Graduate Texts in Mathematics 259

T0178063

For other titles published in this series, go to
http://www.springer.com/series/136

Manfred Einsiedler · Thomas Ward

# Ergodic Theory

with a view towards Number Theory

 Springer

Manfred Einsiedler
ETH Zurich
Departement Mathematik
Rämistrasse 101
8092 Zurich
Switzerland
manfred.einsiedler@math.ethz.ch

Thomas Ward
University of East Anglia
School of Mathematics
NR4 7TJ Norwich
UK
t.ward@uea.ac.uk

ISSN 0072-5285
ISBN 978-1-4471-2591-4          ISBN 978-0-85729-021-2 (eBook)
DOI 10.1007/978-0-85729-021-2
Springer London Dordrecht Heidelberg New York

British Library Cataloguing in Publication Data
A catalogue record for this book is available from the British Library

Mathematics Subject Classification (2010): 37-01, 11-01, 37D40, 05D10, 22D40, 28D15, 37A15, 11J70, 11J71, 11K50

*To the memory of Daniel Jay Rudolph*
*(1949–2010)*

# Preface

Many mathematicians are aware of some of the dramatic interactions between ergodic theory and other parts of the subject, notably Ramsey theory, infinite combinatorics, and Diophantine number theory. These notes are intended to provide a gentle route to a tiny sample of these results. The intended readership is expected to be mathematically sophisticated, with some background in measure theory and functional analysis, or to have the resilience to learn some of this material along the way from other sources.

In this volume we develop the beginnings of ergodic theory and dynamical systems. While the selection of topics has been made with the applications to number theory in mind, we also develop other material to aid motivation and to give a more rounded impression of ergodic theory. Different points of view on ergodic theory, with different kinds of examples, may be found in the monographs of Cornfeld, Fomin and Sinaĭ [60], Petersen [282], or Walters [374]. Ergodic theory is one facet of dynamical systems; for a broad perspective on dynamical systems see the books of Katok and Hasselblatt [182] or Brin and Stuck [44]. An overview of some of the more advanced topics we hope to pursue in a subsequent volume may be found in the lecture notes of Einsiedler and Lindenstrauss [80] in the Clay proceedings of the Pisa Summer school.

Fourier analysis of square-integrable functions on the circle is used extensively. The more general theory of Fourier analysis on compact groups is not essential, but is used in some examples and results. The ergodic theory of commuting automorphisms of compact groups is touched on using a few examples, but is not treated systematically. It is highly developed elsewhere: an extensive treatment may be found in the monograph by Schmidt [332]. Standard background material on measure theory, functional analysis and topological groups is collected in the appendices for convenience.

Among the many *lacunae*, some stand out: Entropy theory; the isomorphism theory of Ornstein, a convenient source being Rudolph [324]; the more advanced spectral theory of measure-preserving systems, a convenient source being Nadkarni [264]; finally Pesin theory and smooth ergodic theory, a source being Barreira and Pesin [19]. Of these omissions, entropy theory is perhaps the most fundamental for applications in number theory, and this was the reason for not including it here. There is simply too much to say about entropy to fit into this volume, so we will treat this important topic, both in general terms and in more detail in the algebraic context needed for number theory, in a subsequent volume. The notion is mentioned in one or two places in this volume, but is never used directly.

No Lie theory is assumed, and for that reason some arguments here may seem laborious in character and limited in scope. Our hope is that seeing the language of Lie theory emerge from explicit matrix manipulations allows a relatively painless route into the ergodic theory of homogeneous spaces. This will be carried further in a subsequent volume, where some of the deeper applications will be given.

## NOTATION AND CONVENTIONS

The symbols $\mathbb{N} = \{1, 2, \ldots\}$, $\mathbb{N}_0 = \mathbb{N} \cup \{0\}$, and $\mathbb{Z}$ denote the natural numbers, non-negative integers and integers; $\mathbb{Q}$, $\mathbb{R}$, $\mathbb{C}$ denote the rational numbers, real numbers and complex numbers; $\mathbb{S}^1$, $\mathbb{T} = \mathbb{R}/\mathbb{Z}$ denote the multiplicative and additive circle respectively. The elements of $\mathbb{T}$ are thought of as the elements of $[0, 1)$ under addition modulo 1. The real and imaginary parts of a complex number are denoted $x = \Re(x + iy)$ and $y = \Im(x + iy)$. The order of growth of real- or complex-valued functions $f, g$ defined on $\mathbb{N}$ or $\mathbb{R}$ with $g(x) \neq 0$ for large $x$ is compared using Landau's notation:

$$f \sim g \text{ if } \left|\frac{f(x)}{g(x)}\right| \longrightarrow 1 \text{ as } x \to \infty;$$

$$f = \mathrm{o}(g) \text{ if } \left|\frac{f(x)}{g(x)}\right| \longrightarrow 0 \text{ as } x \to \infty.$$

For functions $f, g$ defined on $\mathbb{N}$ or $\mathbb{R}$, and taking values in a normed space, we write $f = \mathrm{O}(g)$ if there is a constant $A > 0$ with $\|f(x)\| \leqslant A\|g(x)\|$ for all $x$. In particular, $f = \mathrm{O}(1)$ means that $f$ is bounded. Where the dependence of the implied constant $A$ on some set of parameters $\mathscr{A}$ is important, we write $f = \mathrm{O}_{\mathscr{A}}(g)$. The relation $f = \mathrm{O}(g)$ will also be written $f \ll g$, particularly when it is being used to express the fact that two functions are commensurate, $f \ll g \ll f$. A sequence $a_1, a_2, \ldots$ will be denoted $(a_n)$. Unadorned norms $\|x\|$ will only be used when $x$ lives in a Hilbert space (usually $L^2$) and always refer to the Hilbert space norm. For a topological space $X$, $C(X)$, $C_{\mathbb{C}}(X)$, $C_c(X)$ denote the space of real-valued, complex-valued, compactly supported continuous functions on $X$ respectively, with the supremum norm. For sets $A, B$, denote the set difference by

$$A \smallsetminus B = \{x \mid x \in A, x \notin B\}.$$

Additional specific notation is collected in an index of notation on page 467.

Statements and equations are numbered consecutively within chapters, and exercises are numbered in sections. Theorems without numbers in the main body of the text will not be proved; appendices contain background material in the form of numbered theorems that will not be proved here.

Several of the issues addressed in this book revolve around *measure rigidity*, in which there is a natural measure that other measures are compared with. These natural measures will usually be Haar measure on a compact or locally compact group, or measures constructed from Haar measures, and these will usually be denoted $m$.

We have not tried to be exhaustive in tracing the history of the ideas used here, but have tried to indicate some of the rich history of mathematical developments that have contributed to ergodic theory. Certain references to earlier and to related material is generally collected in endnotes at the end of each chapter; the presence of these references should not be viewed in any way as authoritative. Statements in these notes are informed throughout by a desire to remain rooted in the familiar territory of ergodic theory. The standing assumption is that, unless explicitly noted otherwise, metric spaces are complete and separable, compact groups are metrizable, discrete groups are countable, countable groups are discrete, and measure spaces are assumed to be Borel probability spaces (this assumption is only relevant starting with Sect. 5.3; see Definition 5.13 for the details). A convenient summary of the measure-theoretic background may be found in the work of Royden [320] or of Parthasarathy [280].

## ACKNOWLEDGEMENTS

It is inevitable that we have borrowed ideas and used them inadvertently without citation, and certain that we have misunderstood, misrepresented or misattributed some historical developments; we apologize for any egregious instances of this. We are grateful to several people for their comments on drafts of sections, including Alex Abercrombie, Menny Aka, Sarah Bailey-Frick, Tania Barnett, Vitaly Bergelson, Michael Björklund, Florin Boca, Will Cavendish, Tushar Das, Jerry Day, Jingsong Chai, Alexander Fish, Anthony Flatters, Nikos Frantzikinakis, Jenny George, John Griesmer, Shirali Kadyrov, Cor Kraaikamp, Beverly Lytle, Fabrizio Polo, Christian Röttger, Nimish Shah, Ronggang Shi, Christoph Übersohn, Alex Ustian, Peter Varju and Barak Weiss; the second named author also thanks John and Sandy Phillips for sustaining him with coffee at Le Pas Opton in Summer 2006 and 2009.

We both thank our previous and current home institutions Princeton University, the Clay Mathematics Institute, The Ohio State University, Eidgenössische Technische Hochschule Zürich, and the University of East Anglia, for support, including support for several visits, and for providing the rich mathematical environments that made this project possible. We also thank the National Science Foundation for support under NSF grant DMS-0554373.

Zürich                                                                                    Manfred Einsiedler
Norwich                                                                                       Thomas Ward

# Leitfaden

The dependencies between the chapters is illustrated below, with solid lines indicating logical dependency and dotted lines indicating partial or motivational links.

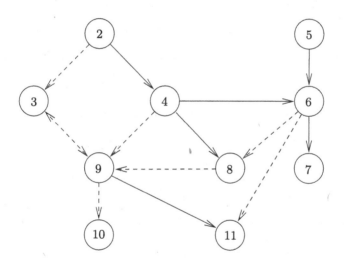

Some possible shorter courses could be made up as follows.

- Chaps. 2 & 4: A gentle introduction to ergodic theory and topological dynamics.
- Chaps. 2 & 3: A gentle introduction to ergodic theory and the continued fraction map (the dotted line indicates that only parts of Chap. 2 are needed for Chap. 3).
- Chaps. 2, 3, & 9: As above, with the connection between the Gauss map and hyperbolic surfaces, and ergodicity of the geodesic flow.
- Chaps. 2, 4, & 8: An introduction to ergodic theory for group actions.

The highlights of this book are Chaps. 7 and 11. Some more ambitious courses could be made up as follows.

- To Chap. 6: Ergodic theory up to conditional measures and the ergodic decomposition.
- To Chap. 7: Ergodic theory including the Furstenberg–Katznelson–Ornstein proof of Szemerédi's theorem.
- To Chap. 11: Ergodic theory and an introduction to dynamics on homogeneous spaces, including equidistribution of horocycle orbits. A minimal path to equidistribution of horocycle orbits on $\mathrm{SL}_2(\mathbb{Z}) \backslash \mathrm{SL}_2(\mathbb{R})$ would include the discussions of ergodicity from Chap. 2, genericity from Chap. 4, Haar measure from Chap. 8, the hyperbolic plane from Chap. 9, and ergodicity and mixing from Chap. 11.

# Contents

1   **Motivation** ............................................... 1
    1.1   Examples of Ergodic Behavior .......................... 1
    1.2   Equidistribution for Polynomials ....................... 3
    1.3   Szemerédi's Theorem .................................. 4
    1.4   Indefinite Quadratic Forms and Oppenheim's Conjecture .... 5
    1.5   Littlewood's Conjecture ................................ 7
    1.6   Integral Quadratic Forms ............................... 8
    1.7   Dynamics on Homogeneous Spaces ...................... 9
    1.8   An Overview of Ergodic Theory ........................ 10

2   **Ergodicity, Recurrence and Mixing** ...................... 13
    2.1   Measure-Preserving Transformations .................... 13
    2.2   Recurrence ........................................... 21
    2.3   Ergodicity ........................................... 23
    2.4   Associated Unitary Operators .......................... 28
    2.5   The Mean Ergodic Theorem ............................ 32
    2.6   Pointwise Ergodic Theorem ............................ 37
          2.6.1   The Maximal Ergodic Theorem .................. 37
          2.6.2   Maximal Ergodic Theorem via Maximal Inequality ... 38
          2.6.3   Maximal Ergodic Theorem via a Covering Lemma .... 40
          2.6.4   The Pointwise Ergodic Theorem ................. 44
          2.6.5   Two Proofs of the Pointwise Ergodic Theorem ....... 45
    2.7   Strong-Mixing and Weak-Mixing ........................ 48
    2.8   Proof of Weak-Mixing Equivalences ..................... 54
          2.8.1   Continuous Spectrum and Weak-Mixing ........... 59
    2.9   Induced Transformations .............................. 61

3   **Continued Fractions** ..................................... 69
    3.1   Elementary Properties ................................. 69
    3.2   The Continued Fraction Map and the Gauss Measure ....... 76

3.3   Badly Approximable Numbers ........................... 87
      3.3.1   Lagrange's Theorem ............................. 88
3.4   Invertible Extension of the Continued Fraction Map ....... 91

**4   Invertible Measures for Continuous Maps** ................ 97
4.1   Existence of Invariant Measures .......................... 98
4.2   Ergodic Decomposition .................................. 103
4.3   Unique Ergodicity ...................................... 105
4.4   Measure Rigidity and Equidistribution ................... 110
      4.4.1   Equidistribution on the Interval .................. 110
      4.4.2   Equidistribution and Generic Points ............... 113
      4.4.3   Equidistribution for Irrational Polynomials ......... 114

**5   Conditional Measures and Algebras** ...................... 121
5.1   Conditional Expectation ................................ 121
5.2   Martingales ............................................ 126
5.3   Conditional Measures ................................... 133
5.4   Algebras and Maps ..................................... 145

**6   Factors and Joinings** ................................... 153
6.1   The Ergodic Theorem and Decomposition Revisited ........ 153
6.2   Invariant Algebras and Factor Maps ..................... 156
6.3   The Set of Joinings ..................................... 158
6.4   Kronecker Systems ..................................... 159
6.5   Constructing Joinings ................................... 163

**7   Furstenberg's Proof of Szemerédi's Theorem** ............. 171
7.1   Van der Waerden ....................................... 172
7.2   Multiple Recurrence .................................... 175
      7.2.1   Reduction to an Invertible System ................. 177
      7.2.2   Reduction to Borel Probability Spaces ............. 177
      7.2.3   Reduction to an Ergodic System ................... 177
7.3   Furstenberg Correspondence Principle .................... 178
7.4   An Instance of Polynomial Recurrence .................... 180
      7.4.1   The van der Corput Lemma ....................... 184
7.5   Two Special Cases of Multiple Recurrence ................ 188
      7.5.1   Kronecker Systems .............................. 188
      7.5.2   Weak-Mixing Systems ............................ 190
7.6   Roth's Theorem ........................................ 192
      7.6.1   Proof of Theorem 7.14 for a Kronecker System ....... 194
      7.6.2   Reducing the General Case to the Kronecker Factor .. 195
7.7   Definitions ............................................. 199
7.8   Dichotomy Between Relatively Weak-Mixing and Compact
      Extensions ............................................. 201

7.9    SZ for Compact Extensions ............................. 207
       7.9.1   SZ for Compact Extensions via van der Waerden ..... 210
       7.9.2   A Second Proof ................................. 212
7.10   Chains of SZ Factors ................................... 216
7.11   SZ for Relatively Weak-Mixing Extensions ................ 218
7.12   Concluding the Proof ................................... 226
7.13   Further Results in Ergodic Ramsey Theory ................ 227
       7.13.1  Other Furstenberg Ergodic Averages ............... 227

8  Actions of Locally Compact Groups ....................... 231
8.1    Ergodicity and Mixing .................................. 231
8.2    Mixing for Commuting Automorphisms ................... 235
       8.2.1   Ledrappier's "Three Dots" Example ............... 236
       8.2.2   Mixing Properties of the ×2, ×3 System ............ 239
8.3    Haar Measure and Regular Representation ................. 243
       8.3.1   Measure-Theoretic Transitivity and Uniqueness ...... 245
8.4    Amenable Groups ...................................... 251
       8.4.1   Definition of Amenability and Existence of Invariant
               Measures ...................................... 251
8.5    Mean Ergodic Theorem for Amenable Groups ............. 254
8.6    Pointwise Ergodic Theorems and Polynomial Growth ....... 257
       8.6.1   Flows ......................................... 257
       8.6.2   Pointwise Ergodic Theorems for a Class of Groups.... 259
8.7    Ergodic Decomposition for Group Actions ................. 266
8.8    Stationary Measures ................................... 272

9  Geodesic Flow on Quotients of the Hyperbolic Plane ..... 277
9.1    The Hyperbolic Plane and the Isometric Action ........... 277
9.2    The Geodesic Flow and the Horocycle Flow ............... 282
9.3    Closed Linear Groups and Left Invariant Riemannian Metric . 288
       9.3.1   The Exponential Map and the Lie Algebra of a
               Closed Linear Group ............................ 289
       9.3.2   The Left-Invariant Riemannian Metric ............. 295
       9.3.3   Discrete Subgroups of Closed Linear Groups ........ 301
9.4    Dynamics on Quotients ................................. 305
       9.4.1   Hyperbolic Area and Fuchsian Groups ............. 306
       9.4.2   Dynamics on $\Gamma \backslash \mathrm{PSL}_2(\mathbb{R})$ ......................... 310
       9.4.3   Lattices in Closed Linear Groups ................. 311
9.5    Hopf's Argument for Ergodicity of the Geodesic Flow ....... 314
9.6    Ergodicity of the Gauss Map ............................ 317
9.7    Invariant Measures and the Structure of Orbits ............ 327
       9.7.1   Symbolic Coding ............................... 327
       9.7.2   Measures Coming from Orbits .................... 328

**10  Nilrotation** ............................................. 331
   10.1 Rotations on the Quotient of the Heisenberg Group ........ 331
   10.2 The Nilrotation ....................................... 333
   10.3 First Proof of Theorem 10.1 ........................... 334
   10.4 Second Proof of Theorem 10.1 ......................... 336
      10.4.1 A Commutative Lemma; The Set $K$ ............... 336
      10.4.2 Studying Divergence; The Set $X_1$ ................. 337
      10.4.3 Combining Linear Divergence
          and the Maximal Ergodic Theorem ................. 339
   10.5 A Non-ergodic Nilrotation ............................. 341
   10.6 The General Nilrotation ............................... 342

**11  More Dynamics on Quotients of the Hyperbolic Plane** .... 347
   11.1 Dirichlet Regions ..................................... 347
   11.2 Examples of Lattices .................................. 357
      11.2.1 Arithmetic and Congruence Lattices in $\mathrm{SL}_2(\mathbb{R})$ ....... 358
      11.2.2 A Concrete Principal Congruence Lattice of $\mathrm{SL}_2(\mathbb{R})$ .. 358
      11.2.3 Uniform Lattices ................................. 361
   11.3 Unitary Representations, Mautner Phenomenon, and
      Ergodicity ............................................ 364
      11.3.1 Three Types of Actions ........................... 364
      11.3.2 Ergodicity ....................................... 366
      11.3.3 Mautner Phenomenon for $\mathrm{SL}_2(\mathbb{R})$ ................. 369
   11.4 Mixing and the Howe–Moore Theorem ................... 370
      11.4.1 First Proof of Theorem 11.22 ..................... 370
      11.4.2 Vanishing of Matrix Coefficients for $\mathrm{PSL}_2(\mathbb{R})$ ........ 372
      11.4.3 Second Proof of Theorem 11.22; Mixing of All Orders . 372
   11.5 Rigidity of Invariant Measures for the Horocycle Flow ....... 378
      11.5.1 Existence of Periodic Orbits; Geometric
          Characterization ................................. 379
      11.5.2 Proof of Measure Rigidity for the Horocycle Flow .... 383
   11.6 Non-escape of Mass for Horocycle Orbits ................. 388
      11.6.1 The Space of Lattices and the Proof of Theorem 11.32
          for $X_2 = \mathrm{SL}_2(\mathbb{Z})\backslash \mathrm{SL}_2(\mathbb{R})$ .......................... 390
      11.6.2 Extension to the General Case .................... 395
   11.7 Equidistribution of Horocycle Orbits ..................... 399

**Appendix A: Measure Theory** ............................. 403
   A.1 Measure Spaces ....................................... 403
   A.2 Product Spaces ....................................... 406
   A.3 Measurable Functions .................................. 407
   A.4 Radon–Nikodym Derivatives ............................ 409
   A.5 Convergence Theorems ................................. 410
   A.6 Well-Behaved Measure Spaces .......................... 411
   A.7 Lebesgue Density Theorem ............................. 412
   A.8 Substitution Rule ..................................... 413

**Appendix B: Functional Analysis** .......................... 417
    B.1  Sequence Spaces....................................... 417
    B.2  Linear Functionals.................................... 418
    B.3  Linear Operators ..................................... 419
    B.4  Continuous Functions ................................. 421
    B.5  Measures on Compact Metric Spaces..................... 422
    B.6  Measures on Other Spaces ............................. 425
    B.7  Vector-valued Integration............................. 425

**Appendix C: Topological Groups** ........................... 429
    C.1  General Definitions ................................... 429
    C.2  Haar Measure on Locally Compact Groups .............. 431
    C.3  Pontryagin Duality ................................... 433

**Hints for Selected Exercises** ............................... 441

**References**................................................. 447

**Author Index**.............................................. 463

**Index of Notation** ......................................... 467

**General Index** ............................................. 471

# Chapter 1
# Motivation

Our main motivation throughout the book will be to understand the applications of ergodic theory to certain problems outside of ergodic theory, in particular to problems in number theory. As we will see, this requires a good understanding of particular examples, which will often be of an algebraic nature. Therefore, we will start with a few concrete examples, and state a few theorems arising from ergodic theory, some of which we will prove within this volume. In Sect. 1.8 we will discuss ergodic theory as a subject in more general terms[1].

## 1.1 Examples of Ergodic Behavior

The *orbit* of a point $x \in X$ under a transformation $T : X \to X$ is the set $\{T^n(x) \mid n \in \mathbb{N}\}$. The structure of the orbit can say a great deal about the original point $x$. In particular, the behavior of the orbit will sometimes detect special properties of the point. A particularly simple instance of this appears in the next example.

*Example 1.1.* Write $\mathbb{T}$ for the quotient group $\mathbb{R}/\mathbb{Z} = \{x + \mathbb{Z} \mid x \in \mathbb{R}\}$, which can be identified with a circle (as a topological space, this can also be obtained as a quotient space of $[0, 1]$ by identifying 0 with 1); there is a natural bijection between $\mathbb{T}$ and the half-open interval $[0, 1)$ obtained by sending the coset $x + \mathbb{Z}$ to the fractional part of $x$. Let $T : \mathbb{T} \to \mathbb{T}$ be defined by $T(x) = 10x \pmod 1$. Then $x \in \mathbb{T}$ is rational if and only if the orbit of $x$ under $T$ is finite. To see this, assume first that $x = \frac{p}{q}$ is rational. In this case the orbit of $x$ is some subset of $\{0, \frac{1}{q}, \dots, \frac{q-1}{q}\}$. Conversely, if the orbit is finite then there must be integers $m, n$ with $1 \leqslant n < m$ for which $T^m(x) = T^n(x)$. It follows that $10^m x = 10^n x + k$ for some $k \in \mathbb{N}$, so $x$ is rational.

Detecting the behavior of the orbit of a given point is usually not so straightforward. Ergodic theory generally has more to say about the orbit of

M. Einsiedler, T. Ward, *Ergodic Theory*, Graduate Texts in Mathematics 259, DOI 10.1007/978-0-85729-021-2_1, © Springer-Verlag London Limited 2011

"typical" points, as illustrated in the next example. Write $\chi_A$ for the indicator function of a set,

$$\chi_A(x) = \begin{cases} 1 \text{ if } x \in A \\ 0 \text{ if } x \notin A. \end{cases}$$

*Example 1.2.* This example recovers a result due to Borel [40]. We shall see later that the map $T : \mathbb{T} \to \mathbb{T}$ defined by $T(x) = 10x \pmod{1}$ preserves Lebesgue measure $m$ on $[0, 1)$ (see Definition 2.1), and is *ergodic* with respect to $m$ (see Definition 2.13). A consequence of the pointwise ergodic theorem (Theorem 2.30) is that for any interval

$$A(j, k) = [\tfrac{j}{10^k}, \tfrac{j+1}{10^k}),$$

we have

$$\frac{1}{N} \sum_{i=0}^{N-1} \chi_{A(j,k)}(T^i x) \longrightarrow \int_0^1 \chi_{A(j,k)}(x) \, dm(x) = \frac{1}{10^k} \qquad (1.1)$$

as $N \to \infty$, for almost every $x$ (that is, for all $x$ in the complement of a set of zero measure, which will be denoted a.e.). For any block $j_1 \ldots j_k$ of $k$ decimal digits, the convergence in (1.1) with $j = 10^{k-1}j_1 + 10^{k-2}j_2 + \cdots + j_k$ shows that the block $j_1 \ldots j_k$ appears with asymptotic frequency $\frac{1}{10^k}$ in the decimal expansion of almost every real number in $[0, 1]$.

Even though the ergodic theorem only concerns the orbital behavior of typical points, there are situations where one is able to describe the orbits for *all* starting points.

*Example 1.3.* We show later that the circle rotation $R_\alpha : \mathbb{T} \to \mathbb{T}$ defined by $R_\alpha(t) = t + \alpha \pmod{1}$ is *uniquely ergodic* if $\alpha$ is irrational (see Definition 4.9 and Example 4.11). A consequence of this is that for any interval $[a, b) \subseteq [0, 1) = \mathbb{T}$,

$$\frac{1}{N} \sum_{n=0}^{N-1} \chi_{[a,b)}(R_\alpha^n(t)) \longrightarrow b - a \qquad (1.2)$$

as $N \to \infty$ for *every* $t \in \mathbb{T}$ (see Theorem 4.10 and Lemma 4.17). As pointed out by Arnol'd and Avez [7] this equidistribution result may be used to find the density of appearance of the digits[2] in the sequence $1, 2, 4, 8, 1, 3, 6, 1, \ldots$ of first digits of the powers of 2:

$$1, 2, 4, 8, 16, 32, 64, 128, 256, 512, 1024, \ldots .$$

A set $A \subseteq \mathbb{N}$ is said to have *density* $\mathbf{d}(A)$ if

$$\mathbf{d}(A) = \lim_{k \to \infty} \frac{1}{k} |A \cap [1, k]|$$

exists. Notice that $2^n$ has first digit $k$ for some $k \in \{1, 2, \ldots, 9\}$ if and only if

$$\log_{10} k \leqslant \{n \log_{10} 2\} < \log_{10}(k+1),$$

where we write $\{t\}$ for the fractional part of the real number $t$.

Since $\alpha = \log_{10} 2$ is irrational, we may apply (1.2) to deduce that

$$\frac{\left|\{n \mid 0 \leqslant n \leqslant N-1, \, 1^{\text{st}} \text{ digit of } 2^n \text{ is } k\}\right|}{N} = \frac{1}{N} \sum_{n=0}^{N-1} \chi_{[\log_{10} k, \log_{10}(k+1))}(R_\alpha^n(0))$$

$$\to \log_{10}\left(\frac{k+1}{k}\right)$$

as $N \to \infty$.

Thus the first digit $k \in \{1, \ldots, 9\}$ appears with density $\log_{10}\left(\frac{k+1}{k}\right)$, and it follows in particular that the digit 1 is the most common leading digit in the sequence of powers of 2.

## Exercises for Sect. 1.1

**Exercise 1.1.1.** A point $x \in X$ is said to be *periodic* for the map $T : X \to X$ if there is some $k \geqslant 1$ with $T^k(x) = x$, and *pre-periodic* if the orbit of $x$ under $T$ is finite. Describe the periodic points and the pre-periodic points for the map $x \mapsto 10x \pmod 1$ from Example 1.1.

**Exercise 1.1.2.** Prove that the orbit of any point $x \in \mathbb{T}$ under the map $R_\alpha$ on $\mathbb{T}$ for $\alpha$ irrational is dense (that is, for any $\varepsilon > 0$ and $t \in \mathbb{T}$ there is some $k \in \mathbb{N}$ for which $T^k x$ lies within $\varepsilon$ of $t$). Deduce that for any finite block of decimal digits, there is some power of 2 that begins with that block of digits.

## 1.2 Equidistribution for Polynomials

A sequence $(a_n)_{n \in \mathbb{N}}$ of numbers in $[0, 1)$ is said to be equidistributed if

$$\mathsf{d}(\{n \in \mathbb{N} \mid a \leqslant a_n < b\}) = b - a$$

for all $a, b$ with $0 \leqslant a < b \leqslant 1$. A classical result of Weyl [381] extends the equidistribution of the numbers $(n\alpha)_{n \in \mathbb{N}}$ modulo 1 for irrational $\alpha$ to the values of any polynomial with an irrational coefficient*.

---

* Numbered theorems like Theorem 1.4 in the main text are proved in this volume, but not necessarily in the chapter in which they first appear.

**Theorem 1.4 (Weyl).** *Let $p(n) = a_k n^k + \cdots + a_0$ be a real polynomial with at least one coefficient among $a_1, \ldots, a_k$ irrational. Then the sequence $(p(n))$ is equidistributed modulo 1.*

Furstenberg extended unique ergodicity to a dynamically defined extension of the irrational circle rotation described in Example 1.3, giving an elegant ergodic-theoretic proof of Theorem 1.4. This approach will be discussed in Sect. 4.4.

## Exercises for Sect. 1.2

**Exercise 1.2.1.** Describe what Theorem 1.4 can tell us about the leading digits of the powers of 2.

## 1.3 Szemerédi's Theorem

Szemerédi, in an intricate and difficult combinatorial argument, proved a long-standing conjecture of Erdős and Turán [85] in his paper [357]. A set $S$ of integers is said to have *positive upper Banach density* if there are sequences $(m_j)$ and $(n_j)$ with $n_j - m_j \to \infty$ as $j \to \infty$ with the property that

$$\lim_{j \to \infty} \frac{|S \cap [m_j, n_j]|}{n_j - m_j} > 0.$$

**Theorem 1.5 (Szemerédi).** *Any subset of the integers with positive upper Banach density contains arbitrarily long arithmetic progressions.*

Furstenberg [102] (see also his book [103] and the article of Furstenberg, Katznelson and Ornstein [107]) showed that Szemerédi's theorem would follow from a generalization of Poincaré's recurrence theorem, and proved that generalization. The connection between recurrence and Szemerédi's theorem will be explained in Sect. 7.3, and Furstenberg's proof of the generalization of Poincaré recurrence needed will be presented in Chap. 7. There are a great many more theorems in this direction which we cannot cover, but it is worth noting that many of these further theorems to date only have proofs using ergodic theory.

More recently, Gowers [122] has given a different proof of Szemerédi's theorem, and in particular has found the following effective form of it*.

**Theorem (Gowers).** *For every integer $s \geqslant 1$ and sufficiently large integer $N$, every subset of $\{1, 2, \ldots, N\}$ with at least*

---

* Theorems and other results that are not numbered will not be proved in this volume, but will also not be used in the main body of the text.

$$N(\log \log N)^{-2^{-2^{s+9}}}$$

elements contains an arithmetic progression of length $s$.

Typically proofs using ergodic theory are not effective: Theorem 1.5 easily implies a finitistic version of Szemerédi's theorem, which states that for every $s$ and constant $c > 0$ and all sufficiently large $N = N(s,c)$, any subset of $\{1, \ldots, N\}$ with at least $cN$ elements contains an arithmetic progression of length $s$. However, the dependence of $N$ on $c$ is not known by this means, nor is it easily deduced from the proof of Theorem 1.5. Gowers' Theorem, proved by different methods, does give an explicit dependence.

We mention Gowers' Theorem to indicate some of the limitations of ergodic theory. While ergodic methods have many advantages, proving quite general theorems which often have no other proofs, they also have disadvantages, one of them being that they tend to be non-effective.

Subsequent development of the combinatorial and arithmetic ideas by Goldston, Pintz and Yıldırım [118][3] and Gowers, and of the ergodic method by Host and Kra [159] and Ziegler [393], has influenced some arguments of Green and Tao [127] in their proof of the following long-conjectured result. This is a good example of how asking for effective or quantitative versions of existing results can lead to new qualitative theorems.

**Theorem (Green and Tao).** The set of primes contains arbitrarily long arithmetic progressions.

## 1.4 Indefinite Quadratic Forms and Oppenheim's Conjecture

Our purpose here is to provide enough background in ergodic theory to quickly reach some understanding of a few deeper results in number theory and combinatorial number theory where ergodic theory has made a contribution. Along the way we will develop a good portion of ergodic theory as well as some other background material. In the rest of this introductory chapter, we mention some more highlights of the many connections between ergodic theory and number theory. The results in this section, and in Sects. 1.5 and 1.6, will not be covered in this book, but we plan to discuss them in a subsequent volume.

The next theorem was conjectured in a weaker form by Oppenheim in 1929 and eventually proved by Margulis in the stronger form stated here in 1986 [247, 250]. In order to state the result, we recall some terminology for quadratic forms.

A *quadratic form* in $n$ variables is a homogeneous polynomial $Q(x_1, \ldots, x_n)$ of degree two. Equivalently, a quadratic form is a polynomial $Q$ for which there is a symmetric $n \times n$ matrix $A_Q$ with

$$Q(x_1, \ldots, x_n) = (x_1, \ldots, x_n) A_Q (x_1, \ldots, x_n)^{\mathrm{t}}.$$

Since $A_Q$ is symmetric, there is an orthogonal matrix $P$ for which $P^{\mathrm{t}} A_Q P$ is diagonal. This means there is a different coordinate system $y_1, \ldots, y_n$ for which

$$Q(x_1, \ldots, x_n) = c_1 y_1^2 + \cdots + c_n y_n^2.$$

The quadratic form is called *non-degenerate* if all the coefficients $c_i$ are non-zero (equivalently, if $\det A_Q \neq 0$), and is called *indefinite* if the coefficients $c_i$ do not all have the same sign. Finally, the quadratic form is said to be *rational* if its coefficients (equivalently, if the entries of the matrix $A_Q$) are rational*.

**Theorem (Margulis).** Let $Q$ be an indefinite non-degenerate quadratic form in $n \geqslant 3$ variables that is not a multiple of a rational form. Then $Q(\mathbb{Z}^n)$ is a dense subset of $\mathbb{R}$.

It is easy to see that two of the stated conditions are necessary for the result: if the form $Q$ is definite then the elements of $Q(\mathbb{Z}^n)$ all have the same sign, and if $Q$ is a multiple of a rational form, then $Q(\mathbb{Z}^n)$ lies in a discrete subgroup of $\mathbb{R}$. The assumption that $Q$ is non-degenerate and $n$ is at least 3 are also necessary, though this is less obvious (requiring in particular the notion of badly approximable numbers from the theory of Diophantine approximation, which will be introduced in Sect. 3.3). This shows that the theorem as stated above is in the strongest possible form. Weaker forms of this result have been obtained by other methods, but the full strength of Margulis' Theorem at the moment requires dynamical arguments (for example, ergodic methods).

Proving the theorem involves understanding the behavior of *orbits* for the action of the subgroup $\mathrm{SO}(2,1) \leqslant \mathrm{SL}_3(\mathbb{R})$ on points $x \in \mathrm{SL}_3(\mathbb{Z}) \backslash \mathrm{SL}_3(\mathbb{R})$ (the space of right cosets of $\mathrm{SL}_3(\mathbb{Z})$ in $\mathrm{SL}_3(\mathbb{R})$); these may be thought of as sets of the form $x \, \mathrm{SO}(2,1)$. As it turns out (a consequence of Raghunathan's conjectures, discussed briefly in Sect. 1.7), such orbits are either closed subsets of $\mathrm{SL}_3(\mathbb{Z}) \backslash \mathrm{SL}_3(\mathbb{R})$ or are dense in $\mathrm{SL}_3(\mathbb{Z}) \backslash \mathrm{SL}_3(\mathbb{R})$. Moreover, the former case happens if and only if the point $x$ corresponds in an appropriate sense to a rational quadratic form.

Margulis' Theorem may be viewed as an extension of Example 1.3 to higher degree in the following sense. The statement that every orbit under the map $R_\alpha(t) = t + \alpha \pmod 1$ is dense in $\mathbb{T}$ is equivalent to the statement that if $L$ is a linear form in two variables that is not a multiple of a rational form, then $L(\mathbb{Z}^2)$ is dense in $\mathbb{R}$.

---

* Note that the rationality of $Q$ cannot be detected using the coefficients $c_1, \ldots, c_n$ after the real coordinate change.

## 1.5 Littlewood's Conjecture

For a real number $t$, write $\langle t \rangle$ for the distance from $t$ to the nearest integer,

$$\langle t \rangle = \min_{q \in \mathbb{Z}} |t - q|.$$

The theory of continued fractions (which will be described in Chap. 3) shows that for any real number $u$, there is a sequence $(q_n)$ with $q_n \to \infty$ such that $q_n \langle q_n u \rangle < 1$ for all $n \geqslant 1$. Littlewood conjectured the following in the 1930s: for any real numbers $u, v$,

$$\liminf_{n \to \infty} n \langle nu \rangle \langle nv \rangle = 0.$$

Some progress was made on this for restricted classes of numbers $u$ and $v$ by Cassels and Swinnerton-Dyer [50], Pollington and Velani [290], and others, but the problem remains open. In 2003 Einsiedler, Katok and Lindenstrauss [79] used ergodic methods to prove that the set of exceptions to Littlewood's conjecture is extremely small.

**Theorem (Einsiedler, Katok & Lindenstrauss).** Let

$$\Theta = \left\{ (u, v) \in \mathbb{R}^2 \mid \liminf_{n \to \infty} n \langle nu \rangle \langle nv \rangle > 0 \right\}.$$

Then the Hausdorff dimension of $\Theta$ is zero.

In fact the result in [79] is a little stronger, showing that $\Theta$ satisfies a stronger property that implies it has Hausdorff dimension zero. The proof relies on a partial classification of certain invariant measures on $\mathrm{SL}_3(\mathbb{Z}) \backslash \mathrm{SL}_3(\mathbb{R})$. This is part of the theory of *measure rigidity*, and the particular type of phenomenon seen has its origins in work of Furstenberg [100], who showed that the natural action $t \mapsto at \pmod{1}$ of the semi-group generated by two multiplicatively independent natural numbers $a_1$ and $a_2$ on $\mathbb{T}$ has, apart from finite sets, no non-trivial closed invariant sets. He asked if this system could have any non-atomic ergodic invariant measures other than Lebesgue measure. Partial results on this and related generalizations led to the formulation of far-reaching conjectures by Margulis [251], by Furstenberg, and by Katok and Spatzier [183, 184]. A special case of these conjectures concerns actions of the group $A$ of positive diagonal matrices in $\mathrm{SL}_k(\mathbb{R})$ for $k \geqslant 3$ on the space $\mathrm{SL}_k(\mathbb{Z}) \backslash \mathrm{SL}_k(\mathbb{R})$: if $\mu$ is an $A$-invariant ergodic probability measure on this space, is there a closed connected group $L \geqslant A$ for which $\mu$ is the unique $L$-invariant measure on a single closed $L$-orbit (that is, is $\mu$ *homogeneous*)?

In the work of Einsiedler, Katok and Lindenstrauss the conjecture stated above is proved under the additional hypothesis that the measure $\mu$ gives positive entropy to some one-parameter subgroup of $A$, which leads to the theorem concerning $\Theta$. A complete classification of these measures without entropy hypotheses would imply the full conjecture of Littlewood.

In this volume we will develop the minimal background needed for the ergodic approach to continued fractions (see Chap. 3) as well as the basic theorems concerning the action of the diagonal subgroup $A$ on the quotient space $\mathrm{SL}_2(\mathbb{Z}) \backslash \mathrm{SL}_2(\mathbb{R})$ (see Chap. 9). We will also describe the connection between these two topics, which will help us to prove results about the continued fraction expansion and about the action of $A$.

## 1.6 Integral Quadratic Forms

An important topic in number theory, both classical and modern, is that of integral quadratic forms. A quadratic form $Q(x_1, \ldots, x_n)$ is said to be *integral* if its coefficients are integers.

A natural problem[4] is to describe the range $Q(\mathbb{Z}^n)$ of an integral quadratic form evaluated on the integers. A classical theorem of Lagrange[5] on the sum of four squares says that $Q_0(\mathbb{Z}^4) = \mathbb{N}_0$ if

$$Q_0(x_1, x_2, x_3, x_4) = x_1^2 + x_2^2 + x_3^2 + x_4^2,$$

solving the problem for a particular form.

More generally, Kloosterman, in his dissertation of 1924, found an asymptotic formula for the number of expressions for an integer in terms of a positive definite quadratic form $Q$ in five or more variables and deduced that any large integer lies in $Q(\mathbb{Z}^n)$ if it satisfies certain congruence conditions. The case of four variables is much deeper, and required him to make new deep developments in analytic number theory; special cases appeared in [201] and the full solution in [202], where he proved that an integral definite quadratic form $Q$ in four variables represents all large enough integers $a$ for which there is no *congruence obstruction*. Here we say that $a \in \mathbb{N}$ has a congruence obstruction for the quadratic form $Q(x_1, \ldots, x_n)$ if $a$ modulo $d$ is not a value of $Q(x_1, \ldots, x_n)$ modulo $d$ for some $d \in \mathbb{N}$.

The methods that are usually applied to prove these theorems are purely number-theoretic. Ellenberg and Venkatesh [83] have introduced a method that combines number theory, algebraic group theory, and ergodic theory to prove results in this field, leading to a different proof of the following special case of Kloosterman's Theorem.

**Theorem (Kloosterman).** Let $Q$ be a positive definite quadratic form with integer coefficients in at least 6 variables. Then all large enough integers that do not fail the congruence conditions can be represented by the form $Q$.

That is, if $a \in \mathbb{N}$ is larger than some constant that depends on $Q$ and for every $d > 0$ there exists some $x_d \in \mathbb{Z}^n$ with $Q(x_d) = a$ modulo $d$, then there exists some $x \in \mathbb{Z}^n$ with $Q(x) = a$. This theorem has purely number-theoretic proofs (see the survey by Schulze-Pillot [335]).

In fact Ellenberg and Venkatesh proved in [83] a different theorem that currently does not have a purely number-theoretic proof. They considered the problem of representing a quadratic form by another quadratic form: If $Q$ is an integral positive definite[6] quadratic form in $n$ variables and $Q'$ is another such form in $m < n$ variables, then one can ask whether there is a subgroup $\Lambda \leqslant \mathbb{Z}^n$ generated by $m$ elements such that when $Q$ is restricted to $\Lambda$ the resulting form is isomorphic to $Q'$. This question has, for instance, been studied by Gauss in the case of $m = 2$ and $n = 3$ in the *Disquisitiones Arithmeticae* [111]. As before, there can be congruence obstructions to this problem, which are best phrased in terms of $p$-adic numbers. Roughly speaking, Ellenberg and Venkatesh show that for a given integral definite quadratic form $Q$ in $n$ variables, every integral definite quadratic form $Q'$ in $m \leqslant n - 5$ variables[7] that does not have small image values can be represented by $Q$, unless there is a congruence obstruction. The assumption that the quadratic form $Q'$ does not have small image means that $\min_{x \in \mathbb{Z}^m \smallsetminus \{0\}} Q'(x)$ should be bigger than some constant that depends on $Q$.

The ergodic theory used in [83] is related to Raghunathan's conjecture mentioned in Sect. 1.4 and discussed again in Sect. 1.7 below, and is the result of work by many people, including Margulis, Mozes, Ratner, Shah, and Tomanov.

## 1.7 Dynamics on Homogeneous Spaces

Let $G \leqslant \mathrm{SL}_n(\mathbb{R})$ be a closed linear group over the reals (or over a local field; see Sect. 9.3 for a precise definition), let $\Gamma < G$ be a discrete subgroup[8], and let $H < G$ be a closed subgroup. For example, the case $G = \mathrm{SL}_3(\mathbb{R})$ and $\Gamma = \mathrm{SL}_3(\mathbb{Z})$ arises in Sect. 1.4 with $H = \mathrm{SO}(2,1)$, and arises in Sect. 1.5 with $H = A$. Dynamical properties of the action of right multiplication by elements of $H$ on the homogeneous space $X = \Gamma \backslash G$ is important for numerous problems[9]. Indeed, all the results in Sects. 1.4–1.6 may be proved by studying concrete instances of such systems. We do not want to go into the details here, but simply mention a few highlights of the theory.

There are many important and general results on the ergodicity and mixing behavior of natural measures on such quotients (see Chap. 2 for the definitions). These results (introduced in Chaps. 9 and 11) are interesting in their own right, but have also found applications to the problem of counting integer (and, more recently, rational) points on groups (or certain other varieties). The first instance of this can be found in Margulis's thesis [252], where this approach is used to find the asymptotics for the number of closed geodesics on compact manifolds of negative curvature. Independently, Eskin and Mc-Mullen [86] found the same method and applied it to a counting problem in certain varieties, which re-proved certain cases of the theorems in the work of Duke, Rudnick and Sarnak [76] in a simpler manner. However, as discussed in Sect. 1.1, the most difficult—and sometimes most interesting—problem is to understand the orbit of a given point rather than the orbit of almost every point. Indeed, the solution of Oppenheim's conjecture in Sect. 1.4 by Margulis involved understanding the $SO(2,1)$-orbit of a point in $SL_3(\mathbb{Z}) \backslash SL_3(\mathbb{R})$ corresponding to the given quadratic form.

We need one more definition before we can state a general theorem in this direction. A subgroup $U < SL_n(\mathbb{R})$ is called a *one-parameter unipotent subgroup* if $U$ is the image of $\mathbb{R}w$ under the exponential map, for some matrix $w \in \text{Mat}_{nn}$ satisfying $w^n = 0$ (that is, $w$ is nilpotent and $\exp(tw)$ has only 1 as an eigenvalue, hence the name). For example, there is an index two subgroup $H \leqslant SO(2,1)$ which is generated by one-parameter unipotent subgroups. However, notice that the diagonal subgroup $A$ is not generated by one-parameter unipotent subgroups.

Raghunathan conjectured that if the subgroup $H$ is generated by one-parameter unipotent subgroups, then the closures of orbits $xH$ are always of the form $xL$ for some closed connected subgroup $L$ of $G$ that contains $H$. This reduces the properties of orbit closures (a dynamical problem) to the algebraic problem of deciding for which closed connected subgroups $L$ the orbit $xL$ is closed.

Ratner [305] proved this important result using methods from ergodic theory. In fact, she deduced Raghunathan's conjecture from Dani's conjecture[10] regarding $H$-invariant measures, which she proved first in the series of papers [302, 303] and [304].

To date there have been numerous applications of the above theorem, and certain extensions of it. To name a few more seemingly unrelated applications, Elkies and McMullen [82] have applied these theorems to obtain the distribution of the gaps in the sequence of fractional parts of $\sqrt{n}$, and Vatsal [367] has studied values of certain $L$-functions using the $p$-adic version of the theorems. There are further applications of the theory too numerous to describe here, but the examples above show again the variety of fields that have connections to ergodic theory.

We will discuss a few special cases of the conjectures of Raghunathan and Dani. Example 1.3, Sect. 4.4, Chap. 10, Sect. 11.5, and Sect. 11.7 treat special cases, some of which were known before the conjectures were formulated.

## 1.8 An Overview of Ergodic Theory

Having seen some statements that qualify as being ergodic in nature, and some of the many important applications of ergodic theory to number theory, in this short section we give a brief overview of ergodic theory. If this is not already clear, notice that it is a rather diffuse subject with ill-defined boundaries[11].

Ergodic theory is the study of long-term behavior in dynamical systems from a statistical point of view. Its origins therefore are intimately connected with the time evolution of systems modeled by measure-preserving actions of the reals or the integers, with the action representing the passage of time. Related approaches, using probabilistic methods to study the evolution of systems, also arose in statistical physics, where other natural symmetries— typically reflected by the presence of a $\mathbb{Z}^d$-action—arise. The rich interaction between arithmetic and geometry present in measure-preserving actions of (lattices in) Lie groups quickly emerged, and it is now natural to view ergodic theory as the study of measure-preserving group actions, containing but not limited to several special branches:

(1) The classical study of single measure-preserving transformations.
(2) Measure-preserving actions of $\mathbb{Z}^d$; more generally of countable amenable groups.
(3) Measure-preserving actions of $\mathbb{R}^d$ and more general amenable groups, called flows.
(4) Measure-preserving and more general actions of groups, in particular of Lie groups and of lattices in Lie groups.

Some of the illuminating results in ergodic theory come from the existence of (counter-)examples. Nonetheless, there are many substantial theorems. In addition to fundamental results (the pointwise and mean ergodic theorems themselves, for example) and structural results (the isomorphism theorem of Ornstein, Krieger's theorem on the existence of generators, the isomorphism invariance of entropy), ergodic theory and its way of thinking have made dramatic contributions to many other fields.

# Notes to Chap. 1

[1](Page 1) The origins of the word 'ergodic' are not entirely clear. Boltzmann coined the word *monode* (unique μὸνος, nature εἶδος) for a set of probability distributions on the phase space that are invariant under the time evolution of a Hamiltonian system, and *ergode* for a monode given by uniform distribution on a surface of constant energy. Ehrenfest and Ehrenfest (in an influential encyclopedia article of 1912, translated as [78]) called a system *ergodic* if each surface of constant energy comprised a single time orbit— a notion called *isodic* by Boltzmann (same ισος, path ὀδός) — and *quasi-ergodic* if each surface has dense orbits. The Ehrenfests themselves suggested that the etymology of the word *ergodic* lies in a different direction (work ἐργον, path ὀδός). This work stimulated interest in the mathematical foundations of statistical mechanics, leading eventually to Birkhoff's formulation of the *ergodic hypothesis* and the notion of systems for which almost every orbit in the sense of measure spends a proportion of time in a given set in the phase space in proportion to the measure of the set.

[2](Page 2) Questions of this sort were raised by Gel'fand; he considered the vector of first digits of the numbers $(2^n, 3^n, 4^n, 5^n, 6^n, 7^n, 8^n, 9^n)$ and asked if (for example) there

is a value of $n > 1$ for which this vector is $(2, 3, 4, 5, 6, 7, 8, 9)$. This circle of problems is related to the classical Poncelet's porism, as explained in an article by King [194]. The influence of Poncelet's book [292] is discussed by Gray [126, Chap. 27].

[3] (Page 5) See also the account with some simplifications by Goldston, Motohashi, Pintz, and Yıldırım [117] and the survey by Goldston, Pintz and Yıldırım [119].

[4] (Page 8) In a more general form, this is the 11th of Hilbert's famous set of problems formulated for the 1900 International Congress of Mathematics.

[5] (Page 8) Bachet conjectured the result, and Diophantus stated it; there are suggestions that Fermat may have known it. The first published proof is that of Lagrange in 1770; a standard proof may be found in [87, Sect. 2.3.1] for example.

[6] (Page 9) For *indefinite* quadratic forms there is a very successful algebraic technique, namely strong approximation for algebraic groups (an account may be found in the monograph [286] of Platonov and Rapinchuk), so ergodic theory does not enter into the discussion.

[7] (Page 9) Under an additional congruence condition on $Q'$ the method also works for $m \leqslant n - 3$.

[8] For some of the statements made here one actually has to assume that $\Gamma$ is a *lattice*; see Sect. 9.4.3.

[9] (Page 9) Further readings from various perspectives on the ergodic theory of homogeneous spaces may be found in the books of Bekka and Mayer [21], Feres [90], Starkov [350], Witte Morris [385, 387] and Zimmer [394].

[10] (Page 10) For linear groups over local fields, and products of such groups, the conjectures of Dani (resp. Raghunathan) have been proved by Margulis and Tomanov [253] and independently by Ratner [306].

[11] (Page 11) Some of the many areas of ergodic theory that we do not treat in a substantial way, and other general sources on ergodic theory, may be found in the following books: the connection with information theory in the work of Billingsley [31] and Shields [342]; a wide-ranging overview of ergodic theory in that of Cornfeld, Fomin and Sinaĭ [60]; ergodic theory developed in the language of joinings in the work of Glasner [116]; more on the theory of entropy and generators in books by Parry [277, 279]; a thorough development of the fundamentals of the measurable theory, including the isomorphism and generator theory, in the book of Rudolph [324].

# Chapter 2
# Ergodicity, Recurrence and Mixing

In this chapter the basic objects studied in ergodic theory, measure-preserving transformations, are introduced. Some examples are given, and the relationship between various mixing properties is described. Background on measure theory appears in Appendix A.

## 2.1 Measure-Preserving Transformations

**Definition 2.1.** Let $(X, \mathscr{B}, \mu)$ and $(Y, \mathscr{C}, \nu)$ be probability spaces. A map* $\phi$ from $X$ to $Y$ is *measurable* if $\phi^{-1}(A) \in \mathscr{B}$ for any $A \in \mathscr{C}$, and is *measure-preserving* if it is measurable and $\mu(\phi^{-1}B) = \nu(B)$ for all $B \in \mathscr{C}$. If in addition $\phi^{-1}$ exists almost everywhere and is measurable, then $\phi$ is called an *invertible measure-preserving map*. If $T : (X, \mathscr{B}, \mu) \to (X, \mathscr{B}, \mu)$ is measure-preserving, then the measure $\mu$ is said to be $T$-*invariant*, $(X, \mathscr{B}, \mu, T)$ is called a *measure-preserving system* and $T$ a *measure-preserving transformation*.

Notice that we work with pre-images of sets rather than images to define measure-preserving maps (just as pre-images of sets are used to define measurability of real-valued functions on a measure space). As pointed out in Example 2.4 and Exercise 2.1.3, it is essential to do this. In order to show that a measurable map is measure-preserving, it is sufficient to check this property on a family of sets whose disjoint unions approximate all measurable sets (see Appendix A for the details).

Most of the examples we will encounter are algebraic or are motivated by algebraic or number-theoretic questions. This is not representative of ergodic theory as a whole, where there are many more types of examples (two non-algebraic classes of examples are discussed on the website [81]).

---

* In this measurable setting, a map is allowed to be undefined on a set of zero measure. Definition 2.7 will give one way to view this: a measurable map undefined on a set of zero measure can be viewed as an everywhere-defined map on an isomorphic measure space.

M. Einsiedler, T. Ward, *Ergodic Theory*, Graduate Texts in Mathematics 259, DOI 10.1007/978-0-85729-021-2_2, © Springer-Verlag London Limited 2011

We define the circle $\mathbb{T} = \mathbb{R}/\mathbb{Z}$ to be the set of cosets of $\mathbb{Z}$ in $\mathbb{R}$ with the quotient topology induced by the usual topology on $\mathbb{R}$. This topology is also given by the metric

$$\mathsf{d}(r + \mathbb{Z}, s + \mathbb{Z}) = \min_{m \in \mathbb{Z}} |r - s + m|,$$

and this makes $\mathbb{T}$ into a compact abelian group (see Appendix C). The interval $[0, 1) \subseteq \mathbb{R}$ is a fundamental domain for $\mathbb{Z}$: that is, every element of $\mathbb{T}$ may be written in the form $t + \mathbb{Z}$ for a unique $t \in [0, 1)$. We will frequently use $[0, 1)$ to define points (and subsets) in $\mathbb{T}$, by identifying $t \in [0, 1)$ with the unique coset $t + \mathbb{Z} \in \mathbb{T}$ defined by $t$.

*Example 2.2.* For any $\alpha \in \mathbb{R}$, define the *circle rotation by* $\alpha$ to be the map

$$R_\alpha : \mathbb{T} \to \mathbb{T}, \ R_\alpha(t) = t + \alpha \pmod 1.$$

We claim that $R_\alpha$ preserves the Lebesgue measure $m_{\mathbb{T}}$ on the circle. By Theorem A.8, it is enough to prove it for intervals, where it is clear. Alternatively, we may note that Lebesgue measure is a Haar measure on the compact group $\mathbb{T}$, which is invariant under any translation by construction (see Sects. 8.3 and C.2).

*Example 2.3.* A generalization of Example 2.2 is a rotation on a compact group. Let $X$ be a compact group, and let $g$ be an element of $X$. Then the map $T_g : X \to X$ defined by $T_g(x) = gx$ preserves the (left) Haar measure $m_X$ on $X$. The Haar measure on a locally compact group is described in Appendix C, and may be thought of as the natural generalization of the Lebesgue measure to a general locally compact group.

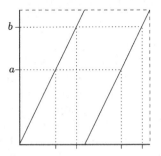

**Fig. 2.1** The pre-image of $[a, b)$ under the circle-doubling map

*Example 2.4.* The *circle-doubling map* is $T_2 : \mathbb{T} \to \mathbb{T}$, $T_2(t) = 2t \pmod 1$. We claim that $T_2$ preserves the Lebesgue measure $m_{\mathbb{T}}$ on the circle. By Theorem A.8, it is sufficient to check this on intervals, so let $B = [a, b) \subseteq [0, 1)$ be any interval. Then it is easy to check that

$$T_2^{-1}(B) = \left[\tfrac{a}{2}, \tfrac{b}{2}\right) \cup \left[\tfrac{a}{2} + \tfrac{1}{2}, \tfrac{b}{2} + \tfrac{1}{2}\right)$$

is a disjoint union (thinking of $a$ and $b$ as real numbers; see Fig. 2.1), so

$$m_{\mathbb{T}}\left(T_2^{-1}(B)\right) = \tfrac{1}{2}(b - a) + \tfrac{1}{2}(b - a) = b - a = m_{\mathbb{T}}(B).$$

Notice that the measure-preserving property cannot be seen by studying forward iterates: if $I$ is a small interval, then $T_2(I)$ is an interval* with total length $2(b - a)$.

*Example 2.5.* Generalizing Example 2.4, let $X$ be a compact abelian group and let $T : X \to X$ be a surjective endomorphism. Then $T$ preserves the Haar measure $m_X$ on $X$ by the following argument. Define a measure $\mu$ on $X$ by $\mu(A) = m_X(T^{-1}A)$. Then, given any $x \in X$ pick $y$ with $T(y) = x$ and notice that

$$\mu(A + x) = m_X(T^{-1}(A + x)) = m_X(T^{-1}A + y) = m_X(T^{-1}A) = \mu(A),$$

so $\mu$ is a translation-invariant Borel probability on $X$ (this just means a probability measure defined on the Borel $\sigma$-algebra). Since the normalized Haar measure is the unique measure with this property, $\mu$ must be $m_X$, which means that $T$ preserves the Haar measure $m_X$ on $X$.

One of the ways in which a measure-preserving transformation may be studied is via its induced action on some natural space of functions. Given any function $f : X \to \mathbb{R}$ and map $T : X \to X$, write $f \circ T$ for the function defined by $(f \circ T)(x) = f(Tx)$. As usual we write $L_\mu^1$ for the space of (equivalence classes of) measurable functions $f : X \to \mathbb{R}$ with $\int |f| \, d\mu < \infty$, $\mathscr{L}^\infty$ for the space of measurable bounded functions and $\mathscr{L}_\mu^1$ for the space of measurable integrable functions (in the usual sense of function, in particular defined everywhere; see Sect. A.3).

**Lemma 2.6.** *A measure $\mu$ on $X$ is $T$-invariant if and only if*

$$\int f \, d\mu = \int f \circ T \, d\mu \qquad (2.1)$$

*for all $f \in \mathscr{L}^\infty$. Moreover, if $\mu$ is $T$-invariant, then (2.1) holds for $f \in L_\mu^1$.*

PROOF. If (2.1) holds, then for any measurable set $B$ we may take $f = \chi_B$ to see that

---

\* We say that a subset of $\mathbb{T}$ is an interval in $\mathbb{T}$ if it is the image of an interval in $\mathbb{R}$. An interval might therefore be represented in our chosen space of coset representatives $[0, 1)$ by the union of two intervals.

$$\mu(B) = \int \chi_B \, d\mu = \int \chi_B \circ T \, d\mu = \int \chi_{T^{-1}B} \, d\mu = \mu(T^{-1}B),$$

so $T$ preserves $\mu$.

Conversely, if $T$ preserves $\mu$ then (2.1) holds for any function of the form $\chi_B$ and hence for any simple function (see Sect. A.3). Let $f$ be a non-negative real-valued function in $\mathscr{L}^1_\mu$. Choose a sequence of simple functions $(f_n)$ increasing to $f$ (see Sect. A.3). Then $(f_n \circ T)$ is a sequence of simple functions increasing to $f \circ T$, and so

$$\int f \circ T \, d\mu = \lim_{n \to \infty} \int f_n \circ T \, d\mu = \lim_{n \to \infty} \int f_n \, d\mu = \int f \, d\mu,$$

showing that (2.1) holds for $f$.                                                     $\square$

One part of ergodic theory is concerned with the structure and classification of measure-preserving transformations. The next definition gives the two basic relationships there may be between measure-preserving transformations[12].

**Definition 2.7.** Let $(X, \mathscr{B}_X, \mu, T)$ and $(Y, \mathscr{B}_Y, \nu, S)$ be measure-preserving systems on probability spaces.

(1) The system $(Y, \mathscr{B}_Y, \nu, S)$ is a *factor* of $(X, \mathscr{B}_X, \mu, T)$ if there are sets $X'$ in $\mathscr{B}_X$ and $Y'$ in $\mathscr{B}_Y$ with $\mu(X') = 1$, $\nu(Y') = 1$, $TX' \subseteq X'$, $SY' \subseteq Y'$ and a measure-preserving map $\phi : X' \to Y'$ with

$$\phi \circ T(x) = S \circ \phi(x)$$

for all $x \in X'$.

(2) The system $(Y, \mathscr{B}_Y, \nu, S)$ is *isomorphic to* $(X, \mathscr{B}_X, \mu, T)$ if there are sets $X'$ in $\mathscr{B}_X$, $Y'$ in $\mathscr{B}_Y$ with $\mu(X') = 1$, $\nu(Y') = 1$, $TX' \subseteq X'$, $SY' \subseteq Y'$, and an invertible measure-preserving map $\phi : X' \to Y'$ with

$$\phi \circ T(x) = S \circ \phi(x)$$

for all $x \in X'$.

In measure theory it is natural to simply ignore null sets, and we will sometimes loosely think of a factor as a measure-preserving map $\phi : X \to Y$ for which the diagram

$$\begin{array}{ccc} X & \xrightarrow{\ T\ } & X \\ \phi \downarrow & & \downarrow \phi \\ Y & \xrightarrow[\ S\ ]{} & Y \end{array}$$

is commutative, with the understanding that the map is not required to be defined everywhere.

A factor map

$$(X, \mathscr{B}_X, \mu, T) \longrightarrow (Y, \mathscr{B}_Y, \nu, S)$$

will also be described as an *extension* of $(Y, \mathscr{B}_Y, \nu, S)$. The factor $(Y, \mathscr{B}_Y, \nu, S)$ is called trivial if as a measure space $Y$ comprises a single element; the extension is called trivial if $\phi$ is an isomorphism of measure spaces.

*Example 2.8.* Define the $(\frac{1}{2}, \frac{1}{2})$ measure $\mu_{(1/2,1/2)}$ on the finite set $\{0, 1\}$ by

$$\mu_{(1/2,1/2)}(\{0\}) = \mu_{(1/2,1/2)}(\{1\}) = \tfrac{1}{2}.$$

Let $X = \{0, 1\}^{\mathbb{N}}$ with the infinite product measure $\mu = \prod_{\mathbb{N}} \mu_{(1/2,1/2)}$ (see Sect. A.2 and Example 2.9 where we will generalize this example). This space is a natural model for the set of possible outcomes of the infinitely repeated toss of a fair coin. The *left shift map* $\sigma : X \to X$ defined by

$$\sigma(x_0, x_1, \dots) = (x_1, x_2, \dots)$$

preserves $\mu$ (since it preserves the measure of the cylinder sets described in Example 2.9). The map $\phi : X \to \mathbb{T}$ defined by

$$\phi(x_0, x_1, \dots) = \sum_{n=0}^{\infty} \frac{x_n}{2^{n+1}}$$

is measure-preserving from $(X, \mu)$ to $(\mathbb{T}, m_{\mathbb{T}})$ and $\phi(\sigma(x)) = T_2(\phi(x))$. The map $\phi$ has a measurable inverse defined on all but the countable set of dyadic rationals $\mathbb{Z}[\frac{1}{2}]/\mathbb{Z}$, where

$$\mathbb{Z}[\tfrac{1}{2}] = \{ \tfrac{m}{2^n} \mid m \in \mathbb{Z}, n \in \mathbb{N} \},$$

so this shows that $(X, \mu, \sigma)$ and $(\mathbb{T}, m_{\mathbb{T}}, T_2)$ are measurably isomorphic.

When the underlying space is a compact metric space, the $\sigma$-algebra is taken to be the Borel $\sigma$-algebra (the smallest $\sigma$-algebra containing all the open sets) unless explicitly stated otherwise. Notice that in both Example 2.8 and Example 2.9 the underlying space is indeed a compact metric space (see Sect. A.2).

*Example 2.9.* The shift map in Example 2.8 is an example of a one-sided Bernoulli shift. A more general[13] and natural two-sided definition is the following. Consider an infinitely repeated throw of a loaded $n$-sided die. The possible outcomes of each throw are $\{1, 2, \dots, n\}$, and these appear with probabilities given by the probability vector $\mathbf{p} = (p_1, p_2, \dots, p_n)$ (probability vector means each $p_i \geqslant 0$ and $\sum_{i=1}^{n} p_i = 1$), so $\mathbf{p}$ defines a measure $\mu_{\mathbf{p}}$ on the finite sample space $\{1, 2, \dots, n\}$, which is given the discrete topology. The sample space for the die throw repeated infinitely often is

$$X = \{1, 2, \ldots, n\}^{\mathbb{Z}}$$
$$= \{x = (\ldots, x_{-1}, x_0, x_1, \ldots) \mid x_i \in \{1, 2, \ldots, n\} \text{ for all } i \in \mathbb{Z}\}.$$

The measure on $X$ is the infinite product measure $\mu = \prod_{\mathbb{Z}} \mu_{\mathbf{p}}$, and the $\sigma$-algebra $\mathscr{B}$ is the Borel $\sigma$-algebra for the compact metric space* $X$, or equivalently is the product $\sigma$-algebra defined below and in Sect. A.2.

A better description of the measure is given via *cylinder sets*. If $I$ is a finite subset of $\mathbb{Z}$, and $\mathbf{a}$ is a map $I \to \{1, 2, \ldots, n\}$, then the cylinder set defined by $I$ and $\mathbf{a}$ is

$$I(\mathbf{a}) = \{x \in X \mid x_j = \mathbf{a}(j) \text{ for all } j \in I\}.$$

It will be useful later to write $x|_I$ for the ordered block of coordinates

$$x_i x_{i+1} \cdots x_{i+s}$$

when $I = \{i, i+1, \ldots, i+s\} = [i, i+s]$. The measure $\mu$ is uniquely determined by the property that

$$\mu\left(I(\mathbf{a})\right) = \prod_{i \in I} p_{a(i)},$$

and $\mathscr{B}$ is the smallest $\sigma$-algebra containing all cylinders (see Sect. A.2 for the details).

Now let $\sigma$ be the (left) shift on $X$: $\sigma(x) = y$ where $y_j = x_{j+1}$ for all $j$ in $\mathbb{Z}$. Then $\sigma$ is $\mu$-preserving and $\mathscr{B}$-measurable. So $(X, \mathscr{B}, \mu, \sigma)$ is a measure-preserving system, called the *Bernoulli scheme* or *Bernoulli shift* based on $\mathbf{p}$. A measure-preserving system measurably isomorphic to a Bernoulli shift is sometimes called a Bernoulli automorphism.

The next example, which we learned from Doug Lind, gives another example of a measurable isomorphism and reinforces the point that being a probability space is a finiteness property of the measure, rather than a metric boundedness property of the space. The measure $\mu$ on $\mathbb{R}$ described in Example 2.10 makes $(\mathbb{R}, \mu)$ into a probability space.

*Example 2.10.* Consider the 2-to-1 map $T : \mathbb{R} \to \mathbb{R}$ defined by

$$T(x) = \frac{1}{2}\left(x - \frac{1}{x}\right)$$

for $x \neq 0$, and $T(0) = 0$. For any $L^1$ function $f$, the substitution $y = T(x)$ shows that

---

* The topology on $X$ is simply the product topology, which is also the metric topology given by the metric defined by $\mathbf{d}(x, y) = 2^{-k}$ where

$$k = \max\{j \mid x_i = y_i \text{ for } |j| \leqslant k\}$$

if $x \neq y$ and $\mathbf{d}(x, x) = 0$. In this metric, points are close together if they agree on a large block of indices around $0 \in \mathbb{Z}$.

$$\int_{-\infty}^{\infty} f(T(x)) \, \frac{dx}{\pi(1+x^2)} = \int_{-\infty}^{\infty} f(y) \frac{dy}{\pi(1+y^2)}$$

(in this calculation, note that $T$ is only injective when restricted to $(0, \infty)$ or $(-\infty, 0)$). It follows by Lemma 2.6 that $T$ preserves the probability measure $\mu$ defined by

$$\mu([a, b]) = \int_a^b \frac{dx}{\pi(1+x^2)}.$$

The map $\phi(x) = \frac{1}{\pi}\arctan(x) + \frac{1}{2}$ from $\mathbb{R}$ to $\mathbb{T}$ is an invertible measure-preserving map from $(\mathbb{R}, \mu)$ to $(\mathbb{T}, m_\mathbb{T})$ where $m_\mathbb{T}$ denotes the Lebesgue measure on $\mathbb{T}$ (notice that the image of $\phi$ is the subset $(0, 1) \subseteq \mathbb{T}$, but this is an invertible map in the measure-theoretic sense).

Define the map $T_2 : \mathbb{T} \to \mathbb{T}$ by $T_2(x) = 2x \pmod 1$ as in Example 2.4. The map $\phi$ is a measurable isomorphism from $(\mathbb{R}, \mu, T)$ to $(\mathbb{T}, m_\mathbb{T}, T_2)$. Example 2.8 shows in turn that $(\mathbb{R}, \mu, T)$ is isomorphic to the one-sided full 2-shift.

It is often more convenient to work with an invertible measure-preserving transformation as in Example 2.9 instead of a non-invertible transformation as in Examples 2.4 and 2.8. Exercise 2.1.7 gives a general construction of an invertible system from a non-invertible one.

## Exercises for Sect. 2.1

**Exercise 2.1.1.** Show that the space $(\mathbb{T}, \mathscr{B}_\mathbb{T}, m_\mathbb{T})$ is isomorphic as a measure space to $(\mathbb{T}^2, \mathscr{B}_{\mathbb{T}^2}, m_{\mathbb{T}^2})$.

**Exercise 2.1.2.** Show that the measure-preserving system $(\mathbb{T}, \mathscr{B}_\mathbb{T}, m_\mathbb{T}, T_4)$, where $T_4(x) = 4x \pmod 1$, is measurably isomorphic to the product system $(\mathbb{T}^2, \mathscr{B}_{\mathbb{T}^2}, m_{\mathbb{T}^2}, T_2 \times T_2)$.

**Exercise 2.1.3.** For a map $T : X \to X$ and sets $A, B \subseteq X$, prove the following.

- $\chi_A(T(x)) = \chi_{T^{-1}(A)}(x)$;
- $T^{-1}(A \cap B) = T^{-1}(A) \cap T^{-1}(B)$;
- $T^{-1}(A \cup B) = T^{-1}(A) \cup T^{-1}(B)$;
- $T^{-1}(A \triangle B) = T^{-1}(A) \triangle T^{-1}(B)$.

Which of these properties also hold with the pre-image under $T^{-1}$ replaced by the forward image under $T$?

**Exercise 2.1.4.** What happens to Example 2.5 if the map $T : X \to X$ is only required to be a continuous homomorphism?

**Exercise 2.1.5.** (a) Find a measure-preserving system $(X, \mathscr{B}, \mu, T)$ with a non-trivial factor map $\phi : X \to X$.
(b) Find an invertible measure-preserving system $(X, \mathscr{B}, \mu, T)$ with a non-trivial factor map $\phi : X \to X$.

**Exercise 2.1.6.** Prove that the circle rotation $R_\alpha$ from Example 2.2 is not measurably isomorphic to the circle-doubling map $T_2$ from Example 2.4.

**Exercise 2.1.7.** Let $\mathsf{X} = (X, \mathscr{B}, \mu, T)$ be any measure-preserving system. A sub-$\sigma$-algebra $\mathscr{A} \subseteq \mathscr{B}_X$ with $T^{-1}\mathscr{A} = \mathscr{A}$ modulo $\mu$ is called a $T$-*invariant sub-$\sigma$-algebra*. Show that the system $\widetilde{\mathsf{X}} = (\widetilde{X}, \widetilde{B}, \widetilde{\mu}, \widetilde{T})$ defined by

- $\widetilde{X} = \{x \in X^{\mathbb{Z}} \mid x_{k+1} = T(x_k) \text{ for all } k \in \mathbb{Z}\}$;
- $(\widetilde{T}(x))_k = x_{k+1}$ for all $k \in \mathbb{Z}$ and $x \in \widetilde{X}$;
- $\widetilde{\mu}(\{x \in \widetilde{X} \mid x_0 \in A\}) = \mu(A)$ for any $A \in \mathscr{B}$, and $\widetilde{\mu}$ is invariant under $\widetilde{T}$;
- $\widetilde{B}$ is the smallest $\widetilde{T}$-invariant $\sigma$-algebra for which the map $\pi : x \mapsto x_0$ from $\widetilde{X}$ to $X$ is measurable;

is an invertible measure-preserving system, and that the map $\pi : x \mapsto x_0$ is a factor map. The system $\widetilde{\mathsf{X}}$ is called the *invertible extension* of $\mathsf{X}$.

**Exercise 2.1.8.** Show that the invertible extension $\widetilde{\mathsf{X}}$ of a measure-preserving system $\mathsf{X}$ constructed in Exercise 2.1.7 has the following universal property. For any extension

$$\phi : (Y, \mathscr{B}_Y, \nu, S) \to (X, \mathscr{B}_X, \mu, T)$$

for which $S$ is invertible, there exists a unique map

$$\widetilde{\phi} : (Y, \mathscr{B}_Y, \nu, S) \to (\widetilde{X}, \widetilde{B}, \widetilde{\mu}, \widetilde{T})$$

for which $\phi = \pi \circ \widetilde{\phi}$.

**Exercise 2.1.9.** (a) Show that the invertible extension of the circle-doubling map from Example 2.4,

$$X_2 = \{x \in \mathbb{T}^{\mathbb{Z}} \mid x_{k+1} = T_2 x_k \text{ for all } k \in \mathbb{Z}\},$$

is a compact abelian group with respect to the coordinate-wise addition defined by $(x + y)_k = x_k + y_k$ for all $k \in \mathbb{Z}$, and the topology inherited from the product topology on $\mathbb{T}^{\mathbb{Z}}$.
(b) Show that the diagonal embedding $\delta(r) = (r, r)$ embeds $\mathbb{Z}[\frac{1}{2}]$ as a discrete subgroup of $\mathbb{R} \times \mathbb{Q}_2$, and that $X_2 \cong \mathbb{R} \times \mathbb{Q}_2 / \delta(\mathbb{Z}[\frac{1}{2}]) \cong \mathbb{R} \times \mathbb{Z}_2 / \delta(\mathbb{Z})$ as compact abelian groups (see Appendix C for the definition of $\mathbb{Q}_p$ and $\mathbb{Z}_p$). In particular, the map $\widetilde{T}_2$ (which may be thought of as the left shift on $X_2$, or as the map that doubles in each coordinate) is conjugate to the map

$$(s, r) + \delta(\mathbb{Z}[\tfrac{1}{2}]) \mapsto (2s, 2r) + \delta(\mathbb{Z}[\tfrac{1}{2}])$$

on $\mathbb{R} \times \mathbb{Q}_2/\delta(\mathbb{Z}[\frac{1}{2}])$. The group $X_2$ constructed in this exercise is a simple example of a *solenoid*.


## 2.2 Recurrence

One of the central themes in ergodic theory is that of *recurrence*, which is a circle of results concerning how points in measurable dynamical systems return close to themselves under iteration. The first and most important of these is a result due to Poincaré [288] published in 1890; he proved this in the context of a natural invariant measure in the "three-body" problem of planetary orbits, before the creation of abstract measure theory[14]. Poincaré recurrence is the pigeon-hole principle for ergodic theory; indeed on a finite measure space it is exactly the pigeon-hole principle.

**Theorem 2.11 (Poincaré Recurrence).** *Let $T : X \to X$ be a measure-preserving transformation on a probability space $(X, \mathscr{B}, \mu)$, and let $E \subseteq X$ be a measurable set. Then almost every point $x \in E$ returns to $E$ infinitely often. That is, there exists a measurable set $F \subseteq E$ with $\mu(F) = \mu(E)$ with the property that for every $x \in F$ there exist integers $0 < n_1 < n_2 < \cdots$ with $T^{n_i} x \in E$ for all $i \geqslant 1$.*

PROOF. Let $B = \{x \in E \mid T^n x \notin E \text{ for any } n \geqslant 1\}$. Then

$$B = E \cap T^{-1}(X \smallsetminus E) \cap T^{-2}(X \smallsetminus E) \cap \cdots,$$

so $B$ is measurable. Now, for any $n \geqslant 1$,

$$T^{-n} B = T^{-n} E \cap T^{-n-1}(X \smallsetminus E) \cap \cdots,$$

so the sets $B, T^{-1}B, T^{-2}B, \ldots$ are disjoint and all have measure $\mu(B)$ since $T$ preserves $\mu$. Thus $\mu(B) = 0$, so there is a set $F_1 \subseteq E$ with $\mu(F_1) = \mu(E)$ and for which every point of $F_1$ returns to $E$ at least once under iterates of $T$. The same argument applied to the transformations $T^2$, $T^3$ and so on defines subsets $F_2, F_3, \ldots$ of $E$ with $\mu(F_n) = \mu(E)$ and with every point of $F_n$ returning to $E$ under $T^n$ for $n \geqslant 1$. The set

$$F = \bigcap_{n \geqslant 1} F_n \subseteq E$$

has $\mu(F) = \mu(E)$, and every point of $F$ returns to $E$ infinitely often.  □

Poincaré recurrence is entirely a consequence of the measure space being of finite measure, as shown in the next example.

*Example 2.12.* The map $T : \mathbb{R} \to \mathbb{R}$ defined by $T(x) = x + 1$ preserves the Lebesgue measure $m_{\mathbb{R}}$ on $\mathbb{R}$. Just as in Definition 2.1, this means that

$$m_{\mathbb{R}}(T^{-1}A) = m_{\mathbb{R}}(A)$$

for any measurable set $A \subseteq \mathbb{R}$. For any bounded set $E \subseteq \mathbb{R}$ and any $x \in E$, the set

$$\{n \geqslant 1 \mid T^n x \in E\}$$

is finite. Thus the map $T$ exhibits no recurrence.

The absence of guaranteed recurrence in infinite measure spaces is one of the main reasons why we restrict attention to probability spaces. There is nonetheless a well-developed ergodic theory of transformations preserving an infinite measure, described in the monograph of Aaronson [1].

Theorem 2.11 may be applied when $E$ is a set in some physical system preserving a finite measure that gives $E$ positive measure. In this case it means that almost every orbit of such a dynamical system returns close to its starting point infinitely often (see Exercise 2.2.3(a)). A much deeper property that a dynamical system may have is that *almost every* orbit returns close to *almost every* point infinitely often, and this property is addressed in Sect. 2.3 (specifically, in Proposition 2.14).

Extending recurrence to multiple recurrence (where the images of a set of positive measure at many different future times is shown to have a non-trivial intersection) is the crucial idea behind the ergodic approach to Szemerédi's theorem (Theorem 1.5). This multiple recurrence generalization of Poincaré recurrence will be proved in Chap. 7.

## Exercises for Sect. 2.2

**Exercise 2.2.1.** Prove the following version of Poincaré recurrence with a weaker hypothesis (finite additivity in place of countable additivity for the measure) and with a stronger conclusion (a bound on the return time). Let $(X, \mathscr{B}, \mu, T)$ be a measure-preserving system with $\mu$ only assumed to be a finitely additive measure (see (A.1)), and let $A \in \mathscr{B}$ have $\mu(A) > 0$. Show that there is some positive $n \leqslant \frac{1}{\mu(A)}$ for which $\mu(A \cap T^{-n}A) > 0$.

**Exercise 2.2.2.** (a) Use Exercise 2.2.1 to show the following. If $A \subseteq \mathbb{N}$ has positive density, meaning that

$$\mathbf{d}(A) = \lim_{k \to \infty} \frac{1}{k} |A \cap [1, k]|$$

exists and is positive, prove that there is some $n \geqslant 1$ with $\overline{\mathbf{d}}(A \cap (A - n)) > 0$ (here $A - n = \{a - n \mid a \in A\}$), where

$$\overline{\mathbf{d}}(B) = \limsup_{k \to \infty} \frac{1}{k} |B \cap [1, k]|.$$

(b) Can you prove this starting with the weaker assumption that the upper density $\bar{\mathbf{d}}(A)$ is positive, and reaching the same conclusion?

**Exercise 2.2.3.** (a) Let $(X, \mathrm{d})$ be a compact metric space and let $T : X \to X$ be a continuous map. Suppose that $\mu$ is a $T$-invariant probability measure defined on the Borel subsets of $X$. Prove that for $\mu$-almost every $x \in X$ there is a sequence $n_k \to \infty$ with $T^{n_k}(x) \to x$ as $k \to \infty$.
(b) Prove that the same conclusion holds under the assumption that $X$ is a metric space, $T : X \to X$ is Borel measurable, and $\mu$ is a $T$-invariant probability measure.

## 2.3 Ergodicity

Ergodicity is the natural notion of indecomposability in ergodic theory[15]. The definition of ergodicity for $(X, \mathscr{B}, \mu, T)$ means that it is impossible to split $X$ into two subsets of positive measure each of which is invariant under $T$.

**Definition 2.13.** A measure-preserving transformation $T : X \to X$ of a probability space $(X, \mathscr{B}, \mu)$ is *ergodic* if for any* $B \in \mathscr{B}$,

$$T^{-1}B = B \implies \mu(B) = 0 \text{ or } \mu(B) = 1. \tag{2.2}$$

When the emphasis is on the map $T : X \to X$, and we are studying different $T$-invariant measures, we will also say that $\mu$ is an ergodic measure for $T$. It is useful to have several different characterizations of ergodicity, and these are provided by the following proposition.

**Proposition 2.14.** *The following are equivalent properties for a measure-preserving transformation $T$ of $(X, \mathscr{B}, \mu)$.*

(1) *$T$ is ergodic.*
(2) *For any $B \in \mathscr{B}$, $\mu(T^{-1}B \triangle B) = 0$ implies that $\mu(B) = 0$ or $\mu(B) = 1$.*
(3) *For $A \in \mathscr{B}$, $\mu(A) > 0$ implies that $\mu\left(\bigcup_{n=1}^{\infty} T^{-n}A\right) = 1$.*
(4) *For $A, B \in \mathscr{B}$, $\mu(A)\mu(B) > 0$ implies that there exists $n \geqslant 1$ with*

$$\mu(T^{-n}A \cap B) > 0.$$

(5) *For $f : X \to \mathbb{C}$ measurable, $f \circ T = f$ almost everywhere implies that $f$ is equal to a constant almost everywhere.*

In particular, for an ergodic transformation and countably many sets of positive measure, almost every point visits all of the sets infinitely often under iterations by the ergodic transformation.

---

* A set $B \in \mathscr{B}$ with $T^{-1}B = B$ is called *strictly invariant* under $T$.

PROOF OF PROPOSITION 2.14. (1) $\implies$ (2): Assume that $T$ is ergodic, so the implication (2.2) holds, and let $B$ be an *almost invariant* measurable set— that is, a measurable set $B$ with $\mu\left(T^{-1}B\triangle B\right) = 0$. We wish to construct an invariant set from $B$, and this is achieved by means of the following limsup construction. Let

$$C = \bigcap_{N=0}^{\infty} \bigcup_{n=N}^{\infty} T^{-n}B.$$

For any $N \geqslant 0$,

$$B\triangle \bigcup_{n=N}^{\infty} T^{-n}B \subseteq \bigcup_{n=N}^{\infty} B\triangle T^{-n}B$$

and $\mu\left(B\triangle T^{-n}B\right) = 0$ for all $n \geqslant 1$, since $B\triangle T^{-n}B$ is a subset of

$$\bigcup_{i=0}^{n-1} T^{-i}B\triangle T^{-(i+1)}B,$$

which has zero measure. Let $C_N = \bigcup_{n=N}^{\infty} T^{-n}B$; the sets $C_N$ are nested,

$$C_0 \supseteq C_1 \supseteq \cdots ,$$

and $\mu(C_N\triangle B) = 0$ for each $N$. It follows that $\mu(C\triangle B) = 0$, so

$$\mu(C) = \mu(B).$$

Moreover,

$$T^{-1}C = \bigcap_{N=0}^{\infty} \bigcup_{n=N}^{\infty} T^{-(n+1)}B = \bigcap_{N=0}^{\infty} \bigcup_{n=N+1}^{\infty} T^{-n}B = C.$$

Thus $T^{-1}C = C$, so by ergodicity $\mu(C) = 0$ or $1$, so $\mu(B) = 0$ or $1$.

(2) $\implies$ (3): Let $A$ be a set with $\mu(A) > 0$, and let $B = \bigcup_{n=1}^{\infty} T^{-n}A$. Then $T^{-1}B \subseteq B$; on the other hand $\mu\left(T^{-1}B\right) = \mu\left(B\right)$ so $\mu(T^{-1}B\triangle B) = 0$. It follows that $\mu(B) = 0$ or $1$; since $T^{-1}A \subseteq B$ the former is impossible, so $\mu(B) = 1$ as required.

(3) $\implies$ (4): Let $A$ and $B$ be sets of positive measure. By (3),

$$\mu\left(\bigcup_{n=1}^{\infty} T^{-n}A\right) = 1,$$

so

$$0 < \mu(B) = \mu\left(\bigcup_{n=1}^{\infty} B\cap T^{-n}A\right) \leqslant \sum_{n=1}^{\infty} \mu\left(B\cap T^{-n}A\right).$$

It follows that there must be some $n \geqslant 1$ with $\mu(B\cap T^{-n}A) > 0$.

(4) $\implies$ (1): Let $A$ be a set with $T^{-1}A = A$. Then

$$0 = \mu(A \cap X \smallsetminus A) = \mu(T^{-n}A \cap X \smallsetminus A)$$

for all $n \geqslant 1$ so, by (4), either $\mu(A) = 0$ or $\mu(X \smallsetminus A) = 0$.

(2) $\implies$ (5): We have seen that if (2) holds, then $T$ is ergodic. Let $f$ be a measurable complex-valued function on $X$, invariant under $T$ in the stated sense. Since the real and the imaginary parts of $f$ must also be invariant and measurable, we may assume without loss of generality that $f$ is real-valued. Fix $k \in \mathbb{Z}$ and $n \geqslant 1$ and write

$$A_n^k = \{x \in X \mid f(x) \in [\tfrac{k}{n}, \tfrac{k+1}{n})\}.$$

Then $T^{-1}A_n^k \triangle A_n^k \subseteq \{x \in X \mid f \circ T(x) \neq f(x)\}$, a null set, so by (2)

$$\mu(A_n^k) \in \{0,1\}.$$

For each $n$, $X$ is the disjoint union $\bigsqcup_{k \in \mathbb{Z}} A_n^k$. It follows that there must be exactly one $k = k(n)$ with $\mu(A_n^{k(n)}) = 1$. Then $f$ is constant on the set

$$Y = \bigcap_{n=1}^{\infty} A_n^{k(n)}$$

and $\mu(Y) = 1$, so $f$ is constant almost everywhere.

(5) $\implies$ (2): If $\mu(T^{-1}B \triangle B) = 0$ then $f = \chi_B$ is a $T$-invariant measurable function, so by (5) $\chi_B$ is a constant almost everywhere. It follows that $\mu(B)$ is either 0 or 1. $\qquad\square$

**Proposition 2.15.** *Bernoulli shifts are ergodic.*

PROOF. Recall the measure-preserving transformation $\sigma$ defined in Example 2.9 on the measure space $X = \{0, 1, \ldots, n\}^{\mathbb{Z}}$ with the product measure $\mu$. Let $B$ denote a $\sigma$-invariant measurable set. Then given any $\varepsilon \in (0, 1)$ there is a finite union of cylinder sets $A$ with $\mu(A \triangle B) < \varepsilon$, and hence with $|\mu(A) - \mu(B)| < \varepsilon$. This means $A$ can be described as

$$A = \{x \in X \mid x|_{[-N,N]} \in F\}$$

for some $N$ and some finite set $F \subseteq \{0, 1, \ldots, n\}^{[-N,N]}$ (for brevity we write $[a, b]$ for the interval of integers $[a, b] \cap \mathbb{Z}$. It follows that for $M > 2N$,

$$\sigma^{-M}(A) = \{x \in X \mid x|_{[M-N,M+N]} \in F\},$$

where we think of $x|_{[M-N,M+N]}$ as a function on $[-N, N]$ in the natural way, is defined by conditions on a set of coordinates disjoint from $[-N, N]$, so

$$\mu(\sigma^{-M}A \smallsetminus A) = \mu(\sigma^{-M}A \cap X \smallsetminus A) = \mu(\sigma^{-M}A)\mu(X \smallsetminus A) = \mu(A)\mu(X \smallsetminus A).$$
$$(2.3)$$

Since $B$ is $\sigma$-invariant, $\mu(B \triangle \sigma^{-1}B) = 0$. Now

$$\mu(\sigma^{-M}A \triangle B) = \mu(\sigma^{-M}A \triangle \sigma^{-M}B)$$
$$= \mu(A \triangle B) < \varepsilon,$$

so $\mu(\sigma^{-M}A \triangle A) < 2\varepsilon$ and therefore

$$\mu(\sigma^{-M}A \triangle A) = \mu(A \smallsetminus \sigma^{-M}A) + \mu(\sigma^{-M}A \smallsetminus A) < 2\varepsilon. \qquad (2.4)$$

Therefore, by (2.3) and (2.4),

$$\mu(B)\mu(X \smallsetminus B) < (\mu(A) + \varepsilon)(\mu(X \smallsetminus A) + \varepsilon)$$
$$= \mu(A)\mu(X \smallsetminus A) + \varepsilon\mu(A) + \varepsilon\mu(X \smallsetminus A) + \varepsilon^2$$
$$< \mu(A)\mu(X \smallsetminus A) + 3\varepsilon < 5\varepsilon.$$

Since $\varepsilon$ was arbitrary, this implies that $\mu(B)\mu(X \smallsetminus B) = 0$, so $\mu(B) = 0$ or $1$ as required. $\qquad \square$

More general versions of this kind of approximation argument appear in Exercises 2.7.3 and 2.7.4.

**Proposition 2.16.** *The circle rotation $R_\alpha : \mathbb{T} \to \mathbb{T}$ is ergodic with respect to the Lebesgue measure $m_{\mathbb{T}}$ if and only if $\alpha$ is irrational.*

PROOF. If $\alpha \in \mathbb{Q}$, then we may write $\alpha = \frac{p}{q}$ in lowest terms, so $R_\alpha^q = I_{\mathbb{T}}$ is the identity map. Pick any measurable set $A \subseteq \mathbb{T}$ with $0 < m_{\mathbb{T}}(A) < \frac{1}{q}$. Then

$$B = A \cup R_\alpha A \cup \cdots \cup R_\alpha^{q-1} A$$

is a measurable set invariant under $R_\alpha$ with $m_{\mathbb{T}}(B) \in (0,1)$, showing that $R_\alpha$ is not ergodic.

If $\alpha \notin \mathbb{Q}$ then for any $\varepsilon > 0$ there exist integers $m, n, k$ with $m \neq n$ and $|m\alpha - n\alpha - k| < \varepsilon$. It follows that $\beta = (m-n)\alpha - k$ lies within $\varepsilon$ of zero but is not zero, and so the set $\{0, \beta, 2\beta, \dots\}$ considered in $\mathbb{T}$ is $\varepsilon$-dense (that is, every point of $\mathbb{T}$ lies within $\varepsilon$ of a point in this set). Thus $(\mathbb{Z}\alpha + \mathbb{Z})/\mathbb{Z} \subseteq \mathbb{T}$ is dense.

Now suppose that $B \subseteq \mathbb{T}$ is invariant under $R_\alpha$. Then for any $\varepsilon > 0$ choose a function $f \in C(\mathbb{T})$ with $\|f - \chi_B\|_1 < \varepsilon$. By invariance of $B$ we have

$$\|f \circ R_\alpha^n - f\|_1 < 2\varepsilon$$

for all $n$. Since $f$ is continuous, it follows that

$$\|f \circ R_t - f\|_1 \leqslant 2\varepsilon$$

for all $t \in \mathbb{R}$. Thus, since $m_{\mathbb{T}}$ is rotation-invariant,

$$\left\| f - \int f(t) \, dt \right\|_1 = \int \left| \int (f(x) - f(x+t)) \, dt \right| \, dx$$

$$\leqslant \int\int |f(x) - f(x+t)| \, dx \, dt \leqslant 2\varepsilon$$

by Fubini's theorem (see Theorem A.13) and the triangle inequality for integrals. Therefore

$$\|\chi_B - \mu(B)\|_1 \leqslant \|\chi_B - f\|_1 + \left\| f - \int f(t) \, dt \right\|_1 + \left\| \int f(t) \, dt - \mu(B) \right\|_1 < 4\varepsilon.$$

Since this holds for every $\varepsilon > 0$ we deduce that $\chi_B$ is constant and therefore $\mu(B) \in \{0,1\}$. Thus for irrational $\alpha$ the transformation $R_\alpha$ is ergodic with respect to Lebesgue measure. $\qquad\square$

**Proposition 2.17.** *The circle-doubling map $T_2 : \mathbb{T} \to \mathbb{T}$ from Example 2.4 is ergodic (with respect to Lebesgue measure).*

PROOF. By Example 2.8, $T_2$ and the Bernoulli shift $\sigma$ on $X = \{0,1\}^{\mathbb{N}}$ together with the fair coin-toss measure are measurably isomorphic. By Proposition 2.15 the latter is ergodic, and it is clear that measurably isomorphic systems are either both ergodic or both not ergodic. $\qquad\square$

Ergodicity (indecomposability in the sense of measure theory) is a universal property of measure-preserving transformations in the sense that every measure-preserving transformation decomposes into ergodic components. This will be shown in Sects. 4.2 and 6.1. In contrast the natural notion of indecomposability in topological dynamics—minimality—does not permit an analogous decomposition (see Exercise 4.2.3).

In Sect. 2.1 we pointed out that in order to check whether a map is measure-preserving it is enough to check this property on a family of sets that generates the $\sigma$-algebra. This is not the case when Definition 2.13 is used to establish ergodicity (see Exercise 2.3.2). Using a different characterization of ergodicity does allow this, as described in Exercise 2.7.3(3).

## Exercises for Sect. 2.3

**Exercise 2.3.1.** Show that ergodicity is not preserved under direct products as follows. Find a pair of ergodic measure-preserving systems $(X, \mathscr{B}_X, \mu, T)$ and $(Y, \mathscr{B}_Y, \nu, S)$ for which $T \times S$ is not ergodic with respect to the product measure $\mu \times \nu$.

**Exercise 2.3.2.** Define a map $R : \mathbb{T} \times \mathbb{T} \to \mathbb{T} \times \mathbb{T}$ by $R(x, y) = (x+\alpha, y+\alpha)$ for an irrational $\alpha$. Show that for any set of the form $A \times B$ with $A, B$ measurable subsets of $\mathbb{T}$ (such a set is called a *measurable rectangle*) has the property of Definition 2.13, but the transformation $R$ is not ergodic, even if $\alpha$ is irrational.

**Exercise 2.3.3.** (a) Find an arithmetic condition on $\alpha_1$ and $\alpha_2$ that is equivalent to the ergodicity of $R_{\alpha_1} \times R_{\alpha_2} : \mathbb{T} \times \mathbb{T} \to \mathbb{T} \times \mathbb{T}$ with respect to $m_\mathbb{T} \times m_\mathbb{T}$.
(b) Generalize part (a) to characterize ergodicity of the rotation

$$R_{\alpha_1} \times \cdots \times R_{\alpha_n} : \mathbb{T}^n \to \mathbb{T}^n$$

with respect to $m_{\mathbb{T}^n}$.

**Exercise 2.3.4.** Prove that any factor of an ergodic measure-preserving system is ergodic.

**Exercise 2.3.5.** Extend Proposition 2.14 by showing that for each $p \in [1, \infty]$ a measure-preserving transformation $T$ is ergodic if and only if for any $L^p$ function $f$, $f \circ T = f$ almost everywhere implies that $f$ is almost everywhere equal to a constant.

**Exercise 2.3.6.** Strengthen Proposition 2.14(5) by showing that a measure-preserving transformation $T$ is ergodic if and only if any measurable function $f : X \to \mathbb{R}$ with $f(Tx) \geqslant f(x)$ almost everywhere is equal to a constant almost everywhere.

**Exercise 2.3.7.** Let $X$ be a compact metric space and let $T : X \to X$ be continuous. Suppose that $\mu$ is a $T$-invariant ergodic probability measure defined on the Borel subsets of $X$. Prove that for $\mu$-almost every $x \in X$ and every $y$ in the support of $\mu$ there exists a sequence $n_k \nearrow \infty$ such that $T^{n_k}(x) \to y$ as $k \to \infty$. Here the support $\mathrm{Supp}(\mu)$ of $\mu$ is the smallest closed subset $A$ of $X$ with $\mu(A) = 1$; alternatively

$$\mathrm{Supp}(\mu) = X \smallsetminus \bigcup_{\substack{O \subseteq X \text{ open,} \\ \mu(O)=0}} O.$$

Notice that $X$ has a countable base for its topology, so the union is still a $\mu$-null set (see p. 406).

## 2.4 Associated Unitary Operators

A different kind of action[16] induced by a measure-preserving map $T$ on a function space is the *associated operator* $U_T : L^2_\mu \to L^2_\mu$ defined by

$$U_T(f) = f \circ T.$$

Recall that $L^2_\mu$ is a Hilbert space, and for any functions $f_1, f_2 \in L^2_\mu$,

$$\langle U_T f_1, U_T f_2 \rangle = \int f_1 \circ T \cdot \overline{f_2 \circ T} \, d\mu$$

$$= \int f_1 \overline{f_2} \, d\mu \qquad \text{(since } \mu \text{ is } T\text{-invariant)}$$

$$= \langle f_1, f_2 \rangle.$$

Here it is natural to think of functions as being complex-valued; it will be clear from the context when members of $L^2_\mu$ are allowed to be complex-valued. Thus $U_T$ is an isometry mapping $L^2_\mu$ into $L^2_\mu$ whenever $(X, \mathscr{B}_X, \mu, T)$ is a measure-preserving system.

If $U : \mathscr{H}_1 \to \mathscr{H}_2$ is a continuous linear operator from one Hilbert space to another then the relation

$$\langle Uf, g \rangle = \langle f, U^* g \rangle$$

defines an associated operator $U^* : \mathscr{H}_2 \to \mathscr{H}_1$ called the *adjoint* of $U$. The operator $U$ is an *isometry* (that is, has $\|Uh\|_{\mathscr{H}_2} = \|h\|_{\mathscr{H}_1}$ for all $h \in \mathscr{H}_1$) if and only if

$$U^* U = I_{\mathscr{H}_1} \tag{2.5}$$

is the identity operator on $\mathscr{H}_1$ and

$$UU^* = P_{\mathrm{Im}\, U} \tag{2.6}$$

is the projection operator onto $\mathrm{Im}\, U$. Finally, an invertible linear operator $U$ is called *unitary* if $U^{-1} = U^*$, or equivalently if $U$ is invertible and

$$\langle Uh_1, Uh_2 \rangle = \langle h_1, h_2 \rangle \tag{2.7}$$

for all $h_1, h_2 \in \mathscr{H}_1$. If $U : \mathscr{H}_1 \to \mathscr{H}_2$ satisfies (2.7) then $U$ is an isometry (even if it is not invertible). Thus for any measure-preserving transformation $T$, the associated operator $U_T$ is an isometry, and if $T$ is invertible then the associated operator $U_T$ is a unitary operator, called the *associated unitary operator* of $T$ or *Koopman operator* of $T$.

A property of a measure-preserving transformation is said to be a *spectral* or *unitary property* if it can be detected by studying the associated operator on $L^2_\mu$.

**Lemma 2.18.** *A measure-preserving transformation $T$ is ergodic if and only if $1$ is a simple eigenvalue of the associated operator $U_T$. Hence ergodicity is a unitary property.*

PROOF. This follows from the proof of the equivalence of (2) and (5) in Proposition 2.14 or via Exercise 2.3.5 applied with $p = 2$: an eigenfunction for the eigenvalue 1 is a $T$-invariant function, and ergodicity is characterized by the property that the only $T$-invariant functions are the constants. $\square$

An isometry $U : \mathscr{H}_1 \to \mathscr{H}_2$ between Hilbert spaces[17] sends the expansion of an element

$$x = \sum_{n=1}^{\infty} c_n e_n$$

in terms of a complete orthonormal basis $\{e_n\}$ for $\mathscr{H}_1$ to a convergent expansion

$$U(x) = \sum_{n=1}^{\infty} c_n U(e_n)$$

in terms of the orthonormal set $\{U(e_n)\}$ in $\mathscr{H}_2$.

We will use this observation to study ergodicity of some of the examples using harmonic analysis rather than the geometrical arguments used earlier in this chapter.

PROOF OF PROPOSITION 2.16 BY FOURIER ANALYSIS. Assume that $\alpha$ is irrational and let $f \in L^2(\mathbb{T})$ be a function invariant under $R_\alpha$. Then $f$ has a Fourier expansion $f(t) = \sum_{n \in \mathbb{Z}} c_n e^{2\pi i n t}$ (both equality and convergence are meant in $L^2(\mathbb{T})$). Now $f$ is invariant, so $\|f \circ R_\alpha - f\|_2 = 0$. By uniqueness of Fourier coefficients, this requires that $c_n = c_n e^{2\pi i n \alpha}$ for all $n \in \mathbb{Z}$. Since $\alpha$ is irrational, $e^{2\pi i n \alpha}$ is only equal to 1 when $n = 0$, so this equation forces $c_n$ to be 0 except when $n = 0$. Thus $f$ is a constant almost everywhere, and hence $R_\alpha$ is ergodic.

If $\alpha \in \mathbb{Q}$ then write $\alpha = \frac{p}{q}$ in lowest terms. The function $g(t) = e^{2\pi i q t}$ is invariant under $R_\alpha$ but is not equal almost everywhere to a constant.    □

Similar methods characterize ergodicity for endomorphisms.

PROOF OF PROPOSITION 2.17 BY FOURIER ANALYSIS. Let $f \in L^2(\mathbb{T})$ be a function with $f \circ T_2 = f$ (equalities again are meant as elements of $L^2(\mathbb{T})$). Then $f$ has a Fourier expansion $f(t) = \sum_{n \in \mathbb{Z}} c_n e^{2\pi i n t}$ with

$$\sum_{n \in \mathbb{Z}} |c_n|^2 = \|f\|_2^2 < \infty. \tag{2.8}$$

By invariance under $T_2$,

$$f(T_2 t) = \sum_{n \in \mathbb{Z}} c_n e^{2\pi i 2 n t} = f(t) = \sum_{n \in \mathbb{Z}} c_n e^{2\pi i n t},$$

so by uniqueness of Fourier coefficients we must have $c_{2n} = c_n$ for all $n \in \mathbb{Z}$. If there is some $n \neq 0$ with $c_n \neq 0$ then this contradicts (2.8), so we deduce that $c_n = 0$ for all $n \neq 0$. It follows that $f$ is constant a.e., so $T_2$ is ergodic.    □

The same argument gives the general abelian case, where Fourier analysis is replaced by character theory (see Sect. C.3 for the background). Notice that for a character $\chi : X \to \mathbb{S}^1$ on a compact abelian group and a continuous homomorphism $T : X \to X$, the map $\chi \circ T : X \to \mathbb{S}^1$ is also a character on $X$.

**Theorem 2.19.** *Let $T : X \to X$ be a continuous surjective homomorphism of a compact abelian group $X$. Then $T$ is ergodic with respect to the Haar measure $m_X$ if and only if the identity $\chi(T^n x) = \chi(x)$ for some $n > 0$ and character $\chi \in \widehat{X}$ implies that $\chi$ is the trivial character with $\chi(x) = 1$ for all $x \in X$.*

PROOF. First assume that there is a non-trivial character $\chi$ with

$$\chi(T^n x) = \chi(x)$$

for some $n > 0$, chosen to be minimal with this property. Then the function

$$f(x) = \chi(x) + \chi(Tx) + \cdots + \chi(T^{n-1}x)$$

is invariant under $T$, and is non-constant since it is a sum of non-trivial distinct characters. It follows that $T$ is not ergodic.

Conversely, assume that no non-trivial character is invariant under a non-zero power of $T$, and let $f \in L^2_{m_X}(X)$ be a function invariant under $T$. Then $f$ has a Fourier expansion in $L^2_{m_X}$,

$$f = \sum_{\chi \in \widehat{X}} c_\chi \chi,$$

with $\sum_\chi |c_\chi|^2 = \|f\|_2^2 < \infty$. Since $f$ is invariant, $c_\chi = c_{\chi \circ T} = c_{\chi \circ T^2} = \cdots$, so either $c_\chi = 0$ or there are only finitely many distinct characters among the $\chi \circ T^i$ (for otherwise $\sum_\chi |c_\chi|^2$ would be infinite). It follows that there are integers $p > q$ with $\chi \circ T^p = \chi \circ T^q$, which means that $\chi$ is invariant under $T^{p-q}$ (the map $\chi \mapsto \chi \circ T$ from $\widehat{X}$ to $\widehat{X}$ is injective since $T$ is surjective), so $\chi$ is trivial by hypothesis. It follows that the Fourier expansion of $f$ is a constant, so $T$ is ergodic. $\qquad\square$

In particular, Theorem 2.19 may be applied to characterize ergodicity for endomorphisms of the torus.

**Corollary 2.20.** *Let $A \in \mathrm{Mat}_{dd}(\mathbb{Z})$ be an integer matrix with $\det(A) \neq 0$. Then $A$ induces a surjective endomorphism $T_A$ of $\mathbb{T}^d = \mathbb{R}^d/\mathbb{Z}^d$ which preserves the Lebesgue measure $m_{\mathbb{T}^d}$. The transformation $T_A$ is ergodic if and only if no eigenvalue of $A$ is a root of unity.*

While harmonic analysis sometimes provides a short and readily understood proof of ergodic or mixing properties, these methods are in general less amenable to generalization than are the more geometric arguments.

## Exercises for Sect. 2.4

**Exercise 2.4.1.** Give a different proof that the circle rotation $R_\alpha : \mathbb{T} \to \mathbb{T}$ is ergodic if $\alpha$ is irrational, using Lebesgue's density theorem (Theorem A.24) as follows. Suppose if possible that $A$ and $B$ are measurable invariant sets with $0 < m_{\mathbb{T}}(A), m_{\mathbb{T}}(B) < 1$ and $A \cap B = \varnothing$, and use the fact that the orbit of a point of density for $A$ is dense to show that $A \cap B$ must be non-empty.

**Exercise 2.4.2.** Prove that an ergodic toral automorphism is not measurably isomorphic to an ergodic circle rotation.

**Exercise 2.4.3.** Extend Proposition 2.16 as follows. If $X$ is a compact abelian group, prove that the group rotation $R_g(x) = gx$ is ergodic with respect to Haar measure if and only if the subgroup $\{g^n \mid n \in \mathbb{Z}\}$ generated by $g$ is dense in $X$.

**Exercise 2.4.4.** In the notation of Corollary 2.20, prove that $A$ is injective if and only if $|\det(A)| = 1$, and in general that $A : \mathbb{T}^d \to \mathbb{T}^d$ is $|\det(A)|$-to-one if $\det(A) \neq 0$. Prove Corollary 2.20 using Theorem 2.19 and the explicit description of characters on the torus from (C.3) on p. 436.

## 2.5 The Mean Ergodic Theorem

Ergodic theorems at their simplest express a relationship between averages taken along the orbit of a point under iteration of a measure-preserving map (in the physical origins of the subject, this represents an average over *time*) and averages taken over the measure space with respect to some invariant measure (an average over *space*). The averages taken are of *observables* in the physical sense, represented in our setting by measurable functions. Much of this way of viewing dynamical systems goes back to the seminal work of von Neumann [268].

We have already seen that ergodicity is a spectral property; the first and simplest ergodic theorem only uses properties of the operator $U_T$ associated to a measure-preserving transformation $T$. Theorem 2.21 is due to von Neumann [267] and predates[18] the pointwise ergodic theorem (Theorem 2.30) of Birkhoff, despite the dates of the published versions.

Write $\xrightarrow[L^p_\mu]{}$ for convergence in the $L^p_\mu$ norm.

**Theorem 2.21 (Mean Ergodic Theorem).** *Let $(X, \mathscr{B}, \mu, T)$ be a measure-preserving system, and let $P_T$ denote the orthogonal projection onto the closed subspace*
$$I = \{g \in L^2_\mu \mid U_T g = g\} \subseteq L^2_\mu.$$
*Then for any $f \in L^2_\mu$,*

$$\frac{1}{N} \sum_{n=0}^{N-1} U_T^n f \xrightarrow[L_\mu^2]{} P_T f.$$

PROOF. Let $B = \{U_T g - g \mid g \in L_\mu^2\}$. We claim that $B^\perp = I$. If

$$U_T f = f,$$

then

$$\langle f, U_T g - g \rangle = \langle U_T f, U_T g \rangle - \langle f, g \rangle = 0,$$

so $f \in B^\perp$. If

$$f \in B^\perp$$

then

$$\langle U_T g, f \rangle = \langle g, f \rangle$$

for all $g \in L_\mu^2$, so

$$U_T^* f = f. \tag{2.9}$$

Thus

$$\begin{aligned}
\|U_T f - f\|_2 &= \langle U_T f - f, U_T f - f \rangle \\
&= \|U_T f\|_2^2 - \langle f, U_T f \rangle - \langle U_T f, f \rangle + \|f\|_2^2 \\
&= 2\|f\|_2^2 - \langle U_T^* f, f \rangle - \langle f, U_T^* f \rangle \\
&= 0 \quad \text{by (2.9)},
\end{aligned}$$

so $f = U_T f$.

It follows that $L_\mu^2 = I \oplus \overline{B}$, so any $f \in L_\mu^2$ decomposes as

$$f = P_T f + h, \tag{2.10}$$

with $h \in \overline{B}$. We claim that

$$\frac{1}{N} \sum_{n=0}^{N-1} U_T^n h \xrightarrow[L_\mu^2]{} 0.$$

This is clear for $h = U_T g - g \in B$, since

$$\begin{aligned}
\left\| \frac{1}{N} \sum_{n=0}^{N-1} U_T^n (U_T g - g) \right\|_2 &= \left\| \frac{1}{N} \left( (U_T g - g) + (U_T^2 g - U_T g) + \cdots \right. \right. \\
&\qquad\qquad \left. \left. + (U_T^N g - U_T^{N-1} g) \right) \right\|_2 \\
&= \frac{1}{N} \left\| U_T^N g - g \right\|_2 \longrightarrow 0 \tag{2.11}
\end{aligned}$$

as $N \to \infty$. All we know is that $h \in \overline{B}$, so let $(g_i)$ be a sequence in $L_\mu^2$ with the property that $h_i = U_T g_i - g_i \to h$ as $i \to \infty$. Then for any $i \geqslant 1$,

$$\left\| \frac{1}{N} \sum_{n=0}^{N-1} U_T^n h \right\|_2 \leqslant \left\| \frac{1}{N} \sum_{n=0}^{N-1} U_T^n (h - h_i) \right\|_2 + \left\| \frac{1}{N} \sum_{n=0}^{N-1} U_T^n h_i \right\|_2. \quad (2.12)$$

Fix $\varepsilon > 0$ and choose, by the convergence (2.11), quantities $i$ and $N$ so large that

$$\| h - h_i \|_2 < \varepsilon$$

and

$$\left\| \frac{1}{N} \sum_{n=0}^{N-1} U_T^n h_i \right\|_2 < \varepsilon.$$

Using these estimates in the inequality (2.12) gives

$$\left\| \frac{1}{N} \sum_{n=0}^{N-1} U_T^n h \right\|_2 \leqslant 2\varepsilon,$$

so

$$\frac{1}{N} \sum_{n=0}^{N-1} U_T^n h \xrightarrow[L_\mu^2]{} 0$$

as $N \to \infty$, for any $h \in \overline{B}$. The theorem follows by (2.10).  $\square$

The quantity studied in Theorem 2.21 is an *ergodic average*, and it will be convenient to fix some notation for these. For a fixed measure-preserving system $(X, \mathscr{B}, \mu, T)$ and a function $f : X \to \mathbb{C}$ the $N$th ergodic average of $f$ is defined to be

$$\mathsf{A}_N = \mathsf{A}_N^f = \mathsf{A}_N(f) = \frac{1}{N} \sum_{n=0}^{N-1} f \circ T^n.$$

It is important to understand that this will be interpreted in several quite different ways.

- In Theorem 2.21 the function $f$ is an element of the Hilbert space $L_\mu^2$ (that is, an equivalence class of measurable functions) and $\mathsf{A}_N^f$ is thought of as an element of $L_\mu^2$.
- In Corollary 2.22 we will want to think of $f$ as an element of $L_\mu^1$, but evaluate the ergodic average $\mathsf{A}_N^f$ at points, sometimes writing $\mathsf{A}_N^f(x)$. Of course in this setting any statement can only be made almost everywhere with respect to $\mu$, since $f$ (and hence $\mathsf{A}_N^f$) is only an equivalence class of functions, with two point functions identified if they agree almost everywhere.

- At times it will be useful to think of $f$ as an element of $\mathscr{L}^p_\mu$ (that is, as a function rather than an equivalence class of functions) in which case $A^f_N$ is defined everywhere. Also, if $f$ is continuous, we will later ask whether the convergence of $A^f_N(x)$ could be uniform across $x \in X$.

**Corollary 2.22.** [19] *Let $(X, \mathscr{B}, \mu, T)$ be a measure-preserving system. Then for any function $f \in L^1_\mu$ the ergodic averages $A^f_N$ converge in $L^1_\mu$ to a $T$-invariant function $f' \in L^1_\mu$.*

PROOF. By the mean ergodic theorem (Theorem 2.21) we know that for any $g \in L^\infty_\mu \subseteq L^2_\mu$, the ergodic averages $A^g_N$ converge in $L^2_\mu$ to some $g' \in L^2_\mu$. We claim that $g' \in L^\infty_\mu$. Indeed, $\|A^g_N\|_\infty \leqslant \|g\|_\infty$ and so

$$|\langle A^g_N, \chi_B \rangle| \leqslant \|g\|_\infty \mu(B)$$

for any $B \in \mathscr{B}$. Since $A^g_N \to g'$ in $L^2_\mu$, this implies that

$$|\langle g', \chi_B \rangle| \leqslant \|g\|_\infty \mu(B)$$

for $B \in \mathscr{B}$, so $\|g'\|_\infty \leqslant \|g\|_\infty$ as required.

Moreover, $\| \cdot \|_1 \leqslant \| \cdot \|_2$, so we deduce that

$$A^g_N \xrightarrow[L^1_\mu]{} g' \in L^\infty_\mu.$$

Thus the corollary holds for the dense set of functions $L^\infty_\mu \subseteq L^1_\mu$.

Let $f \in L^1_\mu$ and fix $\varepsilon > 0$; choose $g \in L^\infty_\mu$ with $\|f - g\|_1 < \varepsilon$. By averaging,

$$\left\| \frac{1}{N} \sum_{n=0}^{N-1} f \circ T^n - \frac{1}{N} \sum_{n=0}^{N-1} g \circ T^n \right\|_1 < \varepsilon,$$

and by the previous paragraph there exists $g'$ and $N_0$ with

$$\left\| \frac{1}{N} \sum_{n=0}^{N-1} g \circ T^n - g' \right\|_1 < \varepsilon$$

for $N \geqslant N_0$. Combining these gives

$$\left\| \frac{1}{N} \sum_{n=0}^{N-1} f \circ T^n - \frac{1}{N'} \sum_{n=0}^{N'-1} f \circ T^n \right\|_1 < 4\varepsilon$$

whenever $N, N' \geqslant N_0$. In other words, the ergodic averages form a Cauchy sequence in $L^1_\mu$, and so they have a limit $f' \in L^1_\mu$ by the Riesz–Fischer theorem (Theorem A.23). Since

$$\left\| \left( \frac{1}{N} \sum_{n=0}^{N-1} f \circ T^n \right) \circ T - \frac{1}{N} \sum_{n=0}^{N-1} f \circ T^n \right\|_1 < \frac{2}{N} \|f\|_1$$

for all $N \geqslant 1$, the limit function $f'$ must be $T$-invariant. ·                                □

## Exercises for Sect. 2.5

**Exercise 2.5.1.** Show that a measure-preserving system $(X, \mathscr{B}, \mu, T)$ is ergodic if and only if, for any $f, g \in L^2_\mu$,

$$\lim_{N \to \infty} \frac{1}{N} \sum_{n=0}^{N-1} \langle U_T^n f, g \rangle = \langle f, 1 \rangle \cdot \langle 1, g \rangle .$$

**Exercise 2.5.2.** Let $(X, \mathscr{B}, \mu, T)$ be a measure-preserving system. For any function $f$ in $L^p_\mu$, $1 \leqslant p < \infty$, prove that

$$\frac{1}{n} \sum_{i=0}^{n-1} f(T^i x) \xrightarrow[L^p_\mu]{} f^*,$$

with $f^* \in L^p_\mu$ a $T$-invariant function.

**Exercise 2.5.3.** Show that a measure-preserving system $(X, \mathscr{B}, \mu, T)$ is ergodic if and only if $\mathsf{A}_N(f) \to \int f \, d\mu$ as $N \to \infty$ for all $f$ in a dense subset of $L^1_\mu$.

**Exercise 2.5.4.** Extend Theorem 2.21 to a uniform mean ergodic theorem as follows. Under the assumptions and with the notation of Theorem 2.21, show that

$$\lim_{N-M \to \infty} \frac{1}{N-M} \sum_{n=M}^{N-1} U_T^n f \to P_T f.$$

**Exercise 2.5.5.** Apply Exercise 2.5.4 to strengthen Poincaré recurrence (Theorem 2.11) as follows. For any set $B$ of positive measure in a measure-preserving system $(X, \mathscr{B}, \mu, T)$,

$$E = \{n \in \mathbb{N} \mid \mu(B \cap T^{-n}B) > 0\}$$

is syndetic: that is, there are finitely many integers $k_1, \ldots, k_s$ with the property that $\mathbb{N} \subseteq \bigcup_{i=1}^s E - k_i$.

**Exercise 2.5.6.** Let $(X, \mathscr{B}, \mu, T)$ be a measure-preserving system. We say that $T$ is *totally ergodic* if $T^n$ is ergodic for all $n \geqslant 1$. Given $K \geqslant 1$ define a space $X^{(K)} = X \times \{1, \ldots, K\}$ with measure $\mu^{(K)} = \mu \times \nu$ defined on the product $\sigma$-algebra $\mathscr{B}^{(K)}$, where $\nu(A) = \frac{1}{K}|A|$ is the normalized counting measure defined on any subset $A \subseteq \{1, \ldots, K\}$, and a $\mu^{(K)}$-preserving transformation $T^{(K)}$ by

$$T^{(K)}(x,i) = \begin{cases} (x, i+1) & \text{if } 1 \leqslant i < K, \\ (Tx, 1) & \text{if } i = K \end{cases}$$

for all $x \in X$. Show that $T^{(K)}$ is ergodic with respect to $\mu^{(K)}$ if and only if $T$ is ergodic with respect to $\mu$, and that $T^{(K)}$ is not totally ergodic if $K > 1$.

## 2.6 Pointwise Ergodic Theorem

The conventional proof of the pointwise ergodic theorem involves two other important results, the maximal inequality and the maximal ergodic theorem. Roughly speaking, the maximal ergodic theorem may be used to show that the set of functions in $L^1_\mu$ for which the pointwise ergodic theorem holds is *closed* as a subset of $L^1_\mu$; one then has to find a *dense* subset of $L^1_\mu$ for which the pointwise ergodic theorem holds. Examples 2.23 and 2.25 give another motivation for the maximal ergodic theorem.

Since the pointwise ergodic theorem involves evaluating a function along the orbit of individual points, it is most naturally phrased in terms of genuine functions (that is, elements of $\mathscr{L}^1_\mu$; see Sect. A.3 for the notation). We will normally apply it to a function in $L^1_\mu$, where the meaning is that for any representative in $\mathscr{L}^1_\mu$ of the equivalence class in $L^1_\mu$ we have convergence almost everywhere.

### 2.6.1 The Maximal Ergodic Theorem

In order to see where the next result comes from, it is useful to ask how likely is it that the orbit of a point spends unexpectedly much time in a given small set (the ergodic theorem says that the orbit of a point spends a predictable amount of time in a given set).

*Example 2.23.* Let $(X, \mathscr{B}_X, \mu, T)$ be a measure-preserving system, and fix a small measurable set $B \in \mathscr{B}_X$ with $\mu(B) = \varepsilon > 0$. Consider the ergodic average

$$A_N^{\chi_B} = \frac{1}{N} \sum_{n=0}^{N-1} \chi_B \circ T^n.$$

Since $T$ preserves $\mu$, $\int_X \chi_B \circ T^n \, d\mu = \mu(B)$ for any $n \geqslant 0$, so

$$\int_X A_N^{\chi_B}\, d\mu = \int_X \chi_B\, d\mu = \mu(B) = \varepsilon.$$

Now ask how likely is it that the orbit of a point $x$ spends more than $\sqrt{\varepsilon} > \varepsilon$ of the time between $0$ and $N-1$ in the set $B$. Notice that

$$\sqrt{\varepsilon}\mu\left(\{x \mid A_N^{\chi_B}(x) > \sqrt{\varepsilon}\}\right) \leqslant \int_X A_N^{\chi_B}\, d\mu = \varepsilon,$$

since

$$\sqrt{\varepsilon}\chi_{\{y|A_N^{\chi_B}(y) > \sqrt{\varepsilon}\}}(x) \leqslant A_N^{\chi_B}(x)$$

for all $x \in X$. Thus on the fixed time scale $[0, N-1]$ the measure of the set $B_\varepsilon^N$ of points that spend in proportion at least $\sqrt{\varepsilon}$ of the time between $0$ and $N-1$ in the set $B$ is no larger than $\sqrt{\varepsilon}$.

We would like to be able to say that one can find a set $B_\varepsilon$ independent of $N$ with similar properties for all $N$; as discussed below, this is a consequence of the maximal ergodic theorem[20].

**Theorem 2.24 (Maximal Ergodic Theorem).**  *Consider the measure-preserving system $(X, \mathscr{B}, \mu, T)$ on a probability space and $g$ a real-valued function in $\mathscr{L}_\mu^1$. Define*

$$E_\alpha = \left\{x \in X \,\Big|\, \sup_{n \geqslant 1} \frac{1}{n}\sum_{i=0}^{n-1} g(T^i x) > \alpha \right\}$$

*for any $\alpha \in \mathbb{R}$. Then*

$$\alpha\mu\left(E_\alpha\right) \leqslant \int_{E_\alpha} g\, d\mu \leqslant \|g\|_1.$$

*Moreover, $\alpha\mu\left(E_\alpha \cap A\right) \leqslant \int_{E_\alpha \cap A} g\, d\mu$ whenever $T^{-1}A = A$.*

*Example 2.25.* We continue the discussion from Example 2.23 by noting that if $B \subseteq X$ has $\mu(B) = \varepsilon > 0$ and $g = \chi_B$ is its characteristic function, then by applying the maximal ergodic theorem (Theorem 2.24) with $\alpha = \sqrt{\varepsilon}$ we get the following statement: There exists a set $B' \subseteq X$ with $\mu(B') \leqslant \sqrt{\varepsilon}$ such that for all $N \geqslant 1$ and all $x \in X \setminus B'$ the orbit of the point $x$ spends at most $\sqrt{\varepsilon}$ in proportion of the times between $0$ and $N-1$ in the set $B$. Thus we have found a set as in Example 2.23, but independently of $N$.

### 2.6.2 Maximal Ergodic Theorem via Maximal Inequality

Notice that the operator $U_T$ associated to a measure-preserving transformation $T$ is a *positive* linear operator on each $L_\mu^p$ space (positive means that $f \geqslant 0$ implies $U_T f \geqslant 0$). A traditional proof of Theorem 2.24 starts with a maximal inequality for positive operators.

**Proposition 2.26 (Maximal Inequality).** *Let $U : L^1_\mu \to L^1_\mu$ be a positive linear operator with $\|U\| \leqslant 1$. For $f \in L^1_\mu$ a real-valued function, define inductively the functions*

$$f_0 = 0$$
$$f_1 = f$$
$$f_2 = f + Uf$$
$$\vdots$$
$$f_n = f + Uf + \cdots + U^{n-1}f$$

*for $n \geqslant 1$, and $F_N = \max\{f_n \mid 0 \leqslant n \leqslant N\}$ (all these functions are defined pointwise). Then*

$$\int_{\{x \mid F_N(x) > 0\}} f \, d\mu \geqslant 0$$

*for all $N \geqslant 1$.*

PROOF. For each $N$, it is clear that $F_N \in L^1_\mu$. Since $U$ is positive and linear, and since

$$F_N \geqslant f_n$$

for $0 \leqslant n \leqslant N$, we have

$$U F_N + f \geqslant U f_n + f = f_{n+1}.$$

Hence

$$U F_N + f \geqslant \max_{1 \leqslant n \leqslant N} f_n.$$

For $x \in P = \{x \mid F_N(x) > 0\}$ we have

$$F_N(x) = \max_{0 \leqslant n \leqslant N} f_n(x) = \max_{1 \leqslant n \leqslant N} f_n(x)$$

since $f_0 = 0$. Therefore,

$$U F_N(x) + f(x) \geqslant F_N(x)$$

for $x \in P$, and so

$$f(x) \geqslant F_N(x) - U F_N(x) \qquad (2.13)$$

for $x \in P$. Now $F_N(x) \geqslant 0$ for all $x$, so $U F_N(x) \geqslant 0$ for all $x$. Hence the inequality (2.13) implies that

$$\int_P f \, \mathrm{d}\mu \geqslant \int_P F_N \, \mathrm{d}\mu - \int_P U F_N \, \mathrm{d}\mu$$

$$= \int_X F_N \, \mathrm{d}\mu - \int_P U F_N \, \mathrm{d}\mu \qquad (\text{since } F_N(x) = 0 \text{ for } x \notin P)$$

$$\geqslant \int_X F_N \, \mathrm{d}\mu - \int_X U F_N \, \mathrm{d}\mu$$

$$= \|F_N\|_1 - \|U F_N\|_1 \geqslant 0,$$

since $\|U\| \leqslant 1$. □

FIRST PROOF OF THEOREM 2.24. Let $f = (g - \alpha)$ and $Uf = f \circ T$ for $f \in \mathscr{L}_\mu^1$ so that, in the notation of Proposition 2.26,

$$E_\alpha = \bigcup_{N=0}^{\infty} \{x \mid F_N(x) > 0\}.$$

It follows that $\int_{E_\alpha} f \, \mathrm{d}\mu \geqslant 0$ and therefore $\int_{E_\alpha} g \, \mathrm{d}\mu \geqslant \alpha \mu(E_\alpha)$. For the last statement, apply the same argument to $f = (g - \alpha)$ on the measure-preserving system $\left(A, \mathscr{B}\big|_A, \frac{1}{\mu(A)}\mu\big|_A, T\big|_A\right)$. □

### 2.6.3 Maximal Ergodic Theorem via a Covering Lemma

In this subsection we use covering properties of intervals in $\mathbb{Z}$ to establish a version of the maximal ergodic theorem (Theorem 2.24). This demonstrates very clearly the strong link between the Lebesgue density theorem (Theorem A.24), whose proof involves the Hardy–Littlewood maximal inequality, and the pointwise ergodic theorem, whose proof involves the maximal ergodic theorem[*]. The material in this section illustrates some of the ideas used in the more extensive results of Bourgain [41]; a little of the history will be given in the note (83) on p. 275.

We will obtain a formally weaker version of Theorem 2.24, by showing that

$$\alpha \mu(E_\alpha) \leqslant 3\|g\|_1 \qquad (2.14)$$

in the notation of Theorem 2.24. This is sufficient for all our purposes. For future applications, we state the covering lemma[(21)] needed in a more general setting.

**Lemma 2.27 (Finite Vitali covering lemma).** *Let* $B_{r_1}(a_1), \ldots, B_{r_K}(a_K)$ *be any collection of balls in a metric space. Then there exists a subcollec-*

---

[*] Additionally, this approach starts to reveal more about what properties of the acting group might be useful for obtaining more general ergodic theorems, and gives a method capable of generalization to ergodic averaging along other sets of integers.

tion $B_{r_{j(1)}}(a_{j(1)}), \ldots, B_{r_{j(k)}}(a_{j(k)})$ of those balls which are disjoint and satisfy

$$B_{r_1}(a_1) \cup \cdots \cup B_{r_K}(a_K) \subseteq B_{3r_{j(1)}}(a_{j(1)}) \cup \cdots \cup B_{3r_{j(k)}}(a_{j(k)}),$$

where in the right-hand side we have tripled the radii of the balls in the sub-collection.

PROOF. By reordering the balls if necessary, we may assume that

$$r_1 \geqslant r_2 \geqslant \cdots \geqslant r_K.$$

Let $j(1) = 1$. We choose the remaining disjoint balls by induction as follows. Assume that we have chosen $j(1), \ldots, j(n)$ from the indices $\{1, \ldots, \ell\}$, discarding those not chosen. If $B_{r_{\ell+1}}(a_{\ell+1})$ is disjoint from

$$B_{r_{j(1)}}(a_{j(1)}) \cup \cdots \cup B_{r_{j(n)}}(a_{j(n)})$$

we choose $j(n+1) = \ell+1$, and if not we discard $\ell+1$, and proceed with studying $\ell + 2$, stopping if $\ell + 1 = K$. Suppose that $B_{r_{j(1)}}(a_{j(1)}), \ldots, B_{r_{j(k)}}(a_{j(k)})$ are the balls chosen from all the balls considered, and let

$$V = B_{3r_{j(1)}}(a_{j(1)}) \cup \cdots \cup B_{3r_{j(k)}}(a_{j(k)}).$$

If $i \in \{j(1), \ldots, j(k)\}$ then $B_{r_i}(a_i) \subseteq B_{3r_i}(a_i) \subseteq V$ by construction. If not, then by the construction there is some $n \in \{1, \ldots, i-1\} \cap \{j(1), \ldots, j(k)\}$ that was selected, such that

$$B_{r_i}(a_i) \cap B_{r_n}(a_n) \neq \varnothing,$$

and $r_n \geqslant r_i$ by the ordering of the indices. By the triangle inequality we therefore have

$$B_{r_i}(a_i) \subseteq B_{3r_n}(a_n) \subseteq V$$

as required. $\qquad \square$

In the integers, the Vitali covering lemma may be formulated as follows (see Exercise 2.6.2).

**Corollary 2.28.** *For any collection of intervals*

$$I_1 = [a_1, a_1 + \ell(1) - 1], \ldots, I_K = [a_K, a_K + \ell(K) - 1]$$

*in $\mathbb{Z}$ there is a disjoint subcollection $I_{j(1)}, \ldots, I_{j(k)}$ such that*

$$I_1 \cup \cdots \cup I_K \subseteq \bigcup_{m=1}^{k} [a_{j(m)} - \ell(j(m)), a_{j(m)} + 2\ell(j(m)) - 1].$$

PROOF OF THE INEQUALITY (2.14). Let $(X, \mathscr{B}, \mu, T)$ be a measure-preserving system, with $g \in \mathscr{L}^1_\mu$, and fix $\alpha > 0$. Define

$$g^*(x) = \sup_{n \geqslant 1} \frac{1}{n} \sum_{i=0}^{n-1} g\left(T^i(x)\right)$$

and $E_\alpha = \{x \in X \mid g^*(x) > \alpha\}$ as before. We will deduce the inequality (2.14) from a similar estimate for the function

$$\phi(j) = \begin{cases} g(T^j x) & \text{for } j = 0, \ldots, J; \\ 0 & \text{for } j < 0 \text{ or } j > J \end{cases} \tag{2.15}$$

for a fixed $x \in X$ and $J \geqslant 1$.

**Lemma 2.29.** *For any $\phi \in \ell^1(\mathbb{Z})$ and $\alpha > 0$, define*

$$\phi^*(a) = \sup_{n \geqslant 1} \frac{1}{n} \sum_{i=0}^{n-1} \phi(a+i),$$

*and*

$$E_\alpha^\phi = \{a \in \mathbb{Z} \mid \phi^*(a) > \alpha\}.$$

*Then $\alpha |E_\alpha^\phi| \leqslant 3\|\phi\|_1$.*

PROOF OF LEMMA 2.29. Let $a_1, \ldots, a_K$ be different elements of $E_\alpha^\phi$, and let $\ell(j)$ for $j = 1, \ldots, K$ be chosen so that

$$\frac{1}{\ell(j)} \sum_{i=0}^{\ell(j)-1} \phi(a_j + i) > \alpha. \tag{2.16}$$

Define the intervals $I_j = [a_j, a_j + \ell(j) - 1]$ for $1 \leqslant j \leqslant K$ and use Corollary 2.28 to construct the subcollection $I_{j(1)}, \ldots, I_{j(k)}$ as in the corollary. Since the intervals $I_{j(1)}, \ldots, I_{j(k)}$ are disjoint, it follows that

$$\sum_{i=1}^{k} \sum_{m \in I_{j(i)}} \phi(m) \leqslant \|\phi\|_1, \tag{2.17}$$

where the left-hand side equals

$$\sum_{i=1}^{k} \ell(j(i)) \frac{1}{\ell(j(i))} \sum_{n=0}^{\ell(j(i))-1} \phi(a_j + n) > \sum_{i=1}^{k} \ell(j(i)) \alpha \tag{2.18}$$

by the choice in (2.16) of the $\ell(j(i))$. However, since

$$\{a_1, \ldots, a_K\} \subseteq \bigcup_{j=1}^{k} [a_{j(i)} - \ell(j(i)), a_{j(i)} + 2\ell(j(i)) - 1]$$

by Corollary 2.28, we therefore have

$$K \leqslant 3 \sum_{i=1}^{k} \ell(j(i)). \tag{2.19}$$

Combining the inequalities (2.19), (2.18), and (2.17) in that order gives

$$\alpha K \leqslant 3 \sum_{i=1}^{k} \ell(j(i))\alpha < 3\|\phi\|_1,$$

which proves the lemma. $\qquad\qquad$ □

Fix now some $M \geqslant 1$ (the parameter $J$ will later be chosen much larger than $M$) and define

$$g_M^*(x) = \sup_{1 \leqslant n \leqslant M} \frac{1}{n} \sum_{i=0}^{n-1} g(T^i x),$$

and

$$E_{\alpha,M}^g = \{x \in X \mid g_M^*(x) > \alpha\}.$$

Using $\phi$ as in (2.15) and, suppressing the dependence on $x$ as before, we also define

$$\phi_M^*(a) = \sup_{1 \leqslant n \leqslant M} \frac{1}{n} \sum_{i=0}^{n-1} \phi(a+i).$$

As $\phi(a+i) = g(T^{a+i}x)$ if $0 \leqslant a < J - M$ and $0 \leqslant i < M$, we have

$$\phi_M^*(a) = g_M^*(T^a x) \tag{2.20}$$

for $0 \leqslant a < J - M$. Also, for any $x \in X$ and $\alpha > 0$ we have

$$\alpha |\{a \in [0, J-1] \mid \phi_M^*(a) > \alpha\}| \leqslant 3\|\phi\|_1$$

by Lemma 2.29. Recalling the definition of $\phi$ and $E_\alpha$ and using (2.20), this may be written in a slightly weaker form as

$$\alpha \sum_{a=0}^{J-M-1} \chi_{E_{\alpha,M}^g}(T^a x) = \alpha \left|\left\{a \in [0, J-M-1] \mid g_M^*(T^a x) > \alpha\right\}\right|$$

$$\leqslant 3 \sum_{i=0}^{J} |g(T^i x)|,$$

which may be integrated over $x \in X$ to obtain

$$(J - M)\alpha\mu\left(E_{\alpha,M}^g\right) \leqslant 3(J+1)\|g\|_1,$$

where we have used the invariance of $\mu$ under $T$. Dividing by $J$ and letting $J \to \infty$ gives $\alpha\mu(E^g_{\alpha,M}) \leqslant 3\|g\|_1$, and finally letting $M \to \infty$ gives inequality (2.14).                                                                                     □

### 2.6.4 The Pointwise Ergodic Theorem

We are now ready to give a proof of Birkhoff's pointwise ergodic theorem [33] using the maximal ergodic theorem[22]. This precisely describes the relationship sought between the space average of a function and the time average along the orbit of a typical point.

**Theorem 2.30 (Birkhoff).** *Let $(X, \mathscr{B}, \mu, T)$ be a measure-preserving system. If $f \in \mathscr{L}^1_\mu$, then*

$$\lim_{n \to \infty} \frac{1}{n} \sum_{j=0}^{n-1} f(T^j x) = f^*(x)$$

*converges almost everywhere and in $L^1_\mu$ to a $T$-invariant function $f^* \in \mathscr{L}^1_\mu$, and*

$$\int f^* \, d\mu = \int f \, d\mu.$$

*If $T$ is ergodic, then*

$$f^*(x) = \int f \, d\mu$$

*almost everywhere.*

*Example 2.31.* [23] In Example 1.2 we explained that almost every real number has the property that any block of length $k$ of digits base 10 appears with asymptotic frequency $\frac{1}{10^k}$, thus almost every number is *normal* base 10. We now have all the material needed to justify this result: By Corollary 2.20, the map $x \mapsto Kx \pmod 1$ on the circle for $K \geqslant 2$ is ergodic, so the pointwise ergodic theorem (Theorem 2.30) may be applied to show that almost every number is normal to each base $K \geqslant 2$, and so (by taking the union of countably many null sets) almost every number is normal in *every* base $K \geqslant 2$.

As with the maximal ergodic theorem (Theorem 2.24), we will give two proofs[24] of the pointwise ergodic theorem. The first is a traditional one while the second is closer to the approach of Bourgain [41] for example, and is better adapted to generalization both of the acting group and of the sequence along which ergodic averages are formed.

Theorem 2.30 will be formulated differently in Theorem 6.1, and will be used in Theorem 6.2 to construct the ergodic decomposition.

### 2.6.5 Two Proofs of the Pointwise Ergodic Theorem

FIRST PROOF OF THEOREM 2.30. Recall that $(X, \mathscr{B}, \mu, T)$ is a measure-preserving system, $\mu(X) = 1$, and $f \in \mathscr{L}^1_\mu$. It is sufficient to prove the result for a real-valued function $f$. Define, for any $x \in X$,

$$f^*(x) = \limsup_{n \to \infty} \frac{1}{n} \sum_{i=0}^{n-1} f(T^i x),$$

$$f_*(x) = \liminf_{n \to \infty} \frac{1}{n} \sum_{i=0}^{n-1} f(T^i x).$$

Then

$$\frac{n+1}{n} \left( \frac{1}{n+1} \sum_{i=0}^{n} f(T^i x) \right) = \frac{1}{n} \sum_{i=0}^{n-1} f(T^i(Tx)) + \frac{1}{n} f(x). \qquad (2.21)$$

By taking the limit along a subsequence for which the left-hand side of (2.21) converges to the limsup, this shows that $f^* \leqslant f^* \circ T$. A limit along a subsequence for which the right-hand side of (2.21) converges to the limsup shows that $f^* \geqslant f^* \circ T$. A similar argument for $f_*$ shows that

$$f^* \circ T = f^*, \quad f_* \circ T = f_*. \qquad (2.22)$$

Now fix rationals $\alpha > \beta$, and write

$$E_\alpha^\beta = \{x \in X \mid f_*(x) < \beta \text{ and } f^*(x) > \alpha\}.$$

By (2.22), $T^{-1} E_\alpha^\beta = E_\alpha^\beta$ and $E_\alpha \supseteq E_\alpha^\beta$ where $E_\alpha$ is the set defined in Theorem 2.24 (with $g = f$). By Theorem 2.24,

$$\int_{E_\alpha^\beta} f \, d\mu \geqslant \alpha \mu \left( E_\alpha^\beta \right). \qquad (2.23)$$

After replacing $f$ by $-f$, a similar argument shows that

$$\int_{E_\alpha^\beta} f \, d\mu \leqslant \beta \mu \left( E_\alpha^\beta \right). \qquad (2.24)$$

Now

$$\{x \mid f_*(x) < f^*(x)\} = \bigcup_{\substack{\alpha, \beta \in \mathbb{Q}, \\ \alpha > \beta}} E_\alpha^\beta,$$

while the inequalities (2.23) and (2.24) show that $\mu(E_\alpha^\beta) = 0$ for $\alpha > \beta$. It follows that

$$\mu \left( \bigcup_{\substack{\alpha,\beta \in \mathbb{Q},\\ \alpha > \beta}} E_\alpha^\beta \right) = 0,$$

so

$$f_*(x) = f^*(x) \text{ a.e.}$$

Thus

$$g_n(x) = \frac{1}{n} \sum_{i=0}^{n-1} f(T^i x) \longrightarrow f^*(x) \text{ a.e.} \tag{2.25}$$

By Corollary 2.22 we also know that

$$g_n \xrightarrow[L^1_\mu]{} f' \in \mathscr{L}^1_\mu. \tag{2.26}$$

By Corollary A.12, this implies that there is a subsequence $n_k \to \infty$ with

$$g_{n_k}(x) \longrightarrow f'(x) \text{ a.e.} \tag{2.27}$$

Putting (2.25), (2.26) and (2.27) together we see that $f^* = f' \in \mathscr{L}^1_\mu$ and that the convergence in (2.25) also happens in $L^1_\mu$. Finally we also get

$$\int f \, d\mu = \int g_n \, d\mu = \int f^* \, d\mu.$$

$\square$

A somewhat different approach is to use the maximal ergodic theorem (Theorem 2.24) to control the gap between mean convergence and pointwise convergence almost everywhere.

SECOND PROOF OF THEOREM 2.30. Assume first that $f_0 \in \mathscr{L}^\infty$. By the mean ergodic theorem in $L^1$ (Corollary 2.22) we know that the ergodic averages

$$\mathsf{A}_N(f_0) = \frac{1}{N} \sum_{n=0}^{N-1} f_0 \circ T^n \to F_0$$

converge in $L^1_\mu$ as $N \to \infty$ to some $T$-invariant function $F_0 \in \mathscr{L}^1_\mu$. Given $\varepsilon > 0$ choose some $M$ such that

$$\|F_0 - \mathsf{A}_M(f_0)\|_1 < \varepsilon^2.$$

By the maximal ergodic theorem (Theorem 2.24) applied to the function

$$g(x) = F_0(x) - \mathsf{A}_M(f_0)$$

we see that

$$\varepsilon \mu \left( \{ x \in X \mid \sup_{N \geq 1} |\mathsf{A}_N (F_0 - \mathsf{A}_M(f_0))| > \varepsilon \} \right) < \varepsilon^2.$$

Clearly $A_N(F_0) = F_0$ since the limit function $F_0$ is $T$-invariant, while if $M$ is fixed and $N \to \infty$ we have (see Exercise 2.6.4)

$$A_N(A_M(f_0)) = \frac{1}{NM} \sum_{n=0}^{N-1} \sum_{m=0}^{M-1} f_0 \circ T^{n+m}$$

$$= A_N(f_0) + O_M\left(\frac{\|f_0\|_\infty}{N}\right). \qquad (2.28)$$

Putting these together, we see that

$$\mu\left(\{x \mid \limsup_{N\to\infty} |F_0 - A_N(f_0)| > \varepsilon\}\right) = \mu\left(\{x \mid \limsup_{N\to\infty} |F_0 - A_N(A_M(f_0))| > \varepsilon\}\right)$$

$$\leqslant \mu\left(\{x \mid \sup_{N\geqslant 1} |A_N(F_0 - A_M(f_0))| > \varepsilon\}\right)$$

$$< \varepsilon,$$

which shows that $A_N(f_0) \to F_0$ almost everywhere.

To prove convergence for any $f \in \mathscr{L}_\mu^1$, fix $\varepsilon > 0$ and choose some $f_0 \in \mathscr{L}^\infty$ with $\|f - f_0\|_1 < \varepsilon^2$. Write $F \in \mathscr{L}_\mu^1$ for the $L^1$-limit of $A_N(f)$ and $F_0 \in \mathscr{L}_\mu^1$ for the $L^1$-limit of $A_N(f_0)$. Since $\|A_N(f) - A_N(f_0)\|_1 \leqslant \|f - f_0\|_1$ we deduce that $\|F - F_0\|_1 < \varepsilon^2$. From this we get

$$\mu\left(\{x \mid \limsup_{N\to\infty} |F - A_N(f)| > 2\varepsilon\}\right)$$

$$\leqslant \mu\left(\{x \mid |F - F_0| + \limsup_{N\to\infty} |F_0 - A_N(f_0)| + \sup_{N\geqslant 1} |A_N(f_0 - f)| > 2\varepsilon\}\right)$$

$$\leqslant \mu\left(\{x \mid |F - F_0| > \varepsilon\}\right) + \mu\left(\{x \mid \sup_{N\geqslant 1} |A_N(f_0 - f)| > \varepsilon\}\right)$$

$$\leqslant \varepsilon^{-1}\|F - F_0\|_1 + \varepsilon^{-1}\|f_0 - f\|_1 \leqslant 2\varepsilon \qquad (2.29)$$

by the maximal ergodic theorem (Theorem 2.24), which shows that $A_N(f)$ converges almost everywhere as $N \to \infty$. □

## Exercises for Sect. 2.6

**Exercise 2.6.1.** Prove the following version of the ergodic theorem for finite permutations (see the book of Nadkarni [263] where this is used to motivate a different approach to ergodic theorems). Let $X = \{x_1, \ldots, x_r\}$ be a finite set, and let $\sigma : X \to X$ be a permutation of $X$. The orbit of $x_j$ under $\sigma$ is the set $\{\sigma^n(x_j)\}_{n\geqslant 0}$, and $\sigma$ is called cyclic if there is an orbit of cardinality $r$.

(1) For a cyclic permutation $\sigma$ and any function $f : X \to \mathbb{R}$, prove that

$$\lim_{n\to\infty} \frac{1}{n} \sum_{j=0}^{n-1} f(\sigma^j x) = \frac{1}{r}\left(f(x_1) + \cdots + f(x_r)\right).$$

(2) More generally, prove that for any permutation $\sigma$ and function $f : X \to \mathbb{R}$,

$$\lim_{n\to\infty} \frac{1}{n} \sum_{j=0}^{n-1} f(\sigma^j x) = \frac{1}{p_x}\left(f(x) + f(\sigma(x)) + \cdots + f(\sigma^{p_x-1}(x))\right)$$

where the orbit of $x$ has cardinality $p_x$ under $\sigma$.

**Exercise 2.6.2.** Mimic the proof of Lemma 2.27 (or give the details of a deduction) to prove Corollary 2.28.

**Exercise 2.6.3.** Let $(X, \mathscr{B}, \mu, T)$ be an invertible measure-preserving system. Prove that, for any $f \in L^1_\mu$,

$$\lim_{N\to\infty} \frac{1}{N} \sum_{n=0}^{N-1} f(T^n x) = \lim_{N\to\infty} \frac{1}{N} \sum_{n=0}^{N-1} f(T^{-n} x)$$

almost everywhere.

**Exercise 2.6.4.** Fill in the details to prove the estimate in (2.28).

**Exercise 2.6.5.** Formulate and prove a pointwise ergodic theorem for a measurable function $f \geqslant 0$ with $\int f \, \mathrm{d}\mu = \infty$, under the assumption of ergodicity.

## 2.7 Strong-Mixing and Weak-Mixing

In this section we step back from thinking of measure-preserving transformations through the functional-analytic prism of their action on $L^p$ spaces to the more fundamental questions discussed in Sects. 2.2 and 2.3. Namely, if $A$ is a measurable set, what can be said about how the set $T^{-n}A$ is spread around the whole measure space for large $n$?

An easy consequence of the mean ergodic theorem is that a measure-preserving system $(X, \mathscr{B}, \mu, T)$ is ergodic if and only if

$$\frac{1}{N} \sum_{n=0}^{N-1} f \circ T^n \xrightarrow[L^2_\mu]{} \int f \, \mathrm{d}\mu$$

as $N \to \infty$ for every $f \in L^2_\mu$. It follows that $(X, \mathscr{B}, \mu, T)$ is ergodic if and only if

$$\frac{1}{N} \sum_{n=0}^{N-1} \langle f \circ T^n, g \rangle \longrightarrow \int f \, \mathrm{d}\mu \int g \, \mathrm{d}\mu \qquad (2.30)$$

as $N \to \infty$ for any $f, g \in L^2_\mu$. The characterization in (2.30) can be cast in terms of the behavior of sets to show that $(X, \mathscr{B}, \mu, T)$ is ergodic if and only if

$$\frac{1}{N} \sum_{n=0}^{N-1} \mu \left( A \cap T^{-n} B \right) \longrightarrow \mu(A)\mu(B) \qquad (2.31)$$

as $N \to \infty$ for all $A, B \in \mathscr{B}$. One direction is clear: if $T$ is ergodic, then the convergence (2.30) may be applied with $g = \chi_A$ and $f = \chi_B$.

Conversely, if $T^{-1}B = B$ then the convergence (2.31) with $A = X \setminus B$ implies that $\mu(X \setminus B)\mu(B) = 0$, so $T$ is ergodic.

There are several ways in which the convergence (2.31) might take place. Recall that measurable sets in $(X, \mathscr{B}, \mu)$ may be thought of as events in the sense of probability, and events $A, B \in \mathscr{B}$ are called *independent* if

$$\mu(A \cap B) = \mu(A)\mu(B).$$

Clearly if the action of $T$ contrives to make $T^{-n}B$ and $A$ become *independent* in the sense of probability for all large $n$, then the convergence (2.31) is assured. It turns out that this is too much to ask (see Exercise 2.7.1), but asking for $T^{-n}B$ and $A$ to become *asymptotically independent* leads to the following non-trivial definition.

**Definition 2.32.** A measure-preserving system $(X, \mathscr{B}, \mu, T)$ is *mixing* if

$$\mu \left( A \cap T^{-n} B \right) \longrightarrow \mu(A)\mu(B)$$

as $n \to \infty$, for all $A, B \in \mathscr{B}$.

Mixing is also sometimes called *strong-mixing*, in contrast to weak-mixing and mild-mixing.

*Example 2.33.* A circle rotation $R_\alpha : \mathbb{T} \to \mathbb{T}$ is not mixing. There is a sequence $n_j \to \infty$ for which $n_j \alpha \pmod 1 \to 0$ (if $\alpha$ is rational we may choose to have $n_j \alpha \pmod 1 = 0$). If $A = B = [0, \frac{1}{2}]$ then $m_{\mathbb{T}}(A \cap R_\alpha^{n_j} A) \to \frac{1}{2}$, so $R_\alpha$ is not mixing.

It is clear that some measure preserving systems make many sets become asymptotically independent as they move apart in time (that is, under iteration), leading to the following natural definition due to Rokhlin [316].

**Definition 2.34.** A measure-preserving system $(X, \mathscr{B}, \mu, T)$ is *k-fold mixing*, *mixing of order k* or *mixing on $k + 1$ sets* if

$$\mu \left( A_0 \cap T^{-n_1} A_1 \cap \cdots \cap T^{-n_k} A_k \right) \longrightarrow \mu(A_0) \cdots \mu(A_k)$$

as

$$n_1, n_2 - n_1, n_3 - n_2, \ldots, n_k - n_{k-1} \longrightarrow \infty$$

for any sets $A_0, \ldots, A_k \in \mathscr{B}$.

Thus mixing coincides with mixing of order 1. One of the outstanding open problems in classical ergodic theory is that it is not known[25] if mixing implies mixing of order $k$ for every $k \geqslant 1$.

Despite the natural definition, mixing turns out to be a rather special property, less useful and less prevalent than a slightly weaker property called weak-mixing introduced by Koopman and von Neumann [209][26]. Nonetheless, many natural examples are mixing of all orders (see the argument in Proposition 2.15 and Exercise 2.7.9 for example).

**Definition 2.35.** A measure-preserving system $(X, \mathscr{B}, \mu, T)$ is *weak-mixing* if

$$\frac{1}{N} \sum_{n=0}^{N-1} \left| \mu(A \cap T^{-n}B) - \mu(A)\mu(B) \right| \longrightarrow 0$$

as $N \to \infty$, for all $A, B \in \mathscr{B}$.

Notice that for any sequence $(a_n)$,

$$\lim_{n \to \infty} a_n = 0 \implies \lim_{n \to \infty} \frac{1}{n} \sum_{i=0}^{n} |a_i| = 0,$$

but the converse does not hold because the second property permits $|a_n|$ to be large along an infinite but thin set of values of $n$. Thus at the level simply of sequences, weak-mixing seems to be strictly weaker than strong-mixing. It turns out that this is also true for measure-preserving transformations—there are weak-mixing transformations that are not mixing[27].

Weak-mixing and its generalizations will turn out to be central to Furstenberg's proof of Szemerédi's theorem presented in Chap. 7. The first intimation that weak-mixing is a natural property comes from the fact that it has many equivalent formulations, and we will start to define and explore some of these in Theorem 2.36 below.

For one of these equivalent properties, it will be useful to recall some terminology concerning the operator $U_T$ on the Hilbert space $L^2_\mu$ associated to a measure-preserving transformation $T$ of $(X, \mathscr{B}, \mu)$. An *eigenvalue* is a number $\lambda \in \mathbb{C}$ for which there is an *eigenfunction* $f \in L^2_\mu$ with $U_T f = \lambda f$ almost everywhere. Notice that 1 is always an eigenvalue, since a constant function $f$ will satisfy $U_T f = f$. Any eigenvalue $\lambda$ lies on $\mathbb{S}^1$, since $U_T$ is an isometry of $L^2_\mu$. A measure-preserving transformation $T$ is said to have *continuous spectrum* if the only eigenvalue of $T$ is 1 and the only eigenfunctions are the constant functions.

Recall that a set $J \subseteq \mathbb{N}$ is said to have *density*

$$\mathbf{d}(J) = \lim_{n \to \infty} \frac{1}{n} \left| \{ j \in J \mid 1 \leqslant j \leqslant n \} \right|$$

if the limit exists.

**Theorem 2.36.** *The following properties of a system $(X, \mathscr{B}, \mu, T)$ are equivalent.*

(1) $T$ *is weakly mixing.*
(2) $T \times T$ *is ergodic with respect to* $\mu \times \mu$.
(3) $T \times T$ *is weakly mixing with respect to* $\mu \times \mu$.
(4) *For any ergodic measure-preserving system* $(Y, \mathscr{B}_Y, \nu, S)$, *the system*

$$(X \times Y, \mathscr{B} \otimes \mathscr{B}_Y, \mu \times \nu, T \times S)$$

*is ergodic.*
(5) *The associated operator* $U_T$ *has no non-constant measurable eigenfunctions (that is, $T$ has continuous spectrum).*
(6) *For every* $A, B \in \mathscr{B}$, *there is a set* $J_{A,B} \subseteq \mathbb{N}$ *with density zero for which*

$$\mu\left(A \cap T^{-n} B\right) \longrightarrow \mu(A)\mu(B)$$

*as* $n \to \infty$ *with* $n \notin J_{A,B}$.
(7) *For every* $A, B \in \mathscr{B}$,

$$\frac{1}{N} \sum_{n=0}^{N-1} \left|\mu(A \cap T^{-n}B) - \mu(A)\mu(B)\right|^2 \longrightarrow 0$$

*as* $N \to \infty$.

The proof of Theorem 2.36 will be given in Sect. 2.8.

**Corollary 2.37.** *If* $(X, \mathscr{B}_X, \mu, T)$ *and* $(Y, \mathscr{B}_Y, \nu, S)$ *are both weak-mixing, then the product system* $(X \times Y, \mathscr{B} \otimes \mathscr{C}, \mu \times \nu, T \times S)$ *is weak-mixing.*

**Corollary 2.38.** *If $T$ is weak-mixing, then for any $k$ the $k$-fold Cartesian product* $T \times \cdots \times T$ *is weak-mixing with respect to* $\mu \times \cdots \times \mu$.

**Corollary 2.39.** *If $T$ is weak-mixing, then for any $n \geqslant 1$, the $n$th iterate $T^n$ is weak-mixing.*

*Example 2.40.* We know that the circle rotation $R_\alpha : \mathbb{T} \to \mathbb{T}$ defined by

$$R_\alpha(t) = t + \alpha \quad (\mathrm{mod}\ 1)$$

is not mixing, but is ergodic if $\alpha \notin \mathbb{Q}$ (cf. Proposition 2.16 and Example 2.33). It is also not weak-mixing; this may be seen using Theorem 2.36(2) since the function $(x, y) \mapsto e^{2\pi i(x-y)}$ from $\mathbb{T} \times \mathbb{T} \to \mathbb{S}^1$ is a non-constant function preserved by $R_\alpha \times R_\alpha$.

## Exercises for Sect. 2.7

**Exercise 2.7.1.** Show that if a measure-preserving system $(X, \mathcal{B}, \mu, T)$ has the property that for any $A, B \in \mathcal{B}$ there exists $N$ such that

$$\mu\left(A \cap T^{-n} B\right) = \mu(A)\mu(B)$$

for all $n \geqslant N$, then it is trivial in the sense that $\mu(A) = 0$ or $1$ for every $A \in \mathcal{B}$.

**Exercise 2.7.2.** [28] Show that if a measure-preserving system $(X, \mathcal{B}, \mu, T)$ has the property that

$$\mu\left(A \cap T^{-n} B\right) \rightarrow \mu(A)\mu(B)$$

uniformly as $n \rightarrow \infty$ for every measurable $A \subseteq B \in \mathcal{B}$, then it is trivial in the sense that $\mu(A) = 0$ or $1$ for every $A \in \mathcal{B}$.

**Exercise 2.7.3.** This exercise generalizes the argument used in the proof of Proposition 2.15 and relates to the material in Appendix A. A collection $\mathscr{A}$ of measurable sets in $(X, \mathcal{B}, \mu)$ is called a *semi-algebra* (cf. Appendix A) if

- $\mathscr{A}$ contains the empty set;
- for any $A \in \mathscr{A}$, $X \smallsetminus A$ is a finite union of pairwise disjoint members of $\mathscr{A}$;
- for any $A_1, \ldots, A_r \in \mathscr{A}$, $A_1 \cap \cdots \cap A_r \in \mathscr{A}$.

The smallest $\sigma$-algebra containing $\mathscr{A}$ is called the $\sigma$-algebra generated by $\mathscr{A}$. Assume that $\mathscr{A}$ is a semi-algebra that generates $\mathcal{B}$, and prove the following characterizations of the basic mixing properties for a measure-preserving system $(X, \mathcal{B}, \mu, T)$:

(1) $T$ is mixing if and only if

$$\mu\left(A \cap T^{-n} B\right) \longrightarrow \mu(A)\mu(B)$$

as $n \rightarrow \infty$ for all $A, B \in \mathscr{A}$.

(2) $T$ is weak-mixing if and only if

$$\frac{1}{N} \sum_{n=0}^{N-1} \left|\mu(A \cap T^{-n} B) - \mu(A)\mu(B)\right| \longrightarrow 0$$

as $N \rightarrow \infty$ for all $A, B \in \mathscr{A}$.

(3) $T$ is ergodic if and only if

$$\frac{1}{N} \sum_{n=0}^{N-1} \mu\left(A \cap T^{-n} B\right) \longrightarrow \mu(A)\mu(B)$$

as $N \rightarrow \infty$ for all $A, B \in \mathscr{A}$.

**Exercise 2.7.4.** Let $\mathscr{A}$ be a generating semi-algebra in $\mathscr{B}$ (cf. Exercise 2.7.3), and assume that for $A \in \mathscr{A}$, $\mu\left(A\triangle T^{-1}A\right) = 0$ implies $\mu(A) = 0$ or 1. Does it follow that $T$ is ergodic?

**Exercise 2.7.5.** Show that a measure-preserving system $(X, \mathscr{B}, \mu, T)$ is mixing if and only if
$$\lim_{n\to\infty} \langle U_T^n f, g \rangle = \langle f, 1 \rangle \cdot \langle 1, g \rangle$$
for all $f$ and $g$ lying in a dense subset of $L_\mu^2$.

**Exercise 2.7.6.** Use Exercise 2.7.5 and the technique from Theorem 2.19 to prove the following.

(1) An ergodic automorphism of a compact abelian group is mixing with respect to Haar measure.
(2) An ergodic automorphism of a compact abelian group is mixing of all orders with respect to Haar measure.

**Exercise 2.7.7.** Show that a measure-preserving system $(X, \mathscr{B}, \mu, T)$ is weak-mixing if and only if
$$\lim_{N\to\infty} \frac{1}{N} \sum_{n=0}^{N-1} |\langle U_T^n f, g \rangle - \langle f, 1 \rangle \cdot \langle 1, g \rangle| = 0$$
for any $f, g \in L_\mu^2$

**Exercise 2.7.8.** Show that a measure-preserving system $(X, \mathscr{B}, \mu, T)$ is weak-mixing if and only if
$$\lim_{N\to\infty} \frac{1}{N} \sum_{n=0}^{N-1} |\langle U_T^n f, f \rangle - \langle f, 1 \rangle \cdot \langle 1, f \rangle| = 0$$
for any $f \in L_\mu^2$.

**Exercise 2.7.9.** Show that a Bernoulli shift (cf. Example 2.9) is mixing of order $k$ for every $k \geqslant 1$.

**Exercise 2.7.10.** Prove the following result due to Rényi [308]: a measure-preserving transformation $T$ is mixing if and only if
$$\mu(A \cap T^{-n}A) \to \mu(A)^2$$
for all $A \in \mathscr{B}$. Deduce that $T$ is mixing if and only if $\langle U_T^n f, f \rangle \to 0$ as $n \to \infty$ for all $f$ in a set of functions dense in the set of all $L^2$ functions of zero integral.

**Exercise 2.7.11.** Prove that a measure-preserving transformation $T$ is weak-mixing if and only if for any measurable sets $A, B, C$ with positive measure, there exists some $n \geqslant 1$ such that $T^{-n}A \cap B \neq \varnothing$ and $T^{-n}A \cap C \neq \varnothing$. (This is a result due to Furstenberg.)

**Exercise 2.7.12.** Write $T^{(k)}$ for the $k$-fold Cartesian product $T \times \cdots \times T$. Prove[29] that $T^{(k)}$ is ergodic for all $k \geqslant 2$ if and only if $T^{(2)}$ is ergodic.

**Exercise 2.7.13.** Let $T$ be an ergodic endomorphism of $\mathbb{T}^d$. The following exponential error rate for the mixing property[30],

$$\left| \langle f_1, U_T^n f_2 \rangle - \int f_1 \int f_2 \right| \leqslant S(f_1) S(f_2) \theta^n$$

for some $\theta < 1$ depending on $T$ and for a pair of constants $S(f_1), S(f_2)$ depending on $f_1, f_2 \in C^\infty(\mathbb{T}^d)$, is known to hold.
(a) Prove an exponential rate of mixing for the map $T_n : \mathbb{T} \to \mathbb{T}$ defined by $T_n(x) = nx \pmod 1$.
(b) Prove an exponential rate of mixing for the automorphism of $\mathbb{T}^2$ defined by $T : \binom{x}{y} \mapsto \binom{y}{x+y}$.
(c) Could an exponential rate of mixing hold for all continuous functions?

## 2.8 Proof of Weak-Mixing Equivalences

Some of the implications in Theorem 2.36 require the development of additional material; after developing it we will end this section with a proof of Theorem 2.36. The first lemma needed is a general one from analysis, due to Koopman and von Neumann [209].

**Lemma 2.41.** *Let $(a_n)$ be a bounded sequence of non-negative real numbers. Then the following are equivalent:*

(1) $\displaystyle \lim_{n \to \infty} \frac{1}{n} \sum_{j=0}^{n-1} a_j = 0$;

(2) *there is a set $J = J((a_n)) \subseteq \mathbb{N}$ with density zero for which $a_n \xrightarrow[n \notin J]{} 0$;*

(3) $\displaystyle \lim_{n \to \infty} \frac{1}{n} \sum_{j=0}^{n-1} a_j^2 = 0$.

PROOF. (1) $\implies$ (2): Let $J_k = \{j \in \mathbb{N} \mid a_j > \frac{1}{k}\}$, so that

$$J_1 \subseteq J_2 \subseteq J_3 \subseteq \cdots . \tag{2.32}$$

For each $k \geqslant 1$,

$$\frac{1}{k} |J_k \cap [0, n)| < \sum_{\substack{i=0,\ldots,n-1, \\ a_i > 1/k}} a_i \leqslant \sum_{i=0}^{n-1} a_i.$$

It follows that

$$\frac{1}{n}|J_k \cap [0,n)| \leqslant k\frac{1}{n}\sum_{i=0}^{n-1}a_i \longrightarrow 0$$

as $n \to \infty$ for each $k \geqslant 1$, so each $J_k$ has zero density. We will construct the set $J$ by taking a union of segments of each set $J_k$. Since each of the sets $J_k$ has zero density, we may inductively choose numbers $0 < \ell_1 < \ell_2 < \cdots$ with the property that

$$\frac{1}{n}|J_k \cap [0,n)| \leqslant \frac{1}{k} \tag{2.33}$$

for $n \geqslant \ell_k$ and any $k \geqslant 1$. Define the set $J$ by

$$J = \bigcup_{k=0}^{\infty}\left(J_k \cap [\ell_k, \ell_{k+1})\right).$$

We claim two properties for the set $J$, namely

- $a_n \xrightarrow[n \notin J]{} 0$ as $n \to \infty$;
- $J$ has density zero.

For the first claim, note that $J_k \cap [\ell_k, \infty) \subseteq J$ by (2.32), so if $J \not\ni n \geqslant \ell_k$ then $n \notin J_k$, and so $a_n \leqslant \frac{1}{k}$. This shows that $a_n \xrightarrow[n \notin J]{} 0$ as claimed.

For the second claim, notice that if $n \in [\ell_k, \ell_{k+1})$ then again by (2.32) $J \cap [0,n) \subseteq J_k \cap [0,n)$ and so

$$\frac{1}{n}|J \cap [0,n)| \leqslant \frac{1}{k}$$

by (2.33), showing that $J$ has density zero.

(2) $\implies$ (1): The sequence $(a_n)$ is bounded, so there is some $R > 0$ with $a_n \leqslant R$ for all $n \geqslant 1$. For each $k \geqslant 1$ choose $N_k$ so that

$$J \not\ni n \geqslant N_k \implies a_n < \frac{1}{k}$$

and so that

$$n \geqslant N_k \implies \frac{1}{n}|J \cap [0,n)| \leqslant \frac{1}{k}.$$

Then for $n \geqslant kN_k$,

$$\frac{1}{n}\sum_{i=0}^{n-1}a_i = \frac{1}{n}\left(\sum_{i=0}^{N_k-1}a_i + \sum_{\substack{i \in J, \\ N_k \leqslant i < n}}a_i + \sum_{\substack{i \notin J, \\ N_k \leqslant i < n}}a_i\right)$$

$$< \frac{1}{n}\left(RN_k + R|J \cap [0,n)| + n\frac{1}{k}\right)$$

$$\leqslant \frac{2R+1}{k},$$

showing (1).

(3) $\iff$ (1): This is clear from the characterization (2) of property (1).
$\square$

PROOF OF THEOREM 2.36. Properties (1), (6) and (7) are equivalent by Lemma 2.41 applied with $a_n = |\mu(A \cap T^{-n}B) - \mu(A)\mu(B)|$.

(6) $\implies$ (3): Given sets $A_1, B_1, A_2, B_2 \in \mathscr{B}$, property (6) gives sets $J_1$ and $J_2$ of density zero with

$$\mu\left(A_1 \cap T^{-n}B_1\right) \xrightarrow[n \notin J_1]{} \mu(A_1)\mu(B_1)$$

and

$$\mu\left(A_2 \cap T^{-n}B_2\right) \xrightarrow[n \notin J_2]{} \mu(A_2)\mu(B_2).$$

Let $J = J_1 \cup J_2$; this still has density zero and

$$\lim_{J \not\ni n \to \infty} \left|(\mu \times \mu)\left((A_1 \times A_2) \cap (T \times T)^{-n}(B_1 \times B_2)\right)\right.$$

$$\left. -(\mu \times \mu)(A_1 \times A_2) \cdot (\mu \times \mu)(B_1 \times B_2)\right|$$

$$= \lim_{J \not\ni n \to \infty} \left|\mu(A_1 \cap T^{-n}B_1) \cdot \mu(A_2 \cap T^{-n}B_2)\right.$$

$$\left. -\mu(A_1)\mu(A_2)\mu(B_1)\mu(B_2)\right|$$

$$= 0,$$

so $T \times T$ is weak-mixing since the measurable rectangles generate $\mathscr{B} \times \mathscr{B}$.

(3) $\implies$ (1): If $T \times T$ is weak-mixing, then property (1) holds in particular for subsets of $X \times X$ of the form $A \times X$ and $B \times X$, which shows that (1) holds for $T$, so $T$ is weak-mixing.

(1) $\implies$ (4): Let $(Y, \mathscr{B}_Y, \nu, S)$ be an ergodic system and assume that $T$ is weak-mixing. For measurable sets $A_1, B_1 \in \mathscr{B}$ and $A_2, B_2 \in \mathscr{B}_Y$,

$$\frac{1}{N}\sum_{n=0}^{N-1}(\mu \times \nu)\left(A_1 \times A_2 \cap (T \times S)^{-n}(B_1 \times B_2)\right)$$

$$= \frac{1}{N}\sum_{n=0}^{N-1}\mu(A_1 \cap T^{-n}B_1)\nu(A_2 \cap S^{-n}B_2)$$

$$= \frac{1}{N}\sum_{n=0}^{N-1}\mu(A_1)\mu(B_1)\nu(A_2 \cap S^{-n}B_2)$$

$$+ \frac{1}{N}\sum_{n=0}^{N-1}\left[\mu(A_1 \cap T^{-n}B_1) - \mu(A_1)\mu(B_1)\right]\nu(A_2 \cap S^{-n}B_2). \quad (2.34)$$

By the characterization in (2.31) and ergodicity of $S$, the expression on the right in (2.34) converges to

$$\mu(A_1)\mu(B_1)\nu(A_2)\nu(B_2).$$

The second term in (2.34) is dominated by

$$\frac{1}{N}\sum_{n=0}^{N-1}\left|\mu(A_1\cap T^{-n}B_1)-\mu(A_1)\mu(B_1)\right|$$

which converges to 0 since $T$ is weak-mixing. It follows that

$$\frac{1}{N}\sum_{n=0}^{N-1}(\mu\times\nu)\left(A_1\times A_2\cap(T\times S)^{-n}(B_1\times B_2)\right)\longrightarrow\mu(A_1)\mu(B_1)\nu(A_2)\nu(B_2)$$

so $T\times S$ is ergodic by the characterization in (2.31).

(4) $\implies$ (2): Let $(Y,\mathscr{B}_Y,\nu,S)$ be the ergodic system defined by the identity map on the singleton $Y=\{y\}$. Then $T\times S$ is isomorphic to $T$, so (4) shows that $T$ is ergodic. Invoking (4) again now shows that $T\times T$ is ergodic, proving (2).

(2) $\implies$ (7): We must show that

$$\frac{1}{N}\sum_{n=0}^{N-1}\left|\mu(A\cap T^{-n}B)-\mu(A)\mu(B)\right|^2\longrightarrow 0$$

as $N\to\infty$, for every $A,B\in\mathscr{B}$. Let $\mu^2$ denote the product measure $\mu\times\mu$ on $(X\times X,\mathscr{B}\otimes\mathscr{B})$. By the ergodicity of $T\times T$,

$$\frac{1}{N}\sum_{n=0}^{N-1}\mu\left(A\cap T^{-n}B\right)=\frac{1}{N}\sum_{n=0}^{N-1}\mu^2\left((A\times X)\cap(T\times T)^{-n}(B\times X)\right)$$

$$\longrightarrow\mu^2\left(A\times X\right)\cdot\mu^2\left(B\times X\right)=\mu(A)\mu(B)$$

and

$$\frac{1}{N}\sum_{n=0}^{N-1}\left(\mu\left(A\cap T^{-n}B\right)\right)^2=\frac{1}{N}\sum_{n=0}^{N-1}\mu^2\left((A\times A)\cap(T\times T)^{-n}(B\times B)\right)$$

$$\longrightarrow\mu^2(A\times A)\cdot\mu^2(B\times B)=\mu(A)^2\mu(B)^2.$$

It follows that

$$\frac{1}{N}\sum_{n=0}^{N-1}\left[\mu\left(A\cap T^{-n}B\right)-\mu(A)\mu(B)\right]^2 = \frac{1}{N}\sum_{n=0}^{N-1}\mu\left(A\cap T^{-n}B\right)^2$$

$$+\mu(A)^2\mu(B)^2$$

$$-2\mu(A)\mu(B)\frac{1}{N}\sum_{n=0}^{N-1}\mu\left(A\cap T^{-n}B\right)$$

$$\to 2\mu(A)^2\mu(B)^2 - 2\mu(A)^2\mu(B)^2 = 0,$$

so (7) holds.

(2) $\implies$ (5): Suppose that $f$ is a measurable eigenfunction for $T$, so

$$U_T f = \lambda f$$

for some $\lambda \in \mathbb{S}^1$. Define a measurable function on $X \times X$ by

$$g(x_1, x_2) = f(x_1)\overline{f(x_2)};$$

then

$$U_{T\times T}g(x,y) = g(Tx, Ty) = \lambda\bar\lambda g(x,y) = g(x,y)$$

so by ergodicity of $T \times T$, $g$ (and hence $f$) must be constant almost everywhere.

All that remains is to prove that (5) $\implies$ (2), and this is considerably more difficult. There are several different proofs, each of which uses a nontrivial result from functional analysis[31]. Assume that $T \times T$ is not ergodic, so there is a non-constant function $f \in L^2_{\mu^2}(X \times X)$ that is almost everywhere invariant under $T \times T$. We would like to have the additional symmetry property $f(x,y) = \overline{f(y,x)}$ for all $(x,y) \in X \times X$. To obtain this additional property, consider the functions

$$(x,y) \mapsto f(x,y) + \overline{f(y,x)}$$

and

$$(x,y) \mapsto i(f(x,y) - \overline{f(y,x)}).$$

Notice that if both of these functions are constant, then $f$ must be constant. It follows that one of them must be non-constant. So without loss of generality we may assume that $f$ satisfies $f(x,y) = \overline{f(y,x)}$. We may further suppose (by subtracting $\int f\,d\mu^2$) that $\int f\,d\mu^2 = 0$. It follows that the operator $F$ on $L^2_\mu$ defined by

$$(F(g))(x) = \int_X f(x,y)g(y)\,d\mu(y)$$

is a non-trivial self-adjoint compact[(32)] operator, and so by Theorem B.3 has at least one non-zero eigenvalue $\lambda$ whose corresponding eigenspace $V_\lambda$ is finite-dimensional. We claim that the finite-dimensional space $V_\lambda \subseteq L^2_\mu$ is invariant under $T$. To see this, assume that $F(g) = \lambda g$. Then

$$\lambda g(Tx) = \int_X f(Tx, y)g(y)\, d\mu(y)$$
$$= \int_X f(Tx, Ty)g(Ty)\, d\mu(y) \quad \text{(since } \mu \text{ is } T\text{-invariant)}$$
$$= \int_X f(x, y)g(Ty)\, d\mu(y),$$

since $f$ is $T \times T$-invariant, so $F(g \circ T) = \lambda(g \circ T)$ and thus $g \circ T \in V_\lambda$. It follows that $U_T$ restricted to $V_\lambda$ is a non-trivial linear map of a finite-dimensional linear space, and therefore has a non-trivial eigenvector. Since $\int f\, d\mu^2 = 0$, any such eigenvector is non-constant. □

## 2.8.1 Continuous Spectrum and Weak-Mixing

A more conventional proof of the difficult step in Theorem 2.36, which may be taken to be (5) $\implies$ (1), proceeds via the Spectral theorem (Theorem B.4) in the following form.

ALTERNATIVE PROOF OF (5) $\implies$ (1) IN THEOREM 2.36. Definition 2.35 is clearly equivalent to the property that

$$\lim_{N \to \infty} \frac{1}{N} \sum_{n=0}^{N-1} |\langle U_T^n f, g \rangle - \langle f, 1 \rangle \cdot \langle 1, g \rangle| = 0$$

for any $f, g \in L^2_\mu$, and by polarization this is in turn equivalent to

$$\lim_{N \to \infty} \frac{1}{N} \sum_{n=0}^{N-1} |\langle U_T^n f, f \rangle - \langle f, 1 \rangle \cdot \langle 1, f \rangle| = 0$$

for any $f \in L^2_\mu$ (see Exercise 2.7.8 and page 441). By subtracting $\int_X f\, d\mu$ from $f$, it is therefore enough to show that if $f \in L^2_\mu$ has $\int_X f\, d\mu = 0$, then

$$\frac{1}{N} \sum_{n=0}^{N-1} |\langle U_T^n f, f \rangle|^2 \longrightarrow 0$$

as $N \to \infty$. By (B.1), it is enough to show that for the non-atomic measure $\mu_f$ on $\mathbb{S}^1$,

$$\frac{1}{N}\sum_{n=0}^{N-1}\left|\int_{\mathbb{S}^1} z^n \, d\mu_f(z)\right|^2 \longrightarrow 0 \qquad (2.35)$$

as $N \to \infty$. Since $\overline{z^n} = z^{-n}$ for $z \in \mathbb{S}^1$ the product in (2.35) may be expanded to give

$$\frac{1}{N}\sum_{n=0}^{N-1}\left|\int_{\mathbb{S}^1} z^n \, d\mu_f(z)\right|^2 = \frac{1}{N}\sum_{n=0}^{N-1}\left(\int_{\mathbb{S}^1} z^n \, d\mu_f(z) \cdot \int_{\mathbb{S}^1} w^{-n} \, d\mu_f(w)\right)$$

$$= \frac{1}{N}\sum_{n=0}^{N-1}\int_{\mathbb{S}^1 \times \mathbb{S}^1} (z/w)^n \, d\mu_f^2(z,w) \quad \text{(by Fubini)}$$

$$= \int_{\mathbb{S}^1 \times \mathbb{S}^1}\left(\frac{1}{N}\sum_{n=0}^{N-1}(z/w)^n\right) d\mu_f^2(z,w).$$

The measure $\mu_f$ is non-atomic so the diagonal set $\{(z,z) \mid z \in \mathbb{S}^1\} \subseteq \mathbb{S}^1 \times \mathbb{S}^1$ has zero $\mu_f^2$-measure. For $z \neq w$,

$$\frac{1}{N}\sum_{n=0}^{N-1}(z/w)^n = \frac{1}{N}\left(\frac{1-(z/w)^N}{1-(z/w)}\right) \longrightarrow 0$$

as $N \to \infty$, so the convergence (2.35) holds by the dominated convergence theorem (Theorem A.18). □

## Exercises for Sect. 2.8

**Exercise 2.8.1.** Is the hypothesis that the sequence $(a_n)$ be bounded necessary in Lemma 2.41?

**Exercise 2.8.2.** Give an alternative proof of (1) $\implies$ (5) in Theorem 2.36 by proving the following statements:

(1) Any factor of a weak-mixing transformation is weak-mixing.
(2) A complex-valued eigenfunction $f$ of $U_T$ has constant modulus.
(3) If $f$ is an eigenfunction of $U_T$, then $x \mapsto \arg\left(f(x)/|f(x)|\right)$ is a factor map from $(X, \mathscr{B}, \mu, T)$ to $(\mathbb{T}, \mathscr{B}_{\mathbb{T}}, m_{\mathbb{T}}, R_\alpha)$ for some $\alpha$.

**Exercise 2.8.3.** Show the following converse to Exercise 2.5.6: if a measure-preserving system $(Y, \mathscr{B}_Y, \nu, S)$ is not totally ergodic then there exists a measure-preserving system $(X, \mathscr{B}, \mu, T)$ and a $K > 1$ with the property that $(Y, \mathscr{B}_Y, \nu, S)$ is measurably isomorphic to the system

$$(X^{(K)}, \mathscr{B}^{(K)}, \mu^{(K)}, T^{(K)})$$

constructed in Exercise 2.5.6.

**Exercise 2.8.4.** Give a different proof[33] of the mean ergodic theorem (Theorem 2.21) as follows. For a measure-preserving system $(X, \mathscr{B}, \mu, T)$ and function $f \in L_\mu^2$, show that the function $n \mapsto \langle U_T^n f, f \rangle$ is positive-definite (see Sect. C.3). Apply the Herglotz–Bochner theorem (Theorem C.9) to translate the problem into one concerned with functions on $\mathbb{S}^1$, and there use the fact that $\frac{1}{N} \sum_{n=1}^{N} \rho^n$ converges for $\rho \in \mathbb{S}^1$ (to zero, unless $\rho = 1$).

## 2.9 Induced Transformations

Poincaré recurrence gives rise to an important inducing construction introduced by Kakutani [172]. Throughout this section, $(X, \mathscr{B}, \mu, T)$ denotes an invertible measure-preserving system[34].

Let $(X, \mathscr{B}, \mu, T)$ be an invertible measure-preserving system, and let $A$ be a measurable set with $\mu(A) > 0$. By Poincaré recurrence, the first return time to $A$, defined by

$$r_A(x) = \inf_{n \geq 1} \{n \mid T^n(x) \in A\} \tag{2.36}$$

exists (that is, is finite) almost everywhere.

**Definition 2.42.** The map $T_A : A \to A$ defined (almost everywhere) by

$$T_A(x) = T^{r_A(x)}(x)$$

is called the transformation *induced* by $T$ on the set $A$.

Notice that both $r_A : X \to \mathbb{N}$ and $T_A : A \to A$ are measurable by the following argument. For $n \geq 1$, write $A_n = \{x \in A \mid r_a(x) = n\}$. Then the sets

$$A_1 = A \cap T^{-1} A,$$
$$A_2 = A \cap T^{-2} A \smallsetminus A_1,$$
$$\vdots$$
$$A_n = A \cap T^{-n} A \smallsetminus \bigcup_{i < n} A_i$$

are all measurable, as is

$$T^n A_n = A \cap T^n A \smallsetminus (TA \cup T^2 A \cup \cdots \cup T^{n-1} A),$$

since $T$ is invertible by assumption.

**Lemma 2.43.** *The induced transformation $T_A$ is a measure-preserving transformation on the space $(A, \mathscr{B}|_A, \mu_A = \frac{1}{\mu(A)} \mu|_A, T_A)$. If $T$ is ergodic with respect to $\mu$ then $T_A$ is ergodic with respect to $\mu_A$.*

The notation means that the $\sigma$-algebra consists of $\mathscr{B}\big|_A = \{B \cap A \mid B \in \mathscr{B}\}$ and the measure is defined for $B \in \mathscr{B}\big|_A$ by $\mu_A(B) = \frac{1}{\mu(A)}\mu(B)$. The effect of $T_A$ is seen in the *Kakutani skyscraper* Fig. 2.2. The original transformation $T$ sends any point with a floor above it to the point immediately above on the next floor, and any point on a top floor is moved somewhere to the base floor $A$. The induced transformation $T_A$ is the map defined almost everywhere on the bottom floor by sending each point to the point obtained by going through all the floors above it and returning to $A$.

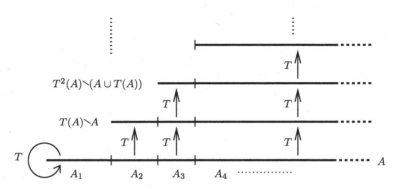

**Fig. 2.2** The induced transformation $T_A$

PROOF OF LEMMA 2.43. If $B \subseteq A$ is measurable, then $B = \bigsqcup_{n \geqslant 1} B \cap A_n$ is a disjoint union so

$$\mu_A(B) = \frac{1}{\mu(A)} \sum_{n \geqslant 1} \mu(B \cap A_n). \qquad (2.37)$$

Now

$$T_A(B) = \bigsqcup_{n \geqslant 1} T_A(B \cap A_n) = \bigsqcup_{n \geqslant 1} T^n(B \cap A_n),$$

so

$$\mu_A(T_A(B)) = \frac{1}{\mu(A)} \sum_{n \geqslant 1} \mu(T^n(B \cap A_n))$$

$$= \frac{1}{\mu(A)} \sum_{n \geqslant 1} \mu(B \cap A_n) \qquad \text{(since } T \text{ preserves } \mu)$$

$$= \mu(B)$$

by (2.37).

If $T_A$ is not ergodic, then there is a $T_A$-invariant measurable set $B \subseteq A$ with $0 < \mu(B) < \mu(A)$; it follows that $\bigcup_{n \geqslant 1} \bigcup_{j=0}^{n-1} T^j(B \cap A_n)$ is a nontrivial $T$-invariant set, showing that $T$ is not ergodic. $\qquad \square$

Poincaré recurrence (Theorem 2.11) says that for any measure-preserving system $(X, \mathscr{B}, \mu, T)$ and set $A$ of positive measure, almost every point on the ground floor of the associated Kakutani skyscraper returns to the ground floor at some point. Ergodicity strengthens this statement to say that almost every point of the entire space $X$ lies on some floor of the skyscraper. This enables a quantitative version of Poincaré recurrence to be found, a result due to Kac [168].

**Theorem 2.44 (Kac).** *Let $(X, \mathscr{B}, \mu, T)$ be an ergodic measure-preserving system and let $A \in \mathscr{B}$ have $\mu(A) > 0$. Then the expected return time to $A$ is $\frac{1}{\mu(A)}$; equivalently*

$$\int_A r_A \, d\mu = 1.$$

PROOF[35]. Referring to Fig. 2.2, each column

$$A_n \sqcup T(A_n) \sqcup \cdots \sqcup T^{n-1}(A_n)$$

comprises $n$ disjoint sets each of measure $\mu(A_n)$, and the entire skyscraper contains almost all of $X$ by ergodicity and Proposition 2.14(3) applied to the transformation $T^{-1}$. It follows that

$$1 = \mu(X) = \sum_{n \geqslant 1} n\mu(A_n) = \int_A r_A \, d\mu$$

by the monotone convergence theorem (Theorem A.16), since $r_A$ is the increasing limit of the functions $\sum_{k=1}^{n} k\chi_{A_k}$ as $n \to \infty$. $\qquad \square$

Kakutani skyscrapers are a powerful tool in ergodic theory. A simple application is to prove the Kakutani–Rokhlin lemma (Lemma 2.45) proved by Kakutani [172] and Rokhlin [315].

**Lemma 2.45 (Kakutani–Rokhlin).** *Let $(X, \mathscr{B}, \mu, T)$ be an invertible ergodic measure-preserving system and assume that $\mu$ is non-atomic (that is, $\mu(\{x\}) = 0$ for all $x \in X$). Then for any $n \geqslant 1$ and $\varepsilon > 0$ there is a set $B \in \mathscr{B}$ with the property that*

$$B, T(B), \ldots, T^{n-1}(B)$$

*are disjoint sets, and*

$$\mu\left(B \sqcup T(B) \sqcup \cdots \sqcup T^{n-1}(B)\right) > 1 - \varepsilon.$$

As the proof will show, the lemma uses only division (constructing a quotient and remainder) and the Kakutani skyscraper.

PROOF OF LEMMA 2.45. Let $A$ be a measurable set with $0 < \mu(A) < \varepsilon/n$ (such a set exists by the assumption that $\mu$ is non-atomic) and form the Kakutani skyscraper over $A$. Then $X$ decomposes into a union of disjoint columns of the form

$$A_k \sqcup T(A_k) \sqcup \cdots \sqcup T^{k-1}(A_k)$$

for $k \geqslant 1$, as in Fig. 2.2. Now let

$$B = \bigsqcup_{k \geqslant n} \bigsqcup_{j=0}^{\lfloor k/n \rfloor - 1} T^{jn}(A_k),$$

the set obtained by grouping together that part of the ground floor made up of the sets $A_k$ with $k \geqslant n$ together with every $n$th floor above that part of the ground floor (stopping before the top of the skyscraper). By construction the sets $B, T(B), \ldots, T^{n-1}(B)$ are disjoint, and together they cover all of $X$ apart from a set comprising no more than $n$ of the floors in each of the towers, which therefore has measure no more than $n \sum_{k=1}^{\infty} \mu(A_k) \leqslant n\mu(A) < \varepsilon$.  □

One often refers to the structure given by Lemma 2.45 as a *Rokhlin tower* of *height* $n$ with *base* $B$ and *residual set of size* $\varepsilon$.

## Exercises for Sect. 2.9

**Exercise 2.9.1.** Show that the inducing construction can be reversed in the following sense. Let $(X, \mathscr{B}, \mu, T)$ be a measure-preserving system, and let $r : X \to \mathbb{N}_0$ be a map in $L_\mu^1$. The *suspension* defined by $r$ is the system $(X^{(r)}, \mathscr{B}^{(r)}, \mu^{(r)}, T^{(r)})$, where:

- $X^{(r)} = \{(x, n) \mid 0 \leqslant n < r(x)\}$;
- $\mathscr{B}^{(r)}$ is the product $\sigma$-algebra of $\mathscr{B}$ and the Borel $\sigma$-algebra on $\mathbb{N}$ (which comprises all subsets);
- $\mu^{(r)}$ is defined by $\mu^{(r)}(A \times N) = \frac{1}{\int r \, d\mu} \mu(A) \times |N|$ for $A \in \mathscr{B}$ and $N \subseteq \mathbb{N}$; and
- $T^{(r)}(x, n) = \begin{cases} (x, n+1) & \text{if } n+1 < r(x); \\ (T(x), 0) & \text{if } n+1 = r(x). \end{cases}$

(a) Verify that this defines a finite measure-preserving system.
(b) Show that the induced map on the set $A = \{(x, 0) \mid x \in X\}$ is isomorphic to the original system $(X, \mathscr{B}, \mu, T)$.

**Exercise 2.9.2.** [36] The hypothesis of ergodicity in Lemma 2.45 can be weakened as follows. An invertible measure-preserving system $(X, \mathscr{B}, \mu, T)$ is called *aperiodic* if $\mu\left(\{x \in X \mid T^k(x) = x\}\right) = 0$ for all $k \in \mathbb{Z} \smallsetminus \{0\}$.
(a) Show that an ergodic transformation on a non-atomic space is aperiodic.
(b) Find an example of an aperiodic transformation on a non-atomic space that is not ergodic.
(c) Prove Lemma 2.45 for an invertible aperiodic transformation on a non-atomic space.

**Exercise 2.9.3.** [37] Show that the Kakutani–Rokhlin lemma (Lemma 2.45) does not hold for arbitrary sequences of iterates of the map $T$. Specifically, show that for an ergodic measure-preserving system $(X, \mathscr{B}, \mu, T)$, sequence $a_1, \ldots, a_n$ of distinct integers, and $\varepsilon > 0$ it is not always possible to find a measurable set $A$ with the properties that $T^{a_1}(A), \ldots, T^{a_n}(A)$ are disjoint and $\mu\left(\bigcup_{i=1}^{n} T^{a_i}(A)\right) > \varepsilon$.

**Exercise 2.9.4.** Use Exercise 2.9.2 above to prove the following result of Steele [351]. Let $(X, \mathscr{B}, \mu, T)$ be an invertible aperiodic measure-preserving system on a non-atomic space. Then, for any $\varepsilon > 0$, there is a set $A \in \mathscr{B}$ with $\mu(A) < \varepsilon$ with the property that for any finite set $F \subseteq X$, there is some $j = j(F)$ with $F \subseteq T^{-j}(A)$.

# Notes to Chap. 2

[12] (Page 16) A measurable isomorphism is also sometimes called a *conjugacy*; conjugacy is also used to describe an isomorphism between the measure algebras that implies isomorphism on sufficiently well-behaved probability spaces. This is discussed in Walters [374, Sect. 2.2] and Royden [320].

[13] (Page 17) The shift maps constructed here are measure-preserving transformations, but they are also homeomorphisms of a compact metric space in a natural way. The study of the dynamics of closed shift-invariant subsets of these systems comprises *symbolic dynamics* and is a rich theory in itself. A gentle introduction may be found in the book of Lind and Marcus [230] or Kitchens [197]; further reading in the collection edited by Berthé, Ferenczi, Mauduit and Siegel [93].

[14] (Page 21) Poincaré's formulation in [288, Th. I, p. 69] is as follows:

"Supposons que le point $P$ reste à distance finie, et que le volume

$$\int \mathrm{d}x_1 \, \mathrm{d}x_2 \, \mathrm{d}x_3$$

soit un invariant intégral; si l'on considère une région $r_0$ quelconque, quelque petite que soite cette région, il y aura des trajectoires qui la traverseront une infinité de fois. [...] En effet le point $P$ restant à distance finie, ne sortira jamais d'une région limitée $R$."

The modern abstract measure-theoretic statement in Theorem 2.11 appears in a paper of Carathéodory [49].

[15] (Page 23) The notion of ergodicity predates the ergodic theorems of the 1930s, in various guises. These include the seminal work of Borel [40], described by Doob as being

"characterized by convenient neglect of error terms in asymptotics, incorrect reasoning, and correct results,"

as well as that of Knopp [205]; a striking remark of Novikoff and Barone [273] is that a result implicit in the work of van Vleck [370] on non-measurable subsets of $[0,1]$ is that any measurable subset of $[0,1]$ invariant under the map $x \mapsto 2x \pmod 1$ has measure zero or one, a prototypical ergodic statement. The general formulation was given by Birkhoff and Smith [35].

$(16)$ (Page 28) These operators are usually called Koopman operators; Koopman [208] used the then-recent development of functional analysis and Hilbert space by von Neumann [266] and Stone [354] to use these operators in the setting of flows arising in classical Hamiltonian mechanics.

$(17)$ (Page 30) Even though this is not necessary here, we assume for simplicity that Hilbert spaces are separable, and as a result that they have countable orthonormal bases. As discussed in Sect. A.6, we only need the separable case.

$(18)$ (Page 32) For a recent account of the history of the relationship between the two results and the account of how they came to be published as and when they did, see Zund [395]. The issue has also been discussed by Ulam [365] and others. The note [25] by Bergelson discusses both the history and how the two results relate to more recent developments.

$(19)$ (Page 35) This result is simply one of many extensions and generalizations of the mean ergodic theorem (Theorem 2.21) to other complete function spaces. It is a special instance of the mean ergodic theorem for Banach spaces, due to Kakutani and Yosida [171, 391, 392].

$(20)$ (Page 38) The maximal ergodic theorem is due to Wiener [382] and was also proved by Yosida and Kakutani [392].

$(21)$ (Page 40) Covering lemmas of this sort were introduced by Vitali [369], and later became important tools in the proof of the Hardy–Littlewood maximal inequality, and thence of the Lebesgue density and differentiation theorems (Theorems A.24 and A.25).

$(22)$ (Page 44) Birkhoff based his proof on a weaker maximal inequality concerning the set of points on which $\limsup_{n \to \infty} \mathsf{A}_n^f \geqslant \alpha$, and initially formulated his result for indicator functions in the setting of a closed analytic manifold with a finite invariant measure. Khinchin [189] showed that Birkhoff's result applies to integrable functions on abstract finite measure spaces, but made clear that the idea of the proof is precisely that used by Birkhoff. A natural question concerning Theorem 2.30, or indeed any convergence result, is whether anything can be said about the rate of convergence. An important special case is the law of the iterated logarithm due to Hartman and Wintner [141]: if $\|f\|_2 = 1$, $\int f \, d\mu = 0$ and the functions $f, U_T f, U_T^2 f, \dots$ are all independent, then

$$\limsup_{n \to \infty} \mathsf{A}_n^f / \sqrt{(2 \log \log n)/n} = 1$$

almost everywhere (and $\liminf = -1$ by symmetry). It follows that

$$\mathsf{A}_n^f = \mathrm{O}\left(\left(\tfrac{1}{n} \log \log n\right)^{1/2}\right)$$

almost everywhere. However, the hypothesis of independence is essential: Krengel [210] showed that for any ergodic Lebesgue measure-preserving transformation $T$ of $[0,1]$ and sequence $(a_n)$ with $a_n \to 0$ as $n \to \infty$, there is a continuous function $f : [0,1] \to \mathbb{R}$ for which

$$\limsup_{n \to \infty} \frac{1}{a_n} \left| \mathsf{A}_n^f - \int f \, dm \right| = \infty$$

almost everywhere, and

$$\limsup_{n \to \infty} \frac{1}{a_n} \left\| \mathsf{A}_n^f - \int f \, dm \right\|_p = \infty$$

for $1 \leqslant p \leqslant \infty$. An extensive treatment of ergodic theorems may be found in the monograph of Krengel [211].

Despite the absence of any general rate bounds in the ergodic theorem, the constructive approach to mathematics has produced rate results in a different sense, which may lead to effective versions of results like the multiple recurrence theorem. Bishop's work [36] included a form of ergodic theorem, and Spitters [348] found constructive characterizations of the ergodic theorem. As an application of 'proof mining', Avigad, Gerhardy and Towsner [12] gave bounds on the rate of convergence that can be explicitly computed in terms of the initial data ($T$ and $f$) under a weak hypotheses, while earlier work of Simic and Avigad [13, 346] showed that, in general, it is impossible to compute such a bound. An overview of this area and its potential may be found in the survey [11] by Avigad.

[23] (Page 44) Despite the impressive result in Example 2.31, the numbers known to be normal to every base have been constructed to meet the definition of normality (with the remarkable exception of Chaitin's constant [53]). Champernowne [54] showed that the specific number $0.123456789101112131415\ldots$ is normal in base 10, and Sierpiński [345] constructed a number normal to every base. Sierpiński's construction was reformulated to be recursive by Becher and Figueira [20], giving a computable number normal to every base. The irrational numbers arising naturally in other fields, like $\pi, e, \zeta(3), \sqrt{2}$, and so on, are not known to be normal to any base.

[24] (Page 44) There are many proofs of the pointwise ergodic theorem; in addition to that of Birkhoff [33] there is a more elementary (though intricate) argument due to Katznelson and Weiss [186], motivated by a paper of Kamae [177]. A different proof is given by Jones [167].

[25] (Page 50) This conjectured result—the "Rokhlin problem"—has been shown in important special cases by Host [158], Kalikow [176], King [193], Ryzhikov [328], del Junco and Yassawi [68, 390] and others, but the general case is open.

[26] (Page 50) The definition used by Koopman and von Neumann is the spectral one that will be given in Theorem 2.36(5), and was called by them the absence of "angle variables"; they also considered flows (measure-preserving actions of $\mathbb{R}$ rather than actions of $\mathbb{Z}$ or $\mathbb{N}$). In physical terms, they characterized lack of ergodicity as barriers that are never passed, and the presence of an angle variable as a clock that never changes, under the dynamics.

[27] (Page 50) Examples of such systems were constructed using Gaussian processes by Maruyama [255]; Kakutani [174] gave a direct combinatorial construction of an example (this example is described in detail in the book of Petersen [282, Sect. 4.5]). Other examples were found by Chacon [51, 52] and Katok and Stepin [185]. Indeed, there is a reasonable way of viewing the collection of all measure-preserving transformations of a fixed space in which a typical transformation is weak-mixing but not mixing (see papers of Rokhlin [315] and Halmos [135] or Halmos' book [138, pp. 77–80]).

[28] (Page 52) This more subtle version of Exercise 2.7.1 appears in a paper of Halmos [136], and is attributed to Ambrose, Halmos and Kakutani in Petersen's book [282].

[29] (Page 54) This is shown in the notes of Halmos [138]. Ergodicity also makes sense for transformations preserving an infinite measure; in that setting Kakutani and Parry [175] used random walk examples of Gillis [115] to show that for any $k \geqslant 1$ there is an infinite measure-preserving transformation $T$ with $T^{(k)}$ ergodic and $T^{(k+1)}$ not ergodic.

[30] (Page 54) This is also known as exponential or effective rate of mixing or decay of correlations; see Baladi [15] for an overview of dynamical settings where it is known.

[31] (Page 58) A more constructive proof of the difficult step in Theorem 2.36 (which may be taken to be (5) $\implies$ (1)) exploiting properties of almost-periodic functions on compact groups, and giving more insight into the structure of ergodic measure-preserving transformations that are not weak-mixing, may be found in Petersen [282, Sect. 4.1].

[32] (Page 59) This is an example of a Hilbert–Schmidt operator [331]; a convenient source for this material is the book of Rudin [321] or Appendix B.

[33] (Page 61) This way of viewing ergodic theorems lies at the start of a sophisticated investigation of ergodic theorems along arithmetic sets of integers by Bourgain [41]. This

exercise already points at a relationship between ergodic theorems and equidistribution on the circle.

[34](Page 61) Notice that the assumption that $(X, \mathscr{B}, \mu, T)$ is invertible also implies that $T$ is *forward measurable*, that is $T(A) \in \mathscr{B}$ for any $A \in \mathscr{B}$. Heinemann and Schmitt [146] prove the Rokhlin lemma for an aperiodic measure-preserving transformation on a Borel probability space using Exercise 5.3.2 and Poincaré recurrence instead of a Kakutani tower (aperiodic is defined in Exercise 2.9.2; for Borel probability space see Definition 5.13). A non-invertible Rokhlin lemma is also developed by Rosenthal [317] in his work on topological models for measure-preserving systems and by Hoffman and Rudolph [155] in their extension of the Bernoulli theory to non-invertible systems.

[35](Page 63) This short proof comes from a paper of Wright [389], in which Kac's theorem is extended to measurable transformations.

[36](Page 64) The extension in Exercise 2.9.2 appears in the notes of Halmos [138, p. 71].

[37](Page 65) Exercise 2.9.3 is taken from a paper of Keane and Michel [188]; they also show that the supremum of $\mu \left( \bigcup_{i=1}^{n} T^{a_i}(A) \right)$ over sets $A$ for which

$$T^{a_1}(A), \dots, T^{a_n}(A)$$

are disjoint is a rational number, and show how this can be computed from the integers $a_1, \dots, a_n$.

# Chapter 3
# Continued Fractions

The continued fraction decomposition of real numbers grows naturally from the Euclidean algorithm, and continued fractions have been used in some form for thousands of years. One goal of this volume is to show how they relate to a natural action on a homogeneous space. To start there would be to willfully reverse their historical development: We start instead with their basic properties[38] from an elementary point of view in Sect. 3.1, then show how continued fractions are related to an explicit measure-preserving transformation in Sect. 3.2. In Chap. 9 we will see how the continued fraction map fits into the more general framework of actions on homogeneous spaces.

Let us mention one result proved in this chapter. We will show that for every irrational $x \in \mathbb{R}$ there is a sequence of 'best rational approxima-tions' $\frac{p_n(x)}{q_n(x)} \in \mathbb{Q}$, defined by the continued fraction expansion of $x$. Moreover, for almost every $x$ we have

$$\lim_{n \to \infty} \frac{1}{n} \log \left| x - \frac{p_n(x)}{q_n(x)} \right| \longrightarrow -\frac{\pi^2}{6 \log 2},$$

which gives a precise description of the expected speed of approximation along this sequence.

## 3.1 Elementary Properties

A (simple) *continued fraction* is a formal expression of the form

$$a_0 + \cfrac{1}{a_1 + \cfrac{1}{a_2 + \cfrac{1}{a_3 + \cfrac{1}{a_4 + \cdots}}}} \tag{3.1}$$

M. Einsiedler, T. Ward, *Ergodic Theory*, Graduate Texts in Mathematics 259,
DOI 10.1007/978-0-85729-021-2_3, © Springer-Verlag London Limited 2011

which we will also denote by

$$[a_0; a_1, a_2, a_3, \ldots]$$

with $a_n \in \mathbb{N}$ for $n \geqslant 1$ and $a_0 \in \mathbb{N}_0$. Also write

$$[a_0; a_1, a_2, \ldots, a_n]$$

for the finite fraction

$$a_0 + \cfrac{1}{a_1 + \cfrac{1}{a_2 + \cdots + \cfrac{1}{a_{n-1} + \cfrac{1}{a_n}}}}.$$

Thus, for example

$$[a_0; a_1, a_2, \ldots, a_n] = a_0 + \frac{1}{[a_1; a_2, \ldots, a_n]}.$$

We will see later that the expression in (3.1)—when suitably interpreted—converges, and therefore defines a real number. The numbers $a_n$ are the *partial quotients* of the continued fraction. The following simple lemma is crucial for many of the basic properties of the continued fraction expansion.

**Lemma 3.1.** *Fix a sequence $(a_n)_{n \geqslant 0}$ with $a_0 \in \mathbb{N}_0$ and $a_n \in \mathbb{N}$ for $n \geqslant 1$. Then the rational numbers*

$$\frac{p_n}{q_n} = [a_0; a_1, a_2, \ldots, a_n] \tag{3.2}$$

*for $n \geqslant 0$ with coprime numerator $p_n \geqslant 1$ and denominator $q_n \geqslant 1$ can be found recursively from the relation*

$$\begin{pmatrix} p_n & p_{n-1} \\ q_n & q_{n-1} \end{pmatrix} = \begin{pmatrix} a_0 & 1 \\ 1 & 0 \end{pmatrix} \begin{pmatrix} a_1 & 1 \\ 1 & 0 \end{pmatrix} \cdots \begin{pmatrix} a_n & 1 \\ 1 & 0 \end{pmatrix} \quad \text{for } n \geqslant 0. \tag{3.3}$$

*In particular, we set $p_{-1} = 1, q_{-1} = 0, p_0 = a_0$, and $q_0 = 1$.*

PROOF. Notice first that the sequence $(a_n)_{n \geqslant 0}$ defines the sequences $(p_n)_{n \geqslant -1}$ and $(q_n)_{n \geqslant -1}$. The claim of the lemma is proved by induction on $n$. Assume that (3.3) holds for $0 \leqslant n \leqslant k - 1$ and $p_n, q_n$ as defined by (3.2) for any sequence $(a_0, a_1, \ldots)$. This is clear for $n = 0$. Thus, on replacing the first $k$ terms of the sequence $(a_n)_{n \geqslant 0}$ with the first $k$ terms of the sequence $(a_n)_{n \geqslant 1}$, we have

$$\frac{x}{y} = [a_1; a_2, \ldots, a_k]$$

as a fraction in lowest terms where $x$ and $y$ are defined by

$$\begin{pmatrix} x & x' \\ y & y' \end{pmatrix} = \begin{pmatrix} a_1 & 1 \\ 1 & 0 \end{pmatrix} \cdots \begin{pmatrix} a_k & 1 \\ 1 & 0 \end{pmatrix}.$$

Then

$$\begin{pmatrix} a_0 & 1 \\ 1 & 0 \end{pmatrix}\begin{pmatrix} x & x' \\ y & y' \end{pmatrix} = \begin{pmatrix} p_k & p_{k-1} \\ q_k & q_{k-1} \end{pmatrix} = \begin{pmatrix} a_0 x + y & a_0 x' + y' \\ x & x' \end{pmatrix},$$

so

$$\frac{p_k}{q_k} = \frac{a_0 x + y}{x} = a_0 + \frac{y}{x} = a_0 + \frac{1}{[a_1; a_2, \ldots, a_k]} = [a_0; a_1, \ldots, a_k],$$

which shows that (3.2) holds for $n = k$ also.                                               □

An immediate consequence of Lemma 3.1 is a pair of recursive formulas

$$p_{n+1} = a_{n+1} p_n + p_{n-1}$$

and

$$q_{n+1} = a_{n+1} q_n + q_{n-1} \tag{3.4}$$

for any $n \geqslant 1$, since

$$\begin{pmatrix} p_{n+1} & p_n \\ q_{n+1} & q_n \end{pmatrix} = \begin{pmatrix} p_n & p_{n-1} \\ q_n & q_{n-1} \end{pmatrix}\begin{pmatrix} a_{n+1} & 1 \\ 1 & 0 \end{pmatrix} = \begin{pmatrix} a_{n+1} p_n + p_{n-1} & p_n \\ a_{n+1} q_n + q_{n-1} & q_n \end{pmatrix}.$$

It follows that

$$1 = q_0 \leqslant q_1 < q_2 < \cdots \tag{3.5}$$

since $a_n \geqslant 1$ for all $n \geqslant 1$; by induction

$$q_n \geqslant 2^{(n-2)/2} \tag{3.6}$$

and similarly

$$p_n \geqslant 2^{(n-2)/2} \tag{3.7}$$

for all $n \geqslant 1$. Taking determinants in (3.3) shows that

$$p_n q_{n-1} - p_{n-1} q_n = (-1)^{n+1} \tag{3.8}$$

and hence $\frac{p_1}{q_1} = a_0 + \frac{1}{q_0 q_1}$, $\frac{p_2}{q_2} = \frac{p_1}{q_1} - \frac{1}{q_1 q_2} = a_0 + \frac{1}{q_0 q_1} - \frac{1}{q_1 q_2}$ and

$$\frac{p_n}{q_n} = \frac{p_{n-1}}{q_{n-1}} + (-1)^{n+1} \frac{1}{q_{n-1} q_n}$$

$$= a_0 + \frac{1}{q_0 q_1} - \frac{1}{q_1 q_2} + \cdots + (-1)^{n+1} \frac{1}{q_{n-1} q_n} \tag{3.9}$$

for all $n \geqslant 1$ by induction.

This shows that an infinite continued fraction is not just a formal object, it in fact converges to a real number. Namely,

$$u = [a_0; a_1, a_2, \ldots] = \lim_{n \to \infty} [a_0; a_1, \ldots, a_n]$$

$$= \lim_{n \to \infty} \frac{p_n}{q_n} = a_0 + \sum_{n=1}^{\infty} \frac{(-1)^{n+1}}{q_{n-1} q_n}, \qquad (3.10)$$

is always convergent (indeed, is absolutely convergent) by the inequality (3.6). Moreover, an immediate consequence of (3.10) and (3.5) is a sequence of inequalities describing how the continued fraction converges: if $a_n \in \mathbb{N}$ for $n \geqslant 1$ then

$$\frac{p_0}{q_0} < \frac{p_2}{q_2} < \cdots < \frac{p_{2n}}{q_{2n}} < \cdots < u < \cdots < \frac{p_{2m+1}}{q_{2m+1}} < \cdots < \frac{p_3}{q_3} < \frac{p_1}{q_1}. \qquad (3.11)$$

We say that $[a_0; a_1, \ldots]$ is the *continued fraction expansion* for $u$. The name suggests that the expansion is (almost) unique and that it always exists. We will see that in fact any irrational number $u$ has a continued fraction expansion, and that it is unique (Lemmas 3.6 and 3.4).

The rational numbers $\frac{p_n}{q_n}$ are called the *convergents* of the continued fraction for $u$ and they provide very rapid rational approximations to $u$. Indeed,

$$u - \frac{p_n}{q_n} = (-1)^n \left[ \frac{1}{q_n q_{n+1}} - \frac{1}{q_{n+1} q_{n+2}} + \cdots \right] \qquad (3.12)$$

so by (3.5) we have[39]

$$\left| u - \frac{p_n}{q_n} \right| < \frac{1}{q_n q_{n+1}}. \qquad (3.13)$$

By (3.4) we deduce that

$$\left| u - \frac{p_n}{q_n} \right| < \frac{1}{a_{n+1} q_n^2} \leqslant \frac{1}{q_n^2}. \qquad (3.14)$$

Recall from Sect. 1.5 that we write

$$\langle t \rangle = \min_{q \in \mathbb{Z}} |t - q|$$

for the distance from $t$ to the nearest integer. The inequality (3.14) gives one explanation* for the comment made on p. 7: using the fact that any irrational has a continued fraction expansion, it follows that for any real number $u$, there is a sequence $(q_n)$ with $q_n \to \infty$ such that $q_n \langle q_n u \rangle < 1$.

---

* This can also be seen more directly as a consequence of the Dirichlet principle (see Exercise 3.1.3).

**Lemma 3.2.** *Let* $a_n \in \mathbb{N}$ *for all* $n \geqslant 0$. *Then the limit in* (3.10) *is irrational.*

PROOF. Suppose that $u = \frac{a}{b} \in \mathbb{Q}$. Then, by (3.14),

$$|q_n a - b p_n| < \frac{b}{a_{n+1} q_n} \leqslant \frac{b}{q_n}.$$

Since $q_n \to \infty$ by the inequality (3.6) and $q_n a - b p_n \in \mathbb{Z}$ we see that

$$q_n a - b p_n = 0$$

and hence $u = \frac{a}{b} = \frac{p_n}{q_n}$ for large enough $n$. However, by Lemma 3.1 $p_n$ and $q_n$ are coprime, so this contradicts the fact that $q_n \to \infty$ as $n \to \infty$. Thus $u$ is irrational. $\qquad\qquad\square$

The continued fraction convergents to a given irrational not only provide good rational approximants. In fact, they provide *optimal* rational approximants in the following sense (see Exercise 3.1.4).

**Proposition 3.3.** *Let* $u = [a_0; a_1, \dots] \in \mathbb{R} \backslash \mathbb{Q}$ *as in* (3.10). *For any* $n > 1$ *and* $p, q$ *with* $0 < q \leqslant q_n$, *if* $\frac{p}{q} \neq \frac{p_n}{q_n}$, *then*

$$|p_n - q_n u| < |p - qu|.$$

*In particular,*

$$\left| \frac{p_n}{q_n} - u \right| < \left| \frac{p}{q} - u \right|.$$

PROOF. Note that $|p_n - q_n u| < |p - qu|$ and $0 < q \leqslant q_n$ together imply that

$$\frac{1}{q} \left| \frac{p_n}{q_n} - u \right| < \frac{1}{q_n} \left| \frac{p}{q} - u \right| \leqslant \frac{1}{q} \left| \frac{p}{q} - u \right|,$$

giving the second statement of the proposition. It is enough therefore to prove the first inequality. Recall from (3.13) that

$$\left| u - \frac{p_n}{q_n} \right| < \frac{1}{q_n q_{n+1}}$$

and

$$\left| u - \frac{p_{n+1}}{q_{n+1}} \right| < \frac{1}{q_{n+1} q_{n+2}}.$$

By the alternating behavior of the convergents in (3.11), each of the three bracketed expressions in the identity

$$\left( u - \frac{p_n}{q_n} \right) = \left( \frac{p_{n+1}}{q_{n+1}} - \frac{p_n}{q_n} \right) - \left( \frac{p_{n+1}}{q_{n+1}} - u \right)$$

is positive (if $n$ is even) or negative (if $n$ is odd). It follows that

$$\left| u - \frac{p_n}{q_n} \right| = \left| \frac{p_{n+1}}{q_{n+1}} - \frac{p_n}{q_n} \right| - \left| \frac{p_{n+1}}{q_{n+1}} - u \right|,$$

so

$$\left| u - \frac{p_n}{q_n} \right| > \frac{1}{q_n q_{n+1}} - \frac{1}{q_{n+1} q_{n+2}} = \frac{q_{n+2} - q_n}{q_n q_{n+1} q_{n+2}} = \frac{a_{n+2}}{q_n q_{n+2}}$$

by (3.4) and (3.14). It follows that

$$\frac{1}{q_{n+2}} < |p_n - q_n u| < \frac{1}{q_{n+1}} \tag{3.15}$$

for $n \geqslant 1$.

By the inequalities (3.15),

$$|q_n u - p_n| < \frac{1}{q_{n+1}} < |q_{n-1} u - p_{n-1}|$$

so we may assume that $q_{n-1} < q \leqslant q_n$ (if not, use downwards induction on $n$).

If $q = q_n$, then $\left| \frac{p_n}{q_n} - \frac{p}{q} \right| \geqslant \frac{1}{q_n}$, while

$$\left| \frac{p_n}{q_n} - u \right| < \frac{1}{q_n q_{n+1}} \leqslant \frac{1}{2 q_n},$$

since $q_{n+1} \geqslant 2$ for all $n \geqslant 1$. Therefore,

$$\left| \frac{p}{q} - u \right| \geqslant \frac{1}{2 q_n} = \frac{1}{2q}$$

and so $|q_n u - p_n| < |q u - p|$.

Assume now that $q_{n-1} < q < q_n$ and write

$$\begin{pmatrix} p_n & p_{n-1} \\ q_n & q_{n-1} \end{pmatrix} \begin{pmatrix} a \\ b \end{pmatrix} = \begin{pmatrix} p \\ q \end{pmatrix},$$

so that $a, b \in \mathbb{Z}$ by (3.8). Clearly $ab \neq 0$ since otherwise $q = q_{n-1}$ or $q = q_n$. Now $q = a q_n + b q_{n-1} < q_n$, so $ab < 0$; by (3.11) we also know that $p_n - q_n u$ and $p_{n-1} - q_{n-1} u$ are of opposite signs. It follows that $a(p_n - q_n u)$ and $b(p_{n-1} - q_{n-1} u)$ are of the same sign, so the fact that

$$p - qu = a(p_n - q_n u) + b(p_{n-1} - q_{n-1} u)$$

implies that

$$|p - qu| > |p_{n-1} - q_{n-1} u| > |p_n - q_n u|$$

as required.                                                                                                                                                      □

We end this section with the uniqueness of the continued fraction expansion.

**Lemma 3.4.** *The map that sends the sequence*

$$(a_0, a_1, \dots) \in \mathbb{N}_0 \times \mathbb{N}^{\mathbb{N}}$$

*to the limit in* (3.10) *is injective.*

PROOF. Let $u = (a_0, a_1, \dots) \in \mathbb{N}_0 \times \mathbb{N}^{\mathbb{N}}$ be given. Then it is clear that

$$u = [a_0; a_1, \dots]$$

is positive. Applying this to $(a_1, a_2, \dots)$ and the inductive relation

$$u = a_0 + \frac{1}{[a_1; a_2, \dots]}$$

we see that

$$u \in \left(a_0, a_0 + \tfrac{1}{a_1}\right) \subseteq (a_0, a_0 + 1).$$

It follows that $u$ uniquely determines $a_0$. Using the inductive relation again, we have

$$[a_1; a_2, \dots] = \frac{1}{u - a_0},$$

which by the argument above shows that $u$ uniquely determines $a_1$. Iterating the procedure shows that all the terms in the continued fraction can be reconstructed from $u$.                                                                 □

The argument used in the proof of Lemma 3.4 also suggests a way to find the continued fraction expansion of a given irrational number $u \in \mathbb{R} \setminus \mathbb{Q}$. This will be pursued further in the next section.

## Exercises for Sect. 3.1

**Exercise 3.1.1.** Show that any positive rational number has exactly two continued fraction expansions, both of which are finite.

**Exercise 3.1.2.** Show that a continued fraction in which some of the digits are allowed to be zero (but that is not allowed to end with infinitely many zeros) can always be rewritten with digits in $\mathbb{N}$.

**Exercise 3.1.3.** [Dirichlet principle] For a given $u \in \mathbb{R}$ and $n \geq 1$ consider the points $0, u, 2u, \dots, nu \pmod 1$ as elements of the circle $\mathbb{T}$. Show that for some $k$, $0 < k < n$ we have $\langle ku \rangle \leqslant \frac{1}{n}$, and deduce that there exists a sequence $q_n \to \infty$ with $q_n \langle q_n u \rangle < 1$.

**Exercise 3.1.4.** Extend Proposition 3.3 in the following way. Given $u$ as in (3.10), and the $n$th convergent $\frac{p_n}{q_n}$, the $(n+1)$th convergent $\frac{p_{n+1}}{q_{n+1}}$ is characterized by being the ratio of the unique pair of positive integers $(p_{n+1}, q_{n+1})$ for which $|p_{n+1} - q_{n+1}u| < |p_n - q_n u|$ with $q_{n+1} > q_n$ minimal. Notice that the same cannot be said when using the expression $\left| u - \frac{p_n}{q_n} \right|$, as becomes clear in the case where $u > \frac{1}{3}$ is very close to $\frac{1}{3}$, in which case the first approximation is not $\frac{1}{2}$.

**Exercise 3.1.5.** Let $u = [a_0; a_1, \dots]$ with convergents $\frac{p_n}{q_n}$. Show that

$$\frac{1}{2q_{n+1}} \leqslant |p_n - q_n u| < \frac{1}{q_{n+1}}.$$

## 3.2 The Continued Fraction Map and the Gauss Measure

Let $Y = [0,1] \setminus \mathbb{Q}$, and define a map $T : Y \to Y$ by

$$T(x) = \frac{1}{x} - \left\lfloor \frac{1}{x} \right\rfloor,$$

where $\lfloor t \rfloor$ denotes the greatest integer less than or equal to $t$. Thus $T(x)$ is the fractional part $\left\{ \frac{1}{x} \right\}$ of $\frac{1}{x}$. The graph of this so-called *continued fraction* or *Gauss map* is shown in Fig. 3.1.

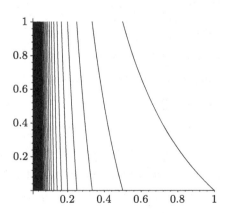

**Fig. 3.1** The Gauss map

Gauss observed in 1845 that $T$ preserves[40] the probability measure given by

$$\mu(A) = \frac{1}{\log 2} \int_A \frac{1}{1+x} \, dx,$$

by showing that the Lebesgue measure of $T^{-n}I$ converges to $\mu(I)$ for each interval $I$.

This map will be studied via a geometric model (for its invertible extension) in Chap. 9; in this section we assemble some basic facts from an elementary point of view, showing that the Gauss measure is $T$-invariant and ergodic. Since the measure defined in Lemma 3.5 is non-atomic, we may extend the map to include the points 0 and 1 in any way without affecting the measurable structure of the system.

**Lemma 3.5.** *The continued fraction map $T(x) = \{\frac{1}{x}\}$ on $(0,1)$ preserves the Gauss measure $\mu$ given by*

$$\mu(A) = \frac{1}{\log 2} \int_A \frac{1}{1+x} \, dx$$

*for any Borel measurable set $A \subseteq [0,1]$.*

A geometric and less formal proof of this will be given on page 93 using basic properties of the invertible extension of the continued fraction map in Proposition 3.15.

PROOF OF LEMMA 3.5. It is sufficient to show that $\mu\left(T^{-1}[0,s]\right) = \mu\left([0,s]\right)$ for every $s > 0$. Clearly

$$T^{-1}[0,s] = \{x \mid 0 \leqslant T(x) \leqslant s\} = \bigsqcup_{n=1}^{\infty} \left[\frac{1}{s+n}, \frac{1}{n}\right]$$

is a disjoint union. It follows that

$$\mu\left(T^{-1}[0,s]\right) = \frac{1}{\log 2} \sum_{n=1}^{\infty} \int_{1/(s+n)}^{1/n} \frac{1}{1+x} \, dx$$

$$= \frac{1}{\log 2} \sum_{n=1}^{\infty} \left(\log(1+\tfrac{1}{n}) - \log(1+\tfrac{1}{s+n})\right)$$

$$= \frac{1}{\log 2} \sum_{n=1}^{\infty} \left(\log(1+\tfrac{s}{n}) - \log(1+\tfrac{s}{n+1})\right) \qquad (3.16)$$

$$= \frac{1}{\log 2} \sum_{n=1}^{\infty} \int_{s/(n+1)}^{s/n} \frac{1}{1+x} \, dx$$

$$= \mu\left([0,s]\right),$$

completing the proof. The identity used in (3.16) amounts to

$$\frac{1 + \frac{s}{n}}{1 + \frac{s}{n+1}} = \frac{1 + \frac{1}{n}}{1 + \frac{1}{s+n}},$$

which may be seen by multiplying numerator and denominator of the left-hand side by $\frac{n+1}{n+s}$, and the interchange of integral and sum is justified by absolute convergence.                                                                □

Thus Lemma 3.5 shows that $\big([0,1], \mathscr{B}_{[0,1]}, \mu, T\big)$ is a measure-preserving system.

Define for $x \in Y = [0,1] \setminus \mathbb{Q}$ and $n \geqslant 1$ the sequence of natural numbers $(a_n) = (a_n(x))$ by

$$\frac{1}{1 + a_n} < T^{n-1}(x) < \frac{1}{a_n}, \tag{3.17}$$

or equivalently by

$$a_n(x) = \left\lfloor \frac{1}{T^{n-1}x} \right\rfloor \in \mathbb{N}. \tag{3.18}$$

For any sequence $(a_n)_{n \geqslant 1}$ of natural numbers we define the continued fraction $[a_1, a_2, \dots]$ just as in (3.1) with $a_0 = 0$.

**Lemma 3.6.** *For any irrational $x \in [0,1] \setminus \mathbb{Q}$ the sequence $(a_n(x))$ defined in (3.18) gives the digits of the continued fraction expansion to $x$. That is,*

$$x = [a_1(x), a_2(x), \dots].$$

PROOF. Define $a_n = a_n(x)$ and let $u = [a_1, a_2, \dots]$ be the limit as in (3.10) with $a_0 = 0$. By (3.11) we have

$$\frac{p_{2n}}{q_{2n}} < u < \frac{p_{2n+1}}{q_{2n+1}}$$

and by (3.8) and the inequality (3.6) we have

$$\frac{p_{2n+1}}{q_{2n+1}} - \frac{p_{2n}}{q_{2n}} = \frac{1}{q_{2n}q_{2n-1}} \leqslant \frac{1}{2^{2n-2}}.$$

We now show by induction that

$$[a_1, \dots, a_{2n}] = \frac{p_{2n}}{q_{2n}} < x < \frac{p_{2n+1}}{q_{2n+1}} = [a_1, \dots, a_{2n+1}], \tag{3.19}$$

which together with the above shows that $u = x$.

Recall that $\frac{p_0}{q_0} = 0$ and $\frac{p_1}{q_1} = \frac{1}{a_1}$, so (3.19) holds for $n = 0$ because of the definition of $a_1$ in (3.18). Now assume that the inequality (3.19) holds for a given $n$ and all $x \in [0,1]$. In particular, we may apply it to $T(x)$ to get

$$[a_2, \dots, a_{2n+1}] < T(x) < [a_2, \dots, a_{2n+2}].$$

Since $T(x) = \frac{1}{x} - a_1$ we get

$$a_1 + [a_2, \ldots, a_{2n+1}] < \frac{1}{x} < a_1 + [a_2, \ldots, a_{2n+2}]$$

and therefore

$$[a_1, \ldots, a_{2n+2}] = \frac{1}{a_1 + [a_2, \ldots, a_{2n+2}]} < x,$$

$$x < \frac{1}{a_1 + [a_2, \ldots, a_{2n+1}]} = [a_1, \ldots, a_{2n+1}]$$

as required.                                                                    □

This gives a description of the continued fraction map as a shift map: the list of digits in the continued fraction expansion of $x \in [0,1] \backslash \mathbb{Q}$ defines a unique element of $\mathbb{N}^{\mathbb{N}}$, and the diagram

$$
\begin{array}{ccc}
\mathbb{N}^{\mathbb{N}} & \xrightarrow{\ \sigma\ } & \mathbb{N}^{\mathbb{N}} \\
\downarrow & & \downarrow \\
(0,1) & \xrightarrow[\ T\ ]{} & (0,1)
\end{array}
$$

commutes, where $\sigma$ is the left shift and the vertical map sends a sequence of digits $(a_n)_{n \geq 1}$ to the real irrational number defined by the continued fraction expansion.

In Corollary 3.8 we will draw some easy consequences[41] of ergodicity for the Gauss measure $\mu$ in terms of properties of the continued fraction expansion for almost every real number. Given a continued fraction expansion, recall that the *convergents* are the terms of the sequence of rationals $\frac{p_n(x)}{q_n(x)}$ in lowest terms defined by

$$\frac{p_n(x)}{q_n(x)} = \cfrac{1}{a_1 + \cfrac{1}{a_2 + \cfrac{1}{a_3 + \cdots + \cfrac{1}{a_n}}}}.$$

**Theorem 3.7.** *The continued fraction map $T(x) = \{\frac{1}{x}\}$ on $(0,1)$ is ergodic with respect to the Gauss measure $\mu$.*

Before proving this[42] we develop some more of the basic identities for continued fractions. Given a continued fraction expansion $u = [a_0; a_1, \ldots]$ of an irrational number $u$, we write $u_n = [a_n; a_{n+1}, \ldots]$ for the $n$th tail of the expansion. By Lemma 3.1 applied twice, we have

$$\begin{pmatrix} p_{n+k} \\ q_{n+k} \end{pmatrix} = \begin{pmatrix} a_0 & 1 \\ 1 & 0 \end{pmatrix} \cdots \begin{pmatrix} a_{n+k} & 1 \\ 1 & 0 \end{pmatrix} \begin{pmatrix} 1 \\ 0 \end{pmatrix}$$

$$= \begin{pmatrix} p_n & p_{n-1} \\ q_n & q_{n-1} \end{pmatrix} \begin{pmatrix} a_{n+1} & 1 \\ 1 & 0 \end{pmatrix} \cdots \begin{pmatrix} a_{n+k} & 1 \\ 1 & 0 \end{pmatrix} \begin{pmatrix} 1 \\ 0 \end{pmatrix}.$$

Writing $p_k(u_{n+1})$ and $q_k(u_{n+1})$ for the numerator and denominator of the $k$th convergents to $u_{n+1}$, we can apply Lemma 3.1 again to deduce that

$$\begin{pmatrix} p_{n+k} \\ q_{n+k} \end{pmatrix} = \begin{pmatrix} p_n & p_{n-1} \\ q_n & q_{n-1} \end{pmatrix} \begin{pmatrix} p_{k-1}(u_{n+1}) & p_{k-2}(u_{n+1}) \\ q_{k-1}(u_{n+1}) & q_{k-2}(u_{n+1}) \end{pmatrix} \begin{pmatrix} 1 \\ 0 \end{pmatrix},$$

so

$$\frac{p_{n+k}}{q_{n+k}} = \frac{p_n \frac{p_{k-1}(u_{n+1})}{q_{k-1}(u_{n+1})} + p_{n-1}}{q_n \frac{p_{k-1}(u_{n+1})}{q_{k-1}(u_{n+1})} + q_{n-1}},$$

which gives

$$u = \frac{p_n u_{n+1} + p_{n-1}}{q_n u_{n+1} + q_{n-1}} \qquad (3.20)$$

in the limit as $k \to \infty$. Notice that the above formulas are derived for a general positive irrational number $u$. If $u = [a_1, \ldots] \in (0,1)$, then $u_{n+1} = (T^n(u))^{-1}$ so that

$$u = \frac{p_n + p_{n-1} T^n(u)}{q_n + q_{n-1} T^n(u)}. \qquad (3.21)$$

PROOF OF THEOREM 3.7. The description of the continued fraction map as a shift on the space $\mathbb{N}^{\mathbb{N}}$ described above suggests the method of proof: the measure $\mu$ corresponds to a rather complicated measure on the shift space, but if we can control the measure of cylinder sets (and their intersections) well enough then we may prove ergodicity along the lines of the proof of ergodicity for Bernoulli shifts in Proposition 2.15. For two expressions $f, g$ we write $f \asymp g$ to mean that there are absolute constants $C_1, C_2 > 0$ such that

$$C_1 f \leqslant g \leqslant C_2 f.$$

Given a vector $\mathbf{a} = (a_1, \ldots, a_n) \in \mathbb{N}^n$ of length $|\mathbf{a}| = n$, define a set

$$I(\mathbf{a}) = \{[x_1, x_2, \ldots] \mid x_i = a_i \text{ for } 1 \leqslant i \leqslant n\}$$

(which may be thought of as an interval in $(0,1)$, or as a cylinder set in $\mathbb{N}^{\mathbb{N}}$).
    The main step towards the proof of the theorem is to show that

$$\mu\left(T^{-n}A \cap I(\mathbf{a})\right) \asymp \mu(A)\mu(I(\mathbf{a})) \qquad (3.22)$$

for any measurable set $A$. Notice that for the proof of (3.22) it is sufficient to show it for any interval $A = [d, e]$; the case of a general Borel set then follows by a standard approximation argument (the set of Borel sets satisfying (3.22) with a fixed choice of constants is easily seen to be a monotone class, so Theorem A.4 may be applied).

Now define $\frac{p_n}{q_n} = [a_1, \ldots, a_n]$ and $\frac{p_{n-1}}{q_{n-1}} = [a_1, \ldots, a_{n-1}]$. Then $u \in I(\mathbf{a})$ if and only if $u = [a_1, \ldots, a_n, a_{n+1}(u), \ldots]$, and so $u \in I(\mathbf{a}) \cap T^{-n}A$ if and only if $u$ can be written as in (3.21), with $T^n(u) \in A = [d, e]$. As $T^n$ restricted to $I(\mathbf{a})$ is continuous and monotone (increasing if $n$ is even, and decreasing if $n$ is odd), it follows that $I(\mathbf{a}) \cap T^{-n}A$ is an interval with endpoints given by

$$\frac{p_n + p_{n-1}d}{q_n + q_{n-1}d}$$

and

$$\frac{p_n + p_{n-1}e}{q_n + q_{n-1}e}.$$

Thus the Lebesgue measure of $I(\mathbf{a}) \cap T^{-n}A$,

$$\left| \frac{p_n + p_{n-1}d}{q_n + q_{n-1}d} - \frac{p_n + p_{n-1}e}{q_n + q_{n-1}e} \right|,$$

expands to

$$\left| \frac{(p_n + p_{n-1}d)(q_n + q_{n-1}e) - (p_n + p_{n-1}e)(q_n + q_{n-1}d)}{(q_n + q_{n-1}d)(q_n + q_{n-1}e)} \right|$$

$$= \left| \frac{p_n q_{n-1}e + p_{n-1}q_n d - p_n q_{n-1}d - p_{n-1}q_n e}{(q_n + q_{n-1}d)(q_n + q_{n-1}e)} \right|$$

$$= (e - d)\frac{|p_n q_{n-1} - p_{n-1}q_n|}{(q_n + q_{n-1}e)(q_n + q_{n-1}f)} = (e - d)\frac{1}{(q_n + q_{n-1}e)(q_n + q_{n-1}f)}$$

by (3.8). On the other hand, the Lebesgue measure of $I(\mathbf{a})$ is

$$\left| \frac{p_n}{q_n} - \frac{p_n + p_{n-1}}{q_n + q_{n-1}} \right| = \frac{|p_n q_{n-1} - p_{n-1}q_n|}{q_n(q_n + q_{n-1})} = \frac{1}{q_n(q_n + q_{n-1})} \qquad (3.23)$$

again by (3.8), which implies that

$$m(I(\mathbf{a}) \cap T^{-n}A) = m(A)m(I(\mathbf{a}))\frac{q_n(q_n + q_{n-1})}{(q_n + q_{n-1}e)(q_n + q_{n-1}f)}$$

$$\asymp m(A)m(I(\mathbf{a})), \qquad (3.24)$$

where $m$ denotes Lebesgue measure on $(0, 1)$. Next notice that

$$\frac{m(B)}{2\log 2} \leqslant \mu(B) \leqslant \frac{m(B)}{\log 2}$$

for any Borel set $B \subseteq (0, 1)$, which together with (3.24) gives (3.22).

Now assume that $A \subseteq (0,1)$ is a Borel set with $T^{-1}A = A$. For such a set, the estimate in (3.22) reads as

$$\mu(A \cap I(\mathbf{a})) \asymp \mu(A)\mu(I(\mathbf{a}))$$

for any interval $I(\mathbf{a})$ defined by $\mathbf{a} = (a_1, \ldots, a_n) \in \mathbb{N}^n$ and any $n$. However, for a fixed $n$ the intervals $I(\mathbf{a})$ partition $(0,1)$ (as $\mathbf{a}$ varies in $\mathbb{N}^n$), and by (3.23)

$$\text{diam}(I(\mathbf{a})) = \frac{1}{q_n(q_n + q_{n-1})}$$
$$\leqslant \frac{1}{2^{n-2}} \quad \text{(by (3.6))},$$

so the lengths of the sets in this partition shrink to zero uniformly as $n \to \infty$. Therefore, the intervals $I(\mathbf{a})$ generate the Borel $\sigma$-algebra, and so

$$\mu(A \cap B) \asymp \mu(A)\mu(B)$$

for any Borel subset $B \subseteq (0,1)$ (again by Theorem A.4). We apply this to the set $B = (0,1) \setminus A$ and obtain $0 \asymp \mu(A)\mu(B)$, which shows that either $\mu(A) = 0$ or $\mu((0,1)\setminus A) = 0$, as needed. $\qquad\square$

We will use the ergodicity of the Gauss map in Corollary 3.8 to deduce statements about the digits of the continued fraction expansion of a typical real number. Just as Borel's normal number theorem (Example 1.2) gives precise statistical information about the decimal expansion of almost every real number, ergodicity of the Gauss map gives precise statistical information about the continued fraction digits of almost every real number. Of course the form of the conclusion is necessarily different. For example, since there are infinitely many different digits in the continued fraction expansion, they cannot all occur with equal frequency, and (3.25) makes precise the way in which small digits occur more frequently than large ones. We also obtain information on the geometric and arithmetic mean of the digits $a_n$ in (3.26) and (3.27), the growth rate of the denominators $q_n$ in (3.28), and the rate at which the convergents $\frac{p_n}{q_n}$ approximate a typical real number in (3.29).

In particular, equations (3.28) and (3.29) together say that the digit $a_{n+1}$ appearing in the estimate (3.14) does not affect the logarithmic rate of approximation of an irrational by the continued fraction partial quotients significantly.

**Corollary 3.8.** *For almost every real number* $x = [a_1, a_2, \ldots] \in (0,1)$, *the digit $j$ appears in the continued fraction with density*

$$\frac{2\log(1+j) - \log j - \log(2+j)}{\log 2}, \tag{3.25}$$

$$\lim_{n\to\infty} (a_1 a_2 \ldots a_n)^{1/n} = \prod_{a=1}^{\infty} \left(\frac{(a+1)^2}{a(a+2)}\right)^{\log a/\log 2}, \qquad (3.26)$$

$$\lim_{n\to\infty} \frac{1}{n}(a_1 + a_2 + \cdots + a_n) = \infty, \qquad (3.27)$$

$$\lim_{n\to\infty} \frac{1}{n}\log q_n(x) = \frac{\pi^2}{12\log 2}, \qquad (3.28)$$

*and*

$$\lim_{n\to\infty} \frac{1}{n}\log\left|x - \frac{p_n(x)}{q_n(x)}\right| \longrightarrow -\frac{\pi^2}{6\log 2}. \qquad (3.29)$$

PROOF. The digit $j$ appears in the first $N$ digits with frequency

$$\frac{1}{N}|\{i \mid i \leqslant N, a_i = j\}| = \frac{1}{N}|\{i \mid i \leqslant N, T^i x \in (\tfrac{1}{j+1}, \tfrac{1}{j})\}|$$

$$\longrightarrow \frac{1}{\log 2}\int_{1/(j+1)}^{1/j} \frac{1}{1+y}\, dy$$

$$= \frac{2\log(1+j) - \log j - \log(2+j)}{\log 2},$$

which proves (3.25).

Define a function $f$ on $(0,1)$ by $f(x) = \log a$ for $x \in (\tfrac{1}{a+1}, \tfrac{1}{a})$. Then

$$\int_0^1 f(x)\, dx = \sum_{a=1}^{\infty} \left(\frac{1}{a} - \frac{1}{a+1}\right)\log a$$

$$\leqslant \sum_{a=1}^{\infty} \frac{1}{a^2}\log a < \infty,$$

so $\int_0^1 f\, d\mu < \infty$ also, since the density $\frac{d\mu}{dx} = \frac{1}{(1+x)\log 2}$ is bounded on $[0,1]$. By the pointwise ergodic theorem (Theorem 2.30) we therefore have, for almost every $x$,

$$\frac{1}{n}\sum_{j=0}^{n-1}\log a_j = \frac{1}{n}\sum_{j=0}^{n-1} f(T^j x) \longrightarrow \int f(x)\, d\mu.$$

This shows (3.26) since

$$\int_0^1 f\, d\mu = \sum_{a=1}^{\infty} \frac{\log a}{\log 2}\int_{1/(1+a)}^{1/a} \frac{1}{1+x}\, dx.$$

Now consider the function $g(x) = e^{f(x)}$ (so $g(x) = a_1$ is the first digit in the continued fraction expansion of $x$). We have

$$\frac{1}{n}(a_1 + \cdots + a_n) = \frac{1}{n}\sum_{j=0}^{n-1} g(T^j x),$$

but the pointwise ergodic theorem cannot be applied to $g$ since $\int_0^1 g\,d\mu = \infty$ (the result needed is Exercise 2.6.5(2); the argument here shows how to do this exercise). However, for any fixed $N$ the truncated function

$$g_N(x) = \begin{cases} g(x) & \text{if } g(x) \leqslant N; \\ 0 & \text{if not} \end{cases}$$

is in $L^1_\mu$ since

$$\int g_N\,d\mu = \frac{1}{\log 2}\sum_{a=1}^{N}\int_{1/(a+1)}^{1/a} a\,dx = \frac{1}{\log 2}\sum_{a=1}^{N}\frac{1}{a+1}.$$

Notice that $\int_0^1 g_N\,d\mu \to \infty$ as $N \to \infty$. By the ergodic theorem,

$$\liminf_{n\to\infty}\frac{1}{n}\sum_{j=0}^{n-1} g(T^j x) \geqslant \lim_{n\to\infty}\frac{1}{n}\sum_{j=0}^{n-1} g_N(T^j x)$$

$$= \int_0^1 g_N\,d\mu \to \infty$$

as $N \to \infty$, showing (3.27).

The proofs of (3.25) and (3.26) were straightforward applications of the ergodic theorem, and (3.27) only required a simple extension to measurable functions. Proving (3.28) and (3.29) takes a little more effort.

First notice that

$$\frac{p_n(x)}{q_n(x)} = \frac{1}{a_1 + [a_2, \ldots, a_n]}$$

$$= \frac{1}{a_1 + \frac{p_{n-1}(Tx)}{q_{n-1}(Tx)}}$$

$$= \frac{q_{n-1}(Tx)}{p_{n-1}(Tx) + q_{n-1}(Tx)a_1},$$

so $p_n(x) = q_{n-1}(Tx)$ since the convergents are in lowest terms. Recall that we always have $p_1 = q_0 = 1$. It follows that

$$\frac{1}{q_n(x)} = \frac{p_n(x)}{q_n(x)} \cdot \frac{p_{n-1}(Tx)}{q_{n-1}(Tx)} \cdots \frac{p_1(T^{n-1}x)}{q_1(T^{n-1}x)},$$

so

$$-\frac{1}{n}\log q_n(x) = \frac{1}{n}\sum_{j=0}^{n-1}\log\left[\frac{p_{n-j}(T^j x)}{q_{n-j}(T^j x)}\right].$$

Let $h(x) = \log x$ (so $h \in L_\mu^1$). Then

$$-\frac{1}{n}\log q_n(x) = \underbrace{\frac{1}{n}\sum_{j=0}^{n-1} h(T^j x)}_{S_n} - \underbrace{\frac{1}{n}\sum_{j=0}^{n-1}\left[\log(T^j x) - \log\left(\frac{p_{n-j}(T^j x)}{q_{n-j}(T^j x)}\right)\right]}_{R_n}$$

gives a splitting of $-\frac{1}{n}\log q_n(x)$ into an ergodic average $S_n = \mathsf{A}_h^n$ and a remainder term $R_n$. By the ergodic theorem,

$$\lim_{n\to\infty}\frac{1}{n}S_n = \frac{1}{\log 2}\int_0^1 \frac{\log x}{1+x}\, dx = -\frac{\pi^2}{12\log 2}.$$

To complete the proof of (3.28), we need to show that $\frac{1}{n}R_n \to 0$ as $n \to \infty$. This will follow from the observation that $\frac{p_{n-j}(T^j x)}{q_{n-j}(T^j x)}$ is a good approximation to $T^j x$ if $(n-j)$ is large enough. Recall from (3.7) and (3.6) that

$$p_k \geqslant 2^{(k-2)/2}, \; q_k \geqslant 2^{(k-1)/2},$$

so, by using the inequality (3.13),

$$\left|\frac{x}{p_k/q_k} - 1\right| = \frac{q_k}{p_k}\left|x - \frac{p_k}{q_k}\right| \leqslant \frac{1}{p_k q_{k+1}} \leqslant \frac{1}{2^{k-1}}.$$

By using this together with the fact that $|\log u| \leqslant 2|u-1|$ whenever $u \in [\frac{1}{2}, \frac{3}{2}]$ (which applies in the sum below with $j \leqslant n-2$), we get

$$|R_n| \leqslant \sum_{j=0}^{n-1}\left|\log\frac{T^j x}{p_{n-j}(T^j x)/q_{n-j}(T^j x)}\right|$$

$$\leqslant 2\underbrace{\sum_{j=0}^{n-2}\left|\frac{T^j x}{p_{n-j}(T^j x)/q_{n-j}(T^j x)} - 1\right|}_{T_n} + \underbrace{\left|\log\frac{T^{n-1} x}{p_1(T^{n-1} x)/q_1(T^{n-1} x)}\right|}_{U_n}.$$

Now

$$T_n \leqslant \sum_{j=0}^{n-2}\frac{2}{2^{n-j-1}} \leqslant 2$$

for all $n$. For the second term, notice that

$$U_n = \left| \log \left[ \left( T^{n-1} x \right) a_1 \left( T^{n-1} x \right) \right] \right|,$$

and by the inequality (3.17) we have

$$1 \geqslant \left( T^{n-1} x \right) a_1 \left( T^{n-1} x \right) \geqslant \frac{a_1 \left( T^{n-1} x \right)}{1 + a_1 \left( T^{n-1} x \right)} \geqslant \frac{1}{2}$$

since $a_1(T^{n-1} x) \geqslant 1$. Therefore,

$$\left| \log \left[ \left( T^{n-1} x \right) a_1 \left( T^{n-1} x \right) \right] \right| \leqslant \log 2,$$

which completes the proof that

$$\frac{1}{n} R_n \to 0$$

as $n \to \infty$, and hence shows (3.28).

Equation (3.29) follows from (3.28), since from the inequalities (3.13) and (3.15) we have

$$\log q_n + \log q_{n+1} \leqslant -\log \left| x - \frac{p_n}{q_n} \right| \leqslant \log q_n + \log q_{n+2}.$$

$\square$

## Exercises for Sect. 3.2

**Exercise 3.2.1.** Use the idea in the proof of (3.27) to extend the pointwise ergodic theorem (Theorem 2.30) to the case of a measurable function $f \geqslant 0$ with $\int_X f \, d\mu = \infty$ without the assumption of ergodicity.

**Exercise 3.2.2.** Show that the map from $\mathbb{N}^{\mathbb{N}}$ to $[0,1] \setminus \mathbb{Q}$ sending $(a_1, a_2, \dots)$ to $[a_1, a_2, \dots]$ is a homeomorphism with respect to the discrete topology on $\mathbb{N}$ and the product topology on $\mathbb{N}^{\mathbb{N}}$.

**Exercise 3.2.3.** Let $\mathbf{p} = (p_1, p_2, \dots)$ be an infinite probability vector (this means that $p_i \geqslant 0$ for all $i$, and $\sum_i p_i = 1$). Show that $\mathbf{p}$ gives rise to a $\sigma$-invariant and ergodic probability measure $\mathbf{p}^{\mathbb{N}}$ on $\mathbb{N}^{\mathbb{N}}$.

**Exercise 3.2.4.** Let $\phi : \mathbb{N}^{\mathbb{N}} \to (0,1) \setminus \mathbb{Q}$ be the map discussed on page 79, and let $\mu$ be the Gauss measure on $[0,1]$. Show that $\phi_*^{-1} \mu$ is not of the form $\mathbf{p}^{\mathbb{N}}$ for any infinite probability vector $\mathbf{p}$.

## 3.3 Badly Approximable Numbers

While Corollary 3.8 gives precise information about the behavior of typical real numbers, it does not say anything about the behavior of all real numbers. In this section we discuss a special class of real numbers that behave very differently to typical real numbers.

**Definition 3.9.** A real number $u = [a_1, a_2, \ldots] \in (0, 1)$ is called *badly approximable* if there is some bound $M$ with the property that $a_n \leqslant M$ for all $n \geqslant 1$.

Clearly a badly approximable number cannot satisfy (3.27). It follows that the set of all badly approximable numbers in $(0, 1)$ is a null set with respect to the Gauss measure, and hence is a null set with respect to Lebesgue measure[43]. The next result explains the terminology: badly approximable numbers cannot be approximated very well by rationals.

**Proposition 3.10.** *A number $u \in (0, 1)$ is badly approximable if and only if there exists some $\varepsilon > 0$ with the property that*

$$\left| u - \frac{p}{q} \right| \geqslant \frac{\varepsilon}{q^2}$$

*for all rational numbers $\frac{p}{q}$.*

PROOF. If $u$ is badly approximable, then (3.4) shows that

$$q_{n+1} \leqslant (M + 1)q_n$$

for all $n \geqslant 0$. For any $q$ there is some $n$ with $q \in (q_{n-1}, q_n]$, and by Proposition 3.3 and (3.15) we therefore have

$$\left| \frac{p}{q} - u \right| > \left| \frac{p_n}{q_n} - u \right| > \frac{1}{q_n q_{n+2}} > \frac{1}{(M+1)^4 q^2}$$

as required.

Conversely, if

$$\left| u - \frac{p}{q} \right| \geqslant \frac{\varepsilon}{q^2}$$

for all rational numbers $\frac{p}{q}$ then, in particular,

$$\frac{\varepsilon}{q_n^2} \leqslant \left| u - \frac{p_n}{q_n} \right| < \frac{1}{q_n q_{n+1}}$$

by (3.13). This implies that

$$a_{n+1}q_n < a_{n+1}q_n + q_{n-1} = q_{n+1} < \frac{1}{\varepsilon}q_n,$$

so $a_{n+1} \leqslant \frac{1}{\varepsilon}$ for all $n \geqslant 1$.                                                                        $\square$

*Example 3.11.* Notice that $\frac{2}{\sqrt{5}-1} = \frac{\sqrt{5}+1}{2} \in (1,2)$ and $\frac{\sqrt{5}+1}{2} - 1 = \frac{\sqrt{5}-1}{2}$. It follows that if

$$\frac{\sqrt{5}-1}{2} = [a_1, a_2, \ldots]$$

then $a_1 + [a_2, a_3, \ldots] \in (1,2)$, so $a_1 = 1$, and hence

$$[a_2, a_3, \ldots] = \frac{\sqrt{5}+1}{2} - 1 = \frac{\sqrt{5}-1}{2} = [a_1, a_2, \ldots].$$

We deduce by the uniqueness of the continued fraction digits that

$$\frac{\sqrt{5}-1}{2} = [1, 1, 1, \ldots],$$

so $\frac{\sqrt{5}-1}{2}$ is badly approximable.

Indeed, the specific number in Example 3.11 is, in a precise sense, the most badly approximable real number in $(0,1)$. In the next section we generalize this example to show that all quadratic irrationals are badly approximable.

### 3.3.1 Lagrange's Theorem

The periodicity of the continued fraction expansion seen in Example 3.11 is a general property of quadratics. A real number $u$ is called a *quadratic irrational* if $u \notin \mathbb{Q}$ and there are integers $a, b, c$ with $au^2 + bu + c = 0$. Notice that $u$ is a quadratic irrational if and only if $\mathbb{Q}(u)$ is a subfield of $\mathbb{R}$ of degree 2 over $\mathbb{Q}$.

**Definition 3.12.** A continued fraction $[a_0; a_1, \ldots]$ is *eventually periodic* if there are numbers $N \geqslant 0$ and $k \geqslant 1$ with $a_{n+k} = a_n$ for all $n \geqslant N$. Such a continued fraction will be written

$$[a_0; a_1, \ldots, a_{N-1}, \overline{a_N, \ldots, a_{N+k}}].$$

The main result describing the special properties of quadratic irrationals is Lagrange's Theorem [218, Sect. 34].

**Theorem 3.13 (Lagrange).** *Let $u$ be an irrational positive real number. Then the continued fraction expansion of $u$ is eventually periodic if and only if $u$ is a quadratic irrational.*

PROOF. Assume first that $u = [\overline{a_0; a_1, \ldots, a_k}]$ has a strictly periodic continued fraction expansion, so that $u_{k+1} = u_0 = u$. Thus

$$u = \frac{up_k + p_{k-1}}{uq_k + q_{k-1}}$$

by (3.20), so

$$u^2 q_k + u(q_{k-1} - p_k) - p_{k-1} = 0$$

and $u$ is a quadratic irrational ($u$ cannot be rational, since it has an infinite continued fraction; alternatively notice that the quadratic equation satisfied by $u$ has discriminant $(q_{k-1} - p_k)^2 + 4q_k p_{k-1} = (q_{k-1} + p_k)^2 - 4(-1)^k$ by (3.8), so cannot be a square).

Now assume that

$$u = [a_0; \ldots, a_{N-1}, \overline{a_N, \ldots, a_{N+k}}].$$

Then, by (3.20),

$$u = \frac{[\overline{a_N; a_{N+1}, \ldots, a_{N+k}}]p_{N-1} + p_{N-2}}{[\overline{a_N; a_{N+1}, \ldots, a_{N+k}}]q_{N-1} + q_{N-2}},$$

so $\mathbb{Q}(u) = \mathbb{Q}([\overline{a_N; a_{N+1}, \ldots, a_{N+k}}])$, and therefore $u$ is a quadratic irrational.

The converse is more involved[44]. Assume now that $u$ is a quadratic irrational, with

$$f_0(u) = \alpha_0 u^2 + \beta_0 u + \gamma_0 = 0$$

for some $\alpha_0, \beta_0, \gamma_0 \in \mathbb{Z}$ and $\delta = \beta_0^2 - 4\alpha_0\gamma_0$ not a square. We claim that for each $n \geq 0$ there is a polynomial

$$f_n(x) = \alpha_n x^2 + \beta_n + \gamma_n$$

with

$$\beta_n^2 - 4\alpha_n\gamma_n = \delta$$

and with the property that $f_n(u_n) = 0$. This claim again follows from the fact that $\mathbb{Q}(u) = \mathbb{Q}(u_n)$, but we will need specific properties of the numbers $\alpha_n, \beta_n, \gamma_n$, so we proceed by induction.

Assume such a polynomial exists for some $n \geq 0$. Since $u_n = a_n + \frac{1}{u_{n+1}}$, we therefore have

$$u_{n+1}^2 f_n\left(a_n + \frac{1}{u_{n+1}}\right) = 0.$$

The resulting relation for $u_{n+1}$ may be written in the form

$$f_{n+1}(x) = \alpha_{n+1}x^2 + \beta_{n+1}x + \gamma_{n+1}$$

where

$$\alpha_{n+1} = a_n^2 \alpha_n + a_n \beta_n + \gamma_n,$$
$$\beta_{n+1} = 2a_n \alpha_n + \beta_n, \qquad\qquad\qquad (3.30)$$
$$\gamma_{n+1} = \alpha_n. \qquad\qquad\qquad\qquad (3.31)$$

It is clear that $\alpha_{n+1}, \beta_{n+1}, \gamma_{n+1} \in \mathbb{Z}$, and a simple calculation shows that

$$\beta_{n+1}^2 - 4\alpha_{n+1}\gamma_{n+1} = \beta_n^2 - 4\alpha_n\gamma_n,$$

proving the claim.

Notice that all the polynomials $f_n$ have the same discriminant $\delta$, which is not a square, so $\alpha_n \neq 0$ for $n \geqslant 0$. If there is some $N$ with $\alpha_n > 0$ for all $n \geqslant N$, then (3.30) shows that the sequence $\beta_N, \beta_{N+1}, \cdots$ is increasing since $a_n > 0$ for $n \geqslant 1$. Thus for large enough $n$, by (3.31), all three of $\alpha_n$, $\beta_n$ and $\gamma_n$ are positive. This is impossible, since $f_n(u_n) = 0$ and $u_n > 0$. A similar argument shows that there is no $N$ with $\alpha_n < 0$ for all $n \geqslant N$. We deduce that $\alpha_n$ must change in sign infinitely often, so in particular there is an infinite set $A \subseteq \mathbb{N}$ with the property that $\alpha_n \alpha_{n-1} < 0$ for all $n \in A$. By (3.31), it follows that $\alpha_n \gamma_n < 0$ for all $n \in A$. Now $\beta_n^2 - 4\alpha_n\gamma_n = \delta$, so for $n \in A$ we must have

$$|\alpha_n| \leqslant \tfrac{1}{4}\delta,$$
$$|\beta_n| < \sqrt{\delta},$$

and

$$|\gamma_n| \leqslant \tfrac{1}{4}\delta.$$

It follows that as $n$ runs through the infinite set $A$ there are only finitely many possibilities for the polynomials $f_n$, so there must be some $n_0 < n_1 < n_2$ with $f_{n_0} = f_{n_1} = f_{n_2}$. Since a quadratic polynomial has only two zeros, and $u_{n_0}, u_{n_1}, u_{n_2}$ are all zeros of the same polynomial, we see that two of them coincide so the continued fraction expansion of $u$ is eventually periodic. $\qquad\square$

**Corollary 3.14.** *Any quadratic irrational is badly approximable.*

PROOF. This is an immediate consequence of Theorem 3.13 and Definition 3.9. $\qquad\square$

It is not known if any other algebraic numbers are badly approximable.

## Exercises for Sect. 3.3

**Exercise 3.3.1.** [45] Show that $\mathbb{Q}(\sqrt{5})$ contains infinitely many elements with a uniform bound on their partial quotients, by checking that the numbers $[\overline{1^{k+1}, 4, 2, 1^k, 3}]$ for $k \geqslant 0$ all lie in $\mathbb{Q}(\sqrt{5})$ (here $1^k$ denotes the string

$1, 1, \ldots, 1$ of length $k$). Can you find a similar pattern in any real quadratic field $\mathbb{Q}(\sqrt{d})$?

**Exercise 3.3.2.** A number $u \in (0, 1)$ is called *very well approximable* if there is some $\delta > 0$ with the property that there are infinitely many rational numbers $\frac{p}{q}$ with $\gcd(p, q) = 1$ for which

$$\left| u - \frac{p}{q} \right| \leqslant \frac{1}{q^{2+\delta}}.$$

(a) Show that $u$ is very well approximable if and only if there is some $\varepsilon > 0$ with the property that $a_{n+1} \geqslant q_n^{\varepsilon}$ for infinitely many values of $n$.
(b) Show that for any very well approximable number the convergence in (3.28) fails.

**Exercise 3.3.3.** Prove Liouville's Theorem[46]: if $u$ is a real algebraic number of degree $d \geqslant 2$, then there is some constant $c(u) > 0$ with the property that

$$\frac{c(u)}{q^d} < \left| u - \frac{p}{q} \right|$$

for any rational number $\frac{p}{q}$.

**Exercise 3.3.4.** Use Liouville's Theorem from Exercise 3.3.3 to show that the number

$$u = \sum_{n=1}^{\infty} 10^{-n!}$$

is transcendental (that is, $u$ is not a zero of any integral polynomial)[47].

**Exercise 3.3.5.** Prove that the theorem of Margulis from p. 6 does not hold for quadratic forms in 2 variables.

## 3.4 Invertible Extension of the Continued Fraction Map

We are interested in finding a geometrically convenient invertible extension of the non-invertible map $T$, and in Sect. 9.6 will re-prove the ergodicity of the Gauss measure in that context.
    Define a set

$$\overline{Y} = \{(y, z) \in [0, 1)^2 \mid 0 \leqslant z \leqslant \frac{1}{1+y}\}$$

(this set is illustrated in Fig. 3.2) and a map $\overline{T} : \overline{Y} \to \overline{Y}$ by

$$\overline{T}(y, z) = (Ty, y(1 - yz)).$$

The map $\overline{T}$ will also be called the Gauss map.

**Proposition 3.15.** *The map* $\overline{T} : \overline{Y} \to \overline{Y}$ *is an area-preserving bijection off a null set. More precisely, there is a countable union $N$ of lines and curves in $\overline{Y}$ with the property that $T|_{\overline{Y} \smallsetminus N} : \overline{Y} \smallsetminus N \to \overline{Y} \smallsetminus N$ is a bijection preserving the Lebesgue measure.*

PROOF. The derivative of the map $\overline{T}$ is

$$\begin{pmatrix} -\frac{1}{y^2} & 0 \\ 1 - 2yz & -y^2 \end{pmatrix},$$

with determinant 1. It follows that $\overline{T}$ preserves area locally. To see that the map is a bijection, define regions $A_n$ and $B_n$ in $\overline{Y}$ by

$$A_n = \{(y, z) \in \overline{Y} \mid \frac{1}{n+1} < y < \frac{1}{n}\}$$

and

$$B_n = \{(y, z) \in \overline{Y} \mid \frac{1}{n+1+y} < z < \frac{1}{n+y} \text{ and } y > 0\}.$$

These sets are shown in Fig. 3.2. Both

$$\{A_n \mid n = 1, 2, \dots\}$$

and

$$\{B_n \mid n = 1, 2, \dots\}$$

define partitions of $\overline{Y}$ after removing countably many vertical lines (or curves in the case of $\{B_n\}$). Since this is a Lebesgue null set, it is enough to show that $\overline{T}|_{A_n} : A_n \to B_n$ is a bijection for each $n \geqslant 1$, for then

$$\overline{T}|_{\cup_{n \geqslant 1} A_n} : \bigcup_{n \geqslant 1} A_n \longrightarrow \bigcup_{n \geqslant 1} B_n$$

is also a bijection, and we can take for the null set $N$ the set of all images and pre-images of

$$\left( \overline{Y} \smallsetminus \bigcup_{n \geqslant 1} A_n \right) \cup \left( \overline{Y} \smallsetminus \bigcup_{n \geqslant 1} B_n \right).$$

Notice that $y > 0$ and $0 < z < \frac{1}{1+y}$ implies that

$$0 < yz < \frac{y}{1+y},$$

$$\frac{1}{1+y} < (1 - yz) < 1,$$

and

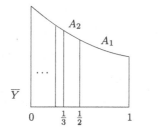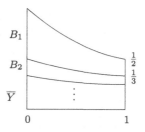

**Fig. 3.2** The Gauss map is a bijection between $\overline{Y}$ and $\overline{Y}$, sending the subset $A_n \subseteq \overline{Y}$ to the subset $B_n \subseteq \overline{Y}$ for each $n \geqslant 1$

$$\frac{y}{1+y} < y(1 - yz) < y. \tag{3.32}$$

If now $(y, z) \in A_n$ for some $n \geqslant 1$ then $y = \frac{1}{n+y_1}$ for $\overline{T}(y, z) = (y_1, z_1)$ and the inequality (3.32) becomes

$$\frac{1}{n+1+y_1} = \frac{y}{1+y} < z_1 = y(1 - yz) < y = \frac{1}{n+y_1},$$

so that $(y_1, z_1) \in B_n$ and therefore $\overline{T}(A_n) \subseteq B_n$. To see that the restriction to $A_n$ is a bijection, fix $(y_1, z_1) \in B_n$. Then $y = \frac{1}{n+y_1}$ is uniquely determined, and the equation $z_1 = y(1 - yz)$ then determines $z$ uniquely. Clearly

$$y \in \left( \frac{1}{n+1}, \frac{1}{n} \right)$$

since $y_1 \in (0, 1)$, and by reversing the argument above (or by a straightforward calculation) we see that

$$\frac{y}{1+y} = \frac{1}{n+1+y_1} < z_1 < \frac{1}{n+y_1} = y$$

implies $0 < z < \frac{1}{1+y}$ so that $(y, z) \in A_n$. □

Lemma 3.5 gives no indication of where the Gauss measure might have came from. The invertible extension, which preserves Lebesgue measure, gives an alternative proof that the Gauss measure is invariant, and gives one explanation of where it might come from.

SECOND PROOF OF LEMMA 3.5. Let $\pi : \overline{Y} \to Y$ be the projection

$$\pi(y, z) = y \tag{3.33}$$

onto $Y$. The Gauss measure $\mu$ on $Y$ is the measure defined* by

$$\mu(B) = m(\pi^{-1}B)$$

* This construction of $\mu$ from $m$ is called the *push-forward* of $m$ by $\pi$.

where $m$ is the normalized Lebesgue measure on $\overline{Y}$. Since $\overline{T} : \overline{Y} \to \overline{Y}$ preserves $m$ by Proposition 3.15 and $\pi \circ \overline{T} = T \circ \pi$, the measure $\mu$ is $T$-invariant.
□

The projection map $\pi : \overline{Y} \to Y$ defined in (3.33) shows that $\overline{T}$ on $\overline{Y}$ is an invertible extension of the non-invertible map $T$ on $Y$.

# Notes to Chap. 3

[38](Page 69) The material in Sect. 3.1 may be found in many places; a convenient source for the path followed here using matrices is a note of van der Poorten [294].

[39](Page 72) In particular, we have Dirichlet's theorem: for any $u \in \mathbb{R}$ and $Q \in \mathbb{N}$, there exists a rational number $\frac{p}{q}$ with $0 < q \leqslant Q$ and $|u - \frac{p}{q}| \leqslant \frac{1}{q(Q+1)}$, which can also be seen via the pigeon-hole principle.

[40](Page 77) A broad overview of continued fractions from an ergodic perspective may be found in the monograph of Iosifescu and Kraaikamp [161]. Kraaikamp and others have suggested ways in which Gauss could have arrived at this measure; see also Keane [187]. Other approaches to the Gauss measure are described in the book of Khinchin [191]. The ergodic approach to continued fractions has a long history. Knopp [205] showed that the Gauss measure is ergodic (in different language); Kuz'min [217] found results on the rate of mixing of the Gauss measure; Doeblin [71] showed ergodicity; Ryll-Nardzewski [326] also showed this (that the Gauss measure is "indecomposable") and used the ergodic theorem to deduce results like (3.26). This had also been shown earlier by Khinchin [190]. Lévy [227] showed (3.25), an implicitly ergodic result, in 1936 (using the language of probability rather than ergodic theory).

[41](Page 79) These results are indeed easily seen given both the ergodic theorem and the ergodicity of the Gauss map; their original proofs by other methods are not easy. For other results on the continued fraction expansion from the ergodic perspective, see Cornfeld, Fomin and Sinaĭ [60, Chap. 7] and from a number-theoretic perspective, see Khinchin [191]. The limit in (3.26), approximately 2.685, is known as Khinchin's constant; the problem of estimating it numerically is considered by Bailey, Borwein and Crandall [14]. Little is known about its arithmetical properties. The (exponential of the) constant appearing in (3.28) is usually called the Khinchin–Lévy constant. Just as in Example 2.31, it is a quite different problem to exhibit any specific number that satisfies these almost everywhere results: Adler, Keane and Smorodinsky exhibit a normal number for the continued fraction map in [2].

[42](Page 79) This is proved here directly, using estimates for conditional measures on cylinder sets; see Billingsley [31] for example. We will re-prove it in Proposition 9.25 on p. 323 using a geometrical argument.

[43](Page 87) Most of this section is devoted to quadratic irrationals, but it is clear there are uncountably many badly approximable numbers; the survey of Shallit [340] describes some of the many settings in which these numbers appear, gives other families of such numbers, and has an extensive bibliography on these numbers (which are also called numbers of *constant type*). For example, Kmošek [203] and Shallit [339] showed that if

$$\sum_{n=0}^{\infty} k^{-2^n} = [a_1^{(k)}, a_2^{(k)}, \dots],$$

then $\sup_{n \geqslant 1}\{a_n^{(2)}\} = 6$ and $\sup_{n \geqslant 1}\{a_n^{(k)}\} = n + 2$ for $k \geqslant 3$.

[44] (Page 89) There are many ways to prove this; we follow the argument of Steinig [352] here.

[45] (Page 90) This remarkable uniformity in Definition 3.9 was shown by Woods [388] for $\mathbb{Q}(\sqrt{5})$ and by Wilson [384] in general, who showed that any real quadratic field $\mathbb{Q}(\sqrt{d})$ contains infinitely many numbers of the form $[\overline{a_1, a_2, \ldots, a_k}]$ with $1 \leqslant a_n \leqslant M_d$ for all $n \geqslant 1$. McMullen [259] has explained these phenomena in terms of closed geodesics; the connection between continued fractions and closed geodesics will be developed in Chap. 9. Exercise 3.3.1 shows that we may take $M_5 = 4$, and the question is raised in [259] of whether there is a tighter bound allowing $M_d$ to be taken equal to 2 for all $d$.

[46] (Page 91) Liouville's Theorem [234, 236] (on Diophantine approximation; there are several important results bearing his name) marked the start of an important series of advances in Diophantine approximation, attempting to sharpen the lower bound. These results may be summarized as follows. The statement that for any algebraic number $u$ of degree $d$ there is a constant $c(u)$ so that for all rationals $p/q$ we have $|u - p/q| > c(u)/q^{\lambda(u)}$ holds: for $\lambda(u) = d$ (Liouville 1844); for any $\lambda(u) > \frac{1}{2}d + 1$ (Thue [360], 1909); for any $\lambda(u) > 2\sqrt{d}$ (Siegel [343], 1921); for any $\lambda(u) > \sqrt{2d}$ (Dyson [77], 1947); finally, and definitively, for any $\lambda(u) > 2$ (Roth [319], 1955).

[47] (Page 91) This observation of Liouville [235] dates from 1844 and seems to be the earliest construction of a transcendental number; in 1874 Cantor [47] used set theory to show that the set of algebraic numbers is countable, deducing that there are uncountably many transcendental real numbers (as pointed out by Herstein and Kaplansky [150, p. 238], and despite what is often taught, Cantor's proof can be used to exhibit many explicit transcendental numbers). In a different direction, many important constants were shown to be transcendental. Examples include: $e$ (Hermite [149], 1873); $\pi$ (Lindemann [232], 1882); $\alpha^\beta$ for $\alpha$ algebraic and not equal to 0 or 1 and $\beta$ algebraic and irrational (Gelfond [113] and Schneider [334], 1934).

# Chapter 4
# Invariant Measures for Continuous Maps

One of the natural ways in which measure-preserving transformations arise is from continuous maps on compact metric spaces. Let $(X, \mathsf{d})$ be a compact metric space, and let $T : X \to X$ be a continuous map. Recall that the dual space $C(X)^*$ of continuous real functionals on the space $C(X)$ of continuous functions $X \to \mathbb{R}$ can be naturally identified with the space of finite signed measures on $X$ equipped with the weak*-topology. Our main interest is in the space $\mathscr{M}(X)$ of Borel probability measures on $X$. The main properties of $\mathscr{M}(X)$ needed are described in Sect. B.5.

Any continuous map $T : X \to X$ induces a continuous map

$$T_* : \mathscr{M}(X) \to \mathscr{M}(X)$$

defined by $T_*(\mu)(A) = \mu(T^{-1}A)$ for any Borel set $A \subseteq X$. Each point $x \in X$ defines a measure $\delta_x$ by

$$\delta_x(A) = \begin{cases} 1 \text{ if } x \in A; \\ 0 \text{ if } x \notin A. \end{cases}$$

We claim that $T_*(\delta_x) = \delta_{T(x)}$ for any $x \in X$. To see this, let $A \subseteq X$ be any measurable set, and notice that

$$(T_* \delta_x)(A) = \delta_x(T^{-1}A) = \delta_{T(x)}(A).$$

This suggests that we should think of the space of measures $\mathscr{M}(X)$ as generalized points, and the transformation $T_* : \mathscr{M}(X) \to \mathscr{M}(X)$ as a natural extension of the map $T$ from the copy $\{\delta_x \mid x \in X\}$ of $X$ to the larger set $\mathscr{M}(X)$. For $f \in C(X)$ and $\mu \in \mathscr{M}(X)$,

$$\int_X f \, \mathrm{d}(T_* \mu) = \int_X f \circ T \, \mathrm{d}\mu,$$

and this property characterizes $T_*$ by (B.2) and Lemma B.12.

M. Einsiedler, T. Ward, *Ergodic Theory*, Graduate Texts in Mathematics 259, DOI 10.1007/978-0-85729-021-2_4, © Springer-Verlag London Limited 2011

The map $T_*$ is continuous and affine, so the set $\mathscr{M}^T(X)$ of $T$-invariant measures is a closed convex subset of $\mathscr{M}(X)$; in the next section[48] we will see that it is always non-empty.

## 4.1 Existence of Invariant Measures

The connection between ergodic theory and the dynamics of continuous maps on compact metric spaces begins with the next result, which shows that invariant measures can always be found.

**Theorem 4.1.** Let $T : X \to X$ be a continuous map of a compact metric space, and let $(\nu_n)$ be any sequence in $\mathscr{M}(X)$. Then any weak*-limit point of the sequence $(\mu_n)$ defined by $\mu_n = \frac{1}{n} \sum_{j=0}^{n-1} T_*^j \nu_n$ is a member of $\mathscr{M}^T(X)$.

An immediate consequence is the following important general statement, which shows that measure-preserving transformations are ubiquitous. It is known as the Kryloff–Bogoliouboff Theorem [214].

**Corollary 4.2 (Kryloff–Bogoliouboff).** Under the hypotheses of Theorem 4.1, $\mathscr{M}^T(X)$ is non-empty.

PROOF. Since $\mathscr{M}(X)$ is weak*-compact, the sequence $(\mu_n)$ must have a limit point. $\qquad\square$

Write $\|f\|_\infty = \sup\{|f(x)| \mid x \in X\}$ as usual.

PROOF OF THEOREM 4.1. Let $\mu_{n(j)} \to \mu$ be a convergent subsequence of $(\mu_n)$ and let $f \in C(X)$. Then, by applying the definition of $T_*\mu_n$, we get

$$
\left| \int f \circ T \, \mathrm{d}\mu_{n(j)} - \int f \, \mathrm{d}\mu_{n(j)} \right| = \frac{1}{n(j)} \left| \int \sum_{i=0}^{n(j)-1} \left( f \circ T^{i+1} - f \circ T^i \right) \mathrm{d}\nu_{n(j)} \right|
$$

$$
= \frac{1}{n(j)} \left| \int \left( f \circ T^{n(j)+1} - f \right) \mathrm{d}\nu_{n(j)} \right|
$$

$$
\leqslant \frac{2}{n(j)} \|f\|_\infty \longrightarrow 0
$$

as $j \to \infty$, for all $f \in C(X)$. It follows that $\int f \circ T \, \mathrm{d}\mu = \int f \, \mathrm{d}\mu$, so $\mu$ is a member of $\mathscr{M}^T(X)$ by Lemma B.12. $\qquad\square$

Thus $\mathscr{M}^T(X)$ is a non-empty compact convex set, since convex combinations of elements of $\mathscr{M}^T(X)$ belong to $\mathscr{M}^T(X)$. It follows that $\mathscr{M}^T(X)$ is an infinite set unless it comprises a single element. For many maps it is difficult to describe the space of invariant measures. The next example has very few ergodic invariant measures, and we shall see later many maps that have only one invariant measure.

*Example 4.3 (North–South map).* Define the stereographic projection $\pi$ from the circle $X = \{z \in \mathbb{C} \mid |z - \mathrm{i}| = 1\}$ to the real axis by continuing the line from 2i through a unique point on $X \setminus \{2\mathrm{i}\}$ until it meets the line $\Im(z) = 0$ (see Fig. 4.1).

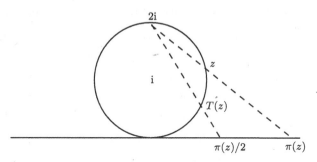

**Fig. 4.1** The North-South map on the circle; for $z \neq 2\mathrm{i}$, $T^n z \to 0$ as $n \to \infty$

The "North–South" map $T : X \to X$ is defined by

$$
T(z) = \begin{cases} 2\mathrm{i} & \text{if } z = 2\mathrm{i}; \\ \pi^{-1}(\pi(z)/2) & \text{if } z \neq 2\mathrm{i} \end{cases}
$$

as shown in Fig. 4.1. Using Poincaré recurrence (Theorem 2.11) it is easy to show that $\mathscr{M}^T(X)$ comprises the measures $p\delta_{2\mathrm{i}} + (1 - p)\delta_0$, $p \in [0, 1]$ that are supported on the two points 2i and 0. Only the measures corresponding to $p = 0$ and $p = 1$ are ergodic.

It is in general difficult to identify measures with specific properties, but the ergodic measures are readily characterized in terms of the geometry of the space of invariant measures.

**Theorem 4.4.** *Let $X$ be a compact metric space and let $T : X \to X$ be a measurable map. The ergodic elements of $\mathscr{M}^T(X)$ are exactly the extreme points of $\mathscr{M}^T(X)$.*

That is, $T$ is ergodic with respect to an invariant probability measure if and only if that measure cannot be expressed as a strict convex combination of two different $T$-invariant probability measures. For any measurable set $A$, define $\mu|_A$ by $\mu|_A(C) = \mu(A \cap C)$. If $T$ is not assumed to be continuous, then we do not know that $\mathscr{M}^T(X) \neq \varnothing$, so without the assumption of continuity Theorem 4.4 may be true but vacuous (see Exercise 4.1.1).

PROOF OF THEOREM 4.4. Let $\mu \in \mathscr{M}^T(X)$ be a non-ergodic measure. Then there is a measurable set $B$ with $\mu(B) \in (0, 1)$ and with $T^{-1}B = B$. It follows that

$$\frac{1}{\mu(B)}\mu\big|_B, \frac{1}{\mu(X\smallsetminus B)}\mu\big|_{X\smallsetminus B} \in \mathscr{M}^T(X),$$

so

$$\mu = \mu(B)\left(\frac{1}{\mu(B)}\mu\big|_B\right) + \mu(X\smallsetminus B)\left(\frac{1}{\mu(X\smallsetminus B)}\mu\big|_{X\smallsetminus B}\right)$$

expresses $\mu$ as a strict convex combination of the invariant probability measures

$$\frac{1}{\mu(B)}\mu\big|_B$$

and

$$\frac{1}{\mu(X\smallsetminus B)}\mu\big|_{X\smallsetminus B},$$

which are different since they give different measures to the set $B$.

Conversely, let $\mu$ be an ergodic measure and assume that

$$\mu = s\nu_1 + (1-s)\nu_2$$

expresses $\mu$ as a strict convex combination of the invariant measures $\nu_1$ and $\nu_2$. Since $s > 0$, $\nu_1 \ll \mu$, so there is a positive function $f \in L^1_\mu$ ($f$ is the Radon–Nikodym derivative $\frac{d\nu_1}{d\mu}$; see Theorem A.15) with the property that

$$\nu_1(A) = \int_A f \, d\mu \tag{4.1}$$

for any measurable set $A$. The set $B = \{x \in X \mid f(x) < 1\}$ is measurable since $f$ is measurable, and

$$\int_{B\cap T^{-1}B} f \, d\mu + \int_{B\smallsetminus T^{-1}B} f \, d\mu = \nu_1(B)$$

$$= \nu_1(T^{-1}B)$$

$$= \int_{B\cap T^{-1}B} f \, d\mu + \int_{(T^{-1}B)\smallsetminus B} f \, d\mu,$$

so

$$\int_{B\smallsetminus T^{-1}B} f \, d\mu = \int_{(T^{-1}B)\smallsetminus B} f \, d\mu. \tag{4.2}$$

By definition, $f(x) < 1$ for $x \in B\smallsetminus(T^{-1}B)$ while $f(x) \geqslant 1$ for $x \in T^{-1}B\smallsetminus B$. On the other hand,

$$\mu((T^{-1}B)\smallsetminus B) = \mu(T^{-1}B) - \mu((T^{-1}B)\cap B)$$

$$= \mu(B) - \mu((T^{-1}B)\cap B)$$

$$= \mu(B\smallsetminus T^{-1}B)$$

so (4.2) implies that $\mu(B\setminus T^{-1}B) = 0$ and $\mu((T^{-1}B)\setminus B) = 0$. Therefore $\mu((T^{-1}B)\triangle B) = 0$, so by ergodicity of $\mu$ we must have $\mu(B) = 0$ or 1. If $\mu(B) = 1$ then

$$\nu_1(X) = \int_X f\,d\mu < \mu(B) = 1,$$

which is impossible. So $\mu(B) = 0$.

A similar argument shows that $\mu(\{x \in X \mid f(x) > 1\}) = 0$, so $f(x) = 1$ almost everywhere with respect to $\mu$. By (4.1), this shows that

$$\nu_1 = \mu,$$

so $\mu$ is an extreme point in $\mathscr{M}^T(X)$. $\qquad\square$

Write $\mathscr{E}^T(X)$ for the set of extreme points in $\mathscr{M}^T(X)$—by Theorem 4.4, this is the set of ergodic measures for $T$.

*Example 4.5.* Let $X = \{1,\ldots,r\}^{\mathbb{Z}}$ and let $T : X \to X$ be the left shift map. In Example 2.9 we defined for any probability vector $\mathbf{p} = (p_1,\ldots,p_r)$ a $T$-invariant probability measure $\mu = \mu_{\mathbf{p}}$ on $X$, and by Proposition 2.15 all these measures are ergodic. Thus for this example the space $\mathscr{E}^T(X)$ of ergodic invariant measures is uncountable. This collection of measures is an inconceivably tiny subset of the set of all ergodic measures—there is no hope of describing all of them.

Measures $\mu_1$ and $\mu_2$ are called *mutually singular* if there exist disjoint measurable sets $A$ and $B$ with $A \cup B = X$ for which $\mu_1(B) = \mu_2(A) = 0$ (see Sect. A.4).

**Lemma 4.6.** *If $\mu_1, \mu_2 \in \mathscr{E}^T(X)$ and $\mu_1 \neq \mu_2$ then $\mu_1$ and $\mu_2$ are mutually singular.*

PROOF. Let $f \in C(X)$ be chosen with $\int f\,d\mu_1 \neq \int f\,d\mu_2$ (such a function exists by Theorem B.11). Then by the ergodic theorem (Theorem 2.30)

$$A_n^f(x) \to \int f\,d\mu_1 \qquad (4.3)$$

for $\mu_1$-almost every $x \in X$, and

$$A_n^f(x) \to \int f\,d\mu_2$$

for $\mu_2$-almost every $x \in X$. It follows that the set $A = \{x \in X \mid (4.3)\text{ holds}\}$ is measurable and has $\mu_1(A) = 1$ but $\mu_2(A) = 0$. $\qquad\square$

Some of the problems for this section make use of the topological analog of Definition 2.7, which will be used later.

**Definition 4.7.** Let $T : X \to X$ and $S : Y \to Y$ be continuous maps
of compact metric spaces (that is, topological dynamical systems). Then a
homeomorphism $\theta : X \to Y$ with $\theta \circ T = S \circ \theta$ is called a *topological conjugacy*,
and if there such a conjugacy then $T$ and $S$ are *topologically conjugate*. A
continuous surjective map $\phi : X \to Y$ with $\phi \circ T = S \circ \phi$ is called a *topological
factor* map, and in this case $S$ is said to be a *factor* of $T$.

## Exercises for Sect. 4.1

**Exercise 4.1.1.** Let $X = \{0, \frac{1}{n} \mid n \geqslant 1\}$ with the compact topology inherited
from the reals. Since $X$ is countable, there is a bijection $\theta : X \to \mathbb{Z}$. Show
that the map $T : X \to X$ defined by $T(x) = \theta^{-1}(\theta(x) + 1)$ is measurable
with respect to the Borel $\sigma$-algebra on $X$ but has no invariant probability
measures.

**Exercise 4.1.2.** Show that a weak*-limit of ergodic measures need not be
an ergodic measure by the following steps. Start with a point $x$ in the full 2-
shift $\sigma : X \to X$ with the property that any finite block of symbols of
length $\ell$ appears in $x$ with asymptotic frequency $\frac{1}{2^{\ell}}$ (such points certainly
exist; indeed the ergodic theorem says that almost every point with respect
to the $(1/2, 1/2)$ Bernoulli measure will do). Write $(x_1 \ldots x_n 0 \ldots 0)^{\infty}$ for the
point $y \in \{0,1\}^{\mathbb{Z}}$ determined by the two conditions

$$y|_{[0,2n-1]} = x_1 \ldots x_n 0 \ldots 0$$

and $\sigma^{2n}(y) = y$. Now for each $n$ construct an ergodic $\sigma$-invariant measure $\mu_n$
supported on the orbit of the periodic point $(x_1 \ldots x_n 0 \ldots 0)^{\infty}$ in which there
are $n$ 0 symbols in every cycle of the periodic point under the shift. Show
that $\mu_n$ converges to some limit $\nu$ and use Theorem 4.4 to deduce that $\nu$ is
not ergodic.

**Exercise 4.1.3.** For a continuous map $T : X \to X$ of a compact metric
space $(X, \mathsf{d})$, define the invertible extension $\widetilde{T} : \widetilde{X} \to \widetilde{X}$ as follows. Let

- $\widetilde{X} = \{x \in X^{\mathbb{Z}} \mid x_{k+1} = Tx_k \text{ for all } k \in \mathbb{Z}\}$;
- $(\widetilde{T}x)_k = x_{k+1}$ for all $k \in \mathbb{Z}$ and $x \in \widetilde{X}$;

with metric $\widetilde{\mathsf{d}}(x, y) = \sum_{k \in \mathbb{Z}} 2^{-|k|} \mathsf{d}(x_k, y_k)$. Write $\pi : \widetilde{X} \to X$ for the map
sending $x$ to $x_0$. Prove the following.

(1) $\widetilde{T}$ is a homeomorphism of a compact metric space, and $\pi : \widetilde{X} \to X$ is a
    topological factor map.
(2) If $(Y, S)$ is any homeomorphism of a compact metric space with the prop-
    erty that there is a topological factor map $(Y, S) \to (X, T)$, then $(\widetilde{X}, \widetilde{T})$
    is a topological factor of $(Y, S)$.

(3) $\pi_* \mathscr{M}^{\widetilde{T}}(\widetilde{X}) = \mathscr{M}^T(X)$.
(4) $\pi_* \mathscr{E}^{\widetilde{T}}(\widetilde{X}) = \mathscr{E}^T(X)$.

**Exercise 4.1.4.** Show that the ergodic Bernoulli measures discussed in Example 4.5 do not exhaust all ergodic measures for the full shift as follows.

(1) Show that any periodic orbit supports an ergodic measure which is not a Bernoulli measure.
(2) Show that there are ergodic measures on the full shift that are neither Bernoulli nor supported on a periodic orbit.

**Exercise 4.1.5.** Give a different proof of Lemma 4.6 using the Radon–Nikodym derivative (Theorem A.15) and the Lebesgue decomposition theorem (Theorem A.14), instead of the pointwise ergodic theorem.

**Exercise 4.1.6.** Prove that the ergodic measures for the circle-doubling map $T_2 : x \mapsto 2x$ (mod 1) are dense in the space of all invariant measures.

## 4.2 Ergodic Decomposition

An important consequence of the fact that $\mathscr{M}^T(X)$ is a compact convex set is that the Choquet representation theorem may be applied[(49)] to it. This generalizes the simple geometrical fact that in a finite-dimensional convex simplex, every point is a unique convex combination of the extreme points, to an infinite-dimensional result. In our setting, this gives a way to decompose any invariant measure into ergodic components.

**Theorem 4.8 (Ergodic decomposition).** *Let $X$ be a compact metric space and $T : X \to X$ a continuous map. Then for any $\mu \in \mathscr{M}^T(X)$ there is a unique probability measure $\lambda$ defined on the Borel subsets of the compact metric space $\mathscr{M}^T(X)$ with the properties that*

(1) $\lambda(\mathscr{E}^T(X)) = 1$, *and*

(2) $\displaystyle\int_X f \, d\mu = \int_{\mathscr{E}^T(X)} \left( \int_X f \, d\nu \right) d\lambda(\nu)$ *for any $f \in C(X)$.*

PROOF. This follows from Choquet's theorem [55] (see also the notes of Phelps [283]). A different proof will be given later (cf. p. 154), and a non-trivial example may be seen in Example 4.13. □

In fact Choquet's theorem is more general than we need: in our setting, $X$ is a compact metric space so $C(X)$ is separable, and hence $\mathscr{M}^T(X)$ is metrizable (see (B.3) for an explicit metric on $\mathscr{M}(X)$ built from a dense set of continuous functions). The picture of the space of invariant measures given by this result is similar to the familiar picture of a finite-dimensional simplex, but in fact few continuous maps[(50)] have a finite-dimensional space of invariant measures.

Indeed, as we have seen in Exercise 4.1.2, the set of ergodic measures is in general not a closed subset of the set of invariant measures.

We will see some non-trivial examples of ergodic decompositions in Sect. 4.3. The existence of the ergodic decomposition is one of the reasons that ergodicity is such a powerful tool: any property that is preserved by the integration in Theorem 4.8(2) which holds for ergodic systems holds for any measure-preserving transformation. A particularly striking case of this general principle will come up in connection with the ergodic proof of Szemerédi's theorem (see Sect. 7.2.3). There is no real topological analog of this decomposition (see Exercises 4.2.3 and 4.2.4).

# Exercises for Sect. 4.2

**Exercise 4.2.1.** A homeomorphism $T : X \to X$ of a compact metric space (a *topological dynamical system* or *cascade*) is called *minimal* if the only non-empty closed $T$-invariant subset of $X$ is $X$ itself.
(a) Show that $(X, T)$ is minimal if and only if the orbit of each point in $X$ is dense.
(b) Show that $(X, T)$ is minimal if and only if $\bigcup_{n \in \mathbb{Z}} T^n O = X$ for every non-empty open set $O \subseteq X$.
(c) Show that any topological dynamical system $(X, T)$ has a *minimal set*: that is, a closed $T$-invariant set $A$ with the property that $T : A \to A$ is minimal.

**Exercise 4.2.2.** Use Exercise 4.2.1(c) to prove Birkhoff's recurrence theorem[51]: every topological dynamical system $(X, T)$ contains a point $x$ for which there is a sequence $n_k \to \infty$ with $T^{n_k} x \to x$ as $k \to \infty$. Such a point is called *recurrent* under $T$.

**Exercise 4.2.3.** Show that in general a topological dynamical system is not a disjoint union of closed minimal subsystems.

**Exercise 4.2.4.** A homeomorphism $T : X \to X$ of a compact metric space is called *topologically ergodic* if every closed proper $T$-invariant subset of $X$ has empty interior. Show that the following properties are equivalent:

- $(X, T)$ is topologically ergodic;
- there is a point in $X$ with a dense orbit;
- for any non-empty open sets $O_1$ and $O_2$ in $X$, there is some $n \geqslant 0$ for which $O_1 \cap T^n O_2 \neq \varnothing$.

Show that in general a topological dynamical system is not a disjoint union of closed topologically ergodic subsystems.

**Exercise 4.2.5.** Let $T : X \to X$ be a continuous map on a compact metric space. Show that the measures in $\mathscr{E}^T(X)$ constrain all the ergodic averages in the following sense. For $f \in C(X)$, define

$$m(f) = \inf_{\mu \in \mathscr{E}^T(X)} \left\{ \int f \, d\mu \right\}$$

and

$$M(f) = \sup_{\mu \in \mathscr{E}^T(X)} \left\{ \int f \, d\mu \right\}.$$

Prove that

$$m(f) \leqslant \liminf_{N \to \infty} \mathsf{A}_N^f(x) \leqslant \limsup_{N \to \infty} \mathsf{A}_N^f(x) \leqslant M(f)$$

for any $x \in X$.

## 4.3 Unique Ergodicity

A natural distinguished class of transformations are those for which there is only one invariant Borel measure. This measure is automatically ergodic, and the uniqueness of this measure has several powerful consequences.

**Definition 4.9.** Let $X$ be a compact metric space and let $T : X \to X$ be a continuous map. Then $T$ is said to be *uniquely ergodic* if $\mathscr{M}^T(X)$ comprises a single measure.

**Theorem 4.10.** *For a continuous map $T : X \to X$ on a compact metric space, the following properties are equivalent.*

(1) *$T$ is uniquely ergodic.*
(2) *$|\mathscr{E}^T(X)| = 1$.*
(3) *For every $f \in C(X)$,*

$$\mathsf{A}_N^f = \frac{1}{N} \sum_{n=0}^{N-1} f(T^n x) \longrightarrow C_f, \tag{4.4}$$

*where $C_f$ is a constant independent of $x$.*
(4) *For every $f \in C(X)$, the convergence (4.4) is uniform across $X$.*
(5) *The convergence (4.4) holds for every $f$ in a dense subset of $C(X)$.*

*Under any of these assumptions, the constant $C_f$ in (4.4) is $\int_X f \, d\mu$, where $\mu$ is the unique invariant measure.*

We will make use of Theorem 4.8 for the equivalence of (1) and (2); the equivalence between (1) and (3)–(5) is independent of it.

PROOF OF THEOREM 4.10. (1) $\iff$ (2): If $T$ is uniquely ergodic and $\mu$ is the only $T$-invariant probability measure on $X$, then $\mu$ must be ergodic by Theorem 4.4. If there is only one ergodic invariant probability measure on $X$, then by Theorem 4.8, it is the only invariant probability measure on $X$.

(1) $\implies$ (3): Let $\mu$ be the unique invariant measure for $T$, and apply Theorem 4.1 to the constant sequence $(\delta_x)$. Since there is only one possible limit point and $\mathscr{M}(X)$ is compact, we must have

$$\frac{1}{N} \sum_{n=0}^{N-1} \delta_{T^n x} \longrightarrow \mu$$

in the weak*-topology, so for any $f \in C(X)$

$$\frac{1}{N} \sum_{n=0}^{N-1} f(T^n x) \longrightarrow \int_X f \, \mathrm{d}\mu.$$

(3) $\implies$ (1): Let $\mu \in \mathscr{M}^T(X)$. Then by the dominated convergence theorem, (4.4) implies that

$$\int_X f \, \mathrm{d}\mu = \int_X \lim_{N \to \infty} \frac{1}{N} \sum_{n=0}^{N-1} f(T^n x) \, \mathrm{d}\mu = C_f$$

for all $f \in C(X)$. It follows that $C_f$ is the integral of $f$ with respect to any measure in $\mathscr{M}^T(X)$, so $\mathscr{M}^T(X)$ can only contain a single measure.

Notice that this also shows $C_f = \int_X f \, \mathrm{d}\mu$ for the unique measure $\mu$.

(1) $\implies$ (4): Let $\mu \in \mathscr{M}^T(X)$, and notice that we must have $C_f = \int f \, \mathrm{d}\mu$ as above. If the convergence is not uniform, then there is a function $g$ in $C(X)$ and an $\varepsilon > 0$ such that for every $N_0$ there is an $N > N_0$ and a point $x_j \in X$ for which

$$\left| \frac{1}{N} \sum_{n=0}^{N-1} g(T^n x_j) - C_g \right| \geq \varepsilon.$$

Let $\mu_N = \frac{1}{N} \sum_{n=0}^{N-1} \delta_{T^n x_j}$, so that

$$\left| \int_X g \, \mathrm{d}\mu_N - C_g \right| \geq \varepsilon. \tag{4.5}$$

By weak*-compactness the sequence $(\mu_N)$ has a subsequence $(\mu_{N(k)})$ with

$$\mu_{N(k)} \to \nu$$

as $k \to \infty$. Then $\nu \in \mathscr{M}^T(X)$ by Theorem 4.1, and

$$\left| \int_X g \, \mathrm{d}\nu - C_g \right| \geq \varepsilon$$

by (4.5). However, this shows that $\mu \neq \nu$, which contradicts (1).

(4) $\implies$ (5): This is clear.

(5) $\implies$ (1): If $\mu, \nu \in \mathscr{E}^T(X)$ then, just as in the proof that (3) $\implies$ (1),

$$\int_X f \, d\nu = C_f = \int_X f \, d\mu$$

for any function $f$ in a dense subset of $C(X)$, so $\nu = \mu$.                    $\square$

The equivalence of (1) and (3) in Theorem 4.10 appeared first in the paper of Kryloff and Bogoliouboff [214] in the context of uniquely ergodic flows.

*Example 4.11.* The circle rotation $R_\alpha : \mathbb{T} \to \mathbb{T}$ is uniquely ergodic if and only if $\alpha$ is irrational. The unique invariant measure in this case is the Lebesgue measure $m_\mathbb{T}$. This may be proved using property (5) of Theorem 4.10 (or using property (1); see Theorem 4.14). Assume first that $\alpha$ is irrational, so $e^{2\pi i k \alpha} = 1$ only if $k = 0$. If $f(t) = e^{2\pi i k t}$ for some $k \in \mathbb{Z}$, then

$$\frac{1}{N} \sum_{n=0}^{N-1} f(R_\alpha^n t) = \frac{1}{N} \sum_{n=0}^{N-1} e^{2\pi i k(t+n\alpha)} = \begin{cases} 1 & \text{if } k = 0; \\ \dfrac{1}{N} e^{2\pi i k t} \dfrac{e^{2\pi i N k \alpha} - 1}{e^{2\pi i k \alpha} - 1} & \text{if } k \neq 0. \end{cases}$$

$$(4.6)$$

Equation (4.6) shows that

$$\frac{1}{N} \sum_{n=0}^{N-1} f(R_\alpha^n t) \longrightarrow \int f \, dm_\mathbb{T} = \begin{cases} 1 \text{ if } k = 0; \\ 0 \text{ if } k \neq 0. \end{cases}$$

By linearity, the same convergence will hold for any trigonometric polynomial, and therefore property (5) of Theorem 4.10 holds. For a curious application of this result, see Example 1.3.

If $\alpha$ is rational, then Lebesgue measure is invariant but not ergodic, so there must be other invariant measures.

Example 4.11 may be used to illustrate the ergodic decomposition of a particularly simple dynamical system.

*Example 4.12.* Let $X = \{z \in \mathbb{C} \mid |z| = 1 \text{ or } 2\}$, let $\alpha$ be an irrational number, and define a continuous map $T : X \to X$ by $T(z) = e^{2\pi i \alpha} z$. By unique ergodicity on each circle, any invariant measure $\mu$ takes the form

$$\mu = s m_1 + (1 - s) m_2,$$

where $m_1$ and $m_2$ denote Lebesgue measures on the two circles comprising $X$. Thus $\mathscr{M}^T(X) = \{s m_1 + (1-s) m_2 \mid s \in [0, 1]\}$, with the two ergodic measures given by the extreme points $s = 0$ and $s = 1$. The decomposition of $\mu$ is described by the measure $\nu = s\delta_{m_1} + (1 - s)\delta_{m_2}$. A convenient notation for this is $\mu = \int_{\mathscr{M}^T(X)} m \, d\nu(m)$.

*Example 4.13.* A more sophisticated version of Example 4.12 is a rotation on the disk. Let $\mathbb{D} = \{z \in \mathbb{C} \mid |z| \leqslant 1\}$, let $\alpha$ be an irrational number, and define a continuous map $T : \mathbb{D} \to \mathbb{D}$ by $T(z) = \mathrm{e}^{2\pi i \alpha} z$. For each $r \in (0,1]$, let $m_r$ denote the normalized Lebesgue measure on the circle $\{z \in \mathbb{C} \mid |z| = r\}$ and let $m_0 = \delta_0$ (these are the ergodic measures). Then the decomposition of $\mu \in \mathscr{M}^T(X)$ is a measure $\nu$ on $\{m_r \mid r \in [0,1]\}$, and

$$\mu(A) = \int_{\mathscr{M}^T(X)} m_r(A) \, \mathrm{d}\nu(m_r).$$

Both Proposition 2.16 and Example 4.11 are special cases of the following more general result about unique ergodicity for rotations on compact groups.

**Theorem 4.14.** *Let $X$ be a compact metrizable group and $R_g(x) = gx$ the rotation by a fixed element $g \in X$. Then the following are equivalent.*

(1) *$R_g$ is uniquely ergodic (with the unique invariant measure being $m_X$, the Haar measure on $X$).*
(2) *$R_g$ is ergodic with respect to $m_X$.*
(3) *The subgroup $\{g^n\}_{n \in \mathbb{Z}}$ generated by $g$ is dense in $X$.*
(4) *$X$ is abelian, and $\chi(g) \neq 1$ for any non-trivial character $\chi \in \widehat{X}$.*

PROOF. (1) $\implies$ (2): This is clear.

(2) $\implies$ (3): Let $Y$ denote the closure of the subgroup generated by $g$. If $Y \neq X$ then there is a continuous non-constant function on $X$ that is constant on each coset of $Y$: in fact if $\mathsf{d}$ is a bi-invariant metric on $X$ giving the topology, then

$$\mathsf{d}_Y(x) = \min\{\mathsf{d}(x,y) \mid y \in Y\}$$

defines such a function (an invariant metric exists by Lemma C.2). Such a function is invariant under $R_g$, showing that $R_g$ is not ergodic.

(3) $\implies$ (1): If $Y = X$ then $X$ is abelian (since it contains a dense abelian subgroup), and any probability measure $\mu$ invariant under $R_g$ is invariant under translation by a dense subgroup. This implies that $\mu$ is invariant under translation by any $y \in X$ by the following argument. Let $f \in C(X)$ be any continuous function, and fix $\varepsilon > 0$. Then for every $\delta > 0$ there is some $n$ with $\mathsf{d}(y, g^n) < \delta$, so by an appropriate choice of $\delta$ we have

$$|f(g^n x) - f(yx)| < \varepsilon$$

for all $x \in X$. Since

$$\int f(x) \, \mathrm{d}\mu(x) = \int f(g^n x) \, \mathrm{d}\mu(x),$$

it follows that

$$\left| \int f(yx) \, \mathrm{d}\mu(x) - \int f(x) \, \mathrm{d}\mu(x) \right| = \left| \int (f(yx) - f(g^n x)) \, \mathrm{d}\mu(x) \right| < \varepsilon$$

for all $\varepsilon > 0$, so $R_y$ preserves $\mu$. Since this holds for all $y \in X$, $\mu$ must be the Haar measure. It follows that $R_g$ is uniquely ergodic.

(4) $\implies$ (2): Assume now that $X$ is abelian and $\chi(g) \neq 1$ for every non-trivial character $\chi \in \widehat{X}$. If $f \in L^2(X)$ is invariant under $R_g$, then the Fourier series

$$f = \sum_{\chi \in \widehat{X}} c_\chi \chi$$

satisfies

$$f = U_{R_g} f = \sum_{\chi \in \widehat{X}} c_\chi \chi(g) \chi,$$

and so $f$ is constant as required.

(2) $\implies$ (4): By (3) it follows that $X$ is abelian. If now $\chi \in \widehat{X}$ is a character with $\chi(g) = 1$, then

$$\chi(R_g x) = \chi(g)\chi(x) = \chi(x)$$

is invariant, which by (2) implies that $\chi$ is itself a constant almost everywhere and so is trivial. $\qquad\square$

**Corollary 4.15.** *Let* $X = \mathbb{T}^\ell$, *and let* $g = (\alpha_1, \alpha_2, \ldots, \alpha_\ell) \in \mathbb{R}^\ell$. *Then the toral rotation* $R_g : \mathbb{T}^\ell \to \mathbb{T}^\ell$ *given by* $R_g(x) = x + g$ *is uniquely ergodic if and only if* $1, \alpha_1, \ldots, \alpha_\ell$ *are linearly independent over* $\mathbb{Q}$.

Theorems 2.19 and 4.14 have been generalized to give characterizations of ergodicity for affine maps on compact abelian groups by Hahn and Parry [131] and Parry [278], and on non-abelian groups by Chu [57].

## Exercises for Sect. 4.3

**Exercise 4.3.1.** Prove that (3) $\implies$ (1) in Theorem 4.14 using Pontryagin duality.

**Exercise 4.3.2.** Show that a surjective homomorphism $T : X \to X$ of a compact group $X$ is uniquely ergodic if and only if $|X| = 1$.

**Exercise 4.3.3.** Extend Theorem 4.14 by using the quotient space $Y \backslash X$ of a compact group $X$ to classify the probability measures on $X$ invariant under the rotation $R_g$ when $Y \neq X$.

**Exercise 4.3.4.** Show that for any Riemann-integrable function $f : \mathbb{T} \to \mathbb{R}$ and $\varepsilon > 0$ there are trigonometric polynomials $p^-$ and $p^+$ such that

$$p^-(t) < f(t) < p^+(t)$$

for all $t \in \mathbb{T}$, and $\int_0^1 (p^+(t) - p^-(t)) \, dt < \varepsilon$. Use this to show that if $\alpha$ is irrational then for any Riemann-integrable function $f : \mathbb{T} \to \mathbb{R}$,

$$\frac{1}{N} \sum_{n=0}^{N-1} f(R_\alpha^n t) \to \int f \, dm_\mathbb{T}$$

for all $t \in \mathbb{T}$.

**Exercise 4.3.5.** Prove Corollary 4.15
(a) using Theorem 4.14;
(b) using Theorem 4.10(5).

**Exercise 4.3.6.** [52] Let $X$ be a compact metric space, and let $T : X \to X$ be a continuous map. Assume that $\mu \in \mathscr{E}^T(X)$, and that for every $x \in X$ there exists a constant $C = C(x)$ such that for every $f \in C(X)$, $f \geqslant 0$,

$$\limsup_{N \to \infty} \frac{1}{N} \sum_{n=0}^{N-1} f(T^n x) \leqslant C \int f \, d\mu.$$

Show that $T$ is uniquely ergodic.

## 4.4 Measure Rigidity and Equidistribution

A natural question in number theory concerns how a sequence of real numbers is distributed when reduced modulo 1. When the terms of the sequence are generated by some dynamical process, then the expressions resemble ergodic averages, and it is natural to expect that ergodic theory will have something to offer.

### 4.4.1 Equidistribution on the Interval

Ergodic theorems give conditions under which all or most orbits in a dynamical system spend a proportion of time in a given set proportional to the measure of the set. In this section we consider a more abstract notion of equidistribution[53] in the specific setting of Lebesgue measure on the unit interval.

**Definition 4.16.** A sequence $(x_n)$ with $x_n \in [0, 1]$ for all $n$ is said to be *equidistributed* or *uniformly distributed* if

$$\lim_{n \to \infty} \frac{1}{n} \sum_{k=1}^{n} f(x_k) = \int_0^1 f(x) \, dx \qquad (4.7)$$

for any $f \in C([0,1])$.

A more intuitive formulation (developed in Lemma 4.17) of equidistribution requires that the terms of the sequence fall in an interval with the correct frequency, just as the pointwise ergodic theorem (Theorem 2.30) says that almost every orbit under an ergodic transformation falls in a measurable set with the correct frequency.

**Lemma 4.17.** [54] *For a sequence $(x_n)$ of elements of $[0,1]$, the following properties are equivalent.*

(1) *The sequence $(x_n)$ is equidistributed.*
(2) *For any $k \neq 0$,*

$$\lim_{n \to \infty} \frac{1}{n} \sum_{j=1}^{n} e^{2\pi i k x_j} = 0.$$

(3) *For any numbers $a, b$ with $0 \leqslant a < b \leqslant 1$,*

$$\frac{1}{n} \left| \{ j \mid 1 \leqslant j \leqslant n, x_j \in [a,b] \} \right| \longrightarrow (b - a)$$

*as $n \to \infty$.*

PROOF. (1) $\iff$ (3): Assume (1) and fix $a, b$ with $0 \leqslant a < b \leqslant 1$. Given a sufficiently small $\varepsilon > 0$, define continuous functions that approximate the indicator function $\chi_{[a,b]}$ by

$$f^+(x) = \begin{cases} 1 & \text{if } a \leqslant x \leqslant b; \\ (x - (a - \varepsilon))/\varepsilon & \text{if } \max\{0, a - \varepsilon\} \leqslant x < a; \\ ((b + \varepsilon) - x)/\varepsilon & \text{if } b < x \leqslant \min\{b + \varepsilon, 1\}; \\ 0 & \text{for other } x, \end{cases}$$

and

$$f^-(x) = \begin{cases} 1 & \text{if } a + \varepsilon \leqslant x \leqslant b - \varepsilon; \\ (x - a)/\varepsilon & \text{if } a \leqslant x < a + \varepsilon; \\ (b - x)/\varepsilon & \text{if } b - \varepsilon < x \leqslant b; \\ 0 & \text{for other } x. \end{cases}$$

Notice that $f^-(x) \leqslant \chi_{[a,b]}(x) \leqslant f^+(x)$ for all $x \in [0,1]$, and

$$\int_0^1 \left( f^+(x) - f^-(x) \right) dx \leqslant 2\varepsilon.$$

For small $\varepsilon$ and $0 < a < b < 1$, these functions are illustrated in Fig. 4.2. It follows that

$$\frac{1}{n} \sum_{j=1}^{n} f^-(x_j) \leqslant \frac{1}{n} \sum_{j=1}^{n} \chi_{[a,b]}(x_j) \leqslant \frac{1}{n} \sum_{j=1}^{n} f^+(x_j).$$

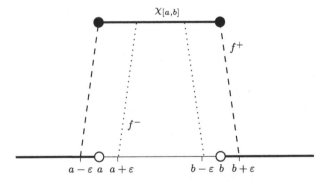

**Fig. 4.2** The function $\chi_{[a,b]}$ and the approximations $f^-$ (dots) and $f^+$ (dashes)

By equidistribution, this implies that

$$b - a - 2\varepsilon \leqslant \int_0^1 f^- \, \mathrm{d}x \leqslant \liminf_{n \to \infty} \frac{1}{n} \sum_{j=1}^n \chi_{[a,b]}(x_j)$$

$$\leqslant \limsup_{n \to \infty} \frac{1}{n} \sum_{j=1}^n \chi_{[a,b]}(x_j) \leqslant \int_0^1 f^+ \, \mathrm{d}x \leqslant b - a + 2\varepsilon.$$

Thus

$$\liminf_{n \to \infty} \frac{1}{n} \sum_{j=1}^n \chi_{[a,b]}(x_j) = \limsup_{n \to \infty} \frac{1}{n} \sum_{j=1}^n \chi_{[a,b]}(x_j) = b - a$$

as required.

Conversely, if (3) holds then (1) holds since any continuous function may be approximated uniformly by a finite linear combination of indicators of intervals*.

(1) $\Longleftrightarrow$ (2): In one direction this is clear; to see that (2) implies (1) it is enough to notice that finite trigonometric polynomials are dense in $C([0,1])$ in the uniform metric.                                                                                      □

Notice that equidistribution of $(x_n)$ does not imply that (4.7) holds for measurable functions (but see Exercise 4.4.7).

*Example 4.18.* [55] A consequence of Theorem 4.10 and Example 4.11 is that for any irrational number $\alpha$, and any initial point $x \in \mathbb{T}$, the orbit $x, R_\alpha x, R_\alpha^2 x, \ldots$ under the circle rotation is an equidistributed sequence. Note that this is proved in Example 4.11 by using property (2) of Lemma 4.17.

---

\* We note that the two implications (3) $\Longrightarrow$ (1) and (2) $\Longrightarrow$ (1) rely on the same argument, which will be explained in detail in the proof of Corollary 4.20.

## 4.4.2 Equidistribution and Generic Points

**Definition 4.19.** If $X$ is a compact metric space, and $\mu$ is a Borel probability measure on $X$, then a sequence $(x_n)$ of elements of $X$ is equidistributed with respect to $\mu$ if for any $f \in C(X)$,

$$\lim_{n \to \infty} \frac{1}{n} \sum_{j=1}^{n} f(x_j) = \int_X f(x) \, d\mu(x).$$

Equivalently, $(x_n)$ is equidistributed if

$$\frac{1}{n} \sum_{j=1}^{n} \delta_{x_j} \longrightarrow \mu$$

in the weak*-topology.

For a continuous transformation $T : X \to X$ and an invariant measure $\mu$ we say that $x \in X$ is *generic* (with respect to $\mu$ and $T$) if the sequence of points along the orbit $(T^n x)$ is equidistributed with respect to $\mu$. Notice that if $x$ is generic with respect to one invariant probability measure for $T$, then $x$ cannot be generic with respect to any other invariant probability measure for $T$. The following is an easy consequence of the ergodic theorem (Theorem 2.30).

**Corollary 4.20.** *Let $X$ be a compact metric space, let $T : X \to X$ be a continuous map, and let $\mu$ be a $T$-invariant ergodic probability measure. Then $\mu$-almost every point in $X$ is generic with respect to $T$ and $\mu$.*

PROOF. Recall that $C(X)$ is a separable metric space with respect to the uniform norm

$$\|f\|_\infty = \sup\{|f(x)| \mid x \in X\}$$

by Lemma B.8. Let $(f_n)_{n \geqslant 1}$ be a dense sequence in $C(X)$. By applying Theorem 2.30 to each of these functions we obtain one set $X' \subseteq X$ of full measure with the property that

$$\frac{1}{N} \sum_{n=0}^{N-1} f_i(T^n x) \longrightarrow \int_X f_i \, d\mu$$

for all $i \geqslant 1$ and $x \in X'$. Now let $f \in C(X)$ be any function and fix $\varepsilon > 0$. By the uniform density of the sequence, we may find an $i \in \mathbb{N}$ for which

$$|f(x) - f_i(x)| < \varepsilon$$

for all $x \in X$. Then

$$\int f \, d\mu - 2\varepsilon \leqslant \liminf_{N \to \infty} \frac{1}{N} \sum_{n=0}^{N-1} f(T^n x) \leqslant \limsup_{N \to \infty} \frac{1}{N} \sum_{n=0}^{N-1} f(T^n x) \leqslant \int f \, d\mu + 2\varepsilon,$$

showing convergence of the ergodic averages for $f$ at any $x \in X'$. The limit must be $\int f \, d\mu$ since $|\int f \, d\mu - \int f_i \, d\mu| \leqslant \varepsilon$, so $x$ is a generic point. □

### 4.4.3 Equidistribution for Irrational Polynomials

Example 4.18 may be thought of as a statement in number theory: for an irrational $\alpha$, the values of the polynomial $p(n) = x + \alpha n$, when reduced modulo 1, form an equidistributed sequence for any value of $x$. Weyl [381] generalized this to more general polynomials, and Furstenberg [98] found that this result could also be understood using ergodic theory. We recall the statement of Weyl's polynomial equidistribution Theorem (Theorem 1.4 on p. 4): Let $p(n) = a_k n^k + \cdots + a_0$ be a real polynomial with at least one coefficient among $a_1, \ldots, a_k$ irrational. Then the sequence $(p(n))$ is equidistributed modulo 1.

As indicated in Example 4.18, the unique ergodicity of irrational circle rotations proves Theorem 1.4 for $k = 1$. More generally, Theorem 4.10 shows that the orbits of any transformation of the circle for which the Lebesgue measure is the unique invariant measure are equidistributed. In order to apply this to the case of polynomials, we turn to a structural result of Furstenberg [99] that allows more complicated transformations to be built up from simpler ones while preserving a dynamical property (in Chap. 7 a similar approach will be used for another application of ergodic theory).

Notice that by Theorem 4.10, orbits of a uniquely ergodic transformation are equidistributed with respect to the unique invariant measure.

**Theorem 4.21 (Furstenberg).** *Let $T : X \to X$ be a uniquely ergodic homeomorphism of a compact metric space with unique invariant measure $\mu$. Let $G$ be a compact group\* with Haar measure $m_G$, and let $c : X \to G$ be a continuous map. Define the skew-product map $S$ on $Y = X \times G$ by*

$$S(x, g) = (T(x), c(x)g).$$

*If $S$ is ergodic with respect to $\mu \times m_G$, then it is uniquely ergodic.*

PROOF. To see that $S$ preserves $\mu \times m_G$, let $f \in C(Y)$. Then, by Fubini's theorem,

---

\* The reader may replace $G$ by a torus $\mathbb{T}^k$ with group operation written additively, together with Lebesgue measure $m_{\mathbb{T}^k}$. Notice that in any case the Haar measure is invariant under multiplication on the right or the left since $G$ is compact (see Sect. C.2).

$$\int_Y f \circ S \, d(\mu \times m_G) = \int_X \int_G f(Tx, c(x)g) \, dm_G(g) \, d\mu(x)$$

$$= \int_X \int_G f(Tx, g) \, dm_G(g) \, d\mu(x)$$

$$= \int_X \int_G f(x, g) \, dm_G(g) \, d\mu(x) = \int_Y f \, d(\mu \times m_G).$$

Assume that $S$ is ergodic. Let

$$E = \{(x, g) \mid (x, g) \text{ is generic w.r.t. } \mu \times m_G\}.$$

By Corollary 4.20, $\mu \times m_G(E) = 1$. We claim that $E$ is invariant under the map $(x, g) \mapsto (x, gh)$. To see this, notice that $(x, g) \in E$ means that

$$\frac{1}{N} \sum_{n=0}^{N-1} f\left(S^n(x, g)\right) \longrightarrow \int f \, d(\mu \times m_G)$$

for all $f \in C(X \times G)$. Writing $f_h(\cdot, g) = f(\cdot, gh)$, it follows that

$$\frac{1}{N} \sum_{n=0}^{N-1} f\left(S^n(x, gh)\right) = \frac{1}{N} \sum_{n=0}^{N-1} f_h\left(S^n(x, g)\right)$$

$$\longrightarrow \int f_h \, d(\mu \times m_G) = \int f \, d(\mu \times m_G)$$

since $m_G$ is invariant under multiplication on the right, so $(x, gh) \in E$ also. It follows that $E = E_1 \times G$ for some set $E_1 \subseteq X, \mu(E_1) = 1$. Now assume that $\nu$ is an $S$-invariant ergodic measure on $Y$. Write $\pi : Y \to X$ for the projection $\pi(x, g) = x$. Then $\pi_* \nu$ is a $T$-invariant measure, so by unique ergodicity $\pi_* \nu = \mu$. In particular, $\nu(E) = \nu(E_1 \times G) = \mu(E_1) = 1$. By Corollary 4.20, $\nu$-almost every point is generic with respect to $\nu$. Thus there must be a point $(x, g) \in E$ generic with respect to $\nu$. By definition of $E$, it follows that $\nu = \mu \times m_G$. $\qquad\square$

**Corollary 4.22.** *Let $\alpha$ be an irrational number. Then the map $S : \mathbb{T}^k \to \mathbb{T}^k$ defined by*

$$S : \begin{pmatrix} x_1 \\ x_2 \\ \vdots \\ x_k \end{pmatrix} \longmapsto \begin{pmatrix} x_1 + \alpha \\ x_2 + x_1 \\ \vdots \\ x_k + x_{k-1} \end{pmatrix}$$

*is uniquely ergodic.*

PROOF. Notice that the transformation $S$ is built up from the irrational circle map by taking $(k-1)$ skew-product extensions as in Theorem 4.21. By Theorem 4.21, it is sufficient to prove that $S$ is ergodic with respect to

Lebesgue measure on $\mathbb{T}^k$. Let $f \in L^2(\mathbb{T}^k)$ be an $S$-invariant function, and write

$$f(\mathbf{x}) = \sum_{\mathbf{n} \in \mathbb{Z}^k} c_{\mathbf{n}} e^{2\pi i \mathbf{n} \cdot \mathbf{x}}$$

for the Fourier expansion of $f$. Then, since $f(\mathbf{x}) = f(S\mathbf{x})$, we have

$$\sum_{\mathbf{n} \in \mathbb{Z}^k} c_{\mathbf{n}} e^{2\pi i \mathbf{n} \cdot S\mathbf{x}} = \sum_{\mathbf{n} \in \mathbb{Z}^k} c_{\mathbf{n}} e^{2\pi i n_1 \alpha} e^{2\pi i S' \mathbf{n} \cdot \mathbf{x}}$$

where

$$S' : \begin{pmatrix} n_1 \\ n_2 \\ \vdots \\ n_{k-1} \\ n_k \end{pmatrix} \longmapsto \begin{pmatrix} n_1 + n_2 \\ n_2 + n_3 \\ \vdots \\ n_{k-1} + n_k \\ n_k \end{pmatrix}$$

is an automorphism of $\mathbb{Z}^k$. By the uniqueness of Fourier coefficients,

$$c_{S'\mathbf{n}} = e^{2\pi i \alpha n_1} c_{\mathbf{n}}, \tag{4.8}$$

and in particular $|c_{S'\mathbf{n}}| = |c_{\mathbf{n}}|$ for all $\mathbf{n}$. Thus for each $\mathbf{n} \in \mathbb{Z}^k$ we either have $\mathbf{n}, S'\mathbf{n}, (S')^2\mathbf{n}, \ldots$ all distinct (in which case $c_{\mathbf{n}} = 0$ since $\sum_{\mathbf{n}} |c_{\mathbf{n}}|^2 < \infty$) or $(S')^p\mathbf{n} = (S')^q\mathbf{n}$ for some $p > q$, so $n_2 = n_3 = \cdots = n_k = 0$ (by downward induction on $k$, for example). Now for $\mathbf{n} = (n_1, 0, \ldots, 0)$, (4.8) simplifies to $c_{\mathbf{n}} = e^{2\pi i n_1 \alpha} c_{\mathbf{n}}$, so $n_1 = 0$ or $c_{\mathbf{n}} = 0$. We deduce that $f$ is constant, so $S$ is ergodic.                                                                                                       □

PROOF OF THEOREM 1.4. Assume that Theorem 1.4 holds for all polynomials of degree strictly less than $k$. If $a_k$ is rational, then $qa_k \in \mathbb{Z}$ for some integer $q$. Then the quantities $p(qn+j)$ modulo 1 for varying $n$ and fixed $j = 0, \ldots, q-1$, coincide with the values of polynomials of degree strictly less than $k$ satisfying the hypothesis of the theorem. It follows that the values of each of those polynomials are equidistributed, so the values of the original polynomial are equidistributed modulo 1 by induction. Therefore, we may assume without loss of generality that the leading coefficient $a_k$ is irrational.

A convenient description of the transformation $S$ in Corollary 4.22 comes from viewing $\mathbb{T}^k$ as $\{\alpha\} \times \mathbb{T}^k$ with a map defined by

$$\begin{pmatrix} 1 & & & & \\ 1 & 1 & & & \\ & 1 & 1 & & \\ & & & \ddots & \\ & & & & 1 & 1 \end{pmatrix} \begin{pmatrix} \alpha \\ x_1 \\ x_2 \\ \vdots \\ x_k \end{pmatrix} = \begin{pmatrix} \alpha \\ x_1 + \alpha \\ x_2 + x_1 \\ \vdots \\ x_k + x_{k-1} \end{pmatrix}.$$

Iterating this map gives

$$
\begin{pmatrix} 1 & & & \\ 1 & 1 & & \\ & 1 & 1 & \\ & & & \ddots & \\ & & & 1 & 1 \end{pmatrix}^n \begin{pmatrix} \alpha \\ x_1 \\ x_2 \\ \vdots \\ x_k \end{pmatrix} = \begin{pmatrix} 1 & & & & \\ n & 1 & & & \\ \binom{n}{2} & n & 1 & & \\ \vdots & & \ddots & \ddots & \\ \binom{n}{k} & \cdots & & n & 1 \end{pmatrix} \begin{pmatrix} \alpha \\ x_1 \\ x_2 \\ \vdots \\ x_k \end{pmatrix}
$$

$$
= \begin{pmatrix} \alpha \\ n\alpha + x_1 \\ \binom{n}{2}\alpha + nx_1 + x_2 \\ \vdots \\ \binom{n}{k}\alpha + \binom{n}{k-1}x_1 + \cdots + nx_{k-1} + x_k \end{pmatrix}.
$$

Now define $\alpha = k! a_k$, and choose points $x_1, \ldots, x_k$ so that

$$
p(n) = \binom{n}{k}\alpha + \binom{n}{k-1}x_1 + \cdots + nx_{k-1} + x_k.
$$

Then by Corollary 4.22, the orbits of this map are equidistributed on $\mathbb{T}^k$, so the same holds for its last component, which coincides with the sequence of values of $p(n)$ reduced modulo 1 in $\mathbb{T}$. □

An alternative approach in the quadratic case will be described in Exercise 7.4.2.

## Exercises for Sect. 4.4

**Exercise 4.4.1.** Consider the circle-doubling map $T_2 : x \mapsto 2x$ (mod 1) on $\mathbb{T}$ with Lebesgue measure $m_{\mathbb{T}}$.
(a) Construct a point that is generic for $m_{\mathbb{T}}$.
(b) Construct a point that is generic for a $T_2$-invariant ergodic measure other than $m_{\mathbb{T}}$.
(c) Construct a point that is generic for a non-ergodic $T_2$-invariant measure.
(d) Construct a point that is not generic for any $T_2$-invariant measure.

**Exercise 4.4.2.** Extend Lemma 4.17 to show that (4.7) holds for Riemann-integrable functions (cf. Exercise 4.3.4). Could it hold for Lebesgue-integrable functions?

**Exercise 4.4.3.** Use Exercise 4.3.4 to show that the fractional parts of the sequence $(n\alpha)$ are uniformly distributed in $[0, 1]$. That is,

$$
\frac{|\{n \mid 0 \leqslant n < N, n\alpha - \lfloor n\alpha \rfloor \in [a, b)\}|}{N} \to (b - a)
$$

as $N \to \infty$, for any $0 \leqslant a < b \leqslant 1$.

**Exercise 4.4.4.** Carry out the procedure used in the proof of Theorem 1.4 to prove that the sequence $(x_n)$ defined by $x_n = \binom{\alpha_1 n}{\alpha_2 n^2}$ is equidistributed in $\mathbb{T}^2$ if and only if $\alpha_1, \alpha_2 \notin \mathbb{Q}$.

**Exercise 4.4.5.** A number $\alpha$ is called a *Liouville number* if there is an infinite sequence $(\frac{p_n}{q_n})_{n \geqslant 1}$ of rationals with the property that

$$\left| \frac{p_n}{q_n} - \alpha \right| < \frac{1}{q_n^n}$$

for all $n \geqslant 1$. Notice that Exercise 3.3.3 shows that algebraic numbers are not Liouville numbers.
(a) Assuming that $\alpha$ is not a Liouville number, prove the following error rate in the equidistribution of the sequence $(x + n\alpha)_{n \geqslant 1}$ modulo 1:

$$\left| \frac{1}{N} \sum_{n=0}^{N-1} f(x + n\alpha) - \int_0^1 f(x)dx \right| \leqslant S(\alpha, f) \frac{1}{N},$$

for $f \in C^\infty(\mathbb{T})$ and some constant $S(\alpha, f)$ depending on $\alpha$ and $f$.
(b) Formulate and prove a generalization to rotations of $\mathbb{T}^d$.

**Exercise 4.4.6.** Use the ideas from Exercise 2.8.4 to prove a mean ergodic theorem along the squares: for a measure-preserving system $(X, \mathscr{B}, \mu, T)$ and $f \in L_\mu^2$, show that

$$\frac{1}{N} \sum_{n=0}^{N-1} U_T^{n^2} f$$

converges in $L_\mu^2$. Under the assumption that $T$ is totally ergodic (see Exercise 2.5.6), show that the limit is $\int f \, d\mu$.

**Exercise 4.4.7.** Let $X$ be a compact metric space, and assume that $\nu_n \to \mu$ in the weak*-topology on $\mathscr{M}(X)$. Show that for a Borel set $B$ with $\mu(\partial B) = 0$,

$$\lim_{n \to \infty} \nu_n(B) = \mu(B).$$

# Notes to Chap. 4

[48] (Page 98) The fact that $\mathscr{M}^T(X)$ is non-empty may also be seen as a result of various fixed-point theorems that generalize the Brouwer fixed point theorem to an infinite-dimensional setting; the argument used in Sect. 4.1 is attractive because it is elementary and is connected directly to the dynamics.
[49] (Page 103) A convenient source for the Choquet representation theorem is the updated lecture notes by Phelps [283]; the original papers are those of Choquet [55, 56].
[50] (Page 103) Notice that the space of invariant measures for a given continuous map is a topological attribute rather than a measurable one: measurably isomorphic systems may

have entirely unrelated spaces of invariant measures. In particular, the Jewett–Krieger theorem shows that any ergodic measure-preserving system $(X, \mathscr{B}, \mu, T)$ on a Lebesgue space is measurably isomorphic to a minimal, uniquely ergodic homeomorphism on a compact metric space (a continuous map on a compact metric space is called *minimal* if every point has a dense orbit; see Exercise 4.2.1). This deep result was found by Jewett [166] for weakly-mixing transformations, and was extended to ergodic systems by Krieger [213] using his proof of the existence of generators [212]. Thus having a model (up to measurable isomorphism) as a uniquely ergodic map on a compact metric space carries no information about a given measurable dynamical system. Among the many extensions and modifications of this important result, Bellow and Furstenberg [22], Hansel and Raoult [140] and Denker [69] gave different proofs; Jakobs [164] and Denker and Eberlein [70] extended the result to flows; Lind and Thouvenot [231] showed that any finite entropy ergodic transformation is isomorphic to a homeomorphism of the torus $\mathbb{T}^2$ preserving Lebesgue measure; Lehrer [222] showed that the homeomorphism can always be chosen to be topologically mixing (a homeomorphism $S : Y \to Y$ of a compact metric space is topologically mixing if for any open sets $U, V \subseteq Y$, there is an $N = N(U,V)$ with $U \cap S^n V \neq \varnothing$ for $n \geqslant N$); Weiss [379] extended to certain group actions and to diagrams of measure-preserving systems; Rosenthal [317] removed the assumption of invertibility. In a different direction, Downarowicz [74] has shown that every possible Choquet simplex arises as the space of invariant measures of a map even in a highly restricted class of continuous maps.

[51] (Page 104) Birkhoff's recurrence theorem may be thought of as a topological analog of Poincaré recurrence (Theorem 2.11), with the essential hypothesis of finite measure replaced by compactness. Furstenberg and Weiss [109] showed that there is also a topological analog of the ergodic multiple recurrence theorem (Theorem 7.4): if $(X, T)$ is minimal and $U \subseteq X$ is open and non-empty, then for any $k > 1$ there is some $n \geqslant 1$ with

$$U \cap T^n U \cap \cdots \cap T^{(k-1)n} U \neq \varnothing.$$

[52] (Page 110) This characterization is due to Pjateckiĭ-Šapiro [285], who showed it as a property characterizing normality for orbits under the map $x \mapsto ax \pmod 1$.

[53] (Page 110) The theory of equidistribution from the viewpoint of number theory is a large and sophisticated one. Extensive overviews of this theory in three different decades may be found in the monographs of Kuipers and Niederreiter [215], Hlawka [154], and Drmota and Tichy [75].

[54] (Page 111) The formulation in (2) is the Weyl criterion for equidistribution; it appears in his paper [381]. Weyl really established the principle that equidistribution can be shown using a sufficiently rich set of test functions; in particular on a compact group it is sufficient to use an appropriate orthonormal basis of $L^2$. Thus a more general formulation of the Weyl criterion is as follows. Let $G$ be a compact metrizable group and let $G^\sharp$ denote the set of conjugacy classes in $G$. Then a sequence $(g_n)$ of elements of $G^\sharp$ is equidistributed with respect to Haar measure if and only if

$$\sum_{j=1}^{n} \mathrm{tr}\,(\pi(g_j)) = \mathrm{o}(n)$$

as $n \to \infty$, for any non-trivial irreducible unitary representation $\pi : G \to \mathrm{GL}_k(\mathbb{C})$. For more about equidistribution in the number-theoretic context, see the monograph of Iwaniec and Kowalski [162, Ch. 21].

[55] (Page 112) This equidistribution result was proved independently by several people, including Weyl [380], Bohl [39] and Sierpiński [344].

# Chapter 5
# Conditional Measures and Algebras

In this chapter we provide some more background in measure theory, which will be used frequently in the rest of the book. One of the most fundamental notions of averaging (in the sense of probability rather than ergodic theory) is afforded by the notion of *conditional expectation*. Recall that in probability the possible events are the measurable sets $A, B, C, \ldots$ in a measure space $(X, \mathscr{B}, \mu)$ with $\mu(X) = 1$. The probability of the event $A$ is $\mu(A)$, and the conditional probability of $A$ given that an event $B$ with $\mu(B) > 0$ has occurred, $\mu(A|B)$, is given by $\frac{\mu(A \cap B)}{\mu(B)}$. It is useful to extend this notion to sub-$\sigma$-algebras of $\mathscr{B}$. This turns out to provide a flexible tool for dealing with probabilities (measures) conditioned on events (measurable sets) that are allowed to be very unlikely. In fact, with some care, we will allow conditioning on events corresponding to null sets.

## 5.1 Conditional Expectation

**Theorem 5.1.** *Let* $(X, \mathscr{B}, \mu)$ *be a probability space, and let* $\mathscr{A} \subseteq \mathscr{B}$ *be a sub-$\sigma$-algebra. Then there is a map*

$$E(\cdot | \mathscr{A}) : L^1(X, \mathscr{B}, \mu) \longrightarrow L^1(X, \mathscr{A}, \mu),$$

*called the conditional expectation, that satisfies the following properties.*

(1) *For* $f \in L^1(X, \mathscr{B}, \mu)$, *the image function* $E(f | \mathscr{A})$ *is characterized almost everywhere by the two properties*

- $E(f | \mathscr{A})$ *is* $\mathscr{A}$-*measurable;*
- *for any* $A \in \mathscr{A}$, $\int_A E(f | \mathscr{A}) \, d\mu = \int_A f \, d\mu$.

M. Einsiedler, T. Ward, *Ergodic Theory*, Graduate Texts in Mathematics 259, DOI 10.1007/978-0-85729-021-2_5, © Springer-Verlag London Limited 2011

(2) $E(\cdot|\mathscr{A})$ is a linear operator of norm 1. Moreover, $E(\cdot|\mathscr{A})$ is positive (that is, $E(f|\mathscr{A}) \geqslant 0$ almost everywhere whenever $f \in L^1(X,\mathscr{B},\mu)$ has $f \geqslant 0$).
(3) For $f \in L^1(X,\mathscr{B},\mu)$ and $g \in L^\infty(X,\mathscr{A},\mu)$,

$$E(gf|\mathscr{A}) = gE(f|\mathscr{A})$$

almost everywhere.
(4) If $\mathscr{A}' \subseteq \mathscr{A}$ is a sub-$\sigma$-algebra, then

$$E\left(E(f|\mathscr{A})|\mathscr{A}'\right) = E(f|\mathscr{A}')$$

almost everywhere.
(5) If $f \in L^1(X,\mathscr{A},\mu)$ then $E(f|\mathscr{A}) = f$ almost everywhere.
(6) For any $f \in L^1(X,\mathscr{B},\mu)$, $|E(f|\mathscr{A})| \leqslant E(|f||\mathscr{A})$ almost everywhere.

For a collection of sets $\{A_\gamma \mid \gamma \in \Gamma\}$, denote by $\sigma\left(\{A_\gamma \mid \gamma \in \Gamma\}\right)$ the $\sigma$-algebra generated by the collection (that is, the smallest $\sigma$-algebra containing all the sets $A_\gamma$). A *partition* $\xi$ of a measure space $X$ is a finite or countable set of disjoint measurable sets whose union is $X$.

*Example 5.2.* If $\mathscr{A} = \sigma(\xi)$ is the finite $\sigma$-algebra generated by a finite partition $\xi = \{A_1,\ldots,A_n\}$ of $X$, then

$$E(f|\mathscr{A})(x) = \frac{1}{\mu(A_i)}\int_{A_i} f\,\mathrm{d}\mu$$

if $x \in A_i$. The $\sigma$-algebra being conditioned on is illustrated in Fig. 5.1 for a partition into $n = 8$ sets; $E(f|\mathscr{A})$ is then a function constant on each element of the partition $\xi$.

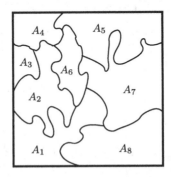

**Fig. 5.1** A partition of $X$ into 8 sets

*Example 5.3.* Let $X = [0,1]^2$ with two-dimensional Lebesgue measure, and let $\mathscr{A} = \mathscr{B}\times\{\varnothing,[0,1]\}$ be the $\sigma$-algebra comprising sets of the form $B\times[0,1]$

for $B$ a measurable subset of $[0,1]$. Then

$$E(f|\mathscr{A})(x_1,x_2) = \int_0^1 f(x_1,t)\,\mathrm{d}t.$$

Notice that every value of $E(f|\mathscr{A})$ is obtained by averaging over a set that is null with respect to the original measure (see Fig. 5.2).

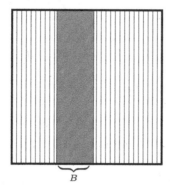

**Fig. 5.2** Conditioning on the $\sigma$-algebra $\mathscr{A}$

Theorem 5.1 has two different assertions, which we will organize as follows. The first step is *existence*, and we will give two different proofs that there is a map $E(\cdot|\mathscr{A})$ with the properties stated in (1). Once existence is established, we will turn to proving that $E(\cdot|\mathscr{A})$ is the unique function with the properties in (1), and that it satisfies the properties (2)–(6).

PROOF OF EXISTENCE WITH (1) IN THEOREM 5.1 VIA MEASURE THEORY. Suppose first that $f \geqslant 0$. Then the measure defined by

$$\mu_f(B) = \int_B f\,\mathrm{d}\mu$$

is a finite, absolutely continuous measure on $(X,\mathscr{B})$. Then $\mu_f\big|_{\mathscr{A}}$ is absolutely continuous with respect to $\mu\big|_{\mathscr{A}}$, so there is a Radon–Nikodym derivative $g$ in $L^1(X,\mathscr{A},\mu)$ (see Sect. A.4) characterized by

$$\mu_f(A) = \int_A g\,\mathrm{d}\mu,$$

so

$$\int_A f\,\mathrm{d}\mu = \int_A g\,\mathrm{d}\mu$$

for any $A \in \mathscr{A}$. The general case follows by decomposing real functions into positive and negative parts and complex functions into real and imaginary parts.                                                                                    $\square$

PROOF OF EXISTENCE WITH (1) IN THEOREM 5.1 VIA FUNCTIONAL ANALYSIS. Let
$$\mathscr{V} = L^2(X, \mathscr{A}, \mu) \subseteq \mathscr{H} = L^2(X, \mathscr{B}, \mu).$$
Then $\mathscr{V}$ is a closed subspace of the Hilbert space $\mathscr{H}$ (closed since $\mathscr{V}$ is itself complete), so there is an orthogonal projection $P : \mathscr{H} \to \mathscr{V}$, with the property that

$$\int_A f \, d\mu = \int \chi_A f \, d\mu = \int \chi_A P f \, d\mu = \int_A P f \, d\mu \qquad (5.1)$$

for $A \in \mathscr{A}$. We claim that the projection $P : \mathscr{H} \to \mathscr{V}$ has a continuous extension to a map
$$L^1(X, \mathscr{B}, \mu) \to L^1(X, \mathscr{A}, \mu),$$
and that this extension is the conditional expectation. To see this, first assume that all functions are real-valued. Notice that $L^2 \subseteq L^1$ is dense, and that for $f \in L^2$ (and hence in $L^1$), the sets

$$\{x \in X \mid Pf(x) > 0\}$$

and
$$\{x \in X \mid Pf(x) < 0\}$$

lie in $\mathscr{A}$, so by (5.1),
$$\|Pf\|_1 \leqslant \|f\|_1.$$

For complex-valued functions, taking real and imaginary parts and applying the same argument to each gives

$$\|Pf\|_1 \leqslant 2\|f\|_1. \qquad (5.2)$$

Equation (5.1) only involves functionals that are continuous in $L^1$, so there is a continuous extension to all of $L^1$ that still satisfies (5.1).          $\square$

PROOF OF THEOREM 5.1. We claim first that the two properties in (1) characterize the conditional expectation of a given function uniquely up to sets of measure zero. In fact if $g_1$ and $g_2$ satisfy both properties then

$$A = \{x \mid g_1(x) < g_2(x)\} \in \mathscr{A}$$

has
$$\int_A g_1 \, d\mu = \int_A f \, d\mu = \int_A g_2 \, d\mu$$

and so $\mu(A) = 0$. Similarly,

$$\mu \left( \{x \mid g_1(x) > g_2(x)\} \right) = 0$$

and so $g_1 = g_2$ almost everywhere.

That $E(\cdot | \mathscr{A})$ is linear follows easily from uniqueness. For positivity, let $f$ be a non-negative function in $L^1(X, \mathscr{B}, \mu)$, and let

$$A = \{x \in X \mid E(f | \mathscr{A}) < 0\}.$$

Then

$$0 \leqslant \int_A f \, d\mu = \int_A E(f | \mathscr{A}) \, d\mu$$

implies that $\mu(A) = 0$.

Property (3) clearly holds for any indicator function $\chi_A$ with $A \in \mathscr{A}$. Any $g \in L^\infty(X, \mathscr{A}, \mu)$ can be approximated by simple functions, so the general case follows from continuity of the conditional expectation operator, which is in turn a consequence of the inequality (5.2).

Now let $A$ be a set in $\mathscr{A}'$. Then

$$\int_A E\left(E(f | \mathscr{A}) | \mathscr{A}'\right) \, d\mu = \int_A E(f | \mathscr{A}) \, d\mu = \int_A f \, d\mu,$$

so property (4) follows by uniqueness.

Property (5) is an easy consequence of (1).

Given $f \in L^1(X, \mathscr{B}, \mu)$ we may find $g \in L^\infty(X, \mathscr{A}, \mu)$ with $|g(x)| = 1$ for $x \in X$ satisfying

$$|E(f | \mathscr{A})| = g \cdot E(f | \mathscr{A}).$$

Then by (3),

$$|E(f | \mathscr{A})| = E(gf | \mathscr{A}),$$

so, for any $A \in \mathscr{A}$,

$$\int_A |E(f | \mathscr{A})| \, d\mu = \int_A E(gf | \mathscr{A}) \, d\mu$$

$$= \int_A gf \, d\mu$$

$$\leqslant \int_A |gf| \, d\mu = \int_A E(|f| | \mathscr{A}) \, d\mu,$$

proving (6). Finally, by integrating (6) we see that $\|E(\cdot | \mathscr{A})\|_{\text{operator}} \leqslant 1$; considering $\mathscr{A}$-measurable functions shows that $\|E(\cdot | \mathscr{A})\|_{\text{operator}} \geqslant 1$, showing the remaining assertion in (2). $\qquad\square$

## 5.2 Martingales

This section[56] provides the basic convergence results for conditional expectations with respect to increasing or decreasing sequences of $\sigma$-algebras. These results are related both in their statements and in their proofs to ergodic theorems[57]: Martingale theorems and the Lebesgue density theorem use analogs of maximal inequalities and the method of subtracting the image under a projection. The increasing martingale theorem (Theorem 5.5) and Lebesgue density theorems involve averaging less and less, so are in some sense opposite to the ergodic theorems; nonetheless, both are useful in ergodic theory.

*Example 5.4.* Consider the partition

$$\xi_n = \{[0, \tfrac{1}{2^n}), [\tfrac{1}{2^n}, \tfrac{2}{2^n}), \dots, [1 - \tfrac{1}{2^n}, 1)\}$$

of $[0, 1)$, with associated finite $\sigma$-algebra $\mathscr{A}_n = \sigma(\xi_n)$ comprising all unions of elements of $\xi_n$. It is clear that

$$\mathscr{A}_1 \subseteq \mathscr{A}_2 \subseteq \cdots,$$

and that the Borel $\sigma$-algebra generated by all of these is the Borel $\sigma$-algebra $\mathscr{B}$ of $[0, 1)$. In what sense does $E(f|\mathscr{A}_n)$ approximate $f = E(f|\mathscr{B})$?

If the measure is Lebesgue, then this can be answered using the Lebesgue density theorem (Theorem A.24). For other measures a different approach is needed.

The next result, a form of Doob's martingale theorem [72], will be used frequently in the sequel to understand limits of $\sigma$-algebras.

**Theorem 5.5 (Increasing martingale theorem).** *Let $(X, \mathscr{B}, \mu)$ be a probability space. Suppose that $\mathscr{A}_n \nearrow \mathscr{A}$ is an increasing sequence of sub-$\sigma$-algebras of $\mathscr{B}$ (the notation $\nearrow$ means that $\mathscr{A}_n \subseteq \mathscr{A}_{n+1}$ for all $n \geqslant 1$ and $\mathscr{A} = \sigma(\bigcup_{n \geqslant 1} \mathscr{A}_n)$). Then, for any $f \in L^1(X, \mathscr{B}, \mu)$,*

$$E(f|\mathscr{A}_n) \longrightarrow E(f|\mathscr{A})$$

*almost everywhere and in $L^1_\mu$.*

Notice that for $f \in L^1(X, \mathscr{B}, \mu)$,

$$\mu\left(\{x \mid E(f|\mathscr{A})(x) > \varepsilon\}\right) \leqslant \frac{\|f\|_1}{\varepsilon}.$$

To see this, let $E = \{x \mid E(f|\mathscr{A})(x) > \varepsilon\}$. Then $E \in \mathscr{A}$ and $\varepsilon \chi_E \leqslant E(f|\mathscr{A})$, so

$$\varepsilon\mu(E) \leqslant \int_E E(f \,|\, \mathscr{A})$$

$$= \int_E f \leqslant \|f\|_1$$

as required. The next lemma will be a very useful generalization of this simple observation, which is an analog of a maximal inequality (Theorem 2.24).

**Lemma 5.6 (Doob's inequality).** *Let* $f \in L^1(X, \mathscr{B}, \mu)$, *let*

$$\mathscr{A}_1 \subseteq \mathscr{A}_2 \subseteq \cdots \subseteq \mathscr{A}_N \subseteq \mathscr{B}$$

*be an increasing list of $\sigma$-algebras, and fix $\lambda > 0$. Let*

$$E = \{x \mid \max_{1 \leqslant i \leqslant N} E(f \,|\, \mathscr{A}_i) > \lambda\}.$$

*Then*

$$\mu(E) \leqslant \frac{1}{\lambda} \|f\|_1.$$

*If $(\mathscr{A}_n)_{n \geqslant 1}$ is an increasing (or decreasing) sequence of $\sigma$-algebras then the same conclusion holds for the set*

$$E = \{x \mid \sup_{i \geqslant 1} E\left(f \,|\, \mathscr{A}_i\right) > \lambda\}.$$

PROOF. Assume that $f \geqslant 0$ (if necessary replacing $f$ by $|f|$, which makes $\mu(E)$ no smaller). Let

$$E_n = \{x \mid E(f \,|\, \mathscr{A}_n) > \lambda \text{ but } E(f \,|\, \mathscr{A}_i) \leqslant \lambda \text{ for } 1 \leqslant i \leqslant n-1\}.$$

Then $E = E_1 \sqcup \cdots \sqcup E_N$ and $E_n \in \mathscr{A}_n$ since $\mathscr{A}_1, \mathscr{A}_2, \ldots, \mathscr{A}_{n-1} \subseteq \mathscr{A}_n$. (In the decreasing case of finitely many $\sigma$-algebras we may reverse the order of the $\sigma$-algebras, since the statement we wish to prove is independent of the order.) It follows that

$$\|f\|_1 \geqslant \int_E f \, \mathrm{d}\mu = \sum_{n=1}^N \int_{E_n} f \, \mathrm{d}\mu$$

$$= \sum_{n=1}^N \int_{E_n} E(f \,|\, \mathscr{A}_n) \, \mathrm{d}\mu$$

$$\geqslant \sum_{n=1}^N \lambda\mu(E_n) = \lambda\mu(E).$$

Taking $N \to \infty$ shows the final remark. $\qquad\square$

PROOF OF THEOREM 5.5. Using Theorem 5.1(4), we may replace the function $f$ by $E(f \,|\, \mathscr{A})$ without changing $E(f \,|\, \mathscr{A}_n)$.

The theorem holds for all $f \in L^1(X, \mathscr{A}_n, \mu)$, $n \geqslant 1$. Now $\bigcup_{n \geqslant 1} L^1(X, \mathscr{A}_n, \mu)$ is dense in $L^1(X, \mathscr{A}, \mu)$. To see this, notice that

$$\{B \in \mathscr{A} \mid \text{for every } \varepsilon > 0 \text{ there exist } m \geqslant 1, A \in \mathscr{A}_m \text{ with } \mu(A \triangle B) < \varepsilon\}$$

is a $\sigma$-algebra by Theorem A.7. Given any $f \in L^1(X, \mathscr{A}, \mu)$ and $\varepsilon > 0$, find $m$ and $g \in L^1(X, \mathscr{A}_m, \mu)$ with $\|f - g\|_1 < \varepsilon$, so that

$$\|E(f \mid \mathscr{A}_n) - f\|_1 \leqslant \|E(f \mid \mathscr{A}_n) - E(g \mid \mathscr{A}_n)\|_1 + \underbrace{\|E(g \mid \mathscr{A}_n) - g\|_1}_{= 0 \text{ for } n \geqslant m} + \|g - f\|_1 < 2\varepsilon$$

for $n \geqslant m$. It follows that

$$\mu\left(\{x \mid \limsup_{n \to \infty} |E(f \mid \mathscr{A}_n) - f| > \sqrt{\varepsilon}\}\right)$$

$$= \mu\left(\{x \mid \limsup_{n \to \infty} |E(f - g \mid \mathscr{A}_n) - (f - g)| > \sqrt{\varepsilon}\}\right)$$

$$\leqslant \mu\left(\{x \mid \sup_{n \geqslant 1} |E(f - g \mid \mathscr{A}_n)| > \tfrac{1}{2}\sqrt{\varepsilon}\}\right) + \mu\left(\{x \mid |f - g| > \tfrac{1}{2}\sqrt{\varepsilon}\}\right)$$

$$\leqslant \tfrac{2}{\sqrt{\varepsilon}}\|f - g\|_1 + \tfrac{2}{\sqrt{\varepsilon}}\|f - g\|_1 \leqslant 4\sqrt{\varepsilon}$$

by Lemma 5.6, so

$$\limsup_{n \to \infty} |E(f \mid \mathscr{A}_n) - f| = 0$$

almost everywhere, showing the almost everywhere convergence.    □

A similar result holds for *decreasing* sequences of $\sigma$-algebras as follows. The notation $\mathscr{A}_n \searrow \mathscr{A}_\infty$ used below means that $\mathscr{A}_{n+1} \subseteq \mathscr{A}_n$ for all $n \geqslant 1$ and

$$\mathscr{A}_\infty = \bigcap_{n \geqslant 1} \mathscr{A}_n.$$

*Example 5.7.* Let $\mathscr{B}$ denote the Borel $\sigma$-algebra on $[0, 1]$ and let

$$\mathscr{A}_n = \{B \in \mathscr{B} \mid B + \tfrac{1}{2^n} = B \pmod 1\}$$

so that $\mathscr{A}_n \searrow \mathscr{N} = \{\varnothing, X\}$ modulo $m$ (meaning that $\bigcap_{n \geqslant 1} \mathscr{A}_n \underset{m}{=} \{\varnothing, X\}$, where $m$ denotes Lebesgue measure on $[0, 1]$). As before, what is the connection between the convergence of $\sigma$-algebras and the convergence of $E(f \mid \mathscr{A}_n)$?

As mentioned at the start of this section, the kind of convergence sought here resembles an ergodic theorem[58]. Indeed, the proof is similar in some ways to the proofs of the ergodic theorems (Theorems 2.21 and 2.30). The usual proof of the decreasing martingale theorem is somewhat opaque because it takes place in $L^1$ rather than in $L^2$, forcing us to replace the geometric methods available in Hilbert space with more flexible methods from functional analysis. To illuminate the different approaches—and the more geometrical approach that working in $L^2$ allows—we give two different arguments for the first part of the proof. Of course the theorem itself is an assertion about $L^1$ convergence, so at some point we must work in $L^1$.

**Theorem 5.8 (Decreasing martingale theorem).** *Let $(X, \mathscr{B}, \mu)$ be a probability space. If $\mathscr{A}_n \searrow \mathscr{A}_\infty$ is a decreasing sequence of sub-$\sigma$-algebras of $\mathscr{B}$ then*

$$E(f|\mathscr{A}_n) \longrightarrow E(f|\mathscr{A}_\infty)$$

*almost everywhere and in $L^1$, for any $f \in L^1(X, \mathscr{B}, \mu)$.*

FIRST PART OF PROOF OF THEOREM 5.8, USING $L^2$. Recall from the proof of Theorem 5.1 that in $L^2(X, \mathscr{B}, \mu)$ the conditional expectation with respect to $\mathscr{A}_n$ (or $\mathscr{A}_\infty$) is precisely the orthogonal projection to $L^2(X, \mathscr{A}_n, \mu)$ (resp. $L^2(X, \mathscr{A}_\infty, \mu)$). Let $V_n = L^2(X, \mathscr{A}_n, \mu)^\perp$ and let $V_* = \bigcup_{n \geqslant 1} V_n$. Notice that for $f \in L^2(X, \mathscr{A}_\infty, \mu) + V_*$ the theorem holds trivially because

$$E(f|\mathscr{A}_n) = E(f|\mathscr{A}_\infty)$$

for sufficiently large $n$. We claim that

$$V = L^2(X, \mathscr{A}_\infty, \mu) + V_*$$

is dense in $L^2(X, \mathscr{B}, \mu)$ with respect to the $L^2_\mu$ norm. To see this, we may use the Riesz representation theorem (see Sect. B.5). If $V$ is not dense in $L^2(X, \mathscr{B}, \mu)$, then there is a continuous non-zero linear functional

$$f \mapsto \int f \bar{h} \, d\mu$$

defined by some $h \in L^2(X, \mathscr{B}, \mu)$ such that

$$\int f \bar{h} \, d\mu = 0$$

for all $f \in V$, and this leads to a contradiction as follows. Clearly

$$h - E\left(h|\mathscr{A}_n\right) \in V_n \subseteq V_*,$$

so

$$\int \left(h - E(h|\mathscr{A}_n)\right) \bar{h} \, d\mu = 0.$$

Since $f \mapsto E(f|\mathscr{A}_n)$ is the orthogonal projection, we also have

$$\int \left(h - E(h|\mathscr{A}_n)\right) \overline{E(h|\mathscr{A}_n)} \, d\mu = 0,$$

which implies that

$$\int \left|h - E(h|\mathscr{A}_n)\right|^2 d\mu = 0$$

and so $h = E(h|\mathscr{A}_n) \in L^2(X, \mathscr{A}_n, \mu)$ for all $n \geqslant 1$. We conclude that

$$h \in L^2(X, \mathscr{A}_\infty, \mu) \subseteq V,$$

and $\int h\bar{h} \, d\mu = 0$, so $h = 0$. This contradiction shows that $V$ is dense in $L^2(X, \mathscr{B}, \mu)$ with respect to the $L^2_\mu$ norm.

Now $\| \cdot \|_1 \leqslant \| \cdot \|_2$ and $L^2(X, \mathscr{B}, \mu) \subseteq L^1(X, \mathscr{B}, \mu)$ is dense with respect to the $L^1_\mu$ norm. It follows that $V$ is also dense in $L^1(X, \mathscr{B}, \mu)$ with respect to the $L^1_\mu$ norm. $\qquad\square$

It might seem unsatisfactory to use $L^2$ arguments in this way to avoid the more complicated theory of the space $L^1$ and its dual $L^\infty$. To give an example of how it is sometimes possible to decompose functions in a way that mimics the orthogonal decomposition available in Hilbert space, we now do the same part of the proof avoiding $L^2$.

FIRST PART OF PROOF OF THEOREM 5.8, USING $L^1$ DIRECTLY. Let

$$V_n = \{ f \in L^1(X, \mathscr{B}, \mu) \mid E\left(f | \mathscr{A}_n\right) = 0 \}$$

for $n \geqslant 1$, so $V_1 \subseteq V_2 \subseteq \cdots$ is an increasing sequence of subspaces of $L^1(X)$. We claim that $V_* = \bigcup_{n \geqslant 1} V_n$ is $L^1$-dense in

$$V_\infty = \{ f \in L^1(X, \mathscr{B}, \mu) \mid E\left(f | \mathscr{A}_\infty\right) = 0 \}.$$

This claim will be crucial for the proof, since it will allow us to split any function $f$ into two parts for which the result will be easier to prove.

By the Hahn–Banach theorem (Theorem B.1), $V_*$ is dense in $V_\infty$ if any continuous linear functional $\Lambda : L^1(X) \to \mathbb{R}$ with $V_* \subseteq \ker \Lambda$ has $V_\infty \subseteq \ker \Lambda$. Any continuous linear functional on $L^1(X)$ has the form

$$\Lambda_h(f) = \int_X fh \, d\mu$$

for some $h \in L^\infty(X)$, and $h$ is uniquely determined by $\Lambda_h$. So suppose that $V_n \subseteq \ker \Lambda_h$ for all $n \geqslant 1$; it follows that

$$\int (f - E(f|\mathscr{A}_n))h \, d\mu = 0$$

for all $f \in L^1(X)$ and $n \geqslant 1$. In particular, we may take $f = h$ (since $L^\infty(X)$ is a subset of $L^1(X)$), so

$$\int (h - E(h|\mathscr{A}_n))h \, d\mu = 0.$$

On the other hand, by Theorem 5.1(3),

$$\int E(h|\mathscr{A}_n)E(h|\mathscr{A}_n) \, d\mu = \int E\left(E(h|\mathscr{A}_n)h | \mathscr{A}_n\right) \, d\mu = \int E\left(h|\mathscr{A}_n\right) h \, d\mu$$

so
$$\int (h - E(h|\mathscr{A}_n))E(h|\mathscr{A}_n)\, d\mu = 0.$$

Now
$$(h - E(h|\mathscr{A}_n))h - (h - E(h|\mathscr{A}_n))E(h|\mathscr{A}_n) = \left(h - E(h|\mathscr{A}_n)\right)^2$$

and therefore
$$\int \left(h - E(h|\mathscr{A}_n)\right)^2 d\mu = 0.$$

It follows that $h = E(h|\mathscr{A}_n) \in L^\infty(X, \mathscr{A}_n, \mu)$, and so $h \in L^\infty(X, \mathscr{A}_\infty, \mu)$. Thus
$$E(f|\mathscr{A}_\infty) = 0$$

implies that
$$\int fh\, d\mu = \int E(fh|\mathscr{A}_\infty)\, d\mu = \int hE(f|\mathscr{A}_\infty)\, d\mu = 0,$$

showing that $\ker \Lambda_h \supseteq V_\infty$ whenever $\ker \Lambda_h \supseteq V_*$ as required.

Clearly the theorem holds for functions in the space
$$V = L^1(X, \mathscr{A}_\infty, \mu) + V_*,$$

which is $L^1$-dense in $L^1(X)$ (to see that this space is dense, write any $f \in L^1$ as $f = E\left(f|\mathscr{A}_\infty\right) + \left(f - E\left(f|\mathscr{A}_\infty\right)\right)$ where the second term belongs to $V_\infty$).  $\square$

The remainder of the proof of Theorem 5.8 of necessity takes place in $L^1(X, \mathscr{B}, \mu)$.

SECOND PART OF PROOF OF THEOREM 5.8. Given $f \in L^1(X)$ and $\varepsilon > 0$, find $g \in V$ with
$$\|f - g\|_1 < \varepsilon.$$

Then
$$\int |E(f|\mathscr{A}_n) - E(f|\mathscr{A}_\infty)|\, d\mu \leqslant \int |E\left((f - g)|\mathscr{A}_n\right) - E\left((f - g)|\mathscr{A}_\infty\right)|\, d\mu$$
$$+ \int |E(g|\mathscr{A}_n) - E(g|\mathscr{A}_\infty)|\, d\mu$$
$$\leqslant 2\int |f - g|\, d\mu + \int |E(g|\mathscr{A}_n) - E(g|\mathscr{A}_\infty)|\, d\mu,$$

so
$$\limsup_{n\to\infty} \int |E(f|\mathscr{A}_n) - E(f|\mathscr{A}_\infty)|\, d\mu \leqslant 2\int |f - g|\, d\mu \leqslant 2\varepsilon,$$

which shows the convergence in $L^1$.

To see the almost everywhere convergence, notice that

$$\mu \left( \{ x \mid \limsup_{n \to \infty} \left| E(f \mid \mathscr{A}_n) - E(f \mid \mathscr{A}_\infty) \right| > \sqrt{\varepsilon} \} \right)$$

$$\leqslant \mu \left( \{ x \mid \limsup_{n \to \infty} \left| E \left( (f - g) \mid \mathscr{A}_n \right) - E \left( (f - g) \mid \mathscr{A}_\infty \right) \right| \right.$$

$$\left. + \limsup_{n \to \infty} \left| E \left( g \mid \mathscr{A}_n \right) - E \left( g \mid \mathscr{A}_\infty \right) \right| > \sqrt{\varepsilon} \} \right)$$

$$\leqslant \mu \left( \{ x \mid \sup_{n \geqslant 1} \left| E \left( (f - g) \mid \mathscr{A}_n \right) - E \left( (f - g) \mid \mathscr{A}_\infty \right) \right| > \sqrt{\varepsilon} \} \right)$$

$$\leqslant \mu \left( \{ x \mid \sup_{n \geqslant 1} \left| E \left( (f - g) \mid \mathscr{A}_n \right) \right| \geqslant \tfrac{1}{2} \sqrt{\varepsilon} \right)$$

$$+ \mu \left( \{ x \mid \sup_{n \geqslant 1} \left| E \left( (f - g) \mid \mathscr{A}_\infty \right) \right| > \tfrac{1}{2} \sqrt{\varepsilon} \} \right)$$

$$\leqslant \tfrac{2}{\sqrt{\varepsilon}} \| f - g \|_1 + \tfrac{2}{\sqrt{\varepsilon}} \| f - g \|_1 \leqslant 4 \sqrt{\varepsilon},$$

by Doob's inequality (Lemma 5.6), so

$$\limsup_{n \to \infty} \left| E(f \mid \mathscr{A}_n) - E(f \mid \mathscr{A}_\infty) \right| = 0$$

almost everywhere.                                                      □

## Exercises for Sect. 5.2

**Exercise 5.2.1.** Use the increasing martingale theorem (Theorem 5.5) to prove the following version of the Borel–Cantelli lemma (Theorem A.9). Suppose that $(X, \mathscr{B}, \mu)$ is a probability space and $(A_n)_{n \geqslant 1}$ is a completely independent sequence of measurable sets (that is, for any finite sequence of indices $i_1 < \cdots < i_\ell$ we have $\mu (A_{i_1} \cap \cdots \cap A_{i_\ell}) = \mu (A_{i_1}) \cdots \mu (A_{i_\ell})$). If additionally

$$\sum_{n=1}^{\infty} \mu(A_n) = \infty,$$

then almost every $x$ is contained in infinitely many of the sets $A_n$; equivalently

$$\mu \left( \bigcap_{N=1}^{\infty} \bigcup_{n=N}^{\infty} A_n \right) = 1.$$

**Exercise 5.2.2.** Use the martingale theorems to prove the following analog of the Lebesgue density theorem (Theorem A.24). Let $m$ be Lebesgue measure on the cube $C = [0,1]^d$. For $n \geqslant 1$ define the partition $\xi_n$ of $C$ into boxes

$$\prod_{i=1}^{d}\left[\frac{j_i}{(1+i)^n}, \frac{j_i+1}{(1+i)^n}\right)$$

for $0 \leqslant j_i < (1+i)^n$, $1 \leqslant i \leqslant d$. For any $x \in C$, let $B_n(x)$ denote the atom of $\xi_n$ containing $x$, and notice that $m(B_n(x)) = \prod_{i=1}^{d}(1+i)^{-n}$. Prove that for any measurable set $A \subseteq C$,

$$\frac{1}{m(B_n(x))} m(A \cap B_n(x)) \to 1$$

as $n \to \infty$ for almost every $x \in A$.

## 5.3 Conditional Measures

Section 5.1 introduced the conditional expectation as a function

$$f \mapsto E(f|\mathscr{A}),$$

and all of its properties, as well as the examples we have seen, suggest that the quantity $E(f|\mathscr{A})(x)$ should be an average of the function $f$ over a part of the measure space, where the part used in the averaging depends on $x$. For well-behaved measure spaces—and in particular for the kind of measure spaces we deal with—this property is reflected in the existence of a measure $\mu_x^{\mathscr{A}}$ with the property that

$$E(f|\mathscr{A})(x) = \int f \, d\mu_x^{\mathscr{A}}$$

for all $f \in L_\mu^1$.

*Example 5.9.* Let $\xi$ be a countable partition of $(X, \mathscr{B}, \mu)$, with $\mathscr{A} = \sigma(\xi)$ the smallest $\sigma$-algebra containing $\xi$. Then

$$\mu_x^{\mathscr{A}} = \frac{1}{\mu(P)} \mu\big|_P$$

for $x \in P \in \mathscr{A}$ defines such a measure for almost every $x$; if $\mu(P) = 0$ then $\mu_x^{\mathscr{A}}$ is not defined for $x \in P$.

*Example 5.10.* Let $\mathscr{A} = \mathscr{B}_{[0,1]} \times \{\varnothing, [0,1]\} \subseteq \mathscr{B}_{[0,1]^2}$. Then

$$\mu_{(x_1,x_2)}^{\mathscr{A}} = \delta_{x_1} \times m_{[0,1]},$$

where as usual $m$ denotes Lebesgue (or Haar) measure (see Fig. 5.2).

As in the last example, a particular difficulty that arises in discussing conditional expectation and conditional measures is that in most non-trivial situations null sets with respect to the original measure become more trou-blesome, since the conditional measures will often be singular with respect to the original measure $\mu$.

Thus we need to pay more attention to null sets—and in particular we will need to specify in a more concrete way with respect to which measure a given set has measure zero. When we say $N$ is a null set we mean that $\mu(N) = 0$; in contrast for $x \in X$ we will say $N$ is a $\mu_x^{\mathscr{A}}$-null set if $\mu_x^{\mathscr{A}}(N) = 0$. Similarly, we will need to make a distinction between the notion of "almost everywhere" (true off a $\mu$-null set) and "$\mu_x^{\mathscr{A}}$-almost everywhere". From now on we will also distinguish more carefully between the space $\mathscr{L}^p(X, \mathscr{B}, \mu)$ of genuine functions and the more familiar space $L^p(X, \mathscr{B}, \mu)$ of equivalence classes of functions; in particular $\mathscr{L}^\infty(X, \mathscr{B})$ denotes the space of bounded measurable functions and will be written $\mathscr{L}^\infty$ if the underlying measure space is clear.

We next formalize our prevailing assumption about the measure spaces we deal with. A *probability space* is any triple $(X, \mathscr{B}, \mu)$ where $\mu$ is a measure on the $\sigma$-algebra $\mathscr{B}$ with $\mu(X) = 1$. It turns out that this definition is too permissive for some—but by no means all—of the natural developments in ergodic theory.

*Example 5.11.* Let $X = \{0, 1\}^{\mathbb{R}}$, with the product topology and the $\sigma$-algebra of Borel sets. The product measure $\mu$ of the $(\frac{1}{2}, \frac{1}{2})$ measure on each of the sets $\{0, 1\}$ makes $X$ into a probability space with the property that there is an uncountable collection $\{A_s\}_{s \in \mathbb{R}}$ of measurable sets with the property that $\mu(A_s) = \frac{1}{2}$ for each $s \in \mathbb{R}$ and the sets are all mutually independent:

$$\mu\left(A_{s_1} \cap \cdots \cap A_{s_n}\right) = \frac{1}{2^n}$$

for any $n$ distinct reals $s_1, \ldots, s_n$. The next definition gives a collection of probability spaces that precludes the possibility of uncountably many inde-pendent sets.

**Definition 5.12.** Let $X$ be a Borel subset of a compact metric space with the restriction of the Borel $\sigma$-algebra $\mathscr{B}$ to $X$. Then the pair $(X, \mathscr{B})$ is a *Borel space*.

**Definition 5.13.** Let $X$ be a dense Borel subset of a compact metric space $\overline{X}$, with a probability measure $\mu$ defined on the restriction of the Borel $\sigma$-algebra $\mathscr{B}$ to $X$. The resulting probability space $(X, \mathscr{B}, \mu)$ is a *Borel probability space*.

For a compact metric space $X$, the space $\mathscr{M}(X)$ of Borel probability measures on $X$ itself carries the structure of a compact metric space with respect to the weak*-topology. In particular, we can define the Borel $\sigma$-algebra $\mathscr{B}_{\mathscr{M}(X)}$ on the space $\mathscr{M}(X)$ in the usual way. If $X$ is a Borel subset of a compact metric space $\overline{X}$, then we define

$$\mathscr{M}(X) = \{\mu \in \mathscr{M}(\overline{X}) \mid \mu(\overline{X} \smallsetminus X) = 0\},$$

and we will see in Lemma 5.23 that $\mathscr{M}(X)$ is a Borel subset of $\mathscr{M}(\overline{X})$.

We are now in a position to state and prove the main result of this chapter. A set is called *conull* if it is the complement of a null set. For $\sigma$-algebras $\mathscr{C}, \mathscr{C}'$ the relation

$$\mathscr{C} \underset{\mu}{\subseteq} \mathscr{C}'$$

means that for any $A \in \mathscr{C}$ there is a set $A' \in \mathscr{C}'$ with $\mu(A \triangle A') = 0$. We also define

$$\mathscr{C} \underset{\mu}{=} \mathscr{C}'$$

to mean that $\mathscr{C} \underset{\mu}{\subseteq} \mathscr{C}'$ and $\mathscr{C}' \underset{\mu}{\subseteq} \mathscr{C}$.

A $\sigma$-algebra $\mathscr{A}$ on $X$ is *countably-generated* if there exists a countable set $\{A_1, A_2, \dots\}$ of subsets of $X$ with the property that $\mathscr{A} = \sigma(\{A_1, A_2, \dots\})$ is the smallest $\sigma$-algebra (that is, the intersection of every) $\sigma$-algebra containing the sets $A_1, A_2, \dots$.

**Theorem 5.14.** *Let $(X, \mathscr{B}, \mu)$ be a Borel probability space, and $\mathscr{A} \subseteq \mathscr{B}$ a $\sigma$-algebra. Then there exists an $\mathscr{A}$-measurable conull set $X' \subseteq X$ and a system $\{\mu_x^{\mathscr{A}} \mid x \in X'\}$ of measures on $X$, referred to as* conditional measures, *with the following properties.*

(1) *$\mu_x^{\mathscr{A}}$ is a probability measure on $X$ with*

$$E(f \mid \mathscr{A})(x) = \int f(y) \, d\mu_x^{\mathscr{A}}(y) \tag{5.3}$$

*almost everywhere for all $f \in \mathscr{L}^1(X, \mathscr{B}, \mu)$. In other words, for any function[*] $f \in \mathscr{L}^1(X, \mathscr{B}, \mu)$ we have that $\int f(y) \, d\mu_x^{\mathscr{A}}(y)$ exists for all $x$ belonging to a conull set in $\mathscr{A}$, that on this set*

$$x \mapsto \int f(y) \, d\mu_x^{\mathscr{A}}(y)$$

*depends $\mathscr{A}$-measurably on $x$, and that*

$$\int_A \int f(y) \, d\mu_x^{\mathscr{A}}(y) \, d\mu(x) = \int_A f \, d\mu$$

*for all $A \in \mathscr{A}$.*

(2) *If $\mathscr{A}$ is countably-generated, then $\mu_x^{\mathscr{A}}([x]_{\mathscr{A}}) = 1$ for all $x \in X'$, where*

$$[x]_{\mathscr{A}} = \bigcap_{x \in A \in \mathscr{A}} A$$

---

[*] Notice that we are forced to work with genuine functions in $\mathscr{L}^1$ in order that the right-hand side of (5.3) is defined. As we said before, $\mu_x^{\mathscr{A}}$ may be singular to $\mu$.

is the atom of $\mathscr{A}$ containing $x$; moreover $\mu_x^{\mathscr{A}} = \mu_y^{\mathscr{A}}$ for $x, y \in X'$ whenever $[x]_{\mathscr{A}} = [y]_{\mathscr{A}}$.

(3) *Property* (1) *uniquely determines* $\mu_x^{\mathscr{A}}$ *for a.e.* $x \in X$. *In fact, property* (1) *for a dense countable set of functions in* $C(\overline{X})$ *uniquely determines* $\mu_x^{\mathscr{A}}$ *for a.e.* $x \in X$.

(4) *If* $\widetilde{\mathscr{A}}$ *is any* $\sigma$*-algebra with* $\mathscr{A} \underset{\mu}{=} \widetilde{\mathscr{A}}$*, then* $\mu_x^{\mathscr{A}} = \mu_x^{\widetilde{\mathscr{A}}}$ *almost everywhere.*

*Remark 5.15.* Theorem 5.14 is rather technical but quite powerful, so we assemble here some comments that will be useful both in the proof and in situations where the results are applied.

(a) For a countably generated $\sigma$-algebra $\mathscr{A} = \sigma\left(\{A_1, A_2, \dots\}\right)$ the atom in (2) is given by

$$[x]_{\mathscr{A}} = \bigcap_{x \in A_i} A_i \cap \bigcap_{x \notin A_i} X \smallsetminus A_i \tag{5.4}$$

and hence is $\mathscr{A}$-measurable (see Exercise 5.3.1). In fact $[x]_{\mathscr{A}}$ is the smallest element of $\mathscr{A}$ containing $x$.

(b) If $N \subseteq X$ is a null set for $\mu$, then $\mu_x^{\mathscr{A}}(N) = 0$ almost everywhere. In other words, for a $\mu$-null set $N$, the set $N$ is also a $\mu_x^{\mathscr{A}}$-null set for $\mu$-almost every $x$. This follows from property (1) applied to the function $f = \chi_N$. In many interesting cases, the atoms $[x]_{\mathscr{A}}$ are null sets with respect to $\mu$, and so $\mu_x^{\mathscr{A}}$ is singular to $\mu$.

(c) The conditional measures constructed in Theorem 5.14 are sometimes said to give a *disintegration of the measure* $\mu$.

(d) Notice that the uniqueness in property (3) (and similarly for (4)) may require switching to smaller conull sets. That is, if $\mu_x^{\mathscr{A}}$ for $x \in X' \subseteq X$ and $\widetilde{\mu}_x^{\mathscr{A}}$ for $x \in \widetilde{X}' \subseteq X$ are two systems of measures as in (1), then the claim is that there exists a conull subset $X'' \subseteq X' \cap \widetilde{X}'$ with $\mu_x^{\mathscr{A}} = \widetilde{\mu}_x^{\mathscr{A}}$ for all $x \in X''$.

(e) We only ever talk about atoms for countably generated $\sigma$-algebras. The first reason for this is that for a general $\sigma$-algebra the expression defined in Theorem 5.14(2) by an uncountable intersection may not be measurable (let alone $\mathscr{A}$-measurable). Moreover, even in those cases where the expression happens to be $\mathscr{A}$-measurable, the definition cannot be used to prove the stated assertions. We also note that it is not true that any sub-$\sigma$-algebra of a countably-generated $\sigma$-algebra is countably generated (but see Lemma 5.17 for a more positive statement). For example, the $\sigma$-algebra of null sets in $\mathbb{T}$ with respect to Lebesgue measure is not countably-generated (but there are more interesting examples, see Exercise 6.1.2).

*Example 5.16.* Let $X = [0,1]^2$ and $\mathscr{A} = \mathscr{B} \times \{\varnothing, [0,1]\}$ as in Example 5.3. In this case Theorem 5.14 claims that any Borel probability measure $\mu$ on $X$ can be decomposed into "vertical components": the conditional measures $\mu^{\mathscr{A}}_{(x_1,x_2)}$ are defined on the vertical line segments $\{x_1\} \times [0,1]$, and these sets are precisely the atoms of $\mathscr{A}$. Moreover,

$$\mu(B) = \int_X \mu^{\mathscr{A}}_{(x_1,x_2)}(B)\, d\mu(x_1,x_2). \tag{5.5}$$

In this example $\mu^{\mathscr{A}}_{(x_1,x_2)} = \nu_{x_1}$ does not depend on $x_2$, so (5.5) may be written as

$$\mu(B) = \int_{[0,1]} \nu_{x_1}(B)\, d\overline{\mu}(x_1) \tag{5.6}$$

where $\overline{\mu} = \pi_* \mu$ is the measure on $[0,1]$ obtained by the projection

$$\pi : [0,1]^2 \longrightarrow [0,1]$$
$$(x_1, x_2) \longmapsto x_1.$$

While (5.6) looks simpler than (5.5), in order to arrive at it a quotient space and a quotient measure has to be constructed (see Sect. 5.4). For simplicity we will often work with expressions like (5.5) in the general context.

Once $\mu$ is known explicitly, the measures $\mu^{\mathscr{A}}_{(x_1,x_2)}$ can often be computed. For example, if $\mu$ is defined by

$$\int f\, d\mu = \tfrac{1}{3} \int f(s,s)\, ds + \int_0^1 \int_0^{\sqrt{s}} f(s,t)\, dt\, ds,$$

then

$$\mu^{\mathscr{A}}_{(x_1,x_2)} = \frac{1}{\sqrt{x_1} + 1/3} \delta_{x_1} \times \left(\tfrac{1}{3}\delta_{x_1} + m_{[0,\sqrt{x_1}]}\right).$$

To see that this equation holds, the reader should use Theorem 5.14(3). However, the real force of Theorem 5.14 lies in the fact that it allows an unknown measure to be decomposed into components which are often easier to work with.

PROOF OF THEOREM 5.14. By assumption, $X$ is contained in a compact metric space $\overline{X}$, which is automatically separable. We note that the statement of the theorem for the ambient compact metric space $\overline{X}$ implies the theorem for $X$ by Remark 5.15(b). Hence we may assume that $X = \overline{X}$ is itself a compact metric space.

Suppose first that $\{\rho_x\}$ and $\{\nu_x\}$ are families of measures defined for almost every $x$ that both satisfy (5.3) for a countable dense subset $\{f_n\}_{n \in \mathbb{N}}$ in $C(X)$. Then for each $n \geqslant 1$ and almost every $x$,

$$\int f_n\, d\rho_x = E(f_n | \mathscr{A}) = \int f_n\, d\nu_x. \tag{5.7}$$

So there is a common null set $N$ with the property that (5.7) holds for all $n \geqslant 1$ and $x \notin N$. By uniform approximation and the dominated convergence theorem (Theorem A.18), this easily extends to show that

$$\int f \, d\rho_x = \int f \, d\nu_x$$

for all $f \in C(X)$ and $x \notin N$. Hence $\rho_x = \nu_x$ for $x \notin N$, which shows that the conditional measures—if they exist—must be unique as claimed in (3).

Now let

$$\widetilde{\mathscr{A}} = \mathscr{A}_\mu$$

and write $\overline{\mathscr{A}}$ for the smallest $\sigma$-algebra containing both $\widetilde{\mathscr{A}}$ and $\mathscr{A}$. Then for any $f \in C(X)$, $g = E(f|\mathscr{A})$ (or $E(f|\widetilde{\mathscr{A}})$) satisfies the characterizing properties of $E(f|\overline{\mathscr{A}})$, so they are equal almost everywhere. Noting this for a countable dense subset of $C(X)$ shows (as in the proof of uniqueness) that $\mu_x^{\mathscr{A}} = \mu_x^{\overline{\mathscr{A}}}$ almost everywhere, showing (4).

Turning to existence, let

$$\mathscr{F} = \{ f_0 \equiv 1, f_1, f_2, \dots \} \subseteq C(X)$$

be a vector space over $\mathbb{Q}$ that is dense* in $C(X)$. For every $i \geqslant 1$, choose an $\mathscr{A}$-measurable function[†] $g_i \in \mathscr{L}_\mu^1$ with $g_i$ representing $E(f_i | \mathscr{A})$. Define $g_0$ to be the constant function 1. Then

- $g_i(x) \geqslant 0$ almost everywhere if $f_i \geqslant 0$;
- $|g_i(x)| \leqslant \|f_i\|_\infty$ almost everywhere;
- if $f_i = \alpha f_j + \beta f_k$ with $\alpha, \beta \in \mathbb{Q}$, then $g_i(x) = \alpha g_j(x) + \beta g_k(x)$ for almost all $x$.

Let $N \in \mathscr{A}$ be the union of all the null sets on the complement of which the properties above hold; since this is a countable union, $N$ is a null set.

For $x \notin N$, define $\Lambda_x(f_i)$ to be $g_i(x)$. Then by the properties above $\Lambda_x$ is a $\mathbb{Q}$-linear map from $\mathscr{F}$ to $\mathbb{R}$ with $\|\Lambda_x\| \leqslant 1$. It follows that $\Lambda_x$ extends uniquely to a continuous positive linear functional

$$\Lambda_x : C(X) \to \mathbb{R}.$$

By the Riesz representation theorem, there is a measure $\mu_x^{\mathscr{A}}$ on $X$ characterized by the property that

---

* Since $X$ is separable we may find a set $\{h_0 \equiv 1, h_1, h_2, \dots\}$ that is dense in $C(X)$. The vector space over $\mathbb{Q}$ spanned by this set is dense and countable, and may be written in the form $\{f_0 \equiv 1, f_1, f_2, \dots\}$.

† Notice that this is a genuine function rather than an equivalence class of functions, so there is a choice involved despite Theorem 5.1(1).

$$\Lambda_x(f) = \int f \, d\mu_x^{\mathscr{A}}$$

for all $f \in C(X)$; moreover $\Lambda_x(1) = 1$, so $\mu_x^{\mathscr{A}}$ is a probability measure.

By our choice of the set $\mathscr{F}$, for any $f \in C(X)$ there is a sequence $(f_{n_i})$ with $f_{n_i} \longrightarrow f$ uniformly. We have already established that

$$x \mapsto \int f_{n_i} \, d\mu_x^{\mathscr{A}}$$

is $\mathscr{A}$-measurable (by Theorem 5.14(1)), and that

$$\int_A \int f_{n_i} \, d\mu_x^{\mathscr{A}} \, d\mu(x) = \int_A f_{n_i} \, d\mu$$

for all $A \in \mathscr{A}$. So, by the dominated convergence theorem (Theorem A.18),

$$\int f_{n_i} \, d\mu_x^{\mathscr{A}} \to \int f \, d\mu_x^{\mathscr{A}} \tag{5.8}$$

is $\mathscr{A}$-measurable as a function of $x$, and

$$\int_A \int f \, d\mu_x^{\mathscr{A}} \, d\mu(x) = \int_A f \, d\mu \tag{5.9}$$

for all $A \in \mathscr{A}$. For any open set $O$ there is a sequence $(f_{n_i})$ with $f_{n_i} \nearrow \chi_O$, so by the monotone convergence theorem (5.8) and (5.9) hold for $\chi_O$. Thus we have (5.8) and (5.9) for the indicator function of any closed $A \subseteq X$, by taking complements. Similarly, these equations extend to any $G_\delta$-set $G$ and any $F_\sigma$-set $F$. Define

$$\mathscr{M} = \{B \in \mathscr{B} \mid f = \chi_B \text{ satisfies (5.8) and (5.9)}\}.$$

By the monotone convergence theorem (Theorem A.16), if $B_1, B_2, \ldots \in \mathscr{M}$ with

$$B_1 \subseteq B_2 \subseteq \cdots,$$

then $\bigcup_{n \geqslant 1} B_n \in \mathscr{M}$ and if $C_1, C_2, \ldots \in \mathscr{M}$ with

$$C_1 \supseteq C_2 \supseteq \cdots,$$

then $\bigcap_{n \geqslant 1} C_n \in \mathscr{M}$. Thus $\mathscr{M}$ is a monotone class (see Definition A.3 and Theorem A.4). Define

$$\mathscr{R} = \left\{ \bigsqcup_{i=1}^n O_i \cap A_i \mid O_i \subseteq X \text{ is open and } A_i \subseteq X \text{ is closed} \right\}$$

for $n \in \mathbb{N}$. We claim that $\mathscr{R}$ is an algebra (that is, $\mathscr{R}$ is closed under complements, finite intersections and finite unions). To see this, notice that the $\sigma$-
algebra $\mathscr{C}$ generated by finitely many open and closed sets has the property
that every element of $\mathscr{C}$ is a disjoint union of atoms of the partition generated
by the same open and closed sets, all of which are precisely of the form $O \cap A$.

Since any set $O \cap A$ is a $G_\delta$-set and (5.8) and (5.9) are linear conditions,
it follows that (5.8) and (5.9) also hold for functions of the form

$$\chi_R = \sum_{i=1}^{n} \chi_{O_i \cap A_i}$$

for all

$$R = \bigsqcup_{i=1}^{n} O_i \cap A_i \in \mathscr{R}.$$

By the monotone class theorem (Theorem A.4), $\mathscr{B} = \sigma(\mathscr{R}) \subseteq \mathscr{M}$. In other
words, for any Borel measurable set $B \in \mathscr{B}$, the characteristic function $\chi_B$
satisfies (5.8) and (5.9). By considering simple functions and applying the
monotone convergence theorem, it follows that (5.8) and (5.9) also hold for
any $\mathscr{B}$-measurable function $f \geqslant 0$.

Finally, given any $\mathscr{B}$-measurable integrable function $f$, we may write

$$f = f^+ - f^-$$

with $f^+, f^-$ non-negative, measurable, and integrable functions. Then, by (5.9),

$$\int f^+ \, \mathrm{d}\mu_x^{\mathscr{A}}, \int f^- \, \mathrm{d}\mu_x^{\mathscr{A}} < \infty$$

almost everywhere. In particular, $f$ is $\mu_x^{\mathscr{A}}$-integrable for almost every $x$, and
where it is $\mu_x^{\mathscr{A}}$-integrable, $\int f \, \mathrm{d}\mu_x^{\mathscr{A}}$ is an $\mathscr{A}$-measurable function of $x$. Finally,
(5.9) holds, proving (1).

Suppose now that $\mathscr{A} = \sigma(\{A_1, A_2, \dots\})$ is countably-generated. Then

$$E(\chi_{A_i} \mid \mathscr{A})(x) = \chi_{A_i}(x)$$
$$= \mu_x^{\mathscr{A}}(A_i)$$

almost everywhere, for any $i \geqslant 1$. Collecting all the null sets arising into a
single null set $N$ gives

$$\mu_x^{\mathscr{A}}(A_i) = \begin{cases} 1 & \text{if } x \in A_i \setminus N; \\ 0 & \text{if } x \in X \setminus (A_i \cup N). \end{cases}$$

Since $\mu_x^{\mathscr{A}}$ is a measure, it follows by (5.4) that

$$\mu_x^{\mathscr{A}}([x]_{\mathscr{A}}) = 1$$

if $x \notin N$. Writing $X'$ for $X \smallsetminus N$, recall that the map

$$X' \ni x \longmapsto \int f \, d\mu_x^{\mathscr{A}}$$

is $\mathscr{A}$-measurable for any $f \in C(X)$. Thus

$$\int f \, d\mu_x^{\mathscr{A}} = \int f \, d\mu_y^{\mathscr{A}}$$

if $x, y \in X'$ and $[x]_{\mathscr{A}} = [y]_{\mathscr{A}}$, so that $[x]_{\mathscr{A}} = [y]_{\mathscr{A}}$ implies that $\mu_x^{\mathscr{A}} = \mu_y^{\mathscr{A}}$. $\square$

One of the many desirable properties of Borel probability spaces is that there is a constraint on the complexity of their sub-$\sigma$-algebras.

**Lemma 5.17.** *If $(X, \mathscr{B}, \mu)$ is a Borel probability space and $\mathscr{A} \subseteq \mathscr{B}$ is a $\sigma$-algebra then there is a countably-generated $\sigma$-algebra $\widetilde{\mathscr{A}}$ with $\mathscr{A} \underset{\mu}{=} \widetilde{\mathscr{A}}$.*

PROOF. Recall that $C(\overline{X})$ is separable for any compact metric space $\overline{X}$ (see Lemma B.8). Since $C(\overline{X})$ is mapped continuously to a dense subspace of $L^1(X, \mathscr{B}, \mu)$, the same holds for $L^1(X, \mathscr{B}, \mu)$. Since subsets of a separable space are separable, it follows that the space

$$\{\chi_A \mid A \in \mathscr{A}\} \subseteq L^1(X, \mathscr{A}, \mu) \subseteq L^1(X, \mathscr{B}, \mu)$$

is separable. Thus there is a set $\{A_1, A_2, \dots\} \subseteq \mathscr{A}$ such that for any $\varepsilon > 0$ and $A \in \mathscr{A}$ there is some $n$ with

$$\mu(A \triangle A_n) = \|\chi_A - \chi_{A_n}\|_1 < \varepsilon.$$

Let $\widetilde{\mathscr{A}} = \sigma(\{A_1, A_2, \dots\})$, so that $\widetilde{\mathscr{A}} \subseteq \mathscr{A}$ and $\{\chi_A \mid A \in \widetilde{\mathscr{A}}\}$ is dense in

$$\{\chi_A \mid A \in \mathscr{A}\}$$

with respect to the $L^1_\mu$ norm. Given $A \in \mathscr{A}$, we can find a sequence $(n_k)$ for which

$$\|\chi_A - \chi_{A_{n_k}}\|_1 < \frac{1}{k}$$

for $k \geqslant 1$. Then the sequence $(\chi_{A_{n_k}})$ is Cauchy in $L^1(X, \widetilde{\mathscr{A}}, \mu) \subseteq L^1(X, \mathscr{A}, \mu)$, so has a limit $f \in L^1(X, \widetilde{\mathscr{A}}, \mu)$. We must have $f = \chi_A$ almost everywhere since the limit is unique, so there is some $\widetilde{A} \in \widetilde{\mathscr{A}}$ with $\mu(A \triangle \widetilde{A}) = 0$. It follows that $\mathscr{A} \underset{\mu}{=} \widetilde{\mathscr{A}}$ as required. $\square$

In the remainder of this section we give extensions and reformulations of Theorem 5.14.

**Lemma 5.18.** *Let $(X, \mathscr{B}, \mu)$ be a Borel probability space and let $\mathscr{A} \subseteq \mathscr{B}$ be a countably-generated $\sigma$-algebra. If $f \in \mathscr{L}^\infty(X, \mathscr{B})$ is constant on atoms of $\mathscr{A}$, then $f|_{X'}$ is $\mathscr{A}$-measurable, where $X'$ is as in Theorem 5.14.*

A set $B$ (or a function $f$) is $\mathscr{A}$-*measurable modulo* $\mu$ if, after removing a $\mu$-null set, $B$ (or $f$) becomes $\mathscr{A}$-measurable. Thus the conclusion of Lemma 5.18 is that $f$ is $\mathscr{A}$-measurable modulo $\mu$.

PROOF OF LEMMA 5.18. By Theorem 5.14(2), on $X'$ we have

$$\int f \, \mathrm{d}\mu_x^{\mathscr{A}} = f(x)$$

since $\mu_x^{\mathscr{A}}([x]_{\mathscr{A}}) = 1$ and, by assumption, $f$ is constant (and equal to $f(x)$) on the set $[x]_{\mathscr{A}}$. By Theorem 5.14(1) we know that $f|_{X'}$ is $\mathscr{A}$-measurable. $\square$

In Theorem 5.14 the conditional measure was characterized in terms of the conditional expectation. The following proposition gives a more geometrical characterization.

**Proposition 5.19.** *Let $(X, \mathscr{B}, \mu)$ be a Borel probability space and let $\mathscr{A}$ be a countably-generated sub-$\sigma$-algebra of $\mathscr{B}$. Suppose that there is a set $X' \in \mathscr{B}$ with $\mu(X') = 1$, and a collection $\{\nu_x \mid x \in X'\}$ of probability measures with the property that*

- *$x \mapsto \nu_x$ is measurable, that is for any $f \in \mathscr{L}^\infty$ we have that $x \mapsto \int f \, \mathrm{d}\nu_x$ is measurable,*
- *$\nu_x = \nu_y$ for $[x]_{\mathscr{A}} = [y]_{\mathscr{A}}$ and $x, y \in X'$,*
- *$\nu_x([x]_{\mathscr{A}}) = 1$, and*
- *$\mu = \int \nu_x \, \mathrm{d}\mu(x)$ in the sense that $\int f \, \mathrm{d}\mu = \int \int f \, \mathrm{d}\nu_x \, \mathrm{d}\mu(x)$ for all $f \in \mathscr{L}^\infty$.*

*Then $\nu_x = \mu_x^{\mathscr{A}}$ for a.e. $x$. The same is true if the properties hold for a dense countable set of functions in $C(\overline{X})$.*

PROOF. First notice that we may assume that $\mu_x^{\mathscr{A}}$ and $\nu_x$ are defined on a common conull set $X''$. Moreover, we may replace $X$ by $X''$ and simultaneously replace $\mathscr{A}$ by $\mathscr{A}|_{X''} = \{A \cap X'' \mid A \in \mathscr{A}\}$. After this replacement, Lemma 5.18 says that any function $f$ which is constant on $\mathscr{A}$-atoms is $\mathscr{A}$-measurable. In order to apply Theorem 5.14(3) we need to check that

$$\int f \, \mathrm{d}\nu_x = E\left(f \mid \mathscr{A}\right)(x) \tag{5.10}$$

almost everywhere, for all $f$ in a countable dense subset of $C(\overline{X})$.

That $x \mapsto \int f \, \mathrm{d}\nu_x$ is measurable is the first assumption on the family of measures in the proposition. Together with Lemma 5.18, the second property shows that $x \mapsto \int f \, \mathrm{d}\nu_x$ is actually $\mathscr{A}$-measurable. This is the first requirement in the direction of showing that (5.10) holds.

To show (5.10), we also need to calculate $\int_A \int f \, d\nu_x \, d\mu(x)$ for any $A \in \mathscr{A}$, as in Theorem 5.1(1). We know that $\chi_A(x)$ is constant $\nu_x$-almost everywhere for any $A \in \mathscr{A}$, by the third property. In fact $\chi_A(x)$ equals 1 $\nu_x$-almost everywhere if $x \in A$ and equals 0 otherwise. Therefore, by the fourth property applied to the function $\chi_A f$, we get

$$
\int_A \int f(z) \, d\nu_x(z) \, d\mu(x) = \int \chi_A(x) \int_{[x]_{\mathscr{A}}} f(z) \, d\nu_x(z) \, d\mu(x)
$$

$$
= \iint \chi_A(z) f(z) \, d\nu_x(z) \, d\mu(x)
$$

$$
= \int \chi_A(z) f(z) \, d\mu(z) = \int_A f \, d\mu
$$

as required. By Theorem 5.14(3) it follows that $\nu_x = \mu_x^{\mathscr{A}}$ almost everywhere.

It remains to prove the last claim of the proposition. So suppose we only assume the first and fourth properties for all functions in a dense countable subset of $C(\overline{X})$. Using dominated convergence, monotone convergence, and the monotone class theorem (Theorems A.18, A.16 and A.4) just as in the proof of Theorem 5.14 on p. 138, we can extend the first and fourth properties in turn to all $f \in C(\overline{X})$, all $f = \chi_B$ for $B$ any open set, any closed set, any $G_\delta$, any $F_\sigma$, any Borel set, and finally to any $f \in L^\infty(\overline{X})$. This implies the last claim.                                                                                      □

**Proposition 5.20.** *Let $(X, \mathscr{B}, \mu)$ be a Borel probability space, and let*

$$
\mathscr{A}' \subseteq \mathscr{A} \subseteq \mathscr{B}
$$

*be countably-generated sub-$\sigma$-algebras. Then $[z]_{\mathscr{A}} \subseteq [z]_{\mathscr{A}'}$ for $z \in X$, and for almost every $z \in X$ the conditional measures for the measure $\mu_z^{\mathscr{A}'}$ with respect to $\mathscr{A}$ are given for $\mu_z^{\mathscr{A}'}$-almost every $x \in [z]_{\mathscr{A}'}$ by $(\mu_z^{\mathscr{A}'})_x^{\mathscr{A}} = \mu_x^{\mathscr{A}}$.*

The proof of this result will reveal that it is a reformulation of Theorem 5.1(4).

PROOF OF PROPOSITION 5.20. We will show that the map $x \mapsto \mu_x^{\mathscr{A}}$ satisfies all the assumptions in Proposition 5.19 with respect to the measure $\mu_z^{\mathscr{A}'}$ for almost every $z \in X$. Let $\mu_z^{\mathscr{A}'}$ be defined on $X'_{\mathscr{A}'} \in \mathscr{A}'$ and let $\mu_x^{\mathscr{A}}$ be defined on $X'_{\mathscr{A}} \in \mathscr{A}$ with all the properties in Theorem 5.14. By Remark 5.15(b), we have $\mu_z^{\mathscr{A}'}(X'_{\mathscr{A}}) = 1$ for $\mu$-almost every $z$. Now fix some $z \in X'_{\mathscr{A}'}$ with $\mu_z^{\mathscr{A}'}(X'_{\mathscr{A}}) = 1$. For $x, y \in X'_{\mathscr{A}}$ we know that $\mu_x^{\mathscr{A}} = \mu_y^{\mathscr{A}}$ if $[x]_{\mathscr{A}} = [y]_{\mathscr{A}}$ and that $\mu_x^{\mathscr{A}}([x]_{\mathscr{A}}) = 1$ by Theorem 5.14(2). Also, if $f \in \mathscr{L}^\infty$, we know that $\int f \, d\mu_x^{\mathscr{A}}$ is measurable by Theorem 5.14(1). Thus we have shown the first three assumptions of Proposition 5.19 on the complement of a single $\mu_z^{\mathscr{A}}$-null set.

It remains to check that

$$\mu_z^{\mathscr{A}'} = \int \mu_x^{\mathscr{A}} \, d\mu_z^{\mathscr{A}'}(x) \tag{5.11}$$

for almost every $z$. Let $f \in C(\overline{X})$. By Theorem 5.14(1), for almost every $z$,

$$\iint f \, d\mu_x^{\mathscr{A}} \, d\mu_z^{\mathscr{A}'}(x) = \int E(f|\mathscr{A})(x) \, d\mu_z^{\mathscr{A}'}(x) = E\left(E(f|\mathscr{A})|\mathscr{A}'\right)(z),$$

which by Theorem 5.1(1) is equal to

$$E(f|\mathscr{A}')(z) = \int f \, d\mu_z^{\mathscr{A}'}$$

for almost every $z$. Using a dense subset $\{f_1, f_2, \dots\} \subseteq C(\overline{X})$, and collecting the countably many null sets arising in these two statements for each $n$ into a single null set, we obtain equality in (5.11) on a conull set $Z$. In other words, we have checked all the requirements of Proposition 5.19 for the family of measures $\nu_x = \mu_x^{\mathscr{A}}$ (and therefore they are equal almost everywhere to $\mu_x^{\mathscr{A}}$) and for the measure $\mu_z^{\mathscr{A}'}$ for $z \in Z \cap X'_{\mathscr{A}}$ with $\mu_z^{\mathscr{A}'}(X'_{\mathscr{A}}) = 1$. □

Theorem 5.14(3) and the more geometric discussion above highlights the significance of the countably-generated hypothesis on the $\sigma$-algebra $\mathscr{A}$, for in that case the conditional measures $\mu_x^{\mathscr{A}}$ can be related to the atoms $[x]_{\mathscr{A}}$. In a Borel probability space it is safe to assume that $\sigma$-algebras are countably-generated by Lemma 5.17.

By combining the increasing and decreasing martingale theorems (Theorems 5.5 and 5.8) with the characterizing properties of the conditional measures we get the following corollary (see Exercise 5.3.5).

**Corollary 5.21.** *If $\mathscr{A}_n \nearrow \mathscr{A}$ or $\mathscr{A}_n \searrow \mathscr{A}$ then $\mu_x^{\mathscr{A}_n} \longrightarrow \mu_x^{\mathscr{A}}$ in the weak\*-topology for $\mu$-almost every $x$.*

This gives an alternative construction of $\mu_x^{\mathscr{A}}$ for a countably-generated $\sigma$-algebra. More concretely, if $\mathscr{A} = \sigma(\{A_1, A_2, \dots\})$ and $\mathscr{A}_n = \sigma(\{A_1, \dots, A_n\})$ is the finite $\sigma$-algebra generated by the first $n$ generators of $\mathscr{A}$, then $\mu_x^{\mathscr{A}_n}$ is readily defined, and $\mu_x^{\mathscr{A}_n} \to \mu_x^{\mathscr{A}}$.

## Exercises for Sect. 5.3

**Exercise 5.3.1.** Prove the equality claimed in (5.4).

**Exercise 5.3.2.** [59] Let $(X, \mathscr{B}, \mu)$ be an aperiodic measure-preserving transformation on a Borel probability space (see Exercise 2.9.2 for the definition of aperiodic). Prove that for any $k \geqslant 1$ there is a set $A \in \mathscr{B}$ with $\mu(A) > 0$ and $\mu(T^{-k}(A) \cap A) = 0$.

**Exercise 5.3.3.** Using the definition from p. 135, show that if $(X, \mathcal{B}, \mu)$ is a Borel probability space, then for countably-generated $\sigma$-algebras $\mathcal{A}, \widetilde{\mathcal{A}} \subseteq \mathcal{B}$ there is a null set $N$ with the property that

$$\mathcal{A}\big|_{X \setminus N} = \widetilde{\mathcal{A}}\big|_{X \setminus N}$$

if and only if $\mathcal{A} \underset{\mu}{=} \mathcal{A}'$.

**Exercise 5.3.4.** In the notation of Example 5.16, find a precise criterion in terms of $\mu^{\mathcal{A}}_{(x_1, x_2)}$ characterizing the property that $\mu$ is a product $\mu_1 \times \mu_2$ of measures $\mu_1, \mu_2$ defined on $[0, 1]$.

**Exercise 5.3.5.** Prove Corollary 5.21, starting with a countable dense set of functions in $C(X)$ and using the appropriate martingale theorem.

## 5.4 Algebras and Maps

Let $X$ and $Y$ be Borel subsets of compact metric spaces $\overline{X}$ and $\overline{Y}$. For a measurable map

$$\phi : X \to Y,$$

write $\phi_* : \mathcal{M}(X) \to \mathcal{M}(Y)$ for the map induced on the space of probability measures by

$$(\phi_*(\mu))(A) = \mu(\phi^{-1}(A))$$

for $A \subseteq Y$ measurable. In this notation, for any integrable function $f : Y \to \mathbb{R}$ and $B \in \mathcal{B}_Y$,

$$\int_{\phi^{-1}B} f \circ \phi \, d\mu = \int_B f \, d\phi_* \mu.$$

In particular, a map $\phi : (X, \mathcal{B}, \mu) \to (Y, \mathcal{B}_Y, \nu)$ between two Borel probability spaces is measure-preserving if and only if $\phi_* \mu = \nu$.

Any measurable function $\phi : X \to Y$ as above defines a $\sigma$-algebra

$$\mathcal{A} = \phi^{-1}(\mathcal{B}_Y)$$

on $X$. The next results, corollaries of Theorem 5.14, show that essentially all $\sigma$-algebras on $X$ arise this way.

**Corollary 5.22.** *Let $(X, \mathcal{B}, \mu)$ be a Borel probability space, and let $\mathcal{A} \subseteq \mathcal{B}$ be a countably-generated $\sigma$-algebra. Then there is a conull set $X' = X \setminus N$ in $\mathcal{A}$, a compact metric space together with its Borel $\sigma$-algebra $(Y, \mathcal{B}_Y)$, and a measurable map $\phi : X' \to Y$ such that*

$$\mathcal{A}\big|_{X'} = \phi^{-1} \mathcal{B}_Y.$$

*Moreover,*

$$[x]_{\mathscr{A}} = \phi^{-1}\left(\phi(x)\right)$$

*for* $x \notin N$, *and* $\mu_x^{\mathscr{A}} = \nu_{\phi(x)}$ *for some measurable map* $y \mapsto \nu_y$ *defined on a* $\phi_*\mu$-*conull subset of* $Y$. *In fact we can take* $Y = \mathscr{M}(\overline{X})$, $\phi(x) = \mu_x^{\mathscr{A}}$, *and* $\nu_y = y$.

This will be proved later; the conclusion described in Corollary 5.22 is depicted in Fig. 5.3.

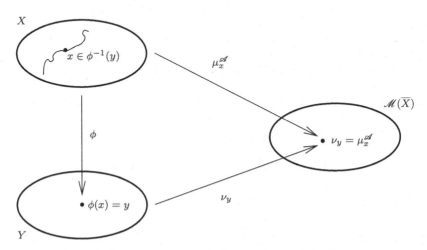

**Fig. 5.3** Each $y \in Y$ determines an atom $\phi^{-1}(y)$ and its conditional measure $\nu_y$

**Lemma 5.23.** *If* $\overline{X}$ *is a compact metric space, and* $f \in \mathscr{L}^{\infty}(\overline{X})$, *then the map*

$$\mathscr{M}(X) \ni \nu \longmapsto \int f \, d\nu$$

*is Borel measurable. In particular, for a Borel subset* $X$ *of* $\overline{X}$, *we have that* $\mathscr{M}(X)$ *is a Borel subset of* $\mathscr{M}(\overline{X})$. *Moreover, if* $\phi : X \to Y$ *is a Borel measurable map between Borel subsets of compact metric spaces, then the induced map* $\phi_* : \mathscr{M}(X) \to \mathscr{M}(Y)$ *is Borel measurable.*

PROOF. Starting with continuous functions, we know that $\int f \, d\nu$ depends continuously on $\nu$ (by definition of the weak*-topology on $\mathscr{M}(X)$). Arguing just as we did on p. 139, this can be extended to show that $\int f \, d\nu$ depends measurably on $\nu$ for all indicator functions of open sets, and thence to show that it does so for indicator functions of Borel measurable sets, and finally for any $f \in \mathscr{L}^{\infty}$. By definition and the argument above, it follows that

$$\mathscr{M}(X) = \{\mu \in \mathscr{M}(\overline{X}) \mid \mu(\overline{X} \smallsetminus X) = 0\}$$

is Borel measurable. Now let $\phi : X \to Y$ be measurable and fix $r \in \mathbb{R}$, $\varepsilon > 0$ and $f \in C(\overline{Y})$. Then

$$O_{f,r,\varepsilon} = \left\{ \mu \in \mathscr{M}(Y) \mid \left| \int f \, d\mu - r \right| < \varepsilon \right\}$$

is an open set in $\mathscr{M}(\overline{Y})$, and clearly

$$\phi_*^{-1} O_{f,r,\varepsilon} = \left\{ \nu \in \mathscr{M}(X) \mid \left| \int f \circ \phi \, d\nu - r \right| < \varepsilon \right\}$$

is measurable in $\mathscr{M}(X)$. Since any open set in $\mathscr{M}(\overline{Y})$ can be written as a countable union of finite intersections of sets of the form $O_{f,r,\varepsilon}$ with $f$ chosen from a dense countable subset of $C(\overline{X})$, $r \in \mathbb{Q}$ and $\varepsilon \in \mathbb{Q}$, the lemma follows. $\square$

PROOF OF COROLLARY 5.22. Let $\mathscr{A} = \sigma(\{A_1, A_2, \dots\})$ be countably-generated. Taking $Y = \mathscr{M}(\overline{X})$ (with the weak*-topology, so that $Y$ is a compact metric space) and $\phi(x) = \mu_x^{\mathscr{A}}$ we can set $\nu_y = y$ and hence $\nu_{\phi(x)} = \mu_x^{\mathscr{A}}$ follows at once. Let $X'$ be a $\mu$-conull set on which all the statements in Theorem 5.14 hold. We claim first that (perhaps after enlarging the complement of $X'$ by null sets countably often), $\mathscr{A} = \phi^{-1} \mathscr{B}_Y$. By Theorem 5.14(2),

$$\chi_{A_i}(x) = \mu_x^{\mathscr{A}}(A_i) \tag{5.12}$$

for almost every $x$, and we may assume that this holds for all $x \in X'$. Since $\{\nu \mid \nu(A_i) = 1\} \in \mathscr{B}_Y$, (5.12) shows that $A_i \cap X' \in \phi^{-1} \mathscr{B}_Y$ and therefore $\mathscr{A}|_{X'} \subseteq \phi^{-1} \mathscr{B}_Y$.

For the reverse direction $\phi^{-1} \mathscr{B}_Y \subseteq \mathscr{A}|_{X'}$, it is sufficient to check this on sets of the form $O_{f,r,\varepsilon}$ since these generate the weak*-topology in a countable manner, and by Theorem 5.14(1) the set

$$\phi^{-1} \left( \left\{ \nu \mid \left| \int f \, d\nu - r \right| < \varepsilon \right\} \right) = \left\{ x \mid \left| \int f \, d\mu_x^{\mathscr{A}} - r \right| < \varepsilon \right\}$$

is $\mathscr{A}$-measurable for any $f \in C(X)$, $r \in \mathbb{R}$ and $\varepsilon > 0$. Hence $\mathscr{A}|_{X'} = \phi^{-1} \mathscr{B}$.

Since $\phi : X' \to Y$ satisfies $\mathscr{A}|_{X'} = \phi^{-1} \mathscr{B}_Y$, and the Borel $\sigma$-algebra $\mathscr{B}_Y$ of $Y$ separates points, $[x]_{\mathscr{A}} = \phi^{-1}(\phi(x))$ follows. $\square$

**Corollary 5.24.** *Let $\phi : (X, \mathscr{B}_X, \mu) \to (Y, \mathscr{B}_Y, \nu)$ be a measure-preserving map between Borel probability spaces, and let $\mathscr{A} \subseteq \mathscr{B}_Y$ be a sub-$\sigma$-algebra. Then*

$$\phi_* \mu_x^{\phi^{-1} \mathscr{A}} = \nu_{\phi(x)}^{\mathscr{A}}$$

*for $\mu$-almost every $x \in X$.*

PROOF. First notice that for any $f \in L^1(Y, \mathscr{B}_Y, \nu)$, $E_\nu(f|\mathscr{A}) \circ \phi$ is $\phi^{-1} \mathscr{A}$-measurable and

$$\int_{\phi^{-1}A} E_\nu(f|\mathscr{A}) \circ \phi \, d\mu = \int_A E_\nu(f|\mathscr{A}) \, d\nu$$
$$= \int_A f \, d\nu$$
$$= \int_{\phi^{-1}A} f \circ \phi \, d\mu.$$

It follows that
$$E_\nu(f|\mathscr{A}) \circ \phi = E_\mu(f \circ \phi|\phi^{-1}\mathscr{A}).$$

Thus for $f \in \mathscr{L}^\infty(Y, \mathscr{B}_Y)$,

$$\int f \, d\nu^{\mathscr{A}}_{\phi(x)} = E_\nu(f|\mathscr{A})(\phi(x))$$
$$= E_\mu(f \circ \phi|\phi^{-1}\mathscr{A})(x)$$
$$= \int f \circ \phi \, d\mu_x^{\phi^{-1}\mathscr{A}}$$
$$= \int f \, d\left(\phi_* \mu_x^{\phi^{-1}\mathscr{A}}\right)$$

(all almost everywhere). Using a dense countable subset of $C(\overline{X})$ completes the proof. □

In the next result we will work with measurability on Borel subsets of compact metric spaces, without reference to a particular measure.

**Lemma 5.25.** *Let $X, Y, Z$ be Borel subsets of compact metric spaces $\overline{X}$, $\overline{Y}$ and $\overline{Z}$ respectively, and let $\phi_Z : X \to Z$ and $\phi_Y : X \to Y$ be measurable maps. Suppose that $\phi_Z$ is $\phi_Y^{-1}(\mathscr{B}_Y)$-measurable. Then there is a measurable map $\psi : Y \to Z$ with $\phi_Z(x) = \psi \circ \phi_Y(x)$ on $X$, as illustrated in Fig. 5.4.*

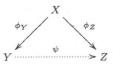

**Fig. 5.4** The map constructed in Lemma 5.25

This is related to Corollary 5.22, in that it allows us to draw the same conclusion that we can write the conditional measure $\mu_x^{\mathscr{A}}$ as a measurable function $\nu_y$ on the image space $Y$ whenever $\mathscr{A} = \phi^{-1}\mathscr{B}_Y$.

PROOF OF LEMMA 5.25. Define $\mathscr{A} = \phi_Y^{-1}(\mathscr{B}_Y)$, which is countably-generated since $\mathscr{B}_Y$ is. Since the Borel $\sigma$-algebra $\mathscr{B}_Y$ of $Y$ separates points,

$$[x]_{\mathscr{A}} = \phi_Y^{-1}(\phi_Y(x))$$

for $x \in X$. By assumption, $\phi_Z$ is $\mathscr{A}$-measurable, and hence $\phi_Z(x) = \phi_Z(x')$ whenever $[x]_{\mathscr{A}} = [x']_{\mathscr{A}}$, or equivalently whenever $\phi_Y(x) = \phi_Y(x')$. So we could define

$$\psi(y) = \phi_Z(x)$$

whenever $y = \phi_Y(x)$ for some $x \in X$, and use some fixed $z_0 \in Z$ to define $\psi(y) = z_0$ for $y \in Y \smallsetminus \phi_Y(X)$. However, it is not clear why this should define a measurable function, so instead we will define $\psi : Y \to Z$ by a limiting process. In order to do this we will cut the target space $Z$ into small metric balls, and ensure that at each finite stage everything is appropriately measurable. For this we start with the additional assumption that $Z = \overline{Z}$ is a compact metric space, and later show that this requirement can be removed.

Since $Z$ is compact, there is a sequence $(\xi_n)$ of finite Borel measurable partitions of $Z$ with the property that

$$\sigma(\xi_n) \subseteq \sigma(\xi_{n+1})$$

for $n \geqslant 1$, and for which every element of $\xi_n$ has diameter less than $\frac{1}{n}$.

We will define related partitions $\xi_n^X$ and $\xi_n^Y$ of $X$ and $Y$. The first of these is defined by taking pre-images as follows. Since $\phi_Z$ is $\mathscr{A}$-measurable we get, for any $P \in \xi_n$, a set

$$P^X = \phi_Z^{-1}(P) \in \mathscr{A},$$

and hence a finite partition $\xi_n^X = \{P^X \mid P \in \xi_n\} \subseteq \mathscr{A}$ of $X$. This is illustrated in Fig. 5.5.

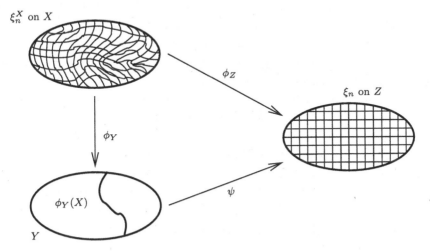

**Fig. 5.5** The partitions $\xi_n$ on $\mathscr{M}(X)$ and $\xi_n^X$ on $X$

We next define the partition $\xi_n^Y$. Note that in the construction of $\xi_n^Y$ care needs to be taken, since the set $\phi_Y(X)$ is not assumed to be measurable. Since $\mathscr{A} = \phi_Y^{-1}\mathscr{B}_Y$, we can choose for every $P^X \in \xi_n^X$ a set $P^Y$ in $\mathscr{B}_Y$ with $P^X = \phi_Y^{-1}P^Y$. Since the various sets $P^X \in \xi_n^X$ are disjoint, we can require the same disjointness for the sets $P^Y$. Indeed if sets $P$ and $Q$ in $\xi_n$ are different, $P^X = \phi_Z^{-1}P = \phi_Y^{-1}P$, and $Q^X = \phi_Z^{-1}Q = \phi_Y^{-1}Q^Y$ with $Q^Y, P^Y \in \mathscr{B}_Y$ not disjoint, then replacing $Q^Y$ by $Q^Y \smallsetminus P^Y$ will not change its properties but will ensure disjointness from $P^Y$. Moreover, using a similar argument inductively we can also insist that $P_{n+1}^Y \subseteq P_n^Y$ whenever $P_{n+1}^X \subseteq P_n^X$ and $P_{n+1}^X \in \xi_{n+1}^X$, $P_n^X \in \xi_n^X$. The sets $P^Y \in \mathscr{B}_Y$ for $P \in \xi_n^X$, together with their common complement $Q_n \in \mathscr{B}_Y$, form a partition $\xi_n^Y$ of $Y$. By construction, $\phi^{-1}Q_n = \varnothing$ and $Q_{n+1} \supseteq Q_n$ for all $n \geqslant 1$. Write

$$Q = \bigcup_{m \geqslant 1} Q_m$$

and define $\psi(y)$ to be some fixed element $z_0 \in Z$ for any $y \in Q$. We define $\psi$ on $Y \smallsetminus Q$ as the limit of the sequence of functions $\psi_n : Y \smallsetminus Q \to Z$ defined as follows. For every $P^Y \in \xi_n^Y$, we choose some $z_P \in P \in \xi_n$ and define $\psi^n(y) = z_P$ for $y \in P^Y \smallsetminus Q$. Clearly $\psi^n : Y \smallsetminus Q \to Z$ is measurable, since the partition $\xi_n^Y$ is measurable by construction. By construction of the partitions $\xi_n$, the diameters of the partition elements $P \in \xi_n$ go to zero, so the function $\psi$ defined by

$$\psi(y) = \lim_{n \to \infty} \psi_n(y)$$

exists for all $y \in Y \smallsetminus Q$; since it is a pointwise limit of a sequence of measurable functions, the map $\psi$ is also measurable. If $y = \phi_Y(x)$ for some $x \in X$, then $\phi_Z(x)$ and $\psi_n(y)$ will belong to the same element of $\xi_n$ for all $n$, so $\phi_Z(x) = \psi(\phi_Y(x))$ as required.

Assume now that $Z$ is only a Borel subset of a compact metric space $\overline{Z}$. The construction above can be used to define a measurable map $\overline{\psi} : Y \to \overline{Z}$ with $\phi_Z = \overline{\psi} \circ \phi_Y$. Since

$$\varnothing = \phi_Z^{-1}\left(\overline{Z} \smallsetminus Z\right) = \phi_Y^{-1}\left(\overline{\psi}^{-1}\left(\overline{Z} \smallsetminus Z\right)\right)$$

we may replace $\overline{\psi}$ by the measurable map

$$\psi(y) = \begin{cases} \overline{\psi}(y) & \text{if } \overline{\psi}(y) \in Z, \text{ and} \\ z_0 \in Z & \text{if not,} \end{cases}$$

for some fixed element $z_0 \in Z$ without affecting the fact that $\phi_Z = \psi \circ \phi_Y$. $\square$

# Notes to Chap. 5

[56](Page 126) The origins of the terminology martingale lie in gambling, where it is related to the so-called St. Petersburg paradox arising from the method of gambling where the stake is doubled after each loss; the etymology is discussed by Mansuy [243] and Mayhew [258]. Lévy [228] introduced their study in probability, and their central importance in probability was developed most significantly by Doob [73]. For the material in this section on martingales we have followed Parry [279, Chap. 2] closely.

[57](Page 126) This was pointed out by Kakutani [173]; a survey of the connections between ergodic and martingale theorems is given by Rao [298] and by Kachurovskiĭ [169]. Jerison [165] showed that ergodic averages for a measure-preserving transformation on a probability space can be represented using martingales in a $\sigma$-finite measure space. Jacobs [163] and Neveu [270] gave proofs of martingale theorems using the method of proof of ergodic theorems.

[58](Page 128) The decreasing martingale theorem may be interpreted as a rather general pointwise ergodic theorem. For example, Theorem 5.8 applied to Example 5.7 gives a pointwise ergodic theorem for the action of the $\mathbb{Z}[\frac{1}{2}]/\mathbb{Z}$ by translations on $\mathbb{T}$, where the ergodic averages are taken over sets of the form $\{\frac{j}{2^n} \mid n \leqslant N\}$. Indeed, Theorem 5.8 may be interpreted in this way as a pointwise ergodic theorem for measure-preserving actions of any group that can be written as a countable increasing union of compact open subgroups.

[59](Page 144) This lemma is used by Heinemann and Schmitt [146] in developing a Rokhlin lemma without the assumption of invertibility (see note (34) to Sect. 2.9 on p. 68).

# Chapter 6
# Factors and Joinings

A central question in ergodic theory is the natural classification one: When are two measure-preserving systems measurably isomorphic? A more refined approach is to ask what kind of internal structures two measure-preserving systems might share. One approach is to look at their *factors*, another more fruitful approach[60] to this problem is to look for their *joinings*, which will be defined and studied in this chapter. In order to discuss these two notions further, we apply some of the measure-theoretic machinery from Chap. 5 to the setting of measure-preserving systems.

## 6.1 The Ergodic Theorem and Decomposition Revisited

We start this chapter by using the results from Chap. 5 to prove the existence of the ergodic decomposition.

Let $(X, \mathscr{B}, \mu, T)$ be a measuring-preserving system on a Borel probability space. Write $\mathscr{E} = \{B \in \mathscr{B} \mid T^{-1}B = B \pmod{\mu}\}$ for the $\sigma$-algebra of (almost) $T$-invariant sets. Ergodicity of $T$ is equivalent to the triviality of $\mathscr{E}$, that is to the property that $\mathscr{E}$ consists only of null and co-null sets. Comparing the pointwise ergodic theorem (Theorem 2.30) with the definition of conditional expectation with respect to $\mathscr{E}$ gives the following reformulation.

**Theorem 6.1.** *Let $(X, \mathscr{B}, \mu, T)$ be a measure-preserving system and $f \in L^1_\mu$. Then*

$$\frac{1}{M} \sum_{n=0}^{M-1} f \circ T^n \longrightarrow E(f \mid \mathscr{E})$$

*almost everywhere and in $L^1$, where $\mathscr{E} = \{B \in \mathscr{B} \mid T^{-1}B \underset{\mu}{=} B\}$. In particular, if $T$ is invertible, then*

M. Einsiedler, T. Ward, *Ergodic Theory*, Graduate Texts in Mathematics 259,
DOI 10.1007/978-0-85729-021-2_6, © Springer-Verlag London Limited 2011

$$\lim_{M \to \infty} \frac{1}{M} \sum_{n=0}^{M-1} f(T^n x) = \lim_{M \to \infty} \frac{1}{M} \sum_{n=0}^{M-1} f(T^{-n} x)$$

*almost everywhere.*

The final conclusion of Theorem 6.1 means that the future and the past of typical points are similar (see Exercise 6.1.1 for a similar conclusion in topological dynamics).

PROOF OF THEOREM 6.1. By Theorem 2.30 the sequence converges to some $f^* \in L^1$ both pointwise and in $L^1$. Now $f^*$ is $\mathscr{E}$-measurable since $f^*$ is $T$-invariant. Moreover, for any set $A \in \mathscr{E}$ with positive measure, we may apply Theorem 2.30 to the restricted measure-preserving system

$$\left( A, \mathscr{B}|_A, \frac{1}{\mu(A)} \mu|_A, T \right)$$

to deduce that $\int_A f^* \, d\mu = \int_A f \, d\mu$. Thus $f^*$ satisfies the two characterizing properties in Theorem 5.1(1), so $f^* = E(f|\mathscr{E})$ almost everywhere. The final assertion follows from the observation that $\mathscr{E}$ can be defined using the map $T$ or the map $T^{-1}$ if $T$ is invertible.

$\square$

The ergodic decomposition result (Theorem 4.8 on p. 103) was seen for a continuous map $T$ as a consequence of Choquet's theorem. We now show how to deduce this important result from properties of conditional measures for any measurable map $T$.

**Theorem 6.2.** *Let* $T : (X, \mathscr{B}, \mu) \to (X, \mathscr{B}, \mu)$ *be a measure-preserving map of a Borel probability space. Then there is a Borel probability space* $(Y, \mathscr{B}_Y, \nu)$ *and a measurable map* $y \mapsto \mu_y$ *for which*

- $\mu_y$ *is a $T$-invariant ergodic probability measure on $X$ for almost every $y$, and*
- $\mu = \int_Y \mu_y \, d\nu(y)$.

*Moreover, we can require that the map* $y \mapsto \mu_y$ *is injective, or alternatively set*

$$(Y, \mathscr{B}_Y, \nu) = (X, \mathscr{B}, \mu)$$

*and* $\mu_x = \mu_x^{\mathscr{E}}$.

PROOF. Choose (by Lemma 5.17) a countably-generated $\sigma$-algebra $\widetilde{\mathscr{E}} \subseteq \mathscr{E}$ with $\widetilde{\mathscr{E}} \overset{\mu}{=} \mathscr{E}$ and write

$$\widetilde{\mathscr{E}} = \sigma \left( \{ E_1, E_2, \dots \} \right).$$

Let

$$N' = \bigcup_{i=1}^{\infty} T^{-1} E_i \triangle E_i = \bigcup_{E \in \widetilde{\mathscr{E}}} T^{-1} E \triangle E,$$

and let $N''$ be a null set off which the conclusions of Theorem 5.14 (with the choices $\mathscr{A} = \widetilde{\mathscr{E}}$ and $\widetilde{\mathscr{A}} = \mathscr{E}$ in the notation of Theorem 5.14(4)) and Corollaries 5.22 (for $\mathscr{A} = \widetilde{\mathscr{E}}$) and 5.24 (for $\mathscr{A} = \widetilde{\mathscr{E}}$ and $\phi = T$) hold. Let

$$N = \bigcup_{n=0}^{\infty} T^{-n} (N' \cup N'') ;$$

then $N$ is a null set containing $N'$ and $N''$ with $T^{-1}N \subseteq N$.

For $x \notin N$, $T^{-1}\mathscr{E} \underset{\mu}{=} \mathscr{E}$ and Corollary 5.24 together show that

$$T_* \mu_x^{\mathscr{E}} = \mu_{Tx}^{\mathscr{E}}.$$

Since $Tx \notin N$, $[x]_{\widetilde{\mathscr{E}}} = [Tx]_{\widetilde{\mathscr{E}}}$, so $\mu_{Tx}^{\mathscr{E}} = \mu_x^{\mathscr{E}}$ by Theorem 5.14. Thus $\mu_x^{\mathscr{E}}$ is $T$-invariant for any $x \notin N$. To prove the ergodicity of $\mu_x^{\mathscr{E}}$ we will use the following general criterion for ergodicity.

**Lemma 6.3.** *Let $(X, \mathscr{B}_X, \nu, T)$ be a measure-preserving system on a Borel probability space, and let $\{f_1, f_2, \dots\}$ be dense in $C(\overline{X})$. Then $\nu$ is ergodic if and only if*

$$\frac{1}{M} \sum_{n=0}^{M-1} f_i(T^n y) \longrightarrow \int f_i \, d\nu \tag{6.1}$$

*for $\nu$-almost every $y$ and all $i \geqslant 1$.*

PROOF. Ergodicity clearly implies the stated property. For the converse, recall that

$$\frac{1}{M} \sum_{n=0}^{M-1} f \circ T^n \xrightarrow[L_\nu^2]{} P_T f,$$

where $P_T$ denotes the projection operator onto the space of $U_T$-invariant functions in $L_\nu^2$ by Theorem 2.21. It follows that if (6.1) holds, we have $P_T(f) = \int f \, d\nu$ for a dense subset of functions $f \in L_\nu^2$, and therefore $P_T$ must be the projection onto the constant functions, which is equivalent to ergodicity. $\square$

Turning to the proof of the convergence (6.1), let $\{f_1, f_2, \dots\}$ be dense in $C(\overline{X})$, and enlarge the set $N$ to ensure that

$$\frac{1}{M} \sum_{n=0}^{M-1} f_i(T^n x) \longrightarrow E(f_i | \mathscr{E})(x) = \int f_i \, d\mu_x^{\mathscr{E}}$$

for $x \notin N$, for each $i \geqslant 1$, by Theorem 6.1. Theorem 5.14 implies that

$$N_1 = N \cup \{x \mid \mu_x^{\mathscr{E}}(N) > 0\}$$

has measure zero (see Remark 5.15(b)). If $[x]_{\widetilde{\mathscr{E}}} = [y]_{\widetilde{\mathscr{E}}}$ for $x \notin N_1, y \notin N$ then $\mu_x^{\mathscr{E}} = \mu_y^{\mathscr{E}}$, so

$$\frac{1}{M} \sum_{n=0}^{M-1} f_i(T^n y) \longrightarrow \int f_i \, \mathrm{d}\mu_x^{\mathscr{E}},$$

which shows the convergence (6.1) for the measure $\nu = \mu_x^{\mathscr{E}}$, since $\mu_x^{\mathscr{E}}(N) = 0$. We deduce that $\mu_x^{\mathscr{E}}$ is ergodic.

Finally, by Corollary 5.22 there is a map $\phi : X \to Y$ and a measure-valued measurable function $\nu_y \in \mathscr{M}(X)$ for $y \in Y$ with $\mu_x^{\mathscr{E}} = \nu_{\phi(x)}$; define $\nu = \phi_* \mu$. The theorem follows since

$$\mu = \int \mu_x^{\mathscr{E}} \, \mathrm{d}\mu(x) = \int \nu_{\phi(x)} \, \mathrm{d}\mu(x) = \int \nu_y \, \mathrm{d}\nu(y).$$

$\square$

## Exercises for Sect. 6.1

**Exercise 6.1.1.** Let $T : X \to X$ be a homeomorphism of a compact metric space, and let $\mu$ be a $T$-invariant Borel probability measure on $X$. Prove that for $\mu$-almost every $x \in X$ the forward and backward orbits of $x$ have the same closures, that is

$$\overline{\{T^n x \mid n \geqslant 1\}} = \overline{\{T^{-n} x \mid n \geqslant 1\}}$$

for almost every $x$, and that the orbit closure contains $x$.

**Exercise 6.1.2.** Let $T : X \to X$ be a measurable transformation on a compact metric space. Assume that there exists a $T$-invariant ergodic probability measure on $X$ without atoms. Let $\mathscr{E} = \{B : T^{-1}B = B\}$ be the $\sigma$-algebra of (strictly) invariant sets. Prove that $\mathscr{E}$ is not countably-generated. This shows that in general the switch from $\mathscr{E}$ to $\widetilde{\mathscr{E}}$ in the proof of Theorem 6.2 is necessary in order to talk about atoms (see Remark 5.15(e)).

## 6.2 Invariant Algebras and Factor Maps

The main result in this section is that a measurable factor of a measure-preserving system determines uniquely, and is uniquely determined by, an invariant sub-$\sigma$-algebra (an invariant sub-$\sigma$-algebra in a measure-preserving system $(X, \mathscr{B}, \mu, T)$ is a sub-$\sigma$-algebra $\mathscr{A} \subseteq \mathscr{B}_X$ with $T^{-1}\mathscr{A} = \mathscr{A}$ modulo $\mu$). It is important not to confuse the notion of an invariant sub-$\sigma$-algebra $\mathscr{A}$ with the specific sub-$\sigma$-algebra of invariant sets $\mathscr{E}$.

In one direction the relationship is clear: Given a factor map, it is easy to construct an invariant $\sigma$-algebra, and we will do this in Lemma 6.4 (the proof of this is Exercise 6.2.1). With a little more effort, it is possible to build a factor from an invariant $\sigma$-algebra. This idea really bears fruit when structural questions are asked: Given a system and a dynamical property, does it make sense to ask for the largest factor with that property?

**Lemma 6.4.** *Let $(X, \mathscr{B}_X, \mu, T)$ and $(Y, \mathscr{B}_Y, \nu, S)$ be invertible systems, and let $\phi : X \to Y$ be a factor map. Then $\mathscr{A} = \phi^{-1}\mathscr{B}_Y \subseteq \mathscr{B}_X$ is an invariant sub-$\sigma$-algebra in the sense that $T^{-1}\mathscr{A} \underset{\mu}{=} \mathscr{A}$.*

**Theorem 6.5.** *Let $(X, \mathscr{B}_X, \mu)$ be a Borel probability space, and let $T$ be a measure-preserving transformation on $X$. Assume furthermore that there is a $T$-invariant sub-$\sigma$-algebra $\mathscr{A} \subseteq \mathscr{B}_X$. Then there is a measure-preserving system $(Y, \mathscr{B}_Y, \nu, S)$ on a Borel probability space and a factor map $\phi : X \to Y$ with $\mathscr{A} = \phi^{-1}\mathscr{B}_Y$ (mod $\mu$). If $T$ is invertible then $S$ may be chosen to be invertible.*

PROOF. We are going to apply Corollary 5.22 with the choice $Y = \mathscr{M}(X)$ appearing in the proof. Let $S : Y \to Y$ be the map defined by $S\nu = T_*\nu$ for any $\nu \in Y$. By Lemma 5.23, $S$ is measurable. Define a map $\phi : X \to Y$ by $\phi(x) = \mu_x^{\mathscr{A}}$, and set $\nu = \phi_*\mu$; $\phi$ is measurable by Theorem 5.14(1). By Corollary 5.24,

$$T_*\mu_x^{T^{-1}\mathscr{A}} = \mu_{Tx}^{\mathscr{A}}$$

almost everywhere, and $T^{-1}\mathscr{A} \underset{\mu}{=} \mathscr{A}$ implies that

$$S(\phi(x)) = \phi(T(x)),$$

almost everywhere, so $\phi$ is a factor map and $S$ preserves $\nu$. Corollary 5.22 shows that

$$\mathscr{A} \underset{\mu}{=} \phi^{-1}\mathscr{B}_Y.$$

If $T$ is invertible, then $S^{-1} = \left(T^{-1}\right)_*$, so $S$ is invertible. ☐

*Example 6.6.* Let $(X, \mathscr{B}, \mu)$ be a Borel probability space, and let $T : X \to X$ be a measure-preserving transformation. The factor corresponding to the $\sigma$-algebra $\mathscr{E}$ of invariant sets is the largest factor on which the induced transformation acts trivially (that is, as the identity) in the sense that any other factor with this property is a factor of this one.

## Exercises for Sect. 6.2

**Exercise 6.2.1.** Prove Lemma 6.4.

**Exercise 6.2.2.** Give the details of the proof that the map $S$ constructed in the proof of Theorem 6.5 is measurable.

**Exercise 6.2.3.** Prove the statement in Example 6.6.

## 6.3 The Set of Joinings

In discussing joinings, it will be useful to emphasize systems instead of transformations, so we will sometimes denote $(X, \mathscr{B}_X, \mu, T)$ by $\mathsf{X}$ and $(Y, \mathscr{B}_Y, \nu, S)$ by $\mathsf{Y}$.

For Borel spaces $(X, \mathscr{B}_X)$ and $(Y, \mathscr{B}_Y)$, denote by

$$(X \times Y, \mathscr{B}_X \otimes \mathscr{B}_Y)$$

the product Borel space; $\mathscr{B}_X \otimes \mathscr{B}_Y$ is the smallest $\sigma$-algebra containing all the measurable rectangles $A \times B$ for $A \in \mathscr{B}_X$ and $B \in \mathscr{B}_Y$.

**Definition 6.7.** Let $(X, \mathscr{B}_X, \mu, T)$ and $(Y, \mathscr{B}_Y, \nu, S)$ be measure-preserving systems on Borel probability spaces. A measure $\rho$ on $(X \times Y, \mathscr{B}_X \otimes \mathscr{B}_Y)$ is a *joining* of the two systems if

- $\rho$ is invariant under $T \times S$;
- the projections of $\rho$ onto the $X$ and $Y$ coordinates are $\mu$ and $\nu$ respectively.

The second property means that

$$\rho(A \times Y) = \mu(A)$$

for all $A \in \mathscr{B}_X$, and

$$\rho(X \times B) = \nu(B)$$

for all $B \in \mathscr{B}_Y$, or equivalently that $(\pi_X)_* (\rho) = \mu$ and $(\pi_Y)_* (\rho) = \nu$ where $\pi_X$ and $\pi_Y$ denote the projections onto $X$ and $Y$.

Denote the set of joinings of $\mathsf{X} = (X, \mathscr{B}_X, \mu, T)$ and $\mathsf{Y} = (Y, \mathscr{B}_Y, \nu, S)$ by $\mathsf{J}(\mathsf{X}, \mathsf{Y})$. Notice that the trivial joining $\mu \times \nu$ is always a member of $\mathsf{J}(\mathsf{X}, \mathsf{Y})$, so the set of joinings is never empty. The ergodic components of any joining of ergodic systems are almost always ergodic, as shown in the next lemma. An alternative proof of the existence of ergodic joinings is given in Exercise 6.3.1.

**Lemma 6.8.** *Let* $\mathsf{X}$ *and* $\mathsf{Y}$ *be ergodic systems with* $\rho \in J(\mathsf{X}, \mathsf{Y})$. *Then almost every ergodic component of* $\rho$ *is an ergodic joining of* $\mathsf{X}$ *and* $\mathsf{Y}$.

Notice that even though Chap. 4 dealt with continuous transformations, the statement and proof of Theorem 4.4 deal with the setting of measure-preserving transformations, so we may use it here.

PROOF OF LEMMA 6.8. Suppose that

$$\rho = \int_Z \rho_z \, d\tau(z)$$

is the ergodic decomposition of $\rho$ from Theorem 6.2, for some probability space $(Z, \mathscr{B}_Z, \tau)$. Recall that $\pi_X : X \times Y \to X$ denotes the projection onto $X$. Then

$$\mu = (\pi_X)_* \rho = \int (\pi_X)_* \rho_z \, d\tau(z) \tag{6.2}$$

is a decomposition of $\mu$. The second equality in (6.2) is a consequence of the definitions

$$\mu(B) = \rho(B \times Y) = \int \rho_z(B \times Y) \, d\tau(z) = \int (\pi_X)_* \rho_z(B) \, d\tau(z).$$

By ergodicity and Theorem 4.4, it follows that $\mu = (\pi_X)_* \rho_z$ for almost every $z \in Z$. By symmetry the same property holds for $Y$, which proves the lemma.                                                                              □

In Sect. 6.5 we will show that the study of joinings is at least as general as the study of factor maps: Every common factor of two systems X and Y gives rise to an element of J(X, Y) (see also Exercise 6.3.3).

## Exercises for Sect. 6.3

**Exercise 6.3.1.** Emulate the proof of Theorem 4.4 to show that J(X, Y) is a convex set, and that the extreme points are ergodic measures for $T \times S$. Deduce that there is always an ergodic joining of two ergodic systems. Give an example in which an ergodic joining cannot be a product of invariant measures on the two systems.

**Exercise 6.3.2.** Suppose X is ergodic and $\rho \in J(X, Y)$. Show that $\rho$ is the trivial joining if and only if $\rho$ is invariant under $T \times I : X \times Y \to X \times Y$ where $(T \times I)(x, y) = (Tx, y)$.

**Exercise 6.3.3.** Show that if Y is a factor of X, then there is a joining of X and Y which gives measure one to the graph of the factor map.

## 6.4 Kronecker Systems

Replacing the word "trivially" in Example 6.6 with other more interesting dynamical properties, and finding the corresponding $\sigma$-algebra, encompasses many of the deepest structural problems in ergodic theory. The first step in this program describes the largest factor on which a measure-preserving system behaves like a rotation of a compact group.

**Lemma 6.9.** *If* $(X, \mathscr{B}, \mu, T)$ *is ergodic then every eigenvalue of* $U_T$ *is simple, and the set of all eigenvalues of* $U_T$ *is a subgroup of* $\mathbb{S}^1$.

PROOF. If $U_T f = \lambda f$ then

$$\langle f, f \rangle = \langle U_T f, U_T f \rangle = \lambda \overline{\lambda} \langle f, f \rangle$$

so $\lambda \overline{\lambda} = 1$. If $f$ is an eigenfunction corresponding to the eigenvalue $\lambda$, then

$$U_T |f| = |U_T f| = |\lambda f| = |f|,$$

so by ergodicity $|f|$ is a non-zero constant almost everywhere.

Thus if $U_T f_1 = \lambda_1 f_1$ and $U_T f_2 = \lambda_2 f_2$ then $U_T (f_1/f_2) = (\lambda_1/\lambda_2)(f_1/f_2)$, so $f_1/f_2$ is also an eigenfunction with eigenvalue $\lambda_1/\lambda_2$. It follows that the set of eigenvalues is a subgroup of $\mathbb{S}^1$. Finally, if $\lambda_1 = \lambda_2$ then, by ergodicity, $f_1/f_2$ is a constant so each eigenvalue is simple. $\square$

**Theorem 6.10 (Kronecker factors).** *Let* $(X, \mathscr{B}, \mu, T)$ *be an invertible ergodic measure-preserving system on a Borel probability space, and let* $\mathscr{A}$ *be the smallest* $\sigma$-*algebra with respect to which all* $L^2_\mu$ *eigenfunctions of* $U_T$ *are measurable. Then the corresponding factor* $(Y, \mathscr{B}_Y, \nu, S)$ *is the largest factor of* $(X, \mathscr{B}, \mu, T)$ *which is isomorphic to a rotation* $R_a(y) = y + a$ *on some compact abelian group* $Y$.

It will transpire that the compact group $Y$ is *monothetic*—that is, it is the closure of the subgroup generated by a single element—and indeed $Y$ is the closure of the subgroup $\{a^n \mid n \in \mathbb{Z}\}$ for some $a$ in the multiplicative infinite torus $(\mathbb{S}^1)^\mathbb{N}$. In particular, $Y$ is automatically a metrizable group, so $(Y, \mathscr{B}_Y, \nu)$ is a Borel probability space.

PROOF OF THEOREM 6.10. Let $\{\chi_i \mid i \in \mathbb{N}\} \subseteq L^2_\mu$ be an enumeration[*] of the eigenfunctions of $U_T$ normalized so that $|\chi_i| = 1$ almost everywhere, and let $U_T \chi_i = \lambda_i \chi_i$. Define a map $\phi : X \to \mathbb{S}^\mathbb{N}$ by

$$\phi(x) = (\chi_1(x), \chi_2(x), \dots),$$

and let $a = (\lambda_1, \lambda_2, \dots)$. Then it is clear that

$$R_a \phi(x) = \phi(Tx)$$

almost everywhere, so $\nu = \phi_* \mu$ is $R_a$-invariant and ergodic (because it is a factor of an ergodic transformation; see Exercise 2.3.4). It follows that $\nu$ is

---

[*] By Lemma B.8 the space $C_{\mathbb{C}}(X)$ is separable, and hence $L^2_\mu$ is separable since the inclusion $C_{\mathbb{C}}(X) \to L^2_\mu$ is continuous with dense image. It follows that the set of eigenvalues for $U_T$ must be countable, since it is an orthonormal set.

the Haar measure for the subgroup $Y = \overline{\langle a^n \rangle} \leqslant \mathbb{S}^{\mathbb{N}}$ (see also Theorem 4.14). It is clear that the resulting factor is the largest with the required property, since it corresponds to the smallest sub-$\sigma$-algebra with the required property, and any ergodic rotation factor will be generated by eigenfunctions. $\square$

Theorem 6.10 suggests that dynamical systems of the sort exhibited— rotations of compact groups—have many special properties. In this section we show that they have a very prescribed measurable structure.

**Definition 6.11.** A measure-preserving system $(X, \mathscr{B}_X, \mu, T)$ with the property that the linear span of the eigenfunctions of $U_T$ in $L^2_\mu$ is dense in $L^2_\mu$ is said to have *discrete spectrum*.

We have seen in Lemma 6.9 that the eigenvalues of the unitary operator associated to an ergodic measure-preserving transformation form a subgroup of $\mathbb{S}^1$. The next lemma shows the converse.

**Lemma 6.12.** *Given any countable subgroup $K \leqslant \mathbb{S}^1$ there is an ergodic measure-preserving system $(X, \mathscr{B}_X, \mu, T)$ on a Borel probability space with the property that $K$ is the group of eigenvalues of $U_T$.*

PROOF. Give $K$ the discrete topology, so that the dual group $X = \widehat{K}$ is a compact metric abelian group; write $\mu = m_X$ for the normalized Haar measure on $X$. The map $\theta : K \to \mathbb{S}^1$ defined by $\theta(\kappa) = \kappa$ is a character on $K$, hence $\theta$ is an element of $X$. Define $T : X \to X$ to be the rotation $T(x) = \theta \cdot x$ and let $\mathscr{B}_X$ be the Borel $\sigma$-algebra on $X$. Then $(X, \mathscr{B}_X, \mu, T)$ is a measure-preserving system, and we claim that it is ergodic (compare this argument with the proof of Proposition 2.16 on p. 26) and that the eigenvalues of $U_T$ comprise $K$.

By Pontryagin's theorem (Theorem C.12) the map $\kappa \mapsto f_\kappa$ defined by

$$f_\kappa(x) = x(\kappa)$$

for $\kappa \in K$, $x \in X$, is an isomorphism from $K$ to $\widehat{X}$.

For any character $f_\kappa \in \widehat{X}$ and $x \in X$,

$$(U_T f_\kappa)(x) = f_\kappa(\theta \cdot x) = f_\kappa(\theta) f_\kappa(x),$$

so $f_\kappa$ is an eigenfunction of $U_T$ with eigenvalue $f_\kappa(\theta)$. Now $f_\kappa(\theta) = \theta(\kappa) = \kappa$, so $K$ is a subgroup of the group of eigenvalues of $U_T$. The set $\widehat{X}$ is a complete orthonormal basis for $L^2_\mu$ (by Theorem C.11), so any eigenfunction $f$ of $U_T$ with eigenvalue $\lambda$ can be written $f = \sum_{\kappa \in K} c_\kappa f_\kappa$ for Fourier coefficients $c_\kappa \in \mathbb{C}$. Then (all equalities are in $L^2_\mu$)

$$(U_T f)(x) = \sum_{\kappa \in K} c_\kappa f_\kappa(\theta \cdot x)$$

$$= \sum_{\kappa \in K} c_\kappa f_\kappa(\theta) f_\kappa(x)$$

$$= \sum_{\kappa \in K} c_\kappa \kappa f_\kappa(x)$$

$$= \lambda f(x) \qquad \text{(since } U_T f = \lambda f)$$

$$= \sum_{\kappa \in K} \lambda c_\kappa f_\kappa(x),$$

so $c_\kappa \kappa = \lambda c_\kappa$ for all $\kappa \in K$. This implies that $c_\kappa = 0$ unless $\kappa = \lambda$, so we must have $f = c_\lambda f_\lambda$. Thus each eigenfunction of $U_T$ is a scalar multiple of a character of $X$. Moreover, each eigenvalue is simple, so $T$ is ergodic. $\qquad\square$

The following theorem, due to Halmos and von Neumann [139], is the simplest classification theorem for a class of measure-preserving systems; also see Exercise 6.4.2. The argument presented here is due to Lemańczyk [225] (see also the article by Thouvenot [359]).

**Theorem 6.13.** *Suppose that $(X, \mathscr{B}_X, \mu, T)$ and $(Y, \mathscr{B}_Y, \nu, S)$ are two ergodic measure-preserving systems with discrete spectrum. Then $T$ and $S$ are measurably isomorphic if and only if they have the same group of eigenvalues.*

PROOF. If $T$ and $S$ are measurably isomorphic then they have the same eigenvalues.

Conversely, let $K$ denote the group of eigenvalues of $U_T$, and assume this is also the group of eigenvalues of $U_S$. We wish to show that any ergodic joining is actually a joining supported on the graph of an isomorphism. By Lemma 6.8 we may choose an ergodic joining $\lambda \in J(\mathsf{X}, \mathsf{Y})$. For each $\kappa \in K$ there are functions $f \in L^2_\mu(X)$ and $g \in L^2_\nu(Y)$ with $U_T f = \kappa f$ and $U_S g = \kappa g$. Write $\overline{f} = f \otimes 1$ for the function on $X \times Y$ defined by $\overline{f}(x, y) = f(x)$ (and similarly define $\overline{g} = 1 \otimes g$). Then $U_{T \times S} \overline{f} = \kappa \overline{f}$ and $U_{T \times S} \overline{g} = \kappa \overline{g}$, so $\overline{f}$ and $\overline{g}$ are eigenfunctions for the ergodic system $(X \times Y, T \times S, \lambda)$ with the same eigenvalue. It follows by Lemma 6.9 that there is some $c \in \mathbb{C}$ with $\overline{f} = c\overline{g}$ modulo $\lambda$. Since the eigenfunctions span a dense set in $L^2_\mu$ and in $L^2_\nu$, this implies that

$$L^2_\mu \otimes \mathbb{C} \underset{\lambda}{=} \mathbb{C} \otimes L^2_\nu \subseteq L^2_\lambda. \tag{6.3}$$

By (6.3) there is a set $G \subseteq X \times Y$ of full $\lambda$-measure with the property that

$$(\{\varnothing, X\} \otimes \mathscr{B}_Y)\big|_G = (\mathscr{B}_X \otimes \{\varnothing, Y\})\big|_G.$$

We claim that $G$ is the graph of an isomorphism between $\mathsf{X}$ and $\mathsf{Y}$. To see this, consider the projection map

$$\pi_Y : X \times Y \to Y,$$

which is $\{\varnothing, X\} \otimes \mathscr{B}_Y$-measurable. Now $\pi_Y|_G$ is $\pi_X^{-1}(\mathscr{B}_X) = \mathscr{B}_X \otimes \{\varnothing, Y\}$-measurable, so by Lemma 5.25 there exists a measurable map $\phi : X \to Y$ such that $\phi \circ \pi_X|_G = \pi_Y|_G$, or equivalently with $G \subseteq \{(x, \phi(x)) \mid x \in X\}$ (as illustrated in Fig. 6.1).

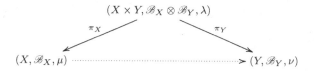

**Fig. 6.1** Constructing the measurable map $\phi$

The argument is symmetrical with respect to $X$ and $Y$, so there is also a measurable map $\psi : Y \to X$ such that $G \subseteq \{(\psi(y), y) \mid y \in Y\}$. The set $\{x \in X \mid \psi(\phi(x)) = x\}$ is clearly measurable, and the pre-image of this set under $\pi_X$ contains $G$ and so has full $\mu$-measure (since $G$ has full $\lambda$-measure and $(\pi_X)_* \lambda = \mu$). Similarly, it follows that $\phi \circ \psi(y) = y$ almost everywhere and that $\phi \circ T = S \circ \phi$ almost everywhere. Thus $\phi$ is an isomorphism between the measure-preserving systems $\mathsf{X}$ and $\mathsf{Y}$. $\qquad\square$

## Exercises for Sect. 6.4

**Exercise 6.4.1.** Prove that a rotation on a compact abelian group has discrete spectrum.

**Exercise 6.4.2.** Prove Theorem 6.13 by using the method in the proof of Theorem 6.10 simultaneously for the two systems.

**Exercise 6.4.3.** Finish the proof of Theorem 6.13 by showing more carefully that $\phi$ is an isomorphism between the measure-preserving systems $\mathsf{X}$ and $\mathsf{Y}$.

## 6.5 Constructing Joinings

One extreme possibility is that a pair of systems may have no joinings apart from the one that always has to exist.

**Definition 6.14.** Measure-preserving systems $\mathsf{X}$ and $\mathsf{Y}$ are *disjoint* if

$$\mathsf{J}(\mathsf{X}, \mathsf{Y}) = \{\mu \times \nu\}.$$

In this case we write $\mathsf{X} \perp \mathsf{Y}$.

In particular, if $X \perp Y$ then $L_0^2(X)$ is orthogonal to $L_0^2(Y)$ as subsets of the Hilbert space $L_0^2(X \times Y, \rho)$ for any joining $\rho$ (as $\mu \times \nu$ is the only joining this is easy to check). Moreover, the sets of eigenvalues of $X$ and of $Y$ are disjoint (apart from the trivial eigenvalue 1 corresponding to the constant functions). This follows from the results of this section: if $\lambda \in \mathbb{S}^1 \setminus \{1\}$ is an eigenvalue of both $X$ and of $Y$, then $X$ and $Y$ have a non-trivial common factor given by the map $x \mapsto \lambda x$ on the closed compact group $\overline{\{\lambda^n \mid n \in \mathbb{Z}\}}$ (see Exercise 6.5.3).

If there is a measurable isomorphism $\phi : X \to Y$ between $X$ and $Y$ then the graph of the isomorphism supports a joining $\rho_\phi$ which is characterized by the property that

$$\rho(B) = \mu \left( \{ x \in X \mid (x, \phi(x)) \in B \} \right)$$
$$= \nu \left( \{ y \in Y \mid (\phi^{-1}(y), y) \in B \} \right)$$

for all $B \in \mathscr{B}_X \otimes \mathscr{B}_Y$.

A more subtle construction of a joining of two systems is the *relatively independent joining* over a common factor. This interpolates between two extremes: the product joining, which is always there and says nothing about any shared measurable structures between the two systems, and at the other extreme the joining induced by the isomorphism between one system and an isomorphic system, which reflects the common structure. The next definition shows how to construct the relatively independent joining, which is illustrated[61] in Fig. 6.2.

For a factor map $\phi_X : X \to Z$ and the $\sigma$-algebra $\mathscr{A}_X = \phi_X^{-1} \mathscr{B}_Z$, we will use the convenient notation $\mu_{\phi_X(x)} = \mu_x^{\mathscr{A}_X}$ using Lemma 5.25 (and the comment after that lemma).

**Definition 6.15.** Let $X = (X, \mathscr{B}_X, \mu, T)$ and $Y = (Y, \mathscr{B}_Y, \nu, S)$ be invertible measure-preserving systems on Borel probability spaces, and assume that $X$ and $Y$ have a common non-trivial measurable factor $Z = (Z, \mathscr{B}_Z, \lambda, R)$. Then the *relatively independent joining*, denoted $\mu \times_\lambda \nu$ or $X \times_Z Y$ is the joining constructed as follows.

Denote the factor maps by $\phi_X : X \to Z$ and $\phi_Y : Y \to Z$, write

$$\mathscr{A}_X = \phi_X^{-1} \mathscr{B}_Z, \quad \mathscr{A}_Y = \phi_Y^{-1} \mathscr{B}_Z.$$

Then $\mathscr{A}_X \subseteq \mathscr{B}_X$ and $\mathscr{A}_Y \subseteq \mathscr{B}_Y$ are invariant $\sigma$-algebras; write

$$\mu_x^{\mathscr{A}_X} = \mu_{\phi_X(x)}$$

and

$$\nu_y^{\mathscr{A}_Y} = \nu_{\phi_Y(y)}$$

for the corresponding conditional measures. Define a measure $\rho$ on $X \times Y$ by

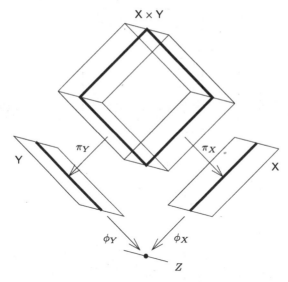

X × Y

$\pi_Y$

$\pi_X$

Y

X

$\phi_Y$

$\phi_X$

Z

**Fig. 6.2** The relatively independent joining $X \times_Z Y$ with projections $\pi_X$ and $\pi_Y$

$$\rho = \int_Z \mu_z \times \nu_z \, d\lambda(z).$$

It is straightforward to check that $\rho \in J(X, Y)$ (cf. Corollaries 5.22 and 5.24): for example, if $B \in \mathscr{B}_X$ then

$$\rho(B \times Y) = \int \mu_z(B) \cdot 1 \, d\lambda(z) = \int \mu_x^{\mathscr{A}}(B) \, d\mu(x) = \mu(B),$$

and if $A \in \mathscr{B}_X, B \in \mathscr{B}_Y$ then

$$\rho((T \times S)^{-1}(A \times B)) = \int_Z \mu_z \times \nu_z(T^{-1}A \times S^{-1}B) \, d\lambda(z)$$

$$= \int_Z \mu_z(T^{-1}A)\nu_z(S^{-1}B) \, d\lambda(z)$$

$$= \int_Z \mu_{Rz}(A)\nu_{Rz}(B) \, d\lambda(z)$$

$$= \rho(A \times B).$$

The basic properties of the relatively independent joining are as follows.

**Proposition 6.16.** *Let $\rho$ be the relatively independent joining of the invertible systems $X$ and $Y$ over a common factor $Z$ as above. Then the following properties hold.*

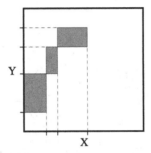

**Fig. 6.3** The union of the rectangles $\{(x,y) \mid \phi_X(x) = \phi_Y(y) = z\}$ for $z \in Z$

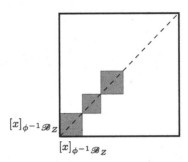

**Fig. 6.4** The relatively independent joining when $X = Y$ and $\phi_X = \phi_Y$

(1) *The relatively independent joining is concentrated on the measurable set*

$$\Phi = \{(x,y) \mid \phi_X(x) = \phi_Y(y)\}.$$

*That is, $\rho(\Phi) = 1$ (cf. Fig. 6.3).*
(2) *If $Z$ is non-trivial (that is, there is no $z_0 \in Z$ with $\lambda(\{z_0\}) = 1$), then $\rho$ is not the trivial joining $\mu \times \nu$.*
(3) *Given functions $f \in L^\infty(X, \mu)$ and $g \in L^\infty(Y, \nu)$, the conditional expectations $E(f | (\phi_X)^{-1} \mathscr{B}_Z)$ and $E(g | (\phi_Y)^{-1} \mathscr{B}_Z)$ can be viewed as functions on $Z$ by Lemma 5.25. In this sense,*

$$\int f(x)g(y)\,\mathrm{d}\rho(x,y) = \int E(f | (\phi_X)^{-1}\mathscr{B}_Z)E(g | (\phi_Y)^{-1}\mathscr{B}_Z)\,\mathrm{d}\lambda.$$

(4) *We have*

$$\mathscr{C} = (\phi_X \pi_X)^{-1}(\mathscr{B}_Z) \underset{\rho}{=} (\phi_Y \pi_Y)^{-1}\mathscr{B}_Z,$$

*the conditional measures for $\mathscr{C}$ are given by*

$$
\begin{aligned}
\rho_{(x,y)}^{\mathscr{C}} &= \mu_x^{(\phi_X)^{-1}\mathscr{B}_Z} \times \nu_y^{(\phi_Y)^{-1}\mathscr{B}_Z} \quad \textit{(for a.e. $(x,y) \in \Phi$)} \\
&= \mu_z \times \nu_z \quad \textit{(for a.e. $z$)} \\
&= \phi_X(x) = \phi_Y(y) \in Z,
\end{aligned}
$$

*and the atoms for $\mathscr{C}$ (after restriction to $\Phi$) are $(\phi_X)^{-1}(z) \times (\phi_Y)^{-1}(z)$ (cf. Figs. 6.3 and 6.4).*

If $X$ and $Y$ are metric spaces and $\phi_X$, $\phi_Y$ are continuous maps, then the set $\Phi$ in Proposition 6.16(1) is closed, and so $\rho$ has support in $\Phi$. In order to follow the proof of Proposition 6.16, the simplified form

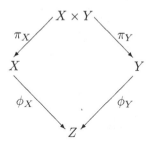

of Fig. 6.2 may be helpful. This diagram clearly commutes if one replaces the set $X \times Y$ by $\Phi$, which is of full $\rho$-measure. Hence it can be viewed as a commutative diagram between the corresponding probability spaces.

PROOF OF PROPOSITION 6.16. Property (1) follows easily by substitution:

$$\rho(\Phi) = \int \mu_z \times \nu_z(\Phi) \, \mathrm{d}\lambda(z) = 1$$

since $\mu_z\left((\phi_X)^{-1}(z)\right) = \nu_z\left((\phi_Y)^{-1}(z)\right) = 1$ for almost every $z \in Z$.

To see (2), let $C \in \mathscr{B}_Z$ be a set with $0 < \lambda(C) < 1$. Then

$$B = (\phi_X)^{-1}(C) \times (\phi_Y)^{-1}(Z \setminus C)$$

has $\rho(B) = 0$ by (1) but $\mu \times \nu(B) = \lambda(C) \cdot (1 - \lambda(C)) > 0$, so $\rho$ is not the trivial joining.

The equation in (3) may be found by integrating $f(x)g(y)$ against the measure

$$\rho = \int \mu_z \times \nu_z \, \mathrm{d}\lambda(z)$$

and interpreting the integral $\int f \, \mathrm{d}\mu_z$ (resp. $\int g \, \mathrm{d}\nu_z$) as $E\left(f \big| (\phi_X)^{-1}\mathscr{B}_Z\right)$ (resp. $E\left(g \big| (\phi_Y)^{-1}\mathscr{B}_Z\right)$). Here we use Lemma 5.25 in the form

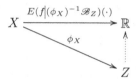

in order to interpret $\int f \, \mathrm{d}\mu_z$.

Finally, the equation

$$(\phi_X \pi_X)^{-1} \mathscr{B}_Z \underset{\rho}{=} (\phi_Y \pi_Y)^{-1} \mathscr{B}_Z$$

follows from (1): if

$$B_X = \{(x,y) \in X \times Y \mid \phi_X(x) \in C\} \in (\phi_X \pi_X)^{-1} \mathscr{B}_Z$$

for some $C \in \mathscr{B}_Z$, then $B_X \cap \Phi = B_Y \cap \Phi$, where

$$B_Y = \{(x,y) \in X \times Y \mid \phi_Y(y) \in C\}.$$

Write $\mathscr{C} = (\phi_X \pi_X)^{-1} \mathscr{B}_Z$. It is clear that

$$[(x,y)]_{\mathscr{C}} = \phi_X^{-1}\phi_X(x) \times Y$$

and that

$$[(x,y)]_{\mathscr{C}} \cap \Phi = \phi_X^{-1}(z) \times \phi_Y^{-1}(z)$$

whenever $(x,y) \in \Phi$ and $\phi_X(x) = \phi_Y(y) = z$.

To see that $\rho_{(x,y)} = \mu_{\phi_X(x)} \times \nu_{\phi_Y(y)}$ for $(x,y) \in \Phi$ defines a choice of the conditional measures of $\rho$ with respect to $\mathscr{C}$ we use Proposition 5.19. First notice that the map $(x,y) \mapsto \mu_{\phi_X(x)} \times \nu_{\phi_Y(y)}$ is measurable, constant on $\mathscr{C}$-atoms, and is supported on the atom $[(x,y)]_{\mathscr{C}}$; moreover

$$\int \mu_{\phi_X(x)} \times \nu_{\phi_Y(y)} \, d\rho(x,y) = \int \mu_{\phi_X(x)} \times \nu_{\phi_Y(y)} \, d(\mu_z \times \nu_z)(x,y) \, d\lambda(z)$$

$$= \int \mu_z \times \nu_z \, d\lambda(z) = \rho,$$

by the definition of $\rho$. This proves the theorem.                          $\square$

## Exercises for Sect. 6.5

**Exercise 6.5.1.** Describe the set of joinings between two circle rotations. When are they disjoint?

**Exercise 6.5.2.** Prove that any ergodic system is disjoint from any identity map.

**Exercise 6.5.3.** Show that if two measure-preserving systems have a common non-trivial eigenvalue, then they have a non-trivial common factor.

**Exercise 6.5.4.** Prove that if $X = (X, \mathscr{B}_X, \mu, T)$ is a weak-mixing system and $Y = (Y, \mathscr{B}_Y, \nu, S)$ is a Kronecker system, then $X \perp Y$.

**Exercise 6.5.5.** Under the hypotheses of Exercise 6.5.4, show that if $x$ is a generic point for $T$ in $X$, then any point $(x, y)$ with $y \in Y$ is a generic point in $\mathsf{X} \times \mathsf{Y}$.

**Exercise 6.5.6.** Use Exercise 6.5.5 to show a Wiener–Wintner ergodic theorem[62] for weak-mixing systems: if $(X, \mathscr{B}, \mu, T)$ is a weak-mixing system then, for $f \in L^1_\mu$ and for $\mu$-almost every $x$,

$$\frac{1}{N} \sum_{n=0}^{N-1} e^{2\pi i \theta n} f(T^n x)$$

converges for every $\theta$.

# Notes to Chap. 6

[60](Page 153) Joinings—and the associated ways of thinking about shared structures between measurable dynamical systems—were introduced by Furstenberg [100]. They were used by Rudolph [323] to produce dramatic examples of new phenomena in measurable dynamics, and they have played a central role ever since. The book of Rudolph [324] develops the theory of joinings, describes many of the important examples that can be constructed using them, and gives proofs of the Krieger generator theorem and the isomorphism theorem for Bernoulli shifts in the language of joinings. Glasner's monograph [116] is an attractive treatment of ergodic theory delivered in the language of joinings.

[61](Page 165) The illustration Fig. 6.2 of how the fiber measure on the square indicated in bold is the direct product is taken from Rudolph [324, Chap. 6].

[62](Page 169) This theorem was shown for ergodic systems by Wiener and Wintner [383]; what makes it a highly non-trivial extension of the pointwise ergodic theorem (Theorem 2.30) is that the null set is independent of $\theta$.

# Chapter 7
# Furstenberg's Proof of Szemerédi's Theorem

In 1927 van der Waerden proved[63], with the help of Artin and Schreier, a conjecture usually attributed to Baudet (or to Schur): if the natural numbers are written as a disjoint union of finitely many sets,

$$\mathbb{N} = C_1 \sqcup C_2 \sqcup \cdots \sqcup C_r,$$

then there must be one set $C_j$ that contains arbitrarily long arithmetic progressions. That is, there is some $j \in \{1, \ldots, r\}$ such that for any $k \geqslant 1$ there are $a \geqslant 1$ and $n \geqslant 1$ with

$$a, a+n, a+2n, \ldots, a+kn \in C_j. \tag{7.1}$$

The original proof appears in van der Waerden's paper [371], and there is a discussion of how he found the proof in [372]. We give an elementary inductive proof in Sect. 7.1.

A set $A \subseteq \mathbb{N}$ is said to have *positive upper Banach density* if

$$\overline{\mathbf{d}}_B(A) = \limsup_{N-M \to \infty} \frac{1}{N-M} \left| \left\{ a \in A \mid M \leqslant a < N \right\} \right| > 0.$$

Erdős and Turán [85] conjectured the stronger statement that any subset of $\mathbb{N}$ with positive upper Banach density must contain arbitrary long arithmetic progressions. This statement was shown for arithmetic progressions of length 3 (the case $k = 2$) by Roth [318] in 1952, then for length 4 ($k = 3$) by Szemerédi [356] in 1969. The general result (Theorem 1.5 on p. 4) was eventually proved[64] by Szemerédi [357] in 1975, in a lengthy and difficult argument. Furstenberg saw that Szemerédi's theorem would follow from a deep extension of Poincaré recurrence and proved that extension [102] (see also the article of Furstenberg, Katznelson and Ornstein [107]). In this chapter we will discuss the latter proof of Szemerédi's theorem, and some related results.

M. Einsiedler, T. Ward, *Ergodic Theory*, Graduate Texts in Mathematics 259, DOI 10.1007/978-0-85729-021-2_7, © Springer-Verlag London Limited 2011

## 7.1 Van der Waerden

We begin by giving a short proof of van der Waerden's theorem following the presentation of Graham and Rothschild [124]. This is not necessary for the proof of Szemerédi's theorem, but does provide an easier alternative proof of one step in the proof of Szemerédi's theorem (see Sect. 7.9.1). The formulation we prove is a finite version of the result (see Exercise 7.1.1).

**Theorem 7.1.** *Given* $\ell, r \geqslant 1$ *there is some* $N(\ell, r)$ *with the property that if* $N \geqslant N(\ell, r)$ *and*

$$\{1, 2, \ldots, N\} = C_1 \sqcup \cdots \sqcup C_r,$$

*then for some* $j$ *the set* $C_j$ *contains an arithmetic progression of length* $\ell$ *as in* (7.1).

In the proof, it will be convenient to write $\{a, a + 1, \ldots, b\}$ as $[a, b]$ and to define a coloring of a set $[1, N]$ into $r$ colors as a map $C : [1, N] \to \{1, \ldots, r\}$.

We define two integer vectors

$$(x_1, \ldots, x_m), (y_1, \ldots, y_m) \in [0, \ell]^m$$

to be *equivalent up to* $\ell$, denoted

$$(x_1, \ldots, x_m) \sim_\ell (y_1, \ldots, y_m)$$

if either $x_i, y_i < \ell$ for all $i = 1, \ldots, m$, or if for some $j \leqslant m$ we have $x_i = y_i$ for each $i \in [1, j]$ with $x_j = y_j = \ell$ and $x_i, y_i < \ell$ for all $i \in [j + 1, m]$. In other words the $\sim_\ell$-equivalence classes are $[0, \ell - 1]^m$ and, for every $j = 1, \ldots, m$ and every $(a_1, \ldots, a_{j-1}) \in [0, \ell]^{j-1}$, the set $\{(a_1, \ldots, a_{j-1}, \ell)\} \times [0, \ell - 1]^{m-j}$. These equivalence classes may be thought of as the index space for multi-dimensional arithmetic progressions.

The proof proceeds by an induction on a more general statement involving two parameters.

The statement $\boldsymbol{V}(\ell, m)$ is defined as follows: *For any* $r \geqslant 1$ *there is some* $N(\ell, m, r) \in \mathbb{N}$ *with the property that for any coloring*

$$C : [1, N(\ell, m, r)] \to \{1, \ldots, r\}$$

*there exist* $a, d_1, \ldots, d_m \in \mathbb{N}$ *with* $a + \sum_{i=1}^{m} \ell d_i \leqslant N(\ell, m, r)$ *and such that the function* $(x_1, \ldots, x_m) \mapsto C(a + \sum_{i=1}^{m} x_i d_i)$ *is constant on each* $\sim_\ell$-*equivalence class in* $[0, \ell]^m$.

Notice that $\boldsymbol{V}(\ell, 1)$ implies that for any coloring of $[1, N(\ell, 1, r)]$ into $r$ colors there is a monochromatic arithmetic progression of length $\ell$ (corresponding to the $\sim_\ell$-equivalence class $[0, \ell - 1]$), and so $\boldsymbol{V}(\ell, 1)$ for all $\ell \geqslant 1$ implies Theorem 7.1. In particular, it is certainly enough to prove $\boldsymbol{V}(\ell, m)$ for all $\ell, m \geqslant 1$.

**Lemma 7.2.** *The properties $V(\ell, 1)$ and $V(\ell, m)$ for some $m \geqslant 1$ together imply $V(\ell, m+1)$.*

PROOF. Fix $r$ and choose $M = N(\ell, m, r)$ and $M' = N(\ell, 1, r^M)$ using properties $V(\ell, m)$ and $V(\ell, 1)$. Let $C : [1, MM'] \to \{1, \ldots, r\}$ be any coloring of $[1, MM']$, and define a coloring

$$C_M : [1, M'] \to \{1, \ldots, r^M\}$$

with the property that $C_M(k) = C_M(k')$ if and only if

$$C(kM - j) = C(k'M - j)$$

for all $j$, $0 \leqslant j < M$. In other words, $C_M$ is the coloring of $[1, M']$ obtained by treating successive blocks of colors in $C$ of length $M$ as one color in $C_M$.

By the choice of $M'$ (which relied on the assumption $V(\ell, 1)$) there are $a', d' > 0$ such that $a' + \ell d' \leqslant M'$ and $C_M(a' + xd')$ is constant as $x$ varies in $[0, \ell - 1]$ (which in the coloring $C$ gives repetitions of a rather big block). Applying $V(\ell, m)$ to the shifted interval $[(a' - 1)M + 1, a'M]$ shows (by the choice of $M$) that there exists $a, d_1, \ldots, d_m > 0$ for which $a \geqslant (a' - 1)M + 1$ and $a + \sum_{i=1}^{m} \ell d_i \leqslant a'M$ and for which $C(a + \sum_{i=1}^{d} x_i d_i)$ is constant on all $\sim_\ell$-equivalence classes of $[0, \ell]^m$. We define $d_{m+1} = d'M$. This gives

$$a + \sum_{i=1}^{m+1} \ell d_i \leqslant a'M + \ell d'M \leqslant M'M$$

as required. Moreover, we claim that $C(a + \sum_{i=1}^{m+1} x_i d_i)$ is constant on all $\sim_\ell$-equivalence classes on $[0, \ell]^{m+1}$, which will give $V(\ell, m+1)$. For this, notice first that for any $b \in [1, M]$ we have

$$C((a' - 1)M + b + x_{m+1}d_{m+1}) = C((a' - 1)M + b)$$

for all $x_{m+1} \in [0, \ell - 1]$ (by our definition of $C_M$ and the choice of $a', d'$). Hence we obtain for $(x_1, \ldots, x_{m+1}) \in [0, \ell - 1]^{m+1}$ that

$$C\left(a + \sum_{i=1}^{m+1} x_i d_i\right) = C\left(a + \sum_{i=1}^{m} x_i d_i\right) = C(a),$$

where the latter equation holds by the choice of $a, d_1, \ldots, d_m$. This deals with the first equivalence class. The argument for the classes of the form

$$\{(a_1, \ldots, a_{j-1}, \ell)\} \times [0, \ell - 1]^{m+1-j}$$

for $j \leqslant m$ is similar. This leaves the equivalence classes of the form

$$\{(a_1, \ldots, a_m, \ell)\},$$

but as those are singletons there is nothing to show for them.                        □

**Lemma 7.3.** $V(\ell, m)$ *for all* $m \geqslant 1$ *implies* $V(\ell + 1, 1)$.

PROOF. Fix $r$, and let

$$C : [1, 2N(\ell, r, r)] \to \{1, \ldots, r\}$$

be given, where $N(\ell, r, r)$ is as in $V(\ell, r)$. Then there exist $a, d_1, \ldots, d_r > 0$ such that

$$a + \sum_{i=1}^{r} \ell d_i \leqslant N(\ell, r, r)$$

and with the property that $C(a + \sum_{i=1}^{r} x_i d_i)$ is constant on $\sim_\ell$ classes for $x_i$ in $[0, \ell]$.

By the pigeon-hole principle there must be some $u, v \in [0, r]$ with $u < v$ such that $C(a + \sum_{i=1}^{u} \ell d_i) = C(a + \sum_{i=1}^{v} \ell d_i)$, so

$$C\left( \left( a + \sum_{i=1}^{u} \ell d_i \right) + x \left( \sum_{i=u+1}^{v} d_i \right) \right)$$

is constant as $x$ varies in $[0, \ell]$. Finally, we also have

$$a + (\ell + 1) \sum_{i=u+1}^{v} d_i \leqslant 2N(\ell, r, r),$$

as required. This proves $V(\ell + 1, 1)$.                                            □

PROOF OF THEOREM 7.1. The statement $V(1, 1)$ is clear, so Lemmas 7.2 and 7.3 show $V(\ell, m)$ for all $\ell, m \geqslant 1$. As mentioned above, $V(\ell, 1)$ for all $\ell \geqslant 1$ gives the theorem.                                              □

# Exercises for Sect. 7.1

**Exercise 7.1.1.** Prove that van der Waerden's theorem as formulated in (7.1) follows from Theorem 7.1.

**Exercise 7.1.2.** Show that in Theorem 7.1 we may take $N(2, r) = r + 1$ for any $r \geqslant 1$, and try to find a value that we may take for $N(3, 2)$.

**Exercise 7.1.3.** Assume that $\mathbb{N} = C_1 \sqcup \cdots \sqcup C_r$, and say that a set $A \subseteq \mathbb{N}$ is monochrome if there is some $j = j(A)$ with $A \subseteq C_j$. Prove that the set of $n \in \mathbb{N}$ with the property that there is some $a \in \mathbb{Z}$ for which the $k$-term arithmetic progression $\{a, a + n, \ldots, a + (k - 1)n\}$ is monochrome is itself syndetic in $\mathbb{N}$ for any $k$ (see Exercise 2.5.5 on p. 36 for the definition of syndetic).

**Exercise 7.1.4.** Let $S = \{a_1, a_2, \ldots\} \subseteq \mathbb{N}$ with $(a_{n+1} - a_n)$ a positive bounded sequence. Show that for any $k \geqslant 1$ there is a $k$-step arithmetic progression in $S$. Find an example to show that such a set $S$ need not contain an infinite arithmetic progression.

## 7.2 Multiple Recurrence

Furstenberg saw that Szemerédi's theorem would be a consequence of a (then unknown) *multiple recurrence* result in ergodic theory, and went on to prove this recurrence property, opening up a significant new chapter in the conversation between ergodic theory and combinatorial number theory.

**Theorem 7.4 (Furstenberg).** *For any system $(X, \mathscr{B}, \mu, T)$ and set $A \in \mathscr{B}$ with $\mu(A) > 0$, and for any $k \in \mathbb{N}$,*

$$\liminf_{N \to \infty} \frac{1}{N} \sum_{n=1}^{N} \mu\left(A \cap T^{-n}A \cap T^{-2n}A \cap \cdots \cap T^{-kn}A\right) > 0. \qquad (7.2)$$

In fact Furstenberg proved that

$$\liminf_{N-M \to \infty} \frac{1}{N-M} \sum_{n=M}^{N-1} \mu\left(A \cap T^{-n}A \cap T^{-2n}A \cap \cdots \cap T^{-kn}A\right) > 0,$$

but the inequality (7.2) is sufficient for Szemerédi's theorem. An immediate consequence of Theorem 7.4 is the following generalization of Poincaré recurrence.

**Theorem 7.5 (Furstenberg).** *For any system $(X, \mathscr{B}, \mu, T)$ and set $A \in \mathscr{B}$ with $\mu(A) > 0$, and for any $k \in \mathbb{N}$ there is some $n \geqslant 1$ with*

$$\mu\left(A \cap T^{-n}A \cap T^{-2n}A \cap \cdots \cap T^{-kn}A\right) > 0. \qquad (7.3)$$

We shall see in the next section that Theorem 7.5 implies* Szemerédi's theorem. Theorem 7.4 is quite straightforward for certain measure-preserving systems.

*Example 7.6.* If $T = R_\alpha$ is an ergodic rotation, then the inequality (7.3) is clear: If $0, n\alpha, 2n\alpha, \ldots, kn\alpha \pmod 1$ are all very close together (which may be arranged for $\alpha \notin \mathbb{Q}$ and any $k \geqslant 1$ by choice of $n$), then the functions

---

* Theorem 7.4 gives more information about the set of possible differences $n$ in arithmetic progressions inside a set of positive density than Theorem 7.5 does; see Exercise 7.3.2.

$$\chi_A, \chi_{A-n\alpha}, \ldots, \chi_{A-kn\alpha}$$

will be close together in $L^2_{m_{\mathrm{T}}}$ for any interval (indeed, for any Borel set) $A$, so the intersection

$$A \cap R_\alpha^{-n} A \cap \cdots \cap R_\alpha^{-kn} A$$

will have measure close to that of $A$. More generally, an ergodic rotation on a compact abelian group satisfies the stronger statement (7.2) (see Proposition 7.12).

*Example 7.7.* If $T$ is a Bernoulli shift (cf. Example 2.9) then the inequality (7.2) is clear by the argument used in Proposition 2.15: the fact that a Bernoulli shift is mixing of order $(k+1)$ means that for any measurable set $A$, if $n$ is large enough then $A, T^{-n}A, \ldots, T^{-kn}A$ are almost independent sets, so their intersection has measure approximately $\mu(A)^{k+1} > 0$. In Proposition 7.13 we will show that this generalizes non-trivially to all weak-mixing systems.

We have seen that ergodic group rotations and Bernoulli shifts are at opposite extremes with respect to their mixing and spectral properties. The fact that Theorem 7.5 holds for two such opposite classes of systems, and for two completely complementary reasons (rotations because they do not move nearby points apart, Bernoulli shifts because they mix collections of sets up so thoroughly as to become asymptotically independent), could be interpreted as a hint that this multiple recurrence is a rather general phenomenon. On the other hand, the fact that multiple recurrence is visible in a circle rotation only along special values of $n$, while in a Bernoulli shift it is seen for all large $n$, already suggests that it is a very subtle phenomenon. Thus Theorem 7.4 holds for two extreme behaviors: discrete spectrum and weak-mixing. The strategy of Furstenberg's proof is to show that an arbitrary system is built up (in a manner to be described below) from these two classes, and the way in which the system is built up preserves the property expressed by the inequality (7.2).

**Definition 7.8.** A measure-preserving system $(X, \mathscr{B}, \mu, T)$ (or a factor system, represented by a $T$-invariant sub-$\sigma$-algebra $\mathscr{A} \subseteq \mathscr{B}$) is said to be SZ if

$$\liminf_{N \to \infty} \frac{1}{N} \sum_{n=1}^{N} \mu\left(A \cap T^{-n}A \cap T^{-2n}A \cap \cdots \cap T^{-kn}A\right) > 0$$

for all $k$ and $A \in \mathscr{B}$ ($A \in \mathscr{A}$ respectively) with $\mu(A) > 0$.

Three immediate simplifications can be made in proving that all measure-preserving systems are SZ; the first two are somewhat technical in that neither is needed for the setting in which Theorem 7.5 is used to prove Szemerédi's theorem (Theorem 1.5). The third is essential, and illustrates once again that the ergodic decomposition makes ergodicity a rather benign hypothesis.

### 7.2.1 Reduction to an Invertible System

The SZ property is preserved by taking the invertible extension of a measure-preserving system. Recall that if $\mathsf{X} = (X, \mathscr{B}, \mu, T)$ is any measure-preserving system, then the system $\widetilde{\mathsf{X}} = (\widetilde{X}, \widetilde{B}, \widetilde{\mu}, \widetilde{T})$ defined by

- $\widetilde{X} = \{x \in X^{\mathbb{Z}} \mid x_{k+1} = Tx_k \text{ for all } k \in \mathbb{Z}\}$;
- $(\widetilde{T}x)_k = x_{k+1}$ for all $k \in \mathbb{Z}$ and $x \in \widetilde{X}$;
- $\widetilde{\mu}(\{x \in X \mid x_0 \in A\}) = \mu(A)$ for any $A \in \mathscr{B}$, and $\widetilde{\mu}$ is invariant under $\widetilde{T}$;
- $\widetilde{\mathscr{B}}$ is the smallest $\widetilde{T}$-invariant $\sigma$-algebra for which the map $x \mapsto x_n$ from $\widetilde{X}$ to $X$ is measurable for all $n \in \mathbb{Z}$;

is called the invertible extension of $\mathsf{X}$. Then $\mathsf{X}$ has property SZ if and only if $\widetilde{\mathsf{X}}$ has property SZ (see Exercise 7.2.5).

### 7.2.2 Reduction to Borel Probability Spaces

Property SZ holds for all measure-preserving systems if it holds for all measure-preserving systems on Borel probability spaces. To see this, let $\mathsf{X}$ be any invertible measure-preserving system and fix $A \in \mathscr{B}$ with $\mu(A) > 0$. Then the factor map $\phi : X \to \{0,1\}^{\mathbb{Z}}$ defined by $\phi(x) = (\chi_A(T^n x))$ gives rise to a Borel probability system, and property SZ for each such factor shows property SZ for $\mathsf{X}$.

### 7.2.3 Reduction to an Ergodic System

Below, and in the rest of this chapter, we will use conditional measures $\mu_x^{\mathscr{A}}$ for invariant sub-$\sigma$-algebras $\mathscr{A}$ and their properties as developed in Sects. 5.3, 5.4 and Chap. 6. By Sect. 7.2.2, we may assume that $(X, \mathscr{B}, \mu)$ is a Borel probability space so that these results apply. In particular, we may apply Theorem 6.2 to show that it suffices to prove the SZ property for ergodic systems as follows. Assume that every ergodic system on a Borel probability space has property SZ, and let $(X, \mathscr{B}, \mu, T)$ be any measure-preserving system on a Borel probability space. By Theorem 6.2, the measure $\mu$ decomposes into ergodic components $\mu_x^{\mathscr{E}}$. Then for any set $A \in \mathscr{B}$ with $\mu(A) > 0$,

$$\mu\left(\{x \in X \mid \mu_x^{\mathscr{E}}(A) > 0\}\right) > 0, \tag{7.4}$$

so (by Fatou's Lemma)

$$\liminf_{N\to\infty} \frac{1}{N} \sum_{n=1}^{N} \mu\left(A \cap T^{-n}A \cap \cdots \cap T^{-kn}A\right)$$

$$= \liminf_{N\to\infty} \int \frac{1}{N} \sum_{n=1}^{N} \mu_x^{\mathscr{E}}\left(A \cap T^{-n}A \cap \cdots \cap T^{-kn}A\right) \, \mathrm{d}\mu(x)$$

$$\geqslant \int \underbrace{\liminf_{N\to\infty} \frac{1}{N} \sum_{n=1}^{N} \mu_x^{\mathscr{E}}\left(A \cap T^{-n}A \cap \cdots \cap T^{-kn}A\right)}_{> \, 0 \text{ on the set of positive measure in (7.4)}} \, \mathrm{d}\mu(x)$$

$$> 0$$

by the SZ property for ergodic systems.

## Exercises for Sect. 7.2

**Exercise 7.2.1.** Assuming Szemerédi's theorem (Theorem 1.5), prove the following finite version of Szemerédi's theorem: For every $k \geqslant 1$ and $\varepsilon > 0$ there is some $N$ with the property that any subset of $\{1, \ldots, N\}$ with more than $\lfloor \varepsilon N \rfloor$ elements contains an arithmetic progression of length $k$.

**Exercise 7.2.2.** [65] Prove the following topological analog of multiple recurrence, due to Furstenberg and Weiss, generalizing Birkhoff's recurrence theorem in Exercise 4.2.2. Let $(X, \mathsf{d})$ be a compact metric space and $T : X \to X$ a continuous map. Prove that for any $k \in \mathbb{N}$ and $\varepsilon > 0$ there is a point $x \in X$ and an $n \in \mathbb{N}$ for which

$$\mathrm{diam}\left(\{x, T^n(x), T^{2n}(x), \ldots, T^{kn}(x)\}\right) < \varepsilon.$$

**Exercise 7.2.3.** Show directly that an ergodic circle rotation is an SZ system.

**Exercise 7.2.4.** Show that if $\mathsf{X}$ is an invertible measure-preserving system, then $\widetilde{\mathsf{X}}$ (the invertible extension) is isomorphic to $\mathsf{X}$ (see Exercise 2.1.7).

**Exercise 7.2.5.** For each of the properties ergodicity, weak-mixing, mixing, and SZ, prove that a measure-preserving system has the property if and only if its invertible extension does (see Exercise 2.1.7).

## 7.3 Furstenberg Correspondence Principle

In this short section we show how Szemerédi's theorem (Theorem 1.5) follows from the multiple recurrence result in Theorem 7.5. This correspondence between multiple recurrence results in ergodic theory and statements in combinatorial number theory holds in great generality; we merely prove it for the case at hand. More general formulations may be found in Bergelson's notes [26] and Furstenberg's monograph [103].

PROOF OF THEOREM 1.5 ASSUMING THEOREM 7.5. Let $S \subseteq \mathbb{Z}$ be a set of positive upper Banach density, and let $k \geqslant 1$ be given. We claim that there exist integers $a$ and $n \geqslant 1$ with $\{a, a + n, \ldots, a + kn\} \subseteq S$.

Let $X_0 = \{0,1\}^{\mathbb{Z}}$ be the full shift on two symbols, given the compact product topology from the discrete topology on $\{0,1\}$, with shift map $\sigma_0$. Define a point $x_S$ in $X_0$ by

$$x_S(\ell) = \begin{cases} 1 \text{ if } \ell \in S; \\ 0 \text{ if } \ell \notin S. \end{cases}$$

Now let $X$ denote the smallest closed* subset of $X_0$ that is invariant under $\sigma_0$ and contains the point $x_S$. Write $\sigma_X$ for the shift $\sigma_0$ restricted to $X$.

Let $A$ denote the cylinder set $\{x \in X \mid x(0) = 1\}$, which is both closed and open (clopen) in $X$. Then

$$\sigma_X^{\ell}(x_S) \in A \iff \left(\sigma_X^{\ell}(x_S)\right)_0 = x_S(\ell) = 1 \iff \ell \in S. \qquad (7.5)$$

The upper Banach density of the set $S$ is positive, so there is a sequence of intervals $[a_1, b_1], [a_2, b_2], \ldots$ with $b_j - a_j \to \infty$ as $j \to \infty$ for which

$$\frac{|S \cap [a_j, b_j]|}{b_j - a_j} \longrightarrow \overline{d}_B(S) > 0$$

as $j \to \infty$. It follows that the measure $\mu_j$ on $X$ defined by

$$\mu_j = \frac{1}{b_j - a_j} \sum_{m=a_j}^{b_j} \delta_{\sigma_X^m(x_S)} \qquad (7.6)$$

has $\mu_j(A) \to \overline{d}_B(S) > 0$ as $j \to \infty$.

By Theorem 4.1 there is a sequence $j_m \to \infty$ with $\mu_{j_m} \to \mu$ in the weak* topology as $m \to \infty$, the measure $\mu$ is invariant under $\sigma_X$, and

$$\mu(A) = \overline{d}_B(S) > 0.$$

Apply Theorem 7.5 to the set $A$ in the system $(X, \mu, \sigma_X)$: there exists $n \geqslant 1$ such that

$$\mu\left(\bigcap_{i=0}^{k} \sigma^{-in}(A)\right) > 0. \qquad (7.7)$$

---

* This set is usually called the *orbit closure* of $x_S$; carrying out this construction for carefully chosen initial points allows for the construction of many important examples in ergodic theory. Notice that an element $x \in X_0$ lies in $X$ if and only if there is a sequence $(n_j)$ of integers with $\sigma_0^{n_j}(x_S) \to x$; equivalently, $x \in X_0$ lies in $X$ if and only if for every $j \geqslant 0$ the block $x_{[-j,j]}$ is seen somewhere in $x_S$.

Now for any clopen measurable set $B$, $\mu(B) > 0$ implies by (7.5) that there is an $a \in \mathbb{Z}$ for which $\sigma_X^a(x_S) \in B$. Thus (7.7) shows there is an $a \in \mathbb{Z}$ for which

$$\sigma_X^a(x_S) \in \bigcap_{i=0}^{k} \sigma_X^{-in}(A),$$

so

$$\sigma_X^{a+in}(x_S) \in A$$

for all $i$, $0 \leqslant i \leqslant k$, and hence $\{a, a+n, \ldots, a+kn\} \subseteq S$ by (7.5), as required.
□

## Exercises for Sect. 7.3

**Exercise 7.3.1.** Prove Theorem 7.5 assuming Theorem 1.5.

**Exercise 7.3.2.** Extend Theorem 1.5 in the following way (assuming Theorem 7.4). Let $S \subseteq \mathbb{Z}$ be a set of positive upper Banach density. Prove that for any $k \geqslant 1$ the set of $n \in \mathbb{N}$ with the property that there is an arithmetic progression of length $k$ in $S$ with common difference $n$ is itself a set of positive upper density.

## 7.4 An Instance of Polynomial Recurrence

To motivate* the deeper results, we begin with a result proved independently by Sárközy [329] and Furstenberg (see [102, Prop. 1.3] or [103, Th. 3.16]).

**Theorem 7.9.** *Let* $E \subseteq \mathbb{N}$ *be a set with positive upper Banach density, and let* $p$ *be a polynomial with integer coefficients with* $p(0) = 0$. *Then there exist* $x, y \in E$ *and an* $n \geqslant 1$ *with* $x - y = p(n)$.

That is, the set of *differences* of a set with positive upper density is such a rich set that it must intersect the range of any polynomial $p$ with $p(0) = 0$[66].
Just as in Sect. 7.3, the Furstenberg correspondence principle may be applied to show that Theorem 7.9 is a consequence of the following dynamical result due[67] to Furstenberg.

---

* This section (with the exception of Theorem 7.11) as well as the following Sects. 7.5 and 7.6 are logically not needed for the proof of Theorem 7.4, but we believe the arguments presented here help to understand the proof of Theorem 7.4 better.

**Theorem 7.10.** *If $p \in \mathbb{Z}[t]$ is a polynomial with $p(0) = 0$ then, for any measure-preserving system $(X, \mathscr{B}, \mu, T)$ and set $A \in \mathscr{B}$ with $\mu(A) > 0$, there is an $n \geqslant 1$ with $\mu(A \cap T^{-p(n)}A) > 0$.*

As we indicated in Sect. 7.2, many of the deeper recurrence results rely on different arguments for orderly systems (Kronecker systems, for example) and more chaotic systems (multiply mixing systems, for example). To obtain a full proof for any system we need to find a way to decompose a general measure-preserving system into orderly and chaotic parts. The details of how this is done depends on the result considered. For Theorem 7.10 this will not be too complicated, but the proof for Theorem 7.4 will need more work (see Sects. 7.10–7.11). Just as in Sect. 7.2, we may assume that $T$ is an ergodic, invertible, transformation of a Borel probability space.

PROOF OF THEOREM 7.10: SPLITTING INTO CHAOTIC AND ORDERLY PARTS. Write $L^2 = L_\mu^2$, and for each $a \geqslant 1$ define

$$\mathscr{H}_a = \{f \in L^2 \mid U_T^a f = f\}$$

and

$$\mathscr{V}_a = \left\{ f \in L^2 \mid \left\| \frac{1}{N} \sum_{n=0}^{N-1} U_T^{an} f \right\|_2 \to 0 \text{ as } N \to \infty \right\}.$$

The decomposition of $L^2$ into the space of invariant functions and its ortho-complement in Theorem 2.21 may be applied to $U_T^a$, giving

$$L^2 = \mathscr{H}_a \oplus \mathscr{V}_a$$

for each $a \geqslant 1$. Notice that the space $\mathscr{H}_a$ detects non-ergodicity of the $a$th iterate of $T$ ($\mathscr{H}_a \supseteq \mathbb{C}$, with equality if and only if $T^a$ is ergodic).

The measure-preserving transformation $T$ is called *totally ergodic* if $T^a$ is ergodic for all $a \geqslant 1$ (see Exercise 7.4.1). Define

$$\mathscr{H}_{\text{rat}} = \overline{\{f \in L^2 \mid U_T^a f = f \text{ for some } a \geqslant 1\}},$$

the *rational spectrum* component and

$$\mathscr{V}_* = \left\{ f \in L^2 \mid \left\| \frac{1}{N} \sum_{n=0}^{N-1} U_T^{an} f \right\|_2 \to 0 \text{ as } N \to \infty \text{ for all } a \geqslant 1 \right\},$$

the *totally ergodic* component. It may be readily checked that

- $\bigcup_{a=1}^{\infty} \mathscr{H}_a$ is dense in $\mathscr{H}_{\text{rat}}$ and $\bigcap_{a=1}^{\infty} \mathscr{V}_a = \mathscr{V}_*$;
- $\mathscr{H}_{\text{rat}}^{\perp} = \mathscr{V}_*$;
- $L^2 = \mathscr{H}_{\text{rat}} \oplus \mathscr{V}_*$.

Let $\chi_A = f + g$ be the unique decomposition with $f \in \mathscr{H}_{\mathrm{rat}}$ and $g \in \mathscr{V}_*$; similarly write $f_a$ for the orthogonal projection of $\chi_A$ onto the subspace $\mathscr{H}_a$. Since $\chi_A$ is a non-negative function with positive integral and $\mathscr{H}_{\mathrm{rat}}$ contains the constants, we have

$$\int f \, \mathrm{d}\mu = \int f_a \, \mathrm{d}\mu = \mu(A) > 0.$$

Moreover, since $f_a = E\left(\chi_A | \mathscr{E}_a\right)$ and $\chi_A \geqslant 0$, we have $f_a \geqslant 0$, where $\mathscr{E}_a$ is the $\sigma$-algebra of $T^a$-invariant sets (cf. Theorem 6.1). Finally, $f \geqslant 0$ since $f_a$ converges to $f$ as $a \to \infty$ by the increasing martingale theorem (Theorem 5.5). As in the case of multiple recurrence, we prove a stronger statement, namely that

$$\lim_{N \to \infty} \frac{1}{N} \sum_{n=0}^{N-1} \mu(A \cap T^{p(n)} A)$$

exists and is positive. Since the orthogonal decomposition $\chi_A = f + g$ is preserved by $U_T$,

$$\mu(A \cap T^{p(n)} A) = \int (f + g) U_T^{p(n)} (f + g) \, \mathrm{d}\mu = \int f U_T^{p(n)} f \, \mathrm{d}\mu + \int g U_T^{p(n)} g \, \mathrm{d}\mu,$$
(7.8)

so it is enough to consider the two components separately. □

PROOF OF THEOREM 7.10: ORDERLY PART. Consider the rational spectrum component $f \in \mathscr{H}_{\mathrm{rat}}$, where we claim that

$$\lim_{N \to \infty} \frac{1}{N} \sum_{n=0}^{N-1} \int f U_T^{p(n)} f \, \mathrm{d}\mu > 0.$$
(7.9)

If $f \in \mathscr{H}_a$ then the sequence of functions $(U_T^{p(n)} f)$ is periodic (since the sequence $(p(n))$ when reduced modulo $a$ is periodic), showing that the limit in (7.9) exists for functions $f$ in a dense subset of $\mathscr{H}_{\mathrm{rat}}$, and hence (by approximation and the Cauchy–Schwartz inequality) for any $f \in \mathscr{H}_{\mathrm{rat}}$.

We must now exclude the possibility that the limit is zero. Choose $a \geqslant 1$ with $\|f - f_a\|_2 < \varepsilon = \frac{1}{4}\mu(A)^2$. For $n \geqslant 1$, $p(an)$ is divisible by $a$ since $p(0) = 0$, and so $U_T^{p(an)} f_a = f_a$. It follows that

$$\int f_a U_T^{p(an)} f_a \, \mathrm{d}\mu = \int f_a^2 \, \mathrm{d}\mu \geqslant \left( \int f_a \cdot 1 \, \mathrm{d}\mu \right)^2 = \mu(A)^2$$

by the Cauchy–Schwartz inequality. Hence

$$\int f U_T^{p(an)} f \, d\mu = \left\langle f, U_T^{p(an)} f \right\rangle$$

$$= \langle f_a, f_a \rangle + \left\langle f_a, U_T^{p(an)} f - U_T^{p(an)} f_a \right\rangle + \left\langle f - f_a, U_T^{p(an)} f \right\rangle$$

$$\geqslant \mu(A)^2 - 2\varepsilon = \tfrac{1}{2}\mu(A)^2 > 0.$$

Recall that $f \geqslant 0$ so $U_T^{p(n)} f \geqslant 0$, which implies that

$$\lim_{N \to \infty} \frac{1}{N} \sum_{n=0}^{N-1} \int f U_T^{p(n)} f \, d\mu \geqslant \tfrac{1}{2a}\mu(A)^2 > 0.$$

$\square$

We complete the proof of Theorem 7.10 by using the Spectral Theorem (Theorem B.4) for the chaotic component.

PROOF OF THEOREM 7.10: CHAOTIC PART USING THE SPECTRAL THEOREM. Consider now the totally ergodic component $g \in \mathscr{V}_*$. We claim that

$$\lim_{N \to \infty} \frac{1}{N} \sum_{n=0}^{N-1} \left\langle U_T^{p(n)} g, g \right\rangle = 0. \tag{7.10}$$

This, together with (7.9) and (7.8), will complete the proof of Theorem 7.10. By (B.1), the claim in (7.10) is equivalent to the statement that

$$\lim_{N \to \infty} \int_{\mathbb{S}^1} \frac{1}{N} \sum_{n=0}^{N-1} z^{p(n)} \, d\mu_g(z) \longrightarrow 0 \tag{7.11}$$

as $N \to \infty$ (see Sect. B.3 for the notation). We will prove (7.11) using Weyl's equidistribution theorem (Theorem 1.4, proved in Sect. 4.4.3). If $z \in \mathbb{S}^1$ is not a root of unity, then $z^m = e^{2\pi i m \theta}$ for some $\theta \in \mathbb{R} \setminus \mathbb{Q}$ and so

$$\frac{1}{N} \sum_{n=0}^{N-1} z^{p(n)} = \frac{1}{N} \sum_{n=0}^{N-1} e^{2\pi i \theta p(n)} \longrightarrow \int_0^1 e^{2\pi i x} \, dx = 0$$

by Theorem 1.4. It is therefore enough to show that $\mu_g(\{z_0\}) = 0$ for any root of unity $z_0 \in \mathbb{S}^1$, for then the dominated convergence theorem (Theorem A.18) implies the claim.

Assume therefore that $\mu_g(\{z_0\}) > 0$ with $z_0 = e^{2\pi i b/a}$ for some $a, b$ integral, $a > 0$. It follows that $L^2_{\mu_g}(\mathbb{S}^1)$ contains the non-zero vector

$$\delta_{z_0}(z) = \begin{cases} 1 & \text{if } z = z_0; \\ 0 & \text{if not,} \end{cases}$$

which is fixed under multiplication by $z^a$, as

$$z^a \delta_{z_0} = z_0^a \delta_{z_0} = \delta_{z_0}.$$

However, by Theorem B.4(2), $L^2_{\mu_g}(\mathbb{S}^1)$ is unitarily isomorphic to the cyclic sub-representation of $L^2_\mu$ generated by $g$ under the unitary map $U_T$. Moreover, by the same theorem this isomorphism conjugates $U_T$ to multiplication by $z$. It follows that there exists some $h \in L^2_\mu$ which is fixed by $U^a_T$, with

$$\langle g, h \rangle_{L^2_\mu} = \langle 1, \delta_{z_0} \rangle_{L^2_{\mu_g}} = \mu_g(\{z_0\}) > 0$$

since the constant function $1 \in L^2_{\mu_g}$ is mapped to $g \in L^2_\mu$ under the unitary isomorphism in Theorem B.4(2). Since $h \in \mathscr{H}_a \subseteq \mathscr{H}_{\mathrm{rat}}$, this contradicts the fact that $g \in \mathscr{V}_* = \mathscr{H}^\perp_{\mathrm{rat}}$ by construction.                                                      □

### 7.4.1 The van der Corput Lemma

In the more complex situation needed for Szemerédi's theorem, we will not be able to use spectral theory (as in the proof above). Instead we will rely on repeated application of the following quite simple argument to handle the chaotic part of the systems arising. This is a version of the *van der Corput Lemma*.

**Theorem 7.11.** *Let $(u_n)$ be a bounded sequence in a Hilbert space $\mathscr{H}$. Define a sequence $(s_h)$ of real numbers by*

$$s_h = \limsup_{N \to \infty} \left| \frac{1}{N} \sum_{n=1}^{N} \langle u_{n+h}, u_n \rangle \right|.$$

*If*

$$\lim_{H \to \infty} \frac{1}{H} \sum_{h=0}^{H-1} s_h = 0,$$

*then*

$$\lim_{N \to \infty} \left\| \frac{1}{N} \sum_{n=1}^{N} u_n \right\| = 0.$$

PROOF. Fix $\varepsilon > 0$ and find $H_0$ such that for $H > H_0$,

$$\frac{1}{H} \sum_{h=0}^{H-1} s_h < \varepsilon. \tag{7.12}$$

Then for $N$ much larger than $H$ (depending on $\varepsilon$), the sums $\frac{1}{N} \sum_{n=1}^{N} u_n$ and $\frac{1}{NH} \sum_{n=1}^{N} \sum_{h=0}^{H-1} u_{n+h}$ differ only in the first few and last few terms, so

$$\left\| \frac{1}{N} \sum_{n=1}^{N} u_n - \frac{1}{N} \frac{1}{H} \sum_{n=1}^{N} \sum_{h=0}^{H-1} u_{n+h} \right\| < \varepsilon. \tag{7.13}$$

This allows us to switch attention and consider the double average instead of the single average. For this double average, notice that by the triangle inequality we have

$$\left\| \frac{1}{N} \frac{1}{H} \sum_{n=1}^{N} \sum_{h=0}^{H-1} u_{n+h} \right\| \leqslant \frac{1}{N} \sum_{n=1}^{N} \underbrace{\left\| \frac{1}{H} \sum_{h=0}^{H-1} u_{n+h} \right\|}_{a_n}.$$

Since the map $a \mapsto a^2$ is convex,

$$\left( \frac{1}{N} \sum_{n=1}^{N} a_n \right)^2 \leqslant \frac{1}{N} \sum_{n=1}^{N} a_n^2.$$

Together these two inequalities give

$$\limsup_{N \to \infty} \left\| \frac{1}{N} \frac{1}{H} \sum_{n=1}^{N} \sum_{h=0}^{H-1} u_{n+h} \right\|^2 \leqslant \limsup_{N \to \infty} \frac{1}{N} \sum_{n=1}^{N} \left\| \frac{1}{H} \sum_{h=0}^{H-1} u_{n+h} \right\|^2$$

$$= \limsup_{N \to \infty} \frac{1}{N} \sum_{n=1}^{N} \frac{1}{H^2} \sum_{h,h'=0}^{H-1} \langle u_{n+h}, u_{n+h'} \rangle$$

$$\leqslant \limsup_{N \to \infty} \frac{1}{H^2} \sum_{h,h'=0}^{H-1} \left| \frac{1}{N} \sum_{n=1}^{N} \langle u_{n+h}, u_{n+h'} \rangle \right|$$

$$\tag{7.14}$$

where the sum has been rearranged and the triangle inequality has been used to give the last line. Notice that for a fixed pair $h, h'$,

$$\limsup_{N \to \infty} \left| \frac{1}{N} \sum_{n=1}^{N} \langle u_{n+h}, u_{n+h'} \rangle \right| = s_{|h-h'|},$$

so the expression in (7.14) is bounded above by

$$\frac{1}{H^2} \sum_{h,h'=0}^{H-1} s_{|h-h'|}.$$

We will split this double average into three terms as indicated in Fig. 7.1.

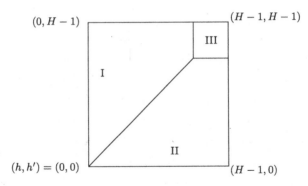

**Fig. 7.1** Decomposing the double average. The lower left corner of the square III has the coordinates $(H - H_0, H - H_0)$

Indeed,

$$\frac{1}{H^2} \sum_{h,h'=0}^{H-1} s_{|h-h'|} = \frac{1}{H} \sum_{h=0}^{H-H_0} \frac{1}{H} \sum_{h'=h}^{H-1} s_{h'-h} \quad \text{(part I)}$$

$$+ \frac{1}{H} \sum_{h'=0}^{H-H_0} \frac{1}{H} \sum_{h=h'+1}^{H-1} s_{h-h'} \quad \text{(part II)}$$

$$+ \frac{1}{H^2} \sum_{h,h'=H-H_0+1}^{H-1} s_{|h-h'|} \quad \text{(part III)}.$$

The first two averages I and II are taken over averages that satisfy the inequality in (7.12), so are less than $\varepsilon$. By initially choosing $H$ much larger than $H_0$ (and using the assumption that $(u_n)$ is bounded), the third term III is bounded by $\varepsilon$ as well. Together with (7.13), this gives

$$\left\| \frac{1}{N} \sum_{n=1}^{N} u_n \right\| < 4\varepsilon$$

as desired.                                                                         □

This gives an alternative approach to the chaotic part of Theorem 7.10.

PROOF OF THEOREM 7.10: CHAOTIC PART USING VAN DER CORPUT. Consider again the totally ergodic component $g \in \mathscr{V}_*$. We claim that

$$\lim_{N \to \infty} \frac{1}{N} \sum_{n=0}^{N-1} \int g U_T^{p(n)} g \, d\mu = 0.$$

In fact, we claim that

$$\lim_{N \to \infty} \left\| \frac{1}{N} \sum_{n=0}^{N-1} U_T^{p(n)} g \right\|_2 = 0 \tag{7.15}$$

holds for any non-constant polynomial $p$ without any condition on $p(0)$. First notice that since $\mathscr{H}_a \subseteq \mathscr{H}_{\mathrm{rat}}$ for all $a \geq 1$, we must have $g \perp \mathscr{H}_a$ and so by the mean ergodic theorem (Theorem 2.21) itself,

$$\lim_{N \to \infty} \left\| \frac{1}{N} \sum_{n=0}^{N-1} U_T^{an} g \right\|_2 = 0$$

if $a \neq 0$, showing (7.15) when $p$ is a linear polynomial with $p(0) = 0$. Also notice that (7.15) for a polynomial $p$ is equivalent to (7.15) for the polynomial $p - p(0)$. The key idea is to use the van der Corput lemma in an inductive argument to extend (7.15) to polynomials of higher degree. Assume therefore that degree$(p) = k$ and we have proved (7.15) for all polynomials $p'$ with $1 \leq$ degree$(p') < k$. Write $u_n = U_T^{p(n)} g$ for $n \geq 1$ and notice that

$$\langle u_{n+h}, u_n \rangle = \left\langle U_T^{p(n+h)} g, U_T^{p(n)} g \right\rangle = \left\langle U_T^{p(n+h) - p(n)} g, g \right\rangle.$$

For any given $h$,

$$\mathrm{degree}\left( p(n+h) - p(n) \right) < \mathrm{degree}\left( p(n) \right).$$

By the inductive hypothesis and the fact that strong convergence in a Hilbert space implies weak convergence, we therefore have for any $h \neq 0$ that

$$s_h = \lim_{N \to \infty} \frac{1}{N} \left| \sum_{n=1}^{N} \langle u_{n+h}, u_n \rangle \right| = \lim_{N \to \infty} \frac{1}{N} \left| \sum_{n=1}^{N} \left\langle U_T^{p(n+h) - p(n)} g, g \right\rangle \right| = 0.$$

Theorem 7.11 may be applied to deduce (7.15), which clearly implies our claim.                                                                           □

## Exercises for Sect. 7.4

**Exercise 7.4.1.** Let $(X, \mathscr{B}, \mu, T)$ be a measure-preserving system. Show that an iterate $T^a$ is not ergodic for some $a \geq 1$ if and only if the operator $U_T$ has an $a$th root of unity as an eigenvalue.

**Exercise 7.4.2.** Re-prove the equidistribution of $\{n^2\alpha\}$ in $\mathbb{T}$ for irrational $\alpha$ (or, more generally, fractional parts of polynomials with an irrational non-constant coefficient as in Theorem 1.4) using $u_n = \mathrm{e}^{2\pi i n^2 \alpha}$ and Theorem 7.11.

**Exercise 7.4.3.** Adapt the proof of the Furstenberg correspondence from Sect. 7.3 to prove Theorem 7.9 assuming Theorem 7.10.

## 7.5 Two Special Cases of Multiple Recurrence

The proof of Theorem 7.4 proceeds by building up a general system from simpler constituents, a process illustrated in Fig. 7.2. As suggested by Fig. 7.2, there are three parts to establishing the SZ property for a general measure-preserving system. First*, it must be shown for certain special systems (Kronecker systems). Second, it must be shown to be preserved by various extensions. Third, the property must be shown to survive taking limits in a sense described later. We will refer to this chain of factors frequently, and will need to establish both the existence of these factors and the fact that all of them have the SZ property. In Sect. 7.12 all the steps used in proving Theorem 7.4 will be summarized in the argument that completes the proof.

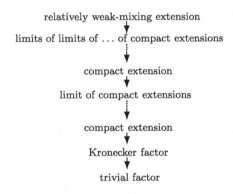

**Fig. 7.2** Building up a measure-preserving system

### 7.5.1 Kronecker Systems

In this section we prove the statement needed to start the induction in Fig. 7.2, by showing that any Kronecker system has the SZ property.

---

\* We could also start with the trivial factor and let the inductive step handle the case of a Kronecker system, but the extra case handled here will show more.

**Proposition 7.12.** *Any Kronecker system has the SZ property.*

That is, if $G$ is a compact abelian metrizable group with Borel $\sigma$-algebra $\mathscr{B}$
(cf. Theorem 6.10), then an ergodic rotation $R : G \to G$ defined by $R(g) = ag$
for a fixed $a \in G$ has the property that

$$\liminf_{N \to \infty} \frac{1}{N} \sum_{n=1}^{N} m_G \left( A \cap R^{-n} A \cap R^{-2n} A \cap \cdots \cap R^{-kn} A \right) > 0 \qquad (7.16)$$

for all $k \geqslant 1$ and $A \in \mathscr{B}$ with $m_G(A) > 0$.

PROOF OF PROPOSITION 7.12. For any function $f$ on $G$ and $g \in G$ write

$$f^{(g)}(h) = f(gh).$$

We claim first that for any $f \in L^{\infty}_{m_G}$ the map $g \mapsto f^{(g)}$ is a continuous
map $G \to L^1_{m_G}$ with respect to the metric d on $G$. To see this, fix $\varepsilon > 0$,
choose $\tilde{f} \in C(G)$ with $\|f - \tilde{f}\|_1 < \varepsilon$ and (by compactness) choose $\delta > 0$ with
the property that $\mathsf{d}(g_1, g_2) < \delta$ implies $|\tilde{f}(g_1 h) - \tilde{f}(g_2 h)| < \varepsilon$ for all $h \in G$.
Then for $\mathsf{d}(g_1, g_2) < \delta$,

$$\|f^{(g_1)} - f^{(g_2)}\|_1 \leqslant \|f^{(g_1)} - \tilde{f}^{(g_1)}\|_1 + \|\tilde{f}^{(g_1)} - \tilde{f}^{(g_2)}\|_1 + \|\tilde{f}^{(g_2)} - f^{(g_2)}\|_1 < 3\varepsilon.$$

Now fix $f \in L^{\infty}_{m_G}$; we know that each map $g \mapsto f^{(g^i)}$ is continuous from $G$
to $L^1_{m_G}$ for $0 \leqslant i \leqslant k$. We claim that the map

$$g \mapsto \int_G f(h) f(gh) \cdots f(g^k h) \, dm_G(h)$$

is continuous. Indeed, for any $\varepsilon > 0$ there exists a $\delta > 0$ such that $\mathsf{d}(g_1, g_2) < \delta^{\bullet}$
implies that $\|f^{(g_1^i)} - f^{(g_2^i)}\|_1 < \varepsilon$ for $0 \leqslant i \leqslant k$; thus

$$\int f(h) f(g_1 h) \cdots f(g_1^k h) \, dm_G(h) - \int f(h) f(g_2 h) \cdots f(g_2^k h) \, dm_G$$

$$= \int f(h) \left( f(g_1 h) - f(g_2 h) \right) f(g_1^2 h) \cdots f(g_1^k h) \, dm_G(h)$$

$$+ \int f(h) f(g_2 h) \left( f(g_1^2 h) - f(g_2^2 h) \right) f(g_1^3 h) \cdots f(g_1^k h) \, dm_G(h)$$

$$\ddots$$

$$+ \int f(h) f(g_2 h) \cdots f(g_2^{k-1} h) \left( f(g_1^k h) - f(g_2^k h) \right) \, dm_G(h),$$

and each integral on the right-hand side is bounded by $\varepsilon \|f\|_\infty^k$.

Now let $f \in L^\infty_{m_G}$ be a non-negative function that is not almost everywhere 0. We claim that

$$\lim_{N \to \infty} \frac{1}{N} \sum_{n=1}^{N} \int_G f(h) f(Rh) \cdots f(R^{kn}h) \, dm_G(h) \qquad (7.17)$$

exists and is positive. By the argument above, the function $\phi$ defined by

$$\phi(a) = \int_G f(h) f(ah) \cdots f(a^k h) \, dm_G(h)$$

is continuous; on the other hand $R$ is uniquely ergodic by Theorem 4.14, and so

$$\lim_{N \to \infty} \frac{1}{N} \sum_{n=1}^{N} \phi(a^n) = \int_G \phi(h) \, dm_G(h),$$

which is the limit in (7.17). By assumption $\phi \geqslant 0$ and $\phi(1_G) > 0$, so the limit is positive since $\phi$ is continuous.      □

In fact much more is true. Returning initially to an ergodic circle rotation $R_\alpha$, for any $\varepsilon > 0$ the set $\{n \in \mathbb{Z} \mid \{n\alpha\} < \varepsilon\}$ is syndetic*. It follows that for fixed $k \geqslant 1$ and any $\varepsilon > 0$ the set

$$\left\{ n \in \mathbb{N} \mid \left| m_{\mathbb{T}}(A \cap R_\alpha^{-n} A \cap \cdots \cap R_\alpha^{-kn} A) - m_{\mathbb{T}}(A) \right| < \varepsilon \right\}$$

is also syndetic for any set $A$ with $m_{\mathbb{T}}(A) > 0$. A general Kronecker system given by a rotation $R : G \to G$ has the property that for any $f \in L^2_{m_G}$ the closure of the orbit $\{U_R^n f \mid n \in \mathbb{Z}\}$ in $L^2_{m_G}$ is compact (a fact that we will be generalizing in order to define compact extensions, see Fig. 7.2 and Sect. 7.9). Moreover, the set $\{n \in \mathbb{N} \mid \|U_R^n f - f\|_2 < \varepsilon\}$ is again syndetic, showing that multiple recurrence occurs along a syndetic set for Kronecker systems.

## 7.5.2 Weak-Mixing Systems

Although logically not necessary, we prove next a multiple recurrence result (in a stronger form, and in particular identifying the limit) for weak-mixing systems. In the proof of Theorem 7.4 we will need to use similar techniques for a relatively weak-mixing extension.

---

* A subset $S$ in a semigroup $G$ is called *syndetic* if there is a finite set $T \subseteq G$ with the property that $\bigcup_{t \in T} S - t = G$, where $S - t = \{g \in G \mid g + t \in S\}$.

**Proposition 7.13.** *Let $(X, \mathscr{B}, \mu, T)$ be a weak-mixing system. Then for any functions $f_1, \ldots, f_k \in L_\mu^\infty$,*

$$\frac{1}{N} \sum_{n=0}^{N-1} U_T^n f_1 U_T^{2n} f_2 \cdots U_T^{kn} f_k \xrightarrow[L_\mu^2]{} \int f_1 \, d\mu \int f_2 \, d\mu \cdots \int f_k \, d\mu. \qquad (7.18)$$

It follows that if $f_0 \in L_\mu^\infty$ is another function then

$$\frac{1}{N} \sum_{n=0}^{N-1} \int_X f_0 U_T^n f_1 U_T^{2n} f_2 \cdots U_T^{kn} f_k \, d\mu \longrightarrow \int f_0 \, d\mu \int f_1 \, d\mu \cdots \int f_k \, d\mu,$$

and, in particular (taking each $f_i$ to be $\chi_A$), we have property SZ.

PROOF OF PROPOSITION 7.13. Since $T$ is weak-mixing it is certainly ergodic, so (7.18) holds for $k = 1$ by the mean ergodic theorem (Theorem 2.21). We proceed by induction on $k$.

Assume that $k = 2$. If either one of $f_1$ or $f_2$ is constant then (7.18) follows from the case $k = 1$. (If $f_2$ is constant this is literally true; if $f_1$ is constant then we need to recall that if $T$ is weak-mixing then so is $T^2$ by Corollary 2.39.) Therefore, we can assume (by subtracting the integral) that $\int_X f_1 \, d\mu = 0$. Write $u_n = U_T^n f_1 U_T^{2n} f_2$ for $n \geqslant 1$ and notice that

$$s_h = \lim_{N \to \infty} \frac{1}{N} \sum_{n=1}^{N} \langle u_{n+h}, u_n \rangle$$

$$= \lim_{N \to \infty} \frac{1}{N} \sum_{n=1}^{N} \int_X U_T^{n+h} f_1 U_T^{2(n+h)} f_2 U_T^n f_1 U_T^{2n} f_2 \, d\mu$$

$$= \lim_{N \to \infty} \frac{1}{N} \sum_{n=1}^{N} \int_X U_T^h f_1 U_T^{n+2h} f_2 f_1 U_T^n f_2 \, d\mu$$

$$(\text{since } T^n \text{ preserves } \mu)$$

$$= \lim_{N \to \infty} \frac{1}{N} \sum_{n=1}^{N} \int_X (f_1 U_T^h f_1) U_T^n (f_2 U_T^{2h} f_2) \, d\mu$$

$$= \int_X f_1 U_T^h f_1 \, d\mu \int_X f_2 U_T^{2h} f_2 \, d\mu$$

$$(\text{by the mean ergodic theorem}).$$

Now $T$ is weak-mixing, so $T^2$ is weak-mixing by Corollary 2.39 and therefore $T \times T^2$ is weak-mixing with respect to $\mu \times \mu$ by Corollary 2.37. Applying the mean ergodic theorem to $T \times T^2$, and writing $f_1 \otimes f_2$ for the function $(x, x') \mapsto f_1(x) f_2(x')$, we have

$$\lim_{H \to \infty} \frac{1}{H} \sum_{h=0}^{H-1} s_h = \lim_{H \to \infty} \frac{1}{H} \sum_{h=0}^{H-1} \int_X f_1 U_T^h f_1 \, d\mu \int_X f_2 U_T^{2h} f_2 \, d\mu$$

$$= \lim_{H \to \infty} \frac{1}{H} \sum_{h=0}^{H-1} \int_{X \times X} (f_1 \otimes f_2) U_{T \times T^2}^h (f_1 \otimes f_2) \, d(\mu \times \mu)$$

$$= \left( \int_{X \times X} f_1 \otimes f_2 \, d(\mu \times \mu) \right)^2 \qquad \text{(by ergodicity of } T \times T^2\text{)}$$

$$= \left( \int_X f_1 \, d\mu \right)^2 \left( \int_X f_2 \, d\mu \right)^2 = 0.$$

Thus we can apply the van der Corput lemma (Theorem 7.11) to the sequence $(u_n)$, giving (7.18) for $k = 2$.

By Corollary 2.39 and Theorem 2.36(3) the map $T \times T^2 \times \cdots \times T^k$ is ergodic, and the same argument (see Exercise 7.5.2) can be used to give (7.18) for any $k$.  $\qquad \square$

## Exercises for Sect. 7.5

**Exercise 7.5.1.** Let $R : G \to G$ be an ergodic group rotation. Prove that for any $\varepsilon > 0$ and $f \in L^2_{m_G}$ the set $\{n \in \mathbb{N} \mid \|R^n f - f\|_{L^2_{m_G}} < \varepsilon\}$ is syndetic. Deduce Proposition 7.12.

**Exercise 7.5.2.** Finish the proof of Proposition 7.13 by giving the details of the inductive step for a general $k$.

## 7.6 Roth's Theorem

In this section we will prove the special case $k = 2$ of Theorem 7.4[68], giving a combinatorial result due to Roth; indeed we will prove the following stronger statement.

**Theorem 7.14.** *Let* $(X, \mathscr{B}, \mu, T)$ *be a measure-preserving system. Then, for any functions* $f_1, f_2 \in L^\infty(X, \mathscr{B}, \mu)$,

$$\frac{1}{N} \sum_{n=1}^{N} U_T^n f_1 U_T^{2n} f_2 \qquad\qquad (7.19)$$

*converges in* $L^2(X, \mathscr{B}, \mu)$. *Moreover, for any* $A \in \mathscr{B}$ *with* $\mu(A) > 0$ *we have*

$$\lim_{N \to \infty} \frac{1}{N} \sum_{n=1}^{N} \mu \left( A \cap T^{-n} A \cap T^{-2n} A \right) > 0.$$

Just as in the proof from Sect. 7.4, we will prove this by decomposing the system (more precisely, by decomposing the space of $L^2$-functions on the space) into an orderly and a chaotic part. We will see below that for this result the appropriate splitting will be given by the Kronecker factor.

Recall that in Sect. 7.5.1 we already proved the last statement of Theorem 7.14 for a Kronecker system $(G, \mathscr{B}_G, m_G, R)$. More precisely, we showed in the proof of Proposition 7.12 that

$$\lim_{N \to \infty} \frac{1}{N} \sum_{n=1}^{N} \int_G f(h) f(Rh) \cdots f(R^{kn} h) \, dm_G(h) > 0 \qquad (7.20)$$

for any $f \in L^\infty_{m_G}$ with $f \geqslant 0$ and $\int f \, dm_G > 0$.

Note however that prior to the statement of Theorem 7.14 we did not claim convergence of averages in $L^2_\mu$ corresponding to multiple recurrence as in (7.19). This convergence is not needed for the application of Theorem 7.14 to give Roth's theorem but is of independent interest*.

In Sect. 7.6.1 we will prove Theorem 7.14 for a Kronecker system. In Sect. 7.6.2 we will prove the following proposition, which reduces the case of a general measure-preserving system to a Kronecker system.

**Proposition 7.15.** *Let $T$ be an invertible ergodic measure-preserving transformation on a Borel probability space $(X, \mathscr{B}, \mu, T)$. Let $\mathscr{K}$ be the $\sigma$-algebra corresponding to the Kronecker factor of $T$, and let $f_1, f_2 \in L^\infty_\mu$. Then*

$$\left\| \frac{1}{N} \sum_{n=1}^{N} U_T^n f_1 U_T^{2n} f_2 - \frac{1}{N} \sum_{n=1}^{N} U_T^n E \left( f_1 \middle| \mathscr{K} \right) U_T^{2n} E \left( f_2 \middle| \mathscr{K} \right) \right\|_2 \to 0.$$

We now show how these two results—the case of Kronecker systems and Proposition 7.15—fit together.

PROOF OF THEOREM 7.14 ASSUMING PROPOSITIONS 7.12 AND 7.15. By Proposition 7.15, the sequence

$$\frac{1}{N} \sum_{n=1}^{N} U_T^n f_1 U_T^{2n} f_2$$

converges to the same limit as

$$\frac{1}{N} \sum_{n=1}^{N} U_T^n E \left( f_1 \middle| \mathscr{K} \right) U_T^{2n} E \left( f_2 \middle| \mathscr{K} \right)$$

---

* The reader interested solely in the application to Roth's theorem may therefore skip Sect. 7.6.1; there will not be an analogous part in the proof of Theorem 7.4.

does, which converges by the Kronecker case (notice that the latter sequence can be viewed as a sequence of functions on the Kronecker factor).

Moreover, if $A \in \mathcal{B}$ has $\mu(A) > 0$, then $f = E\left(\chi_A \big| \mathcal{K}\right)$ has the properties $f \in L^\infty_\mu(X, \mathcal{K})$, $f \geqslant 0$, and $\int_X f \, d\mu > 0$. It follows that

$$\frac{1}{N} \sum_{n=1}^N \mu\left(A \cap T^{-n} A \cap T^{-2n} A\right) = \frac{1}{N} \sum_{n=1}^N \int_X \chi_A U_T^n \chi_A U_T^{2n} \chi_A \, d\mu$$

converges to the same limit as does the sequence

$$\frac{1}{N} \sum_{n=1}^N \int_X \chi_A U_T^n f U_T^{2n} f \, d\mu = \frac{1}{N} \sum_{n=1}^N \int_X f U_T^n f U_T^{2n} f \, d\mu$$

(where we have used Theorem 5.1(3)). By (7.20), the latter limit is positive. □

Finally, let us note that the existence of the limit in Theorem 7.14 would be much harder to establish in Theorem 7.4 (this is discussed further in Sect. 7.13). As we have already seen, the *positivity* of the limit inferior is more than enough to imply Szemerédi's theorem.

### 7.6.1 Proof of Theorem 7.14 for a Kronecker System

In this section we will make use of Theorem 6.10.

PROOF OF THEOREM 7.14 FOR A KRONECKER SYSTEM. Let $G$ be a compact metric abelian group with Borel $\sigma$-algebra $\mathcal{B}$, Haar measure $m_G$ and metric d, and let $R_a(x) = ax$ be an ergodic group rotation defined by some $a \in G$. Let $f_1, f_2 \in L^\infty_\mu$ and write $f_1 \otimes f_2(g_1, g_2) = f_1(g_1) f_2(g_2)$ for $(g_1, g_2) \in G^2$. By the mean ergodic theorem (Theorem 2.21) applied to $R_a \times R_a^2$ (which is not itself ergodic),

$$\widehat{F}_N = \frac{1}{N} \sum_{n=1}^N U_{R_a}^n f_1 \otimes U_{R_a}^{2n} f_2 = \frac{1}{N} \sum_{n=1}^N U_{R_a \times R_a^2}(f_1 \otimes f_2)$$

converges in $L^2(G^2, \mathcal{B} \otimes \mathcal{B}, m_G \times m_G)$ to a limit $\widehat{F}$ which is invariant under the transformation $R_a \times R_{a^2}$. We wish to deduce from this that

$$F_N = \frac{1}{N} \sum_{n=1}^N U_{R_a}^n f_1 U_{R_a}^{2n} f_2$$

converges in $L^2(X, \mathcal{B}, m_G)$ as claimed in the first part of the theorem. Notice that $F_N$ is the restriction of $\widehat{F}_N$ to the null set $\{(x, x) \mid x \in X\}$, so in general there seems little hope of connecting the sequences $(\widehat{F}_N)$ and $(F_N)$. We will use the abelian group structure $U_{R_a}U_{R_t} = U_{R_t}U_{R_a}$ for $t \in G$ to achieve this.

Fix $\varepsilon > 0$. Since $f_2 \in L^\infty_{m_G} \subseteq L^2_{m_G}$ there exists some $\delta > 0$ with the property that $\mathsf{d}(1_G, h) < \delta$ implies that $\|f_2 - U_{R_h}f_2\|_2 < \varepsilon$ (cf. the beginning of the proof of Proposition 7.12).

Assume that $N < M$ and $\mathsf{d}(1_G, h) < \delta$. Then

$$\left\| U^n_{R_a}f_1 U^{2n}_{R_a}f_2 - U^n_{R_a}f_1 U^{2n}_{R_a}U_{R_h}f_2 \right\|_2 = \left\| U^n_{R_a}f_1 U^{2n}_{R_a}(f_2 - U_{R_h}f_2) \right\|_2 < \varepsilon \|f_1\|_\infty$$

and so

$$\|F_N - F_M\|_2 \leqslant \left\| \frac{1}{N}\sum_{n=1}^{N} U^n_{R_a}f_1 U^{2n}_{R_a}U_{R_h}f_2 - \frac{1}{M}\sum_{n=1}^{M} U^n_{R_a}f_1 U^{2n}_{R_a}U_{R_h}f_2 \right\|_2 + 2\varepsilon\|f_1\|_\infty.$$

Thus by averaging the square of these expressions over the set $B$ of points $h$ within $\delta$ of $1_G$ we get (by commutativity)

$$\|F_N - F_M\|_2^2 \leqslant \frac{1}{\mu(B)}\int_B\int_X \left[\widehat{F}_N - \widehat{F}_M\right]^2(g, g+h)\,\mathsf{d}m_G(g)\,\mathsf{d}m_G(h) + O_{f_1,f_2}(\varepsilon).$$

However, the last integral is part of the integral defining $\|\widehat{F}_N - \widehat{F}_M\|_2^2$ (more precisely, it is the restriction of the latter integral to an open neighborhood of the diagonal $\{(g, g) \mid g \in G\}$). Therefore, for $N$ and $M$ sufficiently large we get

$$\|\widehat{F}_N - \widehat{F}_M\|_2^2 < m_G(B)\varepsilon,$$

and so $\|F_N - F_M\|_2^2 \leqslant O_{f_1,f_2}(\varepsilon)$. This shows that $(F_N)$ is a Cauchy sequence in $L^2_{m_G}$ as required. This proves the theorem in the case considered.       □

### 7.6.2 Reducing the General Case to the Kronecker Factor

It remains to prove Proposition 7.15, and this is done once again using the van der Corput lemma (Theorem 7.11).

PROOF OF PROPOSITION 7.15. Notice that

$$\frac{1}{N}\sum_{n=1}^{N} U^n_T f_1 U^{2n}_T f_2 - \frac{1}{N}\sum_{n=1}^{N} U^n_T E\left(f_1 \mid \mathscr{K}\right) U^{2n}_T E\left(f_2 \mid \mathscr{K}\right)$$

$$= \frac{1}{N}\sum_{n=1}^{N} U^n_T\left(f_1 - E\left(f_1 \mid \mathscr{K}\right)\right) U^{2n}_T f_2$$

$$+ \frac{1}{N}\sum_{n=1}^{N} U^n_T E\left(f_1 \mid \mathscr{K}\right) U^{2n}_T\left(f_2 - E\left(f_2 \mid \mathscr{K}\right)\right),$$

so that it is enough to show that

$$\frac{1}{N} \sum_{n=1}^{N} U_T^n f_1 U_T^{2n} f_2 \to 0 \tag{7.21}$$

if either $E\left(f_1 \middle| \mathcal{K}\right) = 0$ or $E\left(f_2 \middle| \mathcal{K}\right) = 0$. Both cases will be proved by applying the van der Corput lemma (Theorem 7.11) with

$$u_n = U_T^n f_1 U_T^{2n} f_2.$$

With this choice,

$$\langle u_{n+h}, u_n \rangle = \int_X U_T^{n+h} f_1 U_T^{2(n+h)} f_2 U_T^n f_1 U_T^{2n} f_2 \, d\mu$$

$$= \int_X f_1 U_T^h f_1 U_T^n (f_2 U_T^{2h} f_2) \, d\mu \tag{7.22}$$

$$= \int_X U_T^{-n} (f_1 U_T^h f_1) f_2 U_T^{2h} f_2 \, d\mu. \tag{7.23}$$

Now assume that $E\left(f_1 \middle| \mathcal{K}\right) = 0$ and apply (7.23) to obtain

$$s_h = \limsup_{N \to \infty} \left| \int_X \left( \frac{1}{N} \sum_{n=1}^{N} U_T^{-n} \left( f_1 U_T^h f_1 \right) \right) f_2 U_T^{2h} f_2 \, d\mu \right|.$$

Notice that

$$\frac{1}{N} \sum_{n=1}^{N} U_T^{-n} (f_1 U_T^h f_1) \to \int_X f_1 U_T^h f_1 \, d\mu$$

by the mean ergodic theorem (Theorem 2.21) applied to $T^{-1}$. Hence

$$s_h \leqslant \|f_2\|_\infty^2 \left| \int_X f_1 U_T^h f_1 \, d\mu \right|,$$

and so we wish to show that

$$\frac{1}{H} \sum_{h=1}^{H} s_h \leqslant \frac{\|f_2\|_\infty^2}{H} \sum_{h=1}^{H} \left| \int_X f_1 U_T^h f_1 \, d\mu \right| \to 0 \tag{7.24}$$

as $H \to \infty$. By Lemma 2.41, this is equivalent to showing that

$$\frac{1}{H}\sum_{h=1}^{H}\left|\int_X f_1 U_T^h \bar f_1 \, \mathrm{d}\mu\right|^2 = \frac{1}{H}\sum_{h=1}^{H}\int_X f_1 \otimes \bar f_1 U_{T\times T}^h(f_1 \otimes \bar f_1)\, \mathrm{d}(\mu \times \mu)$$

$$= \int_X f_1 \otimes \bar f_1 \frac{1}{H}\sum_{h=1}^{H} U_{T\times T}^h(f_1 \otimes \bar f_1)\, \mathrm{d}(\mu \times \mu)$$

converges to zero as $H \to \infty$. By the mean ergodic theorem (Theorem 2.21),

$$F_H = \frac{1}{H}\sum_{h=1}^{H} U_{T\times T}^h(f_1 \otimes \bar f_1)$$

converges in $L^2_{\mu \times \mu}$ to some $F \in L^\infty_{\mu \times \mu}$ which is invariant under $T \times T$. We claim that $F = 0$, which will imply (7.24) and (7.21) by van der Corput's lemma (Theorem 7.11).

The proof of the claim will rely on the argument we used to give the implication (5) $\implies$ (2) in Theorem 2.36 on p. 58. We isolate the statement needed in the form of a lemma (which follows easily from Theorem B.3).

**Lemma 7.16.** *Let $(X, \mathscr{B}, \mu, T)$ be any measure-preserving system on a Borel probability space. Suppose that $K : L^2_\mu \to L^2_\mu$ is a compact self-adjoint operator commuting with $U_T$. Then all eigenspaces of $K$ with non-zero eigenvalue are finite-dimensional, $U_T$-invariant and spanned by eigenfunctions of $U_T$. An operator induced by a $T \times T$-invariant kernel $F$ with $F(y,x) = \overline{F(x,y)}$ satisfies all these assumptions.*

Recall that any kernel in $L^2_{\mu \times \mu}$ defines a compact operator on $L^2_\mu$, hence

$$K_H : g \mapsto \int F_H(x,y)g(y)\, \mathrm{d}\mu(y)$$

and

$$K : g \mapsto \int F(x,y)g(y)\, \mathrm{d}\mu(y)$$

are compact operators on $L^2_\mu$. Moreover, the operator norm of an operator defined by an $L^2_\mu$ kernel is bounded by the $L^2_\mu$-norm of the kernel. In particular, since $F_H$ converges to $F$ in $L^2_{\mu \times \mu}$, the operator $K_H$ converges to $K$ with respect to the operator norm. Finally, notice that if $\zeta \in L^2(X, \mathscr{K})$ is an eigenfunction of $U_T$ (any such eigenfunction is $\mathscr{K}$-measurable by the definition of the Kronecker factor), then

$$\int U_T^h \bar f_1 \zeta \, \mathrm{d}\mu = \int E\left(U_T^h \bar f_1 \zeta \,\middle|\, \mathscr{K}\right)\, \mathrm{d}\mu = \int U_T^h E\left(\bar f_1 \,\middle|\, \mathscr{K}\right) \zeta \, \mathrm{d}\mu = 0$$

since we are assuming that $E\left(f_1 \,\middle|\, \mathscr{K}\right) = 0$. This shows that $K_H \zeta = 0$ and so $K\zeta = 0$ by taking the limit. Since $F_H(y,x) = \overline{F_H(x,y)}$ for all $H$, the same holds for $F$. However, $F$ is $T \times T$-invariant and so by Lemma 7.16 the compact self-adjoint operator $K$ has the property that all its eigenspaces for non-zero eigenvalues are finite-dimensional and generated by eigenfunctions of $U_T$. Since all eigenfunctions of $U_T$ belong to the kernel of $K$ we have $K = 0$, and so $F = 0$ as claimed.

To complete the proof of the proposition we also have to show (7.21) under the assumption that $E\left(f_2 \mid \mathscr{K}\right) = 0$. In this case we can use (7.22) instead of (7.23), and we can proceed as in the first case. However, in this case we need to show that

$$\frac{1}{H} \sum_{h=1}^{H} \left| \int_X f_2 U_T^{2h} f_2 \, \mathrm{d}\mu \right| \to 0$$

as $H \to \infty$, which in turn follows from

$$F_H = \frac{1}{H} \sum_{h=1}^{H} U_{T \times T}^{2h} (f_2 \otimes \bar{f}_2) \to 0$$

in $L^2_{\mu \times \mu}$. As before, we know by the mean ergodic theorem (this time applied for $T^2 \times T^2$ on $L^2_{\mu \times \mu}$) that $F_H$ converges to some $F \in L^2_{\mu \times \mu}$ invariant under $T^2 \times T^2$.

By applying Lemma 7.16 to $T^2$ we know that the eigenspaces of the compact operators are sums of eigenspaces of $U_{T^2}$. We claim that an eigenspace of $U_{T^2}$ is at most the sum of two eigenspaces of $U_T$, if it is not an eigenspace of $U_T$ itself. Let $V_\lambda \subseteq L^2_\mu$ be the eigenspace of $U_{T^2}$ with eigenvalue $\lambda \in \mathbb{S}^1$. Let $\eta \in \mathbb{S}^1$ be a square root of $\lambda$; then for any $v \in V_\lambda$

$$v = \frac{v + \eta^{-1} U_T(v)}{2} + \frac{v - \eta^{-1} U_T(v)}{2},$$

and it is easily checked that

$$\frac{v \pm \eta^{-1} U_T(v)}{2}$$

is an eigenvector of $U_T$ with eigenvalue $\pm \eta$. As in the first case, this together with the assumption $E\left(f_2 \mid \mathscr{K}\right) = 0$ implies that $F = 0$ as needed.     $\square$

## Exercises for Sect. 7.6

**Exercise 7.6.1.** Give a proof of Theorem 7.14 in the case of a Kronecker factor that does not use Theorem 6.10, but uses only the fact that eigenfunctions span a dense subset of $L^2_\mu$.

## 7.7 Definitions

In this section we introduce the two complementary notions of *compact extension* and *weak-mixing extension*; once again Fig. 7.2 summarizes how these will be used in the final induction.

A Kronecker system (that is, a rotation $R$ on a compact group $G$) has the following property: For any $f \in L^2_{m_G}(G)$, the orbit $\{U^n_R f\}_{n \in \mathbb{Z}}$ is a totally bounded subset of $L^2_{m_G}(G)$ (that is, is relatively compact). This is not a property shared by all measure-preserving systems, as shown by the next example.

*Example 7.17.* Let $X = \mathbb{T}^2$ and define[69] a map $T : X \to X$ by

$$T : x = \begin{pmatrix} y \\ z \end{pmatrix} \mapsto \begin{pmatrix} y + \alpha \\ z + y \end{pmatrix}.$$

The iterates of the map $T$ take the form

$$T^n : \begin{pmatrix} y \\ z \end{pmatrix} \mapsto \begin{pmatrix} y + n\alpha \\ z + ny + c(n, \alpha) \end{pmatrix}$$

for some function $c : \mathbb{Z} \times \mathbb{T} \to \mathbb{T}$. It follows that the map $T$ does not have the totally-bounded orbits property of a Kronecker system. For example, the function

$$f : \mathbb{T}^2 \to \mathbb{C}$$

defined by

$$f \begin{pmatrix} y \\ z \end{pmatrix} = e^{2\pi i z}$$

has

$$U^n_T f \begin{pmatrix} y \\ z \end{pmatrix} = e^{2\pi i c(n,\alpha)} e^{2\pi i (z+ny)},$$

so $U^n_T f$ and $U^m_T f$ are orthogonal and hence are distance $\sqrt{2}$ apart in $L^2_{m_{\mathbb{T}^2}}(\mathbb{T}^2)$ for all $m \neq n$.

We will use the same notation from now on: $(X, \mathscr{B}_X, \mu, T)$ is an extension of $(Y, \mathscr{B}_Y, \nu, S)$, and the variable $y$ denotes the image of $x \in X$ under the factor map.

Although the function $f$ appearing in Example 7.17 does not have a totally bounded orbit, the functions arising in the orbit still have a very simple form. This is particularly so when they are viewed as functions of $z$ for fixed $y$. The next definition makes this simplicity more precise.

**Definition 7.18.** Suppose that X is an extension of Y (equivalently, Y is a factor of X).

$$\mathsf{X} = (X, \mathscr{B}_X, \mu, T)$$
$$\downarrow$$
$$\mathsf{Y} = (Y, \mathscr{B}_Y, \nu, S)$$

Then a function $f$ in $L^2_\mu(X)$ is *almost-periodic* (AP) with respect to $\mathsf{Y}$ if, for every $\varepsilon > 0$, there exist $r \geqslant 1$ and functions $g_1, \dots, g_r \in L^2_\mu(X)$ such that

$$\min_{s=1,\dots,r} \|U_T^n f - g_s\|_{L^2_{\mu_y^{\mathscr{A}}}} < \varepsilon$$

for all $n \geqslant 1$, and for almost every $y \in Y$. The extension $\mathsf{X} \to \mathsf{Y}$ is a *compact extension* if the set of functions almost-periodic with respect to $\mathsf{Y}$ is dense in $L^2_\mu(X)$.

*Example 7.19.* The map $T : \mathbb{T}^2 \to \mathbb{T}^2$ constructed in Example 7.17 is a compact extension of the circle rotation $R_\alpha : \mathbb{T} \to \mathbb{T}$. The character

$$\binom{y}{z} \mapsto e^{2\pi i z}$$

is mapped to the function

$$\binom{y}{z} \mapsto \underbrace{C(n, \alpha, y)}_{\text{modulus } 1} e^{2\pi i z}$$

in $L^2_{\mu_y}$, so the orbit is a totally bounded subset of $L^2_{\mu_y}$ since we can take

$$g_i(z) = \lambda_i e^{2\pi i z}$$

for some $\varepsilon$-dense subset $\{\lambda_1, \dots, \lambda_r\} \subseteq \mathbb{S}^1$. This argument applies to all characters and linear combinations of characters, giving a dense subset.

An example of a function that is not AP relative to the rotation $R_\alpha$ for $\alpha$ irrational is

$$f : \binom{y}{z} \mapsto e^{2\pi i k z}$$

for $y \in \left(\frac{1}{k+1}, \frac{1}{k}\right]$, since for a fixed $y$ the set $\{U_T^n f\}$ contains scalar multiples of $z \mapsto e^{2\pi i k z}$ for all $k \geqslant 1$ (where the constant has absolute value one and depends on $y$) and is therefore not totally bounded in $L^2_{\mu_y}$.

Just as weak-mixing and Kronecker systems are opposite extremes to one another, the following notion is opposite to the case of a compact extension.

**Definition 7.20.** Let $(X, \mathscr{B}_X, \mu, T)$ be an ergodic measure-preserving system on a Borel probability space. The extension

$$\mathsf{X} = (X, \mathscr{B}_X, \mu, T)$$
$$\downarrow$$
$$\mathsf{Y} = (Y, \mathscr{B}_Y, \nu, S)$$

is *relatively weak-mixing* (relative to Y) if the system $(X \times X, \mu \times_Y \mu, T \times T)$ is ergodic, where $\mu \times_Y \mu$ is the relatively independent joining over Y. If Y is trivial, then the extension X $\to$ Y is relatively weak-mixing if and only if X is a weak-mixing system.

To understand the motivation for this terminology, notice that if Y is trivial, then the relatively independent joining is precisely the product measure $\mu \times \mu$ on $X \times X$, so by Theorem 2.36 the extension X $\to$ Y is relatively weak-mixing if and only if X is a weak-mixing system if Y is trivial.

## Exercises for Sect. 7.7

**Exercise 7.7.1.** Prove that the systems that appear in the proof of Weyl's equidistribution theorem (Theorem 1.4, proved in Sect. 4.4.3) may be obtained by taking finitely many compact extensions of a circle rotation.

**Exercise 7.7.2.** If X $\to$ Y is a relatively weak-mixing extension, and the system $(Z, U, \mathscr{C}, \lambda)$ is ergodic with $(Y, S)$ as a factor, prove that

$$(X \times Z, T \times U, \mu \times_Y \lambda)$$

is ergodic.

**Exercise 7.7.3.** Let X $\to$ Y be a relatively weak-mixing extension, and consider any measure-preserving system $(Z, U, \mathscr{C}, \lambda)$. Prove that all $T \times S$-invariant sets in the system $(X \times Z, T \times S, \mu \times \lambda)$ are contained in $\mathscr{A} \times \mathscr{C}$.

**Exercise 7.7.4.** Show that if $(X, T) \to (Y, S)$ is a relatively weak-mixing extension* then for any $k \geqslant 1$ the induced extension $(X, T^k) \to (Y, S^k)$ is also relatively weak-mixing.

**Exercise 7.7.5.** Show that the skew-product example in Example 7.17 is not relatively weak-mixing over the rotation $R_\alpha$.

## 7.8 Dichotomy Between Relatively Weak-Mixing and Compact Extensions

It is too much to expect that an extension is either relatively weak-mixing or compact. Indeed it is clear that there are more complicated possibilities (the system in Example 7.17 is neither a compact extension nor a weak-

---

* This exercise requires the definition of relatively weak-mixing to be extended to allow the factor system to be non-ergodic; if Y $= (Y, S)$ is not ergodic, then the extension X $\to$ Y is said to be relatively weak-mixing if all invariant sets in the relatively independent joining X $\times_Y$ X are sets in $Y$ modulo null sets.

mixing extension of the trivial system, for example). However, a weaker form
of this fundamental dichotomy does hold: an extension can only fail to be
weak-mixing if there is an intermediate non-trivial compact extension.

**Theorem 7.21.** *Let $(X, \mathscr{B}, \mu, T)$ be an ergodic measure-preserving system on
a Borel probability space and suppose that*

$$(X, \mathscr{B}, \mu, T)$$
$$\downarrow$$
$$(Y, \mathscr{A}, \nu, S)$$

*is an extension of measure-preserving systems. Then one of the following
holds:*

• X *is a relatively weak-mixing extension of* Y; *or*

• *there exists an intermediate extension* X* *(factors* X → X* → Y*) with the
property that* X* *is a non-trivial compact extension of* Y.

This structure theorem explains the chain of factors in Fig. 7.2. Starting
with the Kronecker factor (which is present only if X is not weak-mixing), one
can find a chain of compact extensions until the original system is a relatively
weak-mixing extension of a factor which is a limit of chains of compact extensions. In the proof, we will make use of the material on relatively independent
joinings from Sect. 6.5 without comment. We will also use $y$ to denote the
image of $x$ in $Y$ corresponding to $[x]_{\mathscr{A}}$, and when the factor system is thought
of in terms of the $T$-invariant sub-$\sigma$-algebra $\mathscr{A}$, we will use $T$ for the factor
map on $Y$ as well as the map on $X$. Similarly we will write $\mu_x^{\mathscr{A}}$ as well as $\mu_y^{\mathscr{A}}$
for the conditional measure on $[x]_{\mathscr{A}}$.

PROOF OF THEOREM 7.21. Let

$$\widetilde{\mathsf{X}} = \left( \widetilde{X} = X \times X, \mathscr{B} \otimes \mathscr{B}, \widetilde{\mu} = \mu \times_{\mathsf{Y}} \mu, \widetilde{T} = T \times T \right)$$

and assume that X → Y is not relatively weak-mixing. Then there is a non-
constant function $H \in \mathscr{L}^{\infty}(\widetilde{X})$ invariant under $\widetilde{T}$. We will use the function $H$
in a similar way to the proof of the implication (5) $\implies$ (2) in Theorem 2.36
(cf. Lemma 7.16) to define the following operator. For $\phi \in \mathscr{L}^2(X, \mathscr{B}, \mu)$ define

$$H * \phi(x) = \int H(x, x')\phi(x')\, d\mu_x^{\mathscr{A}}(x').$$

Notice however that we do not claim that $\phi \mapsto H * \phi$ is a compact operator
on $L^2(X, \mathscr{B}, \mu)$. To see that it defines a bounded operator on $L^2(X, \mathscr{B}, \mu)$
(and indeed, to see that $H * \phi(x)$ is defined almost everywhere) we give an
alternative description that will also be useful later.

Write $\mathscr{N}_X = \{\varnothing, X\}$ for the trivial $\sigma$-algebra on $X$. We claim that

$$H * \phi(x) = E\left(\underbrace{H(x,x')\phi(x')}_{\in L^2(\tilde{X})} \Big| \mathscr{B} \times \mathscr{N}_X\right)(x,\cdot),\tag{7.25}$$

$$\underbrace{\phantom{H * \phi(x) = E\left( H(x,x')\phi(x')\Big| \mathscr{B} \times \mathscr{N}_X\right)(x,\cdot)}}_{\in L^2(X)\subseteq L^2(\tilde{X})}$$

where we indicate in the argument on the right-hand side that the function only depends on the first coordinate.

To see the claim it is sufficient to know that

$$\tilde{\mu}_{(x,x')}^{\mathscr{B}\times\mathscr{N}_X} = \delta_x \times \mu_x^{\mathscr{A}}.\tag{7.26}$$

This follows from Proposition 6.16(4) combined with Proposition 5.20 applied to the $\sigma$-algebras

$$\mathscr{A} \times \mathscr{N}_X \underset{\tilde{\mu}}{=} \mathscr{N}_X \times \mathscr{A} \subseteq \mathscr{B} \times \mathscr{N}_X \subseteq \mathscr{B} \otimes \mathscr{B}.$$

Alternatively, we may appeal to Proposition 5.19 directly as follows. First, the measure $\delta_x \times \mu_x^{\mathscr{A}}$ is independent of $x'$, and therefore is $\mathscr{B} \times \mathscr{N}_X$-measurable. Moreover, $\delta_x \times \mu_x^{\mathscr{A}}$ is supported on $\{x\} \times X$, which is the $\mathscr{B} \times \mathscr{N}_X$-atom of $(x,x')$.

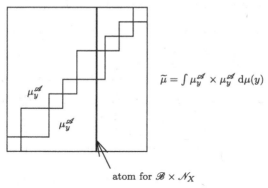

$$\tilde{\mu} = \int \mu_y^{\mathscr{A}} \times \mu_y^{\mathscr{A}}\, d\mu(y)$$

atom for $\mathscr{B} \times \mathscr{N}_X$

**Fig. 7.3** Explanation of $\tilde{\mu}_{(x,x')}^{\mathscr{B}\times\mathscr{N}_X} = \delta_x \times \mu_x^{\mathscr{A}}$

Finally,

$$\int \delta_x \times \mu_x^{\mathscr{A}}\, d\tilde{\mu}(x,x') = \int \delta_x \times \mu_x^{\mathscr{A}}\, d\mu(x)$$

$$= \iint \delta_x \times \mu_x^{\mathscr{A}}\, d\mu_z^{\mathscr{A}}(x)\, d\mu(z)$$

since $\mu = \int \mu_z^{\mathscr{A}}\, d\mu(z)$, and so

$$\int \delta_x \times \mu_x^{\mathscr{A}} \, d\widetilde{\mu}(x,x') = \int \left( \int \delta_x \, d\mu_z^{\mathscr{A}}(x) \right) \times \mu_z^{\mathscr{A}} \, d\mu(z)$$

$$= \int \mu_z^{\mathscr{A}} \times \mu_z^{\mathscr{A}} \, d\mu(z) = \widetilde{\mu},$$

since $\mu_x^{\mathscr{A}} = \mu_z^{\mathscr{A}}$ for $\mu_z^{\mathscr{A}}$-almost every $x \in X$. The claim now follows by Proposition 5.19. The idea behind this is illustrated in Fig. 7.3.

Now, by Corollary 5.24,

$$U_T \left( H * \phi \right)(x) = H * \phi(Tx)$$

$$= \int H(Tx, x') \phi(x') \, d\mu_{Tx}^{\mathscr{A}}(x')$$

$$= \int \underbrace{H(Tx, Tx')}_{=H(x,x')} \phi(Tx') \, d\mu_x^{\mathscr{A}}(x')$$

$$= H * \left( U_T \phi \right)(x). \tag{7.27}$$

If $\phi \in \mathscr{L}^{\infty}(X)$ then $\{U_T^n \phi \mid n \in \mathbb{Z}\} \subseteq L_{\mu_y^{\mathscr{A}}}^{\infty}$ for any fixed $y \in Y$. Recall our convention that $y \in Y$ always stands for the image of $x \in X$ (and hence corresponds to $[x]_{\mathscr{A}}$). Since $\phi \mapsto H * \phi$ is a compact operator $L_{\mu_y^{\mathscr{A}}}^2 \to L_{\mu_y^{\mathscr{A}}}^2$ (see Sect. B.3), it follows from (7.27) that for any fixed $y$ the set

$$\{U_T^n \left( H * \phi \right) \mid n \in \mathbb{Z}\} \subseteq L_{\mu_y^{\mathscr{A}}}^2$$

is totally bounded for any $\phi \in \mathscr{L}^{\infty}(X)$.

Note, however, that the number of functions in $L_{\mu_y^{\mathscr{A}}}^2$ required to $\varepsilon$-cover $\{U_T^n \left( H * \phi \right) \mid n \in \mathbb{Z}\}$ (that is, to have the property that the $\varepsilon$-balls around them together cover the set) may still depend on the point $y$, in contrast to the uniformity in Definition 7.18. We claim that the potential variation in $y$ does not cause a problem, and thus that $H * \phi$ is AP relative to $Y$.

To see this, fix $\varepsilon > 0$ and $y$ in $Y$. Then there exists $M(y)$ such that

$$\{U_{T^j} \left( H * \phi \right) \mid |j| \leqslant M(y)\}$$

is $\varepsilon$-dense in $\{U_{T^n} \left( H * \phi \right)\}$ with respect to $\mu_y^{\mathscr{A}}$. Choosing $M(y)$ to be the smallest such number defines a measurable function $M : Y \to \mathbb{N}$. Notice that $\{y \in Y \mid M(y) \leqslant M\}$ is the set of all $y \in Y$ with the property that for all $m \in \mathbb{Z}$ there is some $j$ with $|j| \leqslant M$ for which

$$\|U_T^m \left( H * \phi \right) - U_T^j \left( H * \phi \right)\|_{L_{\mu_y}^2} < \varepsilon.$$

Therefore, $\{y \in M \mid M(y) \leqslant M\}$ is a measurable set and $M$ is a measurable function. It follows that there is an $M \in \mathbb{N}$ with the property that

$$A = \{y \in Y \mid M(y) \leqslant M\} \in \mathscr{A}$$

has positive measure. For any $j$ with $|j| \leqslant M$ we define the function $g_j$ by the properties

$$g_j(x) = \begin{cases} U_T^j \, (H * \phi) \, (x) = H * (U_T^j \phi)(x) & \text{if } y \in A; \\ g_j(T^m x) & \text{if } y, Ty, \ldots, T^{m-1}y \notin A \text{ and } T^m y \in A. \end{cases}$$

As before, $y \in Y$ corresponds to $[x]_{\mathscr{A}}$ for $x \in X$. Notice that this defines $g_j$ almost everywhere by ergodicity. For $y \in A$, by assumption,

$$\min_{-M \leqslant j \leqslant M} \left\| U_T^n \, (H * \phi) - g_j \right\|_{L^2_{\mu_y^{\mathscr{A}}}} < \varepsilon$$

for all $n \in \mathbb{Z}$. Similarly, if $y, Ty, \ldots, T^{m-1}y \notin A$ and $T^m y \in A$ then we can use the above case for $T^m y$ to get

$$\left\| U_T^n \, (H * \phi) - \underbrace{U_T^m g_j}_{=g_j} \right\|_{L^2_{\mu_y^{\mathscr{A}}}} = \left\| U_T^{n-m} \, (H * \phi) - g_j \right\|_{L^2_{\mu_{T^m y}^{\mathscr{A}}}}$$

so

$$\min_{-M \leqslant j \leqslant M} \left\| U_T^n \, (H * \phi) - g_j \right\|_{L^2_{\mu_y^{\mathscr{A}}}} < \varepsilon$$

for all $n \in \mathbb{Z}$ and almost every $y$. In other words $H * \phi$ is AP relative to Y as claimed earlier. To make use of this we need to know, however, that there exists some $\phi$ for which $H * \phi$ is not a function on $Y$ (that is, is not $\mathscr{A}$-measurable modulo $\mu$).

**Lemma 7.22.** *There is a function $\phi \in L^\infty(X)$ such that $H * \phi \notin L^2(Y)$.*

PROOF. Suppose there is no such function, and choose a sequence $(\mathscr{P}_n)$ of finite partitions of $X$ with the property that

$$\sigma \left( \bigcup_{n \geqslant 1} \sigma(\mathscr{P}_n) \right) = \mathscr{B}.$$

Then for $x_2 \in P \in \mathscr{P}_n$, the definition of conditional expectations implies that

$$E \left( H \middle| \mathscr{B} \times \sigma \left( \mathscr{P}_n \right) \right) (x_1, x_2) = \frac{E \left( H \cdot \chi_{X \times P} \middle| \mathscr{B} \times \mathscr{N}_X \right) (x_1, x_2)}{E \left( \chi_{X \times P} \middle| \mathscr{B} \times \mathscr{N}_X \right) (x_1, x_2)}. \tag{7.28}$$

By the earlier reformulation of the operator $\phi \mapsto H * \phi$ in (7.25) and our assumption, the numerator

$$E \left( H \cdot \chi_{X \times P} \middle| \mathscr{B} \times \mathscr{N}_X \right) \tag{7.29}$$

is $\mathscr{A} \times \mathscr{N}_X$-measurable. The same applies to the denominator

$$E\left(\chi_{X \times P} \big| \mathscr{B} \times \mathscr{N}_X\right)(x_1, x_2) = \mu_{x_1}^{\mathscr{A}}(P) \qquad (7.30)$$

(using the fact that $\widetilde{\mu}_{(x_1,x_2)}^{\mathscr{B} \times \mathscr{N}_X} = \delta_{x_1} \times \mu_{x_1}^{\mathscr{A}}$ proved after (7.26)). Since the formula in (7.28) holds under the assumption that $x_2 \in P \in \mathscr{P}_n$, and both of the expressions (7.29) and (7.30) are $\mathscr{A} \times \mathscr{N}_X$-measurable, it follows that $E\left(H \big| \mathscr{B} \times \sigma(\mathscr{P}_n)\right)$ is $\mathscr{A} \times \sigma(\mathscr{P}_n)$-measurable. By the increasing martingale theorem (Theorem 5.5) applied to the sequence

$$\mathscr{B} \times \sigma(\mathscr{P}_n) \nearrow \mathscr{B} \otimes \mathscr{B},$$

we deduce that $H$ is $\mathscr{A} \times \mathscr{B}$-measurable. However, by Proposition 6.16(4) we have $\mathscr{A} \times \mathscr{N}_X = \mathscr{N}_X \times \mathscr{A}$ modulo $\widetilde{\mu}$, so $H$ is actually $\mathscr{N}_X \times \mathscr{B}_X$-measurable. Thus $H$ is a function of $x_2 \in X$ alone. Since $H$ is a non-constant invariant function, this contradicts the ergodicity of $T$.                                   □

Thus the set

$$\mathscr{F} = \{f \in L^\infty(X) \mid f \text{ is AP relative to } \mathsf{Y}\}$$

contains functions that are not $\mathscr{A}$-measurable. We claim that $\mathscr{F}$ is an algebra of functions; the only (slightly) difficult part of this assertion is to show that $\mathscr{F}$ is closed under multiplication.

Suppose that $f_1, f_2 \in \mathscr{F}$, $g_j \in L^2(X)$ for $j = 1, \ldots, J$, and $h_k \in L^2(X)$ for $k = 1, \ldots, K$, are functions with the properties that

$$\min_j \|U_T^n f_1 - g_j\|_{L^2_{\mu_y^{\mathscr{A}}}} < \frac{\varepsilon}{\|f_2\|_\infty} \qquad (7.31)$$

and

$$\min_k \|U_T^n f_2 - h_k\|_{L^2_{\mu_y^{\mathscr{A}}}} < \frac{\varepsilon}{\|f_1\|_\infty} \qquad (7.32)$$

for almost every $y \in Y$. Notice that we can modify $g_j$ if necessary to ensure that $\|g_j\|_\infty \leqslant \|f_1\|_\infty$ for all $j = 1, \ldots, J$. Then

$$\|U_T^n(f_1 f_2) - g_j h_k\|_{L^2_{\mu_y^{\mathscr{A}}}} < \|U_T^n f_2 \left(U_T^n f_1 - g_j\right)\|_{L^2_{\mu_y^{\mathscr{A}}}}$$

$$+ \|g_j \left(U_T^n f_2 - h_k\right)\|_{L^2_{\mu_y^{\mathscr{A}}}}$$

$$< 2\varepsilon$$

if $j$ and $k$ are chosen so as to minimize the expressions in (7.31) and (7.32). This implies that $f_1 f_2 \in \mathscr{F}$, so $\mathscr{F}$ is an algebra.

Let $\mathscr{B}^*$ denote the smallest $\sigma$-algebra with respect to which the members of $\mathscr{F}$ are measurable. Since $\mathscr{F}$ is invariant under $U_T$, $\mathscr{B}^*$ is invariant under $T$.

We will show that $\mathscr{F} \subseteq L^2(X, \mathscr{B}^*)$ is dense, from which it follows that $\mathscr{B}^*$ defines a factor of X that is a non-trivial compact extension $X^*$ of Y.

To see that $\mathscr{F}$ is dense in $L^2(X, \mathscr{B}^*)$, fix $f \in \mathscr{F}$, $\varepsilon > 0$ and an interval $[a, b] \subseteq \mathbb{R}$. By the Stone–Weierstrass Theorem, $\chi_{[a,b]}$ may be approximated arbitrarily closely by a polynomial $p \in \mathbb{R}[t]$ on $[-\|f\|_\infty, \|f\|_\infty]$. Thus there is a polynomial $p$ with

$$\|\chi_{[a,b]} - p\|_{L^2_{f_* \mu}} < \varepsilon$$

or equivalently

$$\|\chi_{f^{-1}[a,b]} - p(f)\|_2 < \varepsilon.$$

This shows that all the generators of $\mathscr{B}^*$ can be approximated by elements of the algebra $\mathscr{F}$. Let

$$\mathscr{C} = \{B \in \mathscr{B}^* \mid \chi_B \text{ belongs to the } L^2\text{-closure of } \mathscr{F}\}.$$

Since $f \in \mathscr{F}$ implies that $1 - f \in \mathscr{F}$, $C \in \mathscr{C}$ implies that $X \setminus C \in \mathscr{C}$. Suppose now that $C_1, C_2 \in \mathscr{C}$ and $\varepsilon > 0$, then there exists $f_1 \in \mathscr{F}$ with

$$\|\chi_{C_1} - f_1\|_2 < \frac{\varepsilon}{\|f_2\|_\infty}$$

and $f_2 \in \mathscr{F}$ with

$$\|\chi_{C_2} - f_2\|_2 < \frac{\varepsilon}{\|f_1\|_\infty}.$$

This implies that $\|\chi_{C_1}\chi_{C_2} - f_1 f_2\|_2 < 2\varepsilon$, showing that $C_1 \cap C_2 \in \mathscr{C}$. Thus the functions $\chi_{C_1}, \chi_{C_2}, \chi_{C_1 \cap C_2}$ can all be approximated, so the same holds for

$$\chi_{C_1 \cup C_2} = \chi_{C_1} + \chi_{C_2} - \chi_{C_1 \cap C_2},$$

which implies that $C_1 \cup C_2 \in \mathscr{C}$. Thus $\mathscr{C}$ is an algebra; since every countable union can be approximated by a finite union $\mathscr{C}$ is in fact a $\sigma$-algebra. It follows that $\mathscr{C} = \mathscr{B}^*$, so by considering simple functions we see that $\mathscr{F}$ is dense in $L^2_\mu(X^*)$ as claimed. $\square$

## 7.9 SZ for Compact Extensions

We have already established that every Kronecker system satisfies the SZ property. By Theorem 7.21 from the last section there is, for a given system $X = (X, \mathscr{B}, \mu, T)$, either an intermediate compact extension $X^*$ of the Kronecker factor Y, or X is a relatively weak-mixing extension of Y. In the former case we want to prove property SZ for $X^*$ first and proceed induc-

tively through the sequence of factors* in Fig. 7.2. Thus we need to prove the
following proposition regarding compact extensions.

**Proposition 7.23.** *Let*

$$\mathsf{X} = (X, \mathscr{B}, \mu, T)$$
$$\downarrow$$
$$\mathsf{Y} = (Y, \mathscr{A}, \nu, S)$$

*be a compact extension of invertible measure-preserving systems on Borel
probability spaces. If* $\mathsf{Y}$ *is SZ, then so is* $\mathsf{X}$.

As seen in Example 7.19, it is not true that any function $f \in L^\infty(X)$ in
the compact extension $\mathsf{X}$ of $\mathsf{Y}$ is automatically AP relative to $\mathsf{Y}$. The next
lemma will be used to circumvent this problem in the case of characteristic
functions.

**Lemma 7.24.** *In the notation of Proposition 7.23, let* $B \in \mathscr{B}$ *have* $\mu(B) > 0$.
*Then there exists a set* $\widetilde{B} \subseteq B$ *with* $\mu(\widetilde{B}) > 0$ *such that*

- $\chi_{\widetilde{B}}$ *is AP relative to* $\mathsf{Y}$, *and*
- $\mu_y^{\mathscr{A}}(\widetilde{B}) > \frac{1}{2}\mu(\widetilde{B})$ *or* $\mu_y^{\mathscr{A}}(\widetilde{B}) = 0$ *for all* $y \in Y$.

PROOF. Both properties of the lemma will be achieved by removing an ele-
ment $A \in \mathscr{A}$ from $B$ (in other words, by removing a collection of entire $\mathscr{A}$-
atoms). We will do this in two stages, first defining $B' \subseteq B$ which satisfies
the second property, and then defining $\widetilde{B} \subseteq B'$. Notice that if we indeed
have $\widetilde{B} = B' \smallsetminus A$ for some $A \in \mathscr{A}$, and $B'$ satisfies the second property, then
so does $\widetilde{B}$.
Let

$$B' = \{x \in B \mid \mu_y^{\mathscr{A}}(B) > \tfrac{1}{2}\mu(B)\}.$$

Hence, if $y \notin B'$ then we have some $x \in B \smallsetminus B'$ that is mapped to $y$, and

$$\mu_y^{\mathscr{A}}(B \smallsetminus B') = \mu_y^{\mathscr{A}}(B) \leqslant \tfrac{1}{2}\mu(B),$$

and integration over $y \in Y$ yields $\mu(B \smallsetminus B') \leqslant \tfrac{1}{2}\mu(B)$.
Moreover, since $\mu(B) > 0$ we have $\mu(B') > 0$ and

$$\mu_x^{\mathscr{A}}(B') > \tfrac{1}{2}\mu(B) \geqslant \tfrac{1}{2}\mu(B')$$

for every $x \in B'$, so the set $B'$ satisfies the second property of the lemma.
We now proceed to define $\widetilde{B}$. Choose a decreasing sequence $(\varepsilon_\ell)_{\ell \geqslant 1}$
with $\varepsilon_\ell > 0$ for all $\ell \geqslant 1$ and with

---

* Even though the result in this section is the next logical step, the reader may wish to look
at the argument for sequences of factors in the next section, which is both independent
and easier, and may help the reader to prepare for the current argument.

$$\sum_{\ell=1}^{\infty} \varepsilon_\ell < \tfrac{1}{2}\mu(B').$$

For every $\ell \geqslant 1$, there is an AP function $f_\ell$ such that

$$\|\chi_{B'} - f_\ell\|_{L_\mu^2}^2 = \int |\chi_{B'} - f_\ell|^2 \, d\mu < \varepsilon_\ell^2$$

by density. Let

$$A_\ell = \{y \in Y \mid \|\chi_{B'} - f_\ell\|_{L^2_{\mu_y^{\mathscr{A}}}}^2 \geqslant \varepsilon_\ell\},$$

then

$$\mu(A_\ell) \leqslant \frac{1}{\varepsilon_\ell} \int_{A_\ell} \|\chi_{B'} - f_\ell\|_{L^2_{\mu_y^{\mathscr{A}}}}^2 \, d\nu(y) \leqslant \varepsilon_\ell.$$

Let

$$\widetilde{B} = B' \smallsetminus \bigcup_{\ell \geqslant 1} A_\ell,$$

so in particular $\mu(\widetilde{B}) \geqslant \tfrac{1}{2}\mu(B')$. As explained earlier, either

$$\mu_y^{\mathscr{A}}(\widetilde{B}) > \tfrac{1}{2}\mu(\widetilde{B})$$

or

$$\mu_y^{\mathscr{A}}(\widetilde{B}) = 0,$$

since $\bigcup_\ell A_\ell \in \mathscr{A}$. Finally, for the AP property, fix $\varepsilon > 0$ and choose some $\ell$ with $\varepsilon_\ell < \tfrac{1}{2}\varepsilon$. Then if $T^n y \notin \bigcup_\ell A_\ell$,

$$\|U_T^n \chi_{\widetilde{B}} - U_T^n f_\ell\|_{L^2_{\mu_y^{\mathscr{A}}}} = \|\chi_{\widetilde{B}} - f_\ell\|_{L^2_{\mu_{T^n y}^{\mathscr{A}}}}$$
$$= \|\chi_{B'} - f_\ell\|_{L^2_{\mu_{T^n y}^{\mathscr{A}}}}$$
$$< \varepsilon_\ell < \tfrac{1}{2}\varepsilon.$$

On the other hand, if $T^n y \in \bigcup_\ell A_\ell$ then

$$\|U_T^n \chi_{\widetilde{B}}\|_{L^2_{\mu_y^{\mathscr{A}}}} = \|\chi_{\widetilde{B}}\|_{L^2_{\mu_{T^n y}^{\mathscr{A}}}} = 0.$$

Since $f_\ell$ is AP relative to $Y$, there exist functions $g_1, \ldots, g_m$ with

$$\min_{1 \leqslant j \leqslant m} \|U_T^n f_\ell - g_j\|_{L^2_{\mu_y^{\mathscr{A}}}} < \tfrac{1}{2}\varepsilon$$

almost everywhere; set $g_0 = 0$ to deduce that

$$\min_{0 \leqslant j \leqslant m} \|U_T^n \chi_{\widetilde{B}} - g_j\|_{L^2_{\mu_y^{\mathscr{A}}}} < \varepsilon,$$

showing the AP property. □

### 7.9.1 SZ for Compact Extensions via van der Waerden

The following approach to the SZ property for compact extensions (Proposition 7.23) is taken from the notes of Bergelson [26], and uses van der Waerden's theorem (Theorem 7.1).

PROOF OF PROPOSITION 7.23 USING VAN DER WAERDEN. By Lemma 7.24, we may assume that $f = \chi_B$ is AP and that there exists some set $A \in \mathscr{A}$ of positive measure with $\mu_y^{\mathscr{A}}(B) > \frac{1}{2}\mu(B)$ for $y \in A$. We will use SZ for $A$ (for arithmetic progressions of quite large length $K$) to show SZ for $B$ (for arithmetic progressions of length $k$). Given $\varepsilon = \frac{\mu(B)}{6(k+1)} > 0$ we may find (using the AP property of $f$) functions $g_1, \ldots, g_r$ such that

$$\min_{1 \leqslant s \leqslant r} \|U_T^n f - g_s\|_{L^2_{\mu_y^{\mathscr{A}}}} < \varepsilon$$

for all $n \in \mathbb{Z}$ and almost every $y \in Y$. We may assume that $\|g_s\|_\infty \leqslant 1$ for $s = 1, \ldots, r$. By Theorem 7.1 we may choose a $K$ for which any coloring of $\{1, \ldots, K\}$ with $r$ colors contains a monochrome arithmetic progression of length $k + 1$. There is some $c_1 > 0$ for which

$$R_K = \{n \in \mathbb{N} \mid \nu\left(A \cap S^{-n}A \cap \cdots S^{-Kn}A\right) > c_1\}$$

has positive lower density. In fact this follows from SZ for $A$: If

$$\liminf_{N \to \infty} \frac{1}{N} \sum_{n=1}^N \nu\left(A \cap S^{-n}A \cap \cdots \cap S^{-Kn}A\right) \geqslant c_0 > 0,$$

then, for large enough $N$ and some $c_1 > 0$ depending only on $c_0$, we must have

$$\frac{1}{N}|R_K \cap [1, N]| > c_1.$$

Let $n \in R_K$. Then, for every $y \in A \cap S^{-n}A \cap \cdots \cap S^{-Kn}A$, we have

$$\min_{1 \leqslant s \leqslant r} \|U_T^{in} f - g_s\|_{L^2_{\mu_y^{\mathscr{A}}}} < \varepsilon$$

for $1 \leqslant i \leqslant K$. We may use this to choose a coloring $c(i)$ on $[1, K]$ with $r$ colors (that depends on $y$ and $n$) by requiring that

$$\|U_T^{in} f - g_{c(i)}\|_{L^2_{\mu_y^{\mathscr{A}}}} < \varepsilon.$$

By van der Waerden's theorem there is a monochrome progression

$$\{i, i + d, \ldots, i + kd\} \subseteq \{1, \ldots, K\}$$

and so there is some $g_{s(y)} = g_*$ for which

$$\|U_T^{(i+jd)n} f - g_*\|_{L^2_{\mu_y^{\mathscr{A}}}} < \varepsilon$$

for $j = 0, \ldots, k$. Writing $\widetilde{g} = U_T^{-in} g_*$, this means that

$$\|U_T^{jdn} f - \widetilde{g}\|_{L^2_{\mu_{S^{in}y}^{\mathscr{A}}}} < \varepsilon$$

for $j = 0, \ldots, k$. Since $j = 0$ is allowed here, we also get

$$\|U_T^{jdn} f - f\|_{L^2_{\mu_{S^{in}y}^{\mathscr{A}}}} < 2\varepsilon \tag{7.33}$$

for $j = 1, \ldots, k$. This shows that, for a given $n$, the set

$$A \cap S^{-n} A \cap \cdots \cap S^{-Kn} A$$

is partitioned into finitely many sets $D_{n,1}, \ldots, D_{n,M}$, where $M$ is the number of arithmetic progressions of length $(k+1)$ in $[1, K]$, with the property that $i, d, \widetilde{g}$ do not change within such a given set. In particular, if $n \in R_K$, and so

$$\mu(A \cap S^{-n} A \cap \cdots \cap S^{-Kn} A) > c_1,$$

then for one of these sets $D$ (with corresponding progression

$$\{i, i+d, \ldots, i+kd\}$$

in $[1, K]$), we have $\mu(D) > \frac{c_1}{M}$.

Since $D \subseteq A \cap S^{-n} A \cap \cdots \cap S^{-Kn} A$, we have

$$\mu_{S^{in}y}^{\mathscr{A}}(B) > \tfrac{1}{2}\mu(B).$$

Thus, by the choice of $\varepsilon$, we have

$$\mu_{S^{in}y}^{\mathscr{A}}(B \cap T^{-dn} B \cap \cdots \cap T^{-k(dn)} B) = \int f U_T^{dn} f \cdots U_T^{kdn} f \, d\mu_{S^{in}y}^{\mathscr{A}}$$
$$> \int f^{k+1} \, d\mu_{S^{in}y}^{\mathscr{A}} - (k+1)\varepsilon$$
$$\text{(by (7.33))}$$
$$> \tfrac{1}{2}\mu(B) - (k+1)\varepsilon = \tfrac{1}{3}\mu(B).$$

Recall that this holds for all $y \in D$, and that on this set $i$ is constant, hence we may integrate the above inequality (with respect to $S^{in}y \in S^{in}D$) to get

$$\mu\left(B \cap T^{-dn} B \cap \cdots \cap T^{-k(dn)} B\right) > \tfrac{1}{3}\mu(B)\nu(D) \geqslant \tfrac{c_1}{3M}\mu(B)$$

which holds for all $n \in R_K$. Note that the number $d \in [1, K]$ may depend on $n$. To summarize, let $R'$ be the set of those $n$ for which

$$\mu\left(B \cap T^{-n}B \cap \cdots \cap T^{kn}\right) \geqslant \frac{c_1}{3M}\mu(B).$$

Then we have shown that for every $n \in R_K$ there is some $d \in [1, K]$ for which $dn \in R'$. This implies that the lower density of $R'$ is positive, in fact that

$$\liminf_{N \to \infty} \frac{|R' \cap [1, N]|}{N} \geqslant \frac{c_1}{2K^2}.$$

To see this, suppose that $N$ is large, so we have more than $\frac{c_1}{2K}N$ elements $n$ in $R_K \cap [1, \frac{N}{K}]$, and for each of those $n$ there exists a $d \in [1, K]$ with $dn \in R'$. As there are at most $K$ many $n'$ which can give rise to the same $dn$, the claim follows. This gives the theorem. $\qquad\qquad\qquad\qquad\qquad\qquad\qquad\qquad\square$

### 7.9.2 A Second Proof

The following argument, taken from the survey [107] by Furstenberg, Katznelson and Ornstein, does not use van der Waerden's theorem. It has the advantage that in showing the SZ property for $k$-term arithmetic progressions in the extension, it only uses the SZ property for $k$-term arithmetic progressions in the factor. There is a price to pay for this, in that the argument is a little more complicated.

SECOND PROOF OF PROPOSITION 7.23. By Lemma 7.24 we may assume that the function $f = \chi_B$ is AP, and that there exists a set $A \in \mathscr{A}$ of positive measure with $\mu_y^{\mathscr{A}}(B) > \frac{1}{2}\mu(B)$ for $y \in A$. By assumption, the factor $\mathsf{Y}$ satisfies the SZ property, which we will apply to a subset $\widetilde{A}$ of $A$ found below. We will then use the AP property for $f = \chi_B$ to prove the SZ property for $B$.

In fact we claim that for any $\varepsilon > 0$ there exists a subset $\widetilde{A} \subseteq A$ of positive measure and a finite set $F \subseteq \mathbb{Z}$ such that for any $n \geqslant 0$ there is some $m \in F$ with

$$\|f - U_T^{\ell(n+m)}f\|_{L^2_{\mu_y^{\mathscr{A}}}} \leqslant \varepsilon \tag{7.34}$$

for $\ell = 1, \ldots, k$, whenever $y \in \bigcap_{\ell=0}^{k} T^{-\ell n}\widetilde{A}$ (though we will use the claim only for $\varepsilon = \frac{1}{4}\mu(B)$.) Roughly speaking, this statement is desirable as there will be many such points $y \in \widetilde{A}$ and $n \geqslant 0$ (by property SZ applied to the set $\widetilde{A}$) and the restriction that $m$ belongs to a fixed finite set $F$ will allow us to deduce that there must be elements $x \in \bigcap_{\ell=0}^{k} T^{-\ell p}B$ for infinitely many $p \geqslant 0$. The inequality (7.34) may be viewed as the relative version of the fact that in a Kronecker system the set

$$\{n \in \mathbb{Z} \mid \|f - U_T^n f\|_2 < \varepsilon\}$$

is syndetic. As we will see, the claim (7.34) implies the main result rather quickly.

To prove the claim we introduce some notation. Let

$$\mathscr{L}(y) = \{(f, U_T^m f, \ldots, U_T^{km} f) \mid m \in \mathbb{Z}\} \subseteq \left(L_{\mu_y^{\mathscr{A}}}^2\right)^{k+1}$$

where we equip the space $\left(L_{\mu_y^{\mathscr{A}}}^2\right)^{k+1}$ with the norm

$$\| (f_0, f_1, \ldots, f_k) \| = \max_{\ell=0,\ldots,k} \|f_\ell\|_{L_{\mu_y^{\mathscr{A}}}^2}.$$

Clearly $\mathscr{L}(y)$ is a totally bounded subset of $\left(L_{\mu_y^{\mathscr{A}}}^2\right)^{k+1}$ with respect to this norm. It follows that the same holds for the subset

$$\mathscr{L}^*(y) = \{(f, U_T^m f, \ldots, U_T^{km} f) \mid m \in \mathbb{Z}, y \in \bigcap_{\ell=0}^{k} T^{-\ell m} A\}.$$

The additional condition in the definition of $\mathscr{L}^*(y)$ forces every component of the vector

$$(f, U_T^m f, \ldots, U_T^{km} f)$$

to be a non-zero element of $L_{\mu_y^{\mathscr{A}}}^2$. (Recall that $f = \chi_B$ and $\mu_x^{\mathscr{A}}(B) > \frac{1}{2}\mu(B)$ for $y \in A$.)

Since $\mathscr{L}^*(y)$ is totally bounded, for every $\varepsilon > 0$ there exists a finite maximal $\varepsilon$-separated subset of $\mathscr{L}^*(y)$. A set is called $\varepsilon$-*separated* if any pair of elements are at distance greater than $\varepsilon$ from each other.

For any finite set $F \subseteq \mathbb{Z}$ define $g_F(y) = 0$ if $y \notin \bigcap_{\ell=0}^{k} \bigcap_{m \in F} T^{-\ell m} A$ and define

$$g_F(y) = \min_{m \neq m' \in F} \max_{\ell=0,\ldots,k} \|U_T^{\ell m} f - U_T^{\ell m'} f\|_{L_{\mu_y^{\mathscr{A}}}^2}$$

if $y \in \bigcap_{\ell=0}^{k} \bigcap_{m \in F} T^{-\ell m} A$. Thus $g_F(y)$ measures how widely separated the vectors corresponding to $F \subseteq \mathbb{Z}$ are in $\mathscr{L}^*(y)$.

Also define

$$\mathrm{Sep}_\varepsilon(F) = \{y \mid g_F(y) > \varepsilon \text{ but } g_{\widetilde{F}} \leqslant \varepsilon \text{ whenever } |\widetilde{F}| > |F|\}$$

$$= \left\{ y \in \bigcap_{\ell=0}^{k} \bigcap_{m \in F} T^{-\ell m} A \mid F \text{ defines a maximal } \varepsilon\text{-separated} \right.$$

$$\left. \text{subset of } \mathscr{L}^*(y) \right\}.$$

The sets $\mathrm{Sep}_\varepsilon(F) \in \mathscr{A}$ for all finite subsets $F \subseteq \mathbb{Z}$ cover $A$: for almost every $y$ in $A$ the set $\mathscr{L}^*(y)$ is non-empty (because, for example, we may set $m = 0$ in the definition of $\mathscr{L}^*(y)$), and since $f$ is AP there exists a set $F$ with the property that $y \in \mathrm{Sep}_\varepsilon(F)$. Therefore, there exists a set $F$ such that $\mathrm{Sep}_\varepsilon(F)$ has positive measure. Moreover, there exists $\eta > 0$ such that

$$\{y \mid g_F(y) > \varepsilon + \eta \text{ but } g_{\widetilde{F}}(y) \leqslant \varepsilon \text{ for all } |\widetilde{F}| > |F|\} \tag{7.35}$$

has positive measure. Let $\widetilde{A} \in \mathscr{A}$ be a subset of the set defined by (7.35)—and hence also of $A$—with $\mu(\widetilde{A}) > 0$ and with the property that for all $m, m' \in F$ the function

$$\|U_T^{\ell m} f - U_T^{\ell m'} f\|_{L^2_{\mu_y^{\mathscr{A}}}} \tag{7.36}$$

changes its values inside $\widetilde{A}$ by less than $\eta$. Such a set may be found by decomposing the set (7.35) into finitely many measurable subsets each satisfying the property. We now assume that $y \in \bigcap_{\ell=0}^k T^{-\ell n} \widetilde{A}$ for some integer $n \geqslant 0$ (the existence of sufficiently many $y$ and $n$ with this property follows from property SZ for the set $\widetilde{A}$). Since $T^{\ell n} y \in \widetilde{A}$ we have $g_F(T^{\ell n} y) > 0$ by the construction of $\widetilde{A}$. From the definition of $g_F$, it follows that

$$T^{\ell(m+n)} y = T^{\ell m}(T^{\ell n} y) \in A$$

for all $m \in F$. The next lemma is the main step towards the proof of the claim in inequality (7.34), and hence Proposition 7.23.

**Lemma 7.25.** *For any $y \in \bigcap_{\ell=0}^k T^{-\ell n} \widetilde{A}$, the set*

$$B = \{(f, U_T^{n+m} f, \dots, U_T^{k(n+m)} f) \mid m \in F\} \qquad \bullet$$

*is an $\varepsilon$-separated subset of $\mathscr{L}^*(y)$.*

PROOF. Since $T^{\ell(n+m)} y \in A$ for $\ell = 0, \dots, k$ and $m \in F$, the set $B$ is in $\mathscr{L}^*(y)$. Take $m, m' \in F$, $m \neq m'$. Since $y \in \widetilde{A}$ belongs to the set in (7.35), there exists $\ell \leqslant k$ with

$$\|U_T^{\ell m} f - U_T^{\ell m'} f\|_{L^2_{\mu_y^{\mathscr{A}}}} > \varepsilon + \eta.$$

Since, moreover, $T^{n\ell} y \in \widetilde{A}$ we conclude from the small variance in (7.36) inside $\widetilde{A}$ that

$$\|U_T^{\ell m} f - U_T^{\ell m'} f\|_{L^2_{\mu_{T^{\ell n} y}^{\mathscr{A}}}} > \varepsilon,$$

and so

$$\|U_T^{\ell(m+n)} f - U_T^{\ell(m'+n)} f\|_{L^2_{\mu_y^{\mathscr{A}}}} > \varepsilon,$$

which proves the lemma. $\qquad\square$

However, the set corresponding to $\widetilde{F} = \{0\} \cup (F + n)$, that is

$$\{(f, f, \dots, f)\} \cup \{(f, U_T^{n+m} f, \dots, U_T^{k(n+m)} f) \mid m \in F\},$$

cannot be $\varepsilon$-separated by (7.35).

Thus, for any $y \in \bigcap_{\ell=0}^{k} T^{-\ell n} \widetilde{A}$, there is an $m \in F$ such that $(f, \ldots, f)$ is $\varepsilon$-close to $(f, U_T^{n+m} f, \ldots, U_T^{k(n+m)} f)$, so

$$\| f - U_T^{\ell(n+m)} f \|_{L^2_{\mu_y^{\mathscr{A}}}} \leqslant \varepsilon \tag{7.37}$$

for $\ell = 1, \ldots, k$. This was precisely our claim.

We now apply the claim with $\varepsilon = \mu(B)/4k$ to prove the proposition. Since $f = \chi_B$ we have $\| f - U_T^i f \|_\infty \leqslant 1$ for any $i$. Write $\underset{\varepsilon}{\approx}$ for equality to within error $\varepsilon$ (as in (7.37)). Assume once more that $y$ lies in $\bigcap_{\ell=0}^{k} T^{-\ell n} \widetilde{A}$, and let $m$ be as in (7.37). Then

$$\mu_y^{\mathscr{A}} \left( \bigcap_{\ell=0}^{k} T^{-\ell(n+m)} B \right) = \int f \underbrace{U_T^{n+m} f}_{\underset{\varepsilon}{\approx} f} \cdots \underbrace{U_T^{k(n+m)} f}_{\underset{\varepsilon}{\approx} f} \, d\mu_y^{\mathscr{A}}$$

$$= \int f^{k+1} \, d\mu_y^{\mathscr{A}} +$$

$$+ \sum_{\ell=1}^{k} \int f^\ell \underbrace{\left( U_T^{\ell(n+m)} f - f \right)}_{\| \cdot \|_{L^2_{\mu_y^{\mathscr{A}}}} < \varepsilon} U_T^{(\ell+1)(n+m)} f \cdots U_T^{k(n+m)} f \, d\mu_y^{\mathscr{A}}$$

$$\geqslant \mu_y^{\mathscr{A}}(B) - k\varepsilon \geqslant \tfrac{1}{2}\mu(B) - \tfrac{1}{4}\mu(B) = \tfrac{1}{4}\mu(B),$$

where we used the second property of $B$ ensured by Lemma 7.25 and our choice of $\varepsilon$.

Note that here $m$ depends on $y$, but the set $F$ is fixed. Thus

$$\sum_{m \in F} \mu_y^{\mathscr{A}} \left( \bigcap_{\ell=0}^{k} T^{-\ell(n+m)} B \right) \geqslant \frac{1}{4} \mu(B). \tag{7.38}$$

Integrating the inequality (7.38) over $\bigcap_{\ell=0}^{k} T^{-\ell n} \widetilde{A}$ gives

$$\sum_{m \in F} \mu \left( \bigcap_{\ell=0}^{k} T^{-\ell(n+m)} B \right) \geqslant \frac{1}{4} \mu(B) \mu \left( \bigcap_{\ell=0}^{k} T^{-\ell n} \widetilde{A} \right),$$

so

$$|F| \liminf_{N \to \infty} \frac{1}{N} \sum_{p=1}^{N} \mu \left( \bigcap_{\ell=0}^{k} T^{-\ell p} B \right) \geqslant \frac{1}{4} \mu(B) \liminf_{N \to \infty} \frac{1}{N} \sum_{n=1}^{N} \mu \left( \bigcap_{\ell=0}^{k} T^{-\ell n} \widetilde{A} \right) > 0$$

as required. $\qquad\qquad\qquad\qquad\qquad\qquad\qquad\qquad\qquad\qquad\qquad\qquad\qquad\square$

## 7.10 Chains of SZ Factors

We already showed that the Kronecker factor and any factor that can be obtained by a finite number of compact extensions of the Kronecker factor satisfy property SZ. However, there seems to be no reason why it should be possible to attain the original system in this manner. Thus we need to show that property SZ survives taking limits in the following sense (cf. Theorem 6.5).

**Proposition 7.26.** *Let* $X = (X, \mathscr{B}, \mu, T)$ *be invertible measure-preserving system on a Borel probability space, and let* $\mathscr{A}_1 \subseteq \mathscr{A}_2 \subseteq \cdots$ *be an increasing chain of factors (that is, $T$-invariant sub-$\sigma$-algebras of $\mathscr{B}$). Assume that $\mathscr{A}_n$ is SZ for all $n \geqslant 1$. Then the factor $\sigma\left(\bigcup_{n \geqslant 1} \mathscr{A}_n\right)$ is SZ also.*

PROOF. Let $\mathscr{A} = \sigma\left(\bigcup_{n \geqslant 1} \mathscr{A}_n\right)$ and let $A \in \mathscr{A}$. Then for any $\varepsilon > 0$ there exists some $n \geqslant 1$ and a set $A_1 \in \mathscr{A}_n$ with

$$\mu\left(A \triangle A_1\right) < \varepsilon.$$

Fix $k \geqslant 1$, let $\eta = \frac{1}{2(k+1)}$ and use the construction above for $\varepsilon = \frac{1}{4}\eta\mu(A)$. We define

$$A_0 = \{y \in A_1 \mid \mu_y^{\mathscr{A}_n}(A) \geqslant 1 - \eta\},$$

and claim that $\mu(A_0) > \frac{1}{2}\mu(A)$. This set will allow us to transport property SZ from $\mathscr{A}_n$ to $\mathscr{A}$ (see Fig. 7.4 for an indication of why the set $A_0$ could be helpful). To see the claim, notice that

$$
\begin{aligned}
\varepsilon = \tfrac{1}{4}\eta\mu(A) &> \mu(A_1 \smallsetminus A) && \text{(by definition of } A_1) \\
&= \int_{A_1} \mu_y^{\mathscr{A}_n}(A_1 \smallsetminus A)\, d\mu(y) && \text{(since } A_1 \in \mathscr{A}_n) \\
&\geqslant \int_{A_1 \smallsetminus A_0} \underbrace{\left(1 - \mu_y^{\mathscr{A}_n}(A)\right)}_{\geqslant \eta}\, d\mu(y) \\
&\geqslant \eta\mu(A_1 \smallsetminus A_0), && \text{(by definition of } A_0),
\end{aligned}
$$

and so

$$\mu(A_1 \smallsetminus A_0) < \tfrac{1}{4}\mu(A).$$

This implies that

$$\mu\left(A_0\right) = \mu\left(A_1\right) - \mu\left(A_1 \smallsetminus A_0\right) > \tfrac{3}{4}\mu\left(A\right) - \tfrac{1}{4}\mu\left(A\right)$$

by the construction of $A_1$. It follows that $\mu(A_0) > \frac{1}{2}\mu(A)$ as claimed.

We now show that $k$-multiple recurrence for the set $A_0 \in \mathscr{A}_n$ proves $k$-multiple recurrence for the given set $A$. More concretely, we claim that

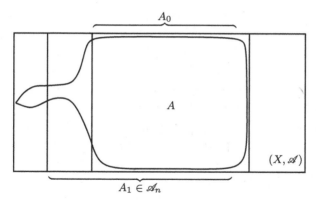

**Fig. 7.4** The set $A \in \mathscr{A}$ is approximated by $A_0 \in \mathscr{A}_n$ in a stronger sense than it is by $A_1 \in \mathscr{A}_n$

$$\mu\left(A \cap T^{-n}A \cap \cdots \cap T^{-kn}A\right) \geqslant \tfrac{1}{2}\mu\left(A_0 \cap T^{-n}A_0 \cap \cdots \cap T^{-kn}A_0\right). \quad (7.39)$$

To see this, suppose that

$$y \in A_0 \cap T^{-n}A_0 \cap \cdots \cap T^{-kn}A_0.$$

Then $y \in A_0$, so

$$\mu_y^{\mathscr{A}_n}(A) \geqslant 1 - \eta.$$

Similarly, $y \in T^{-jn}A_0$ for $0 \leqslant j \leqslant k$, so $\mu_{T^{jn}y}^{\mathscr{A}_n}(A) \geqslant 1 - \eta$ and hence

$$\mu_y^{\mathscr{A}_n}(T^{-jn}A) \geqslant 1 - \eta$$

for all $j$, $0 \leqslant j \leqslant k$. It follows that

$$\mu_y^{\mathscr{A}_n}\left(A \cap T^{-n}A \cap \cdots \cap T^{-kn}A\right) \geqslant 1 - (k+1)\eta = \tfrac{1}{2},$$

which shows the inequality (7.39) by integration.

Finally, by property SZ for the factor $\mathscr{A}_n$,

$$\liminf_{N \to \infty} \frac{1}{N} \sum_{n=1}^{N} \mu\left(A \cap T^{-n}A \cap \cdots \cap T^{-kn}A\right)$$

$$\geqslant \frac{1}{2} \liminf_{N \to \infty} \frac{1}{N} \sum_{n=1}^{N} \mu\left(A_0 \cap T^{-n}A_0 \cap \cdots \cap T^{-kn}A_0\right) > 0,$$

as required. $\qquad\qquad\qquad\qquad\qquad\qquad\qquad\qquad\qquad\qquad\qquad\qquad\square$

A consequence of Proposition 7.26 is that the SZ property can be transported up the diagram in Fig. 7.2: the assumption that $(X, \mathscr{B})$ is a Borel probability space means it is sufficient to only consider sequences of factors rather than totally ordered families of factors.

## 7.11 SZ for Relatively Weak-Mixing Extensions

The following result (and its second corollary below) comprise the last main step towards the proof of Theorem 1.5 (see Fig. 7.2 and Sect. 7.12).

**Theorem 7.27.** *Let*

$$\mathsf{X} = (X, \mathscr{B}, \mu, T)$$
$$\downarrow$$
$$\mathsf{Y} = (Y, \mathscr{A}, \nu, S)$$

*be a relatively weak-mixing extension of measure-preserving transformations of Borel probability spaces. Then for any $k \geqslant 1$ and sets $B_0, \ldots, B_k \in \mathscr{B}$,*

$$\lim_{N \to \infty} \frac{1}{N} \sum_{n=1}^{N} \int \Big[ \mu_y^{\mathscr{A}} (B_0 \cap \cdots \cap T^{-kn} B_k)$$
$$- \mu_y^{\mathscr{A}} (B_0) \mu_y^{\mathscr{A}} (T^{-n} B_1) \cdots \mu_y^{\mathscr{A}} (T^{-kn} B_k) \Big]^2 \, \mathrm{d}\mu = 0. \tag{7.40}$$

*Equivalently, for a given $\varepsilon > 0$ there exists some $N_0$ such that, for $N \geqslant N_0$,*

$$\left| \mu_y^{\mathscr{A}} (B_0 \cap T^{-n} B_1 \cap \cdots \cap T^{-kn} B_k) - \mu_y^{\mathscr{A}} (B_0) \mu_y^{\mathscr{A}} \cdots \mu_y^{\mathscr{A}} (T^{-kn} B_k) \right| < \varepsilon \tag{7.41}$$

*for all $(n, y) \in [1, N] \times Y$ except possibly for a subset of $[1, N] \times Y$ of relative measure $\varepsilon$ (with respect to the product of normalized counting measure on $[1, N]$ and $\nu$).*

The (7.41) in the theorem could also be written in the shorthand form

$$\mu_y^{\mathscr{A}} (B_0 \cap T^{-n} B_1 \cap \cdots \cap T^{-kn} B_k) \underset{\varepsilon}{\approx} \mu_y^{\mathscr{A}} (B_0) \mu_y^{\mathscr{A}} (T^{-n} B_1) \cdots \mu_y^{\mathscr{A}} (T^{-kn} B_k)$$

for large $N$.

Notice that for $k = 1$ and $\mathsf{Y}$ trivial, (7.40) is one of the equivalent characterizations of weak-mixing for $\mathsf{X}$ by Theorem 2.36.

For $k \geqslant 1$ Theorem 7.27 gives the following immediate corollary\*, which colloquially means that weak-mixing implies weak-mixing of all orders.

**Corollary 7.28.** *If $(X, \mathscr{B}, \mu, T)$ is a weak-mixing measure-preserving system, then for $B_0, \ldots, B_k \in \mathscr{B}$,*

---

\* Just as in Sects. 7.2.1 and 7.2.2, we do not have to assume that the space is a Borel space and the transformation is invertible here.

$$\lim_{N \to \infty} \frac{1}{N} \sum_{n=1}^{N} \left[ \mu(B_0 \cap T^{-n}B_1 \cap \cdots \cap T^{-kn}B_k) \right.$$

$$\left. - \mu(B_0)\mu(B_1)\cdots\mu(B_k) \right]^2 = 0.$$

**Corollary 7.29.** *Let*

$$\mathsf{X} = (X, \mathscr{B}, \mu, T)$$
$$\downarrow$$
$$\mathsf{Y} = (Y, \mathscr{A}, \nu, S)$$

*be a relatively weak-mixing extension of invertible measure-preserving transformations on Borel probability spaces. If* $\mathsf{Y}$ *satisfies property SZ, then so does* $\mathsf{X}$.

Szemerédi's theorem will follow from Corollary 7.29 by the argument in Sect. 7.12. Corollary 7.28, which says that weak-mixing implies weak-mixing of all orders, is an immediate consequence of Theorem 7.27; we next prove Corollary 7.29 assuming Theorem 7.27, whose proof begins on p. 224 after some more preparatory material.

PROOF OF COROLLARY 7.29 ASSUMING THEOREM 7.27. Let $B \in \mathscr{B}$ be a set with $\mu(B) > 0$, and for each $a > 0$ set

$$A = \{y \mid \mu_y^{\mathscr{A}}(B) > a\}.$$

Choose and fix a value of $a > 0$ for which $\mu(A) > 0$ (there must be such an $a$ by Theorem 5.14(1)). Then, given $\varepsilon > 0$,

$$\frac{1}{N} \sum_{n=1}^{N} \mu\left(B \cap \cdots \cap T^{-kn}B\right) = \frac{1}{N} \sum_{n=1}^{N} \int \mu_y^{\mathscr{A}}\left(B \cap \cdots \cap T^{-kn}B\right) \, d\mu(y)$$

$$\geqslant \frac{1}{N} \sum_{n=1}^{N} \int \left(\mu_y^{\mathscr{A}}(B)\cdots \mu_y^{\mathscr{A}}(T^{-kn}B) - \varepsilon\right) d\mu(y) - \varepsilon$$

$$(7.42)$$

by Theorem 7.27, if $N$ is large enough. Indeed, (7.41) shows that

$$\mu_y^{\mathscr{A}}\left(B \cap \cdots \cap T^{-kn}B\right) \geqslant \mu_y^{\mathscr{A}}(B)\cdots \mu_y^{\mathscr{A}}(T^{-kn}B) - \varepsilon$$

for most $(n, y) \in [1, N] \times Y$; more precisely, for all but $\varepsilon$ in proportion of all the possible pairs

$$(n, y) \in [1, N] \times Y.$$

Moreover, for those pairs $(n, y)$ for which we do not have (7.41), we do have

$$\mu_y^{\mathscr{A}}\left(B \cap \cdots \cap T^{-kn}B\right) \geqslant 0 \geqslant \mu_y^{\mathscr{A}}(B)\cdots \mu_y^{\mathscr{A}}(T^{-kn}B) - \varepsilon - 1,$$

so integrating over all of $[1, N] \times Y$ gives (7.42). Now notice that for $y \in A \cap T^{-n} A \cap \cdots \cap T^{-kn} A$ we have

$$\mu_y^{\mathscr{A}} \left( T^{-\ell n} B \right) = \mu_{T^{\ell n} y}^{\mathscr{A}} (B) > a$$

for $0 \leqslant \ell \leqslant k$. So if we restrict, for each $n$ individually, the integral to this intersection we get from (7.42)

$$\frac{1}{N} \sum_{n=1}^{N} \mu \left( B \cap \cdots \cap T^{-kn} B \right) \geqslant (a^{k+1} - \varepsilon) \frac{1}{N} \sum_{n=1}^{N} \mu \left( A \cap \cdots \cap T^{-kn} A \right) - \varepsilon.$$

It follows that

$$\liminf_{N \to \infty} \frac{1}{N} \sum_{n=1}^{N} \mu \left( B \cap \cdots \cap T^{-kn} B \right) \geqslant a^{k+1} \liminf_{N \to \infty} \sum_{n=1}^{N} \mu \left( A \cap \cdots \cap T^{-kn} A \right) > 0$$

as required, by property SZ for $A \in \mathscr{A}$.                                      □

It remains to prove Theorem 7.27, and we start with the case $k = 1$ which we will use in the general case.

**Proposition 7.30.** *Let*

$$(X, \mathscr{B}, \mu, T)$$
$$\downarrow$$
$$(Y, \mathscr{A}, \nu, S)$$

*be a relatively weak-mixing extension of measure-preserving transformations on Borel probability spaces. Then for $f, g \in L^\infty(X, \mathscr{B}, \mu)$,*

$$\lim_{N \to \infty} \frac{1}{N} \sum_{n=1}^{N} \left\| E \left( f U_T^n g \big| \mathscr{A} \right) - E \left( f \big| \mathscr{A} \right) U_T^n E \left( g \big| \mathscr{A} \right) \right\|_2 = 0. \qquad (7.43)$$

*Equivalently, given $\varepsilon > 0$, for large enough $n$,*

$$E \left( f U_T^n g \big| \mathscr{A} \right) \underset{\varepsilon}{\approx} E \left( f \big| \mathscr{A} \right) U_T^n E \left( g \big| \mathscr{A} \right) \qquad (7.44)$$

*for all but $\varepsilon$ in measure of points $(n, y) \in [1, N] \times X$, for large enough $N$.*

Notice that by Lemma 2.41 on p. 54, (7.43) is equivalent to the same statement for the square of the $L^2$ norm. The equivalence of (7.44) and (7.43) is a consequence of the maximal inequality

$$m \left( \{ z \mid F(z) > \varepsilon \} \right) < \frac{\|F\|_1}{\varepsilon},$$

applied to the function

$$F(n, x) = \left| E \left( f U_T^n g \big| \mathscr{A} \right)(x) - E \left( f \big| \mathscr{A} \right)(x) U_T^n E \left( g \big| \mathscr{A} \right)(x) \right|^2$$

on the probability space $[1, N] \times X$. Notice that this implies that the result holds for $f = f_1 + f_2$ and $g$ whenever it holds for the two pairs $f_1, g$ and $f_2, g$. Also recall from Proposition 6.16(3) the formula

$$\int_X E\left(f \middle| \mathscr{A}\right) E\left(g \middle| \mathscr{A}\right) \, d\mu = \int_{\widehat{X}} f \otimes g \, d\widehat{\mu} \tag{7.45}$$

where $\widehat{X} = X \times X$ and $\widehat{\mu} = \mu \times_{\mathsf{Y}} \mu$ is the relatively independent joining of $\mu$ with itself over the factor $\mathsf{Y}$.

PROOF OF PROPOSITION 7.30. For $f_1 \in L^\infty(\mathscr{A})$,

$$E(f_1 U_T^n g \middle| \mathscr{A}) = f_1 E(U_T^n g \middle| \mathscr{A}) = E(f_1 \middle| \mathscr{A}) U_T^n E(g \middle| \mathscr{A}),$$

so in this case the statement holds trivially. Since

$$f = E(f \middle| \mathscr{A}) + (f - E(f \middle| \mathscr{A})),$$

we may assume without loss of generality that $E(f_1 \middle| \mathscr{A}) = 0$. By assumption $\widehat{T} = T \times T$ acts ergodically on $\widehat{X}$ with respect to $\widehat{\mu}$. Using the mean ergodic theorem (Theorem 2.21) and (7.45) twice we see that

$$\lim_{N \to \infty} \frac{1}{N} \sum_{n=1}^{N} \int \left[E\left(fU_T^n g \middle| \mathscr{A}\right)\right]^2 \, d\mu = \lim_{N \to \infty} \int f \otimes f \underbrace{\frac{1}{N} \sum_{n=1}^{N} U_{\widehat{T}}^n (g \otimes g)}_{\to C \text{ in } L^2} \, d\widehat{\mu}$$

$$= \int f \otimes f \cdot C \, d\widehat{\mu}$$

$$= C \int E\left(f \middle| \mathscr{A}\right)^2 \, d\mu = 0.$$

$\square$

We are now ready to start the proof of Theorem 7.27 for a general $k \geqslant 1$.

**Proposition 7.31.** *Let*

$$(X, \mathscr{B}, \mu, T)$$
$$\downarrow$$
$$(Y, \mathscr{A}, \nu, S)$$

*be a relatively weak-mixing extension of invertible measure-preserving systems on Borel probability spaces. Then for $f_1, \ldots, f_k \in L^\infty(X)$,*

$$\left\| \frac{1}{N} \sum_{n=1}^{N} U_T^n f_1 \cdots U_T^{kn} f_k - \frac{1}{N} \sum_{n=1}^{N} U_T^n E\left(f_1 \middle| \mathscr{A}\right) \cdots U_T^{kn} E\left(f_k \middle| \mathscr{A}\right) \right\|_2 \longrightarrow 0$$

*as $N \to \infty$.*

In contrast to Theorem 7.27, Proposition 7.31 has the average inside the norm, which appears to be a much weaker statement. Also notice that inside the norm only one average involves the expectation operator $E\left(\cdot\middle|\mathscr{A}\right)$.

PROOF OF PROPOSITION 7.31. This is proved by induction on $k$ and van der Corput's Lemma (Theorem 7.11). The case $k = 1$ is a consequence of the mean ergodic theorem (Theorem 2.21) as in this case both averages inside the $L^2$-norm converge in $L^2$ to $\int f_1\,\mathrm{d}\mu$.

For the inductive step we write the difference inside the norm in the form

$$\frac{1}{N}\sum_{n=1}^{N}U_T^n f_1\cdots U_T^{kn}f_k - \frac{1}{N}\sum_{n=1}^{N}U_T^n E(f_1\middle|\mathscr{A})\cdots U_T^{kn}E(f_k\middle|\mathscr{A})$$

$$= \frac{1}{N}\sum_{n=1}^{N}U_T^n\left(f_1 - E(f_1\middle|\mathscr{A})\right)U_T^{2n}f_2\cdots U_T^{kn}f_k$$

$$+\cdots+\frac{1}{N}\sum_{n=1}^{N}U_T^n E(f_1\middle|\mathscr{A})\cdots U_T^{kn}\left(f_k - E(f_k\middle|\mathscr{A})\right) \quad (7.46)$$

as a telescoping sum. Each individual sum on the right-hand side of (7.46) has the same shape

$$\frac{1}{N}\sum_{n=1}^{N}U_T^n h_1\cdots U_T^{kn}h_k$$

for various choices of functions $h_1,\ldots,h_k$ with the property that in each case there is one function $h_\ell$ with $E(h_\ell\middle|\mathscr{A}) = 0$. We can therefore assume without loss of generality that for some $\ell$ we have $E(f_\ell\middle|\mathscr{A}) = 0$.

The van der Corput lemma may be applied as follows. Let

$$u_n = U_T^n f_1\cdots U_T^{kn}f_k.$$

Then

$$\frac{1}{N}\sum_{n=1}^{N}\langle u_n, u_{n+h}\rangle = \frac{1}{N}\sum_{n=1}^{N}\int U_T^n f_1\cdots U_T^{kn}f_k U_T^{n+h}f_1\cdots U_T^{k(n+h)}f_k\,\mathrm{d}\mu$$

$$= \frac{1}{N}\sum_{n=1}^{N}\int f_1 U_T^h f_1 U_T^n(f_2 U_T^{2h}f_2)\cdots U_T^{(k-1)n}(f_k U_T^{kh}f_k)\,\mathrm{d}\mu$$

$$= \int f_1 U_T^h f_1 \frac{1}{N}\sum_{n=1}^{N}U_T^n(f_2 U_T^h f_2)\cdots U_T^{(k-1)n}(f_k U_T^{kh}f_k)\,\mathrm{d}\mu.$$

However, for the average inside the integral we may use the inductive hypothesis as follows. With an error (for large enough $N$) of at most $\varepsilon\|f_1\|_\infty^2$

we have

$$\frac{1}{N}\sum_{n=1}^{N}\langle u_n, u_{n+h}\rangle \underset{O_{f_1}(\varepsilon)}{\approx} \int f_1 U_T^h f_1 \frac{1}{N}\sum_{n=1}^{N} U_T^n E(f_2 U_T^h f_2 | \mathscr{A})$$

$$\cdots U_T^{(k-1)n} E(f_k U_T^{kh} f_k | \mathscr{A}) \, d\mu$$

$$= \frac{1}{N}\sum_{n=1}^{N} \int E(f_1 U_T^h f_1 | \mathscr{A}) U_T^n \left( E(f_2 U_T^{2h} f_2 | \mathscr{A}) \right)$$

$$\cdots U_T^{(k-1)n} E(f_k U_T^{kh} f_k | \mathscr{A}) \, d\mu. \qquad (7.47)$$

Recall that we assume without loss of generality that $E(f_\ell | \mathscr{A}) = 0$. It follows that each individual integral in the average (7.47) is bounded in absolute value by $C\|E(f_\ell U_T^{\ell h} f_\ell | \mathscr{A})\|_2$, for some constant $C$ depending on $f_1, \ldots, f_k$, so (in the notation of Theorem 7.11)

$$s_h \leqslant C \left\| E\left(f_\ell U_T^{\ell h} f_\ell \big| \mathscr{A}\right) \right\|_2.$$

It follows that

$$\frac{1}{H}\sum_{h=0}^{H-1} s_h \leqslant \frac{C}{H}\sum_{h=0}^{H-1} \left\| E\left(f_\ell U_T^{\ell h} f_\ell \big| \mathscr{A}\right) \right\|_2.$$

Proposition 7.30 implies that the last average converges to zero as $H \to \infty$ by the assumption that $E(f_\ell | \mathscr{A}) = 0$. Thus, by Theorem 7.11,

$$\left\| \frac{1}{N}\sum_{n=1}^{N} u_n \right\|_2 = \left\| \frac{1}{N}\sum_{n=1}^{N} U_T^n f_1 \cdots U_T^{kn} f_k \right\|_2 \longrightarrow 0$$

as required. $\qquad\qquad\qquad\qquad\qquad\qquad\qquad\qquad\qquad\qquad\qquad\square$

Recall from Sect. 6.5 that the relatively independent joining $\mathsf{X} \times_{\mathscr{A}} \mathsf{X}$ is also an extension of $\mathsf{Y}$. To obtain Theorem 7.27 from Proposition 7.31 we first need to prove an analog of the implication (1) $\implies$ (3) in Theorem 2.36.

**Lemma 7.32.** *If* $\mathsf{X} \to \mathsf{Y} = (Y, \mathscr{A})$ *is a relatively weak-mixing extension, then so is* $\mathsf{X} \times_{\mathscr{A}} \mathsf{X} \to \mathsf{Y}$.

PROOF. Let $\widehat{X} = X \times X$, $\widehat{T} = T \times T$, and

$$\widehat{\mu} = \mu \times_{\mathscr{A}} \mu = \int \mu_y^{\mathscr{A}} \times \mu_y^{\mathscr{A}} \, d\mu(y).$$

By the hypothesis, the system $\widehat{\mathsf{X}} = (\widehat{X}, \widehat{T}, \widehat{\mu})$ is ergodic. This characterization of relative weak-mixing in terms of ergodicity of a product will be used for $\widehat{\mathsf{X}} \times_{\mathscr{A}} \widehat{\mathsf{X}}$ itself, which involves the system $\mathsf{X}^4$.

Write $\widetilde{X} = \widehat{X} \times \widehat{X}$, $\widetilde{T} = \widehat{T} \times \widehat{T}$, and

$$\widetilde{\mu} = \widehat{\mu} \times_{\mathscr{A}} \widehat{\mu} = \int \widehat{\mu}^{\mathscr{A}}_{(x_1,x_2)} \times \widehat{\mu}^{\mathscr{A}}_{(x_1,x_2)} \, d\widehat{\mu}(x_1,x_2)$$
$$= \int (\mu^{\mathscr{A}}_y)^4 \, d\mu(y),$$

where we have used Sect. 6.5. We wish to show that $(\widetilde{X}, \widetilde{T}, \widetilde{\mu})$ is ergodic.

Write $F = f_1 \otimes f_2 \otimes f_3 \otimes f_4$ and $G = g_1 \otimes g_2 \otimes g_3 \otimes g_4$ for $f_i, g_j \in L^\infty(X)$. Now for given $\varepsilon > 0$, if $N$ is large enough, Proposition 7.30 shows (a more careful argument using the reasoning from (7.42) would allow the $O(\varepsilon)$ error to be made more explicit) that for large enough $N$

$$\frac{1}{N} \sum_{n=1}^{N} \int F U_{\widetilde{T}}^n G \, d\widetilde{\mu} = \frac{1}{N} \sum_{n=1}^{N} \int E\left(f_1 U_T^n g_1 \big| \mathscr{A}\right) \cdots E\left(f_4 U_T^n g_4 \big| \mathscr{A}\right) d\mu$$

$$\underset{O(\varepsilon)}{\approx} \frac{1}{N} \sum_{n=1}^{N} \int E\left(f_1 \big| \mathscr{A}\right) U_T^n E\left(g_1 \big| \mathscr{A}\right)$$
$$\cdots E\left(f_4 \big| \mathscr{A}\right) U_T^n E\left(g_4 \big| \mathscr{A}\right) d\mu$$

$$= \frac{1}{N} \sum_{n=1}^{N} \int E\left(f_1 \big| \mathscr{A}\right) \cdots E\left(f_4 \big| \mathscr{A}\right)$$
$$U_T^n E\left(g_1 \big| \mathscr{A}\right) \cdots U_T^n E\left(g_4 \big| \mathscr{A}\right) d\mu$$

$$\longrightarrow \int E\left(f_1 \big| \mathscr{A}\right) \cdots E\left(f_4 \big| \mathscr{A}\right) d\mu$$
$$\int E\left(g_1 \big| \mathscr{A}\right) \cdots E\left(g_4 \big| \mathscr{A}\right) d\mu$$
$$\text{(since } (T, \mu) \text{ is ergodic)}$$

$$= \int F \, d\widetilde{\mu} \int G \, d\widetilde{\mu},$$

so $(\widetilde{X}, \widetilde{T}, \widetilde{\mu})$ is ergodic.                                        □

We are now in a position to prove Theorem 7.27, which is the last step towards Szemerédi's theorem.

PROOF OF THEOREM 7.27. We wish to show, by induction on $k$, that for functions

$$f_0, \ldots, f_k \in L^\infty(X),$$

we have

$$\lim_{N \to \infty} \frac{1}{N} \sum_{n=1}^{N} \int \left[E\left(f_0 U_T^n f_1 \cdots U_T^{kn} f_k \big| \mathscr{A}\right)\right.$$
$$\left. - E\left(f_0 \big| \mathscr{A}\right) U_T^n E\left(f_1 \big| \mathscr{A}\right) \cdots U_T^{kn} E\left(f_k \big| \mathscr{A}\right)\right]^2 d\mu = 0. \quad (7.48)$$

Note that applying this claim for the characteristic functions $f_\ell = \chi_{B_\ell}$ gives the first statement in Theorem 7.27. We also note that (7.48) is equivalent to the following claim (which is the generalization of (7.41) in the statement of Theorem 7.27): For any $\varepsilon > 0$ there is an $N_0$ such that, for all $N > N_0$,

$$\left| E\left(f_0 U_T^n f_1 \cdots U_T^{kn} f_k \big| \mathscr{A}\right) - E(f_0 | \mathscr{A}) U_T^n E(f_1 | \mathscr{A}) \cdots U_T^{kn} E(f_k | \mathscr{A}) \right| (y) < \varepsilon$$

for all $(n, y) \in [1, N] \times Y$ except possibly for a set of measure $\varepsilon$. The equivalence follows from the argument immediately after Proposition 7.30. Assume first that $f_k \in L^\infty(Y, \mathscr{A})$, in which case

$$E\left(f_0 U_T^n f_1 \cdots U_T^{kn} f_k \big| \mathscr{A}\right) = E\left(f_0 U_T^n f_1 \cdots U_T^{(k-1)n} f_{k-1}\right) U_T^{kn} E(f_k | \mathscr{A})$$

by Theorem 5.1(3). Using the second formulation of the claim for the functions $f_0, \ldots, f_{k-1}$ we deduce the same claim for $f_0, \ldots, f_k$ in the case considered.

Expressing, in the general case, the function $f_k$ as

$$E(f_k | \mathscr{A}) + \left(f_k - E(f_k | \mathscr{A})\right),$$

it follows that it is enough to consider the case $E(f_k | \mathscr{A}) = 0$ (the linearity needed for this reduction is easily seen in the second formulation of the claim).

By Lemma 7.32 we know that $(X \times X, T \times T, \mu \times_Y \mu)$ is a relatively weak-mixing extension of $Y$. Apply Proposition 7.31 to this extension and the functions $f_1 \otimes f_1, \ldots, f_k \otimes f_k$ to see that

$$\left\| \frac{1}{N} \sum_{n=1}^N U_{\widehat{T}}^n (f_1 \otimes f_1) U_{\widehat{T}}^{2n}(f_2 \otimes f_2) \cdots U_{\widehat{T}}^{kn}(f_k \otimes f_k) \right\|_2 \longrightarrow 0. \qquad (7.49)$$

For this also notice that

$$E\left(f_k \otimes f_k \big| \mathscr{A}\right) = E\left(f_k | \mathscr{A}\right) \otimes E\left(f_k | \mathscr{A}\right) = 0$$

by our assumption on $f_k$ and Proposition 6.16. Recalling (7.45) and taking the inner product of (7.49) with $f_0 \otimes f_0$ gives

$$\int \left( \frac{1}{N} \sum_{n=0}^{N-1} (f_0 \otimes f_0) U_{\widehat{T}}^n (f_1 \otimes f_1) U_{\widehat{T}}^{2n}(f_2 \otimes f_2) \cdots U_{\widehat{T}}^{kn}(f_k \otimes f_k) \right) d\widehat{\mu}$$

$$= \frac{1}{N} \sum_{n=0}^{N-1} \int E\left(f_0 U_T^n f_1 \cdots U_T^{kn} f_k \big| \mathscr{A}\right)^2 d\mu \longrightarrow 0,$$

which concludes the inductive step and hence the proof of Theorem 7.27. $\square$

## 7.12 Concluding the Proof

Proposition 7.12 shows in particular that any Kronecker system $Y_0$ is SZ (this can also be deduced from Proposition 7.23). Moreover, Proposition 7.23 shows that any compact extension $Y_1 \to Y_0$ of a Kronecker system is also SZ; similarly a system obtained from a finite number of compact extensions,

$$Y_n \to Y_{n-1} \to \cdots \to Y_1 \to Y_0$$

is also SZ. Proposition 7.26 takes this further: if the $\sigma$-algebra of $X_\infty$ is generated by all the $\sigma$-algebras of factors (cf. Theorem 6.5 in Chap. 6 identifying factors with invariant sub-$\sigma$-algebras)

$$X_\infty \to \cdots \to Y_{n+1} \to Y_n \to \cdots \to Y_1 \to Y_0,$$

and $Y_{n+1} \to Y_n$ is a compact extension for all $n$, then $X_\infty$ is SZ.

Now let $X$ be a measure-preserving system on a Borel probability space. We claim that there exists a factor $X \to Y$ such that $Y$ is SZ and $X \to Y$ is relatively weak-mixing by the following argument, which will complete the proof of Theorem 7.4 by Corollary 7.29. If $X$ is weak-mixing, then we may take the trivial factor for $Y$. If $X$ is not weak-mixing, then it has a Kronecker factor $Y_0$, which is SZ as above. The claim is proved in this case by transfinite induction[70].

Suppose we have already found an ordinal number $\alpha$ with the following property. For every $\beta < \alpha$ there is a factor $Y_\beta$ of $X$ with property SZ such that, if $\beta + 1 < \alpha$, then $Y_{\beta+1} \to Y_\beta$ is a proper compact extension, and, if $\gamma < \alpha$ is a limit ordinal, then the $\sigma$-algebra corresponding to the factor $Y_\gamma$ is generated by the $\sigma$-algebras of $Y_\beta$ for $\beta < \gamma$.

We prove the inductive step as follows. If $\alpha = \beta + 1$ is a successor, then there are two possibilities. Either the extension $X \to Y_\beta$ is relatively weak-mixing; if so the claim is proved. If not, then by Theorem 7.21 there exists an intermediate system $X \to Y_\alpha \to Y_\beta$ such that $Y_\alpha \to Y_\beta$ is a proper compact extension. By Proposition 7.23, this implies that $Y_\alpha$ is SZ and this case of the inductive step is concluded. Suppose now that $\alpha$ is a limit ordinal. By assumption, the extension $Y_{\beta+1} \to Y_\beta$ is a proper extension for every $\beta < \alpha$, so $L^2(X) \supseteq L^2(Y_{\beta+1}) \supsetneq L^2(Y_\beta)$. Since $X$ is a Borel probability space, $L^2(X)$ is separable and so this chain of closed subspaces has to be countable. Thus $\alpha$ is a countable ordinal and we may write $\alpha = \lim_{n \to \infty} \beta_n$ for some sequence $(\beta_n)$ of ordinals with $\beta_n < \alpha$ for all $n$. Let $Y_\alpha$ be the factor corresponding to the $\sigma$-algebra generated by the $\sigma$-algebras of $Y_{\beta_n}$ (or, equivalently, of $Y_\beta$ for $\beta < \alpha$). Then Proposition 7.26 shows that $Y_\alpha$ is SZ, which concludes the inductive step. Moreover, the inductive step has also shown that if the inductive hypothesis holds for a limit ordinal $\alpha$, then $\alpha$ must be countable. Let $\omega_1$ be the first uncountable ordinal. Then the construction of the factors has to stop at some $\beta < \omega_1$, for otherwise $\omega_1$ would fulfill the inductive hypothesis, which would contradict the assumption that $X$ is itself a Borel space as we have seen. However, the only way for the construction to end is with the proof of the claim that $\mathsf{Y}_\beta$ is SZ and $\mathsf{X} \to \mathsf{Y}_\beta$ is relatively weak-mixing.

# 7.13 Further Results in Ergodic Ramsey Theory

Ergodic Ramsey theory, the study of combinatorial, geometrical, and arithmetical structures preserved in sufficiently large subsets of a structure using methods and results from multiple recurrence in ergodic theory, is a large area with many profound and novel applications. This is an area which continues to see dramatic progress, and we merely mention some of the ideas and results that arise; an attractive overview may be found in Bergelson's notes [26].

The next result, due to Bergelson and Leibman [29], subsumes many earlier results of Szemerédi type[71]. The notion of positive upper density extends easily to subsets of groups like $\mathbb{Z}^d$, just by counting the proportion of elements in the set inside a $d$-dimensional box of side $n$ and letting $n$ go to infinity.

**Theorem (Bergelson and Leibman).** For any $r, \ell \in \mathbb{N}$ and multi-variable polynomial $P : \mathbb{Z}^r \to \mathbb{Z}^\ell$ with $P(0) = 0$, any set $S \subseteq \mathbb{Z}^\ell$ with positive upper density, and any finite set $F \subseteq \mathbb{Z}^r$, there is an $n \in \mathbb{N}$ and a $u \in \mathbb{Z}^\ell$ for which $u + P(nF) \subseteq S$.

### 7.13.1 Other Furstenberg Ergodic Averages

The ergodic theorems in Chap. 2 concern averages of the form

$$\mathsf{A}_N^f = \frac{1}{N} \sum_{n=0}^{N-1} U_T^n f$$

and they describe the limit and mode of convergence in great detail. Furstenberg's proof of Szemerédi's theorem involves properties of averages of the form

$$\frac{1}{N} \sum_{n=0}^{N-1} \mu \left( A \cap T^{-n} A \cap \cdots \cap T^{-kn} A \right)$$

or more generally of the form

$$\frac{1}{N} \sum_{n=0}^{N-1} U_T^n f_1 \cdots U_T^{kn} f_k,$$

but does not say whether the limits exist, nor does it describe the limit when it does exist. Apart from some special cases, the existence of such limits was an open problem until quite recently. Host and Kra [159], and independently Ziegler [393], proved the existence of the limit

$$\lim_{N \to \infty} \frac{1}{N} \sum_{n=0}^{N-1} U_T^n f_1 \cdots U_T^{kn} f_k \tag{7.50}$$

in the $L^2$ sense for $f_1, \ldots, f_k \in L^\infty$. In both approaches, a fundamental role is played by *characteristic factors*. A factor $\mathscr{A}$ is called characteristic for the expression (7.50) if

$$\lim_{N \to \infty} \left\| \frac{1}{N} \sum_{n=0}^{N-1} \left( U_T^n f_1 \cdots U_T^{kn} f_k - U_T^n E\left(f_1 \middle| \mathscr{A}\right) \cdots U_T^{kn} E\left(f_k \middle| \mathscr{A}\right) \right) \right\|_2 = 0.$$

The strategy is to find characteristic factors that are manageable and minimal; where this is achieved it makes sense to speak of "the" characteristic factor. Having found a good description of the characteristic factor, the question of the existence of the limit then reduces to a more concrete problem for that factor. To see how this might proceed, we briefly outline the first few cases.

For $k = 1$ this corresponds to the formulation of the ergodic theorem in Theorem 6.1, where the characteristic factor is

$$\mathscr{E} = \{B \in \mathscr{B} \mid T^{-1}B = B\},$$

the $\sigma$-algebra of invariant sets, by the mean ergodic theorem.

For $k = 2$ the convergence was shown by Furstenberg [102]. Assuming that X is an ergodic system, the characteristic factor for

$$\lim_{N \to \infty} \frac{1}{N} \sum_{n=0}^{N-1} U_T^n f_1 U_T^{2n} f_2$$

is the Kronecker factor. This allows the convergence question to be reduced to the case of a rotation on a compact group considered in Sect. 7.5.1, which was our approach in Sect. 7.6.

For $k = 3$ the problem is much deeper. Conze and Lesigne [58, 59] showed that the characteristic factor is a 2-step nilpotent system $(M, T)$, in which $M$ is a compact homogeneous space of a 2-step nilpotent group $G$ with Haar measure, and the map $T$ is translation by an element of $G$. If $G$ is a Lie group, then $M$ is a nilmanifold; if $G$ is a projective limit of nilpotent Lie groups, then $M$ is called a pro-nilmanifold. The characteristic factor in this case turns out to be a pro-nilmanifold. Conze and Lesigne used this to show the existence of the limit under some technical hypotheses, a result re-proved by Furstenberg and Weiss [110].

The general case ends up resembling the case $k = 3$, though a great many new ideas are involved: The characteristic factor lives on a $(k - 1)$-step pronilmanifold. Once the existence and structure of the characteristic factor is established, it remains to check the existence of the limit for $(k-1)$-nilsystems on nilmanifolds. The factors that appear in this setting are also characteristic factors for many other averaging schemes of the same *degree* (the degree is the number of times the van der Corput lemma has to be applied).

# Notes to Chap. 7

[63] (Page 171) Schur made this conjecture in his work on the distribution of quadratic residues in $\mathbb{Z}/p\mathbb{Z}$; van der Waerden heard of the conjecture through Baudet, a student at Göttingen. Research by Soifer [347, Chap. 34] suggests that Baudet and Schur may have independently formulated the conjecture; this book also contains an account of the lives of Schur, Baudet and van der Waerden. In van der Waerden's own account [372] of how the proof was found, he describes in detail how many of the ideas were worked out at a blackboard with Artin and Schreier. More details on the history of this conjecture and its proof may be found in several places; a particularly accessible source is Brauer's review [43] of a book by Khinchin [192] which includes a proof of van der Waerden's theorem as one of three "pearls" of number theory. The result has been generalized by Rado [296] and others; a particularly short proof is given by Graham and Rothschild [124], and this will be presented in Sect. 7.1. An extended and carefully motivated account of the proof may be found in the monograph of Graham, Rothschild and Spencer [125], which also presents a different approach due to Shelah [341] in which the enormously rapidly growing function $n \mapsto N(\ell, m, r)$ appearing in the proof of Theorem 7.1 is replaced by a much more slowly growing function appearing from an induction in one variable. Furstenberg and Weiss [109] developed topological versions of multiple recurrence, and using these found proofs of van der Waerden's theorem and many similar results using topological dynamics.

[64] (Page 171) Szemerédi's theorem asserts the existence of a finite structure resembling that of the integers (an arithmetic progression) in any set sufficiently thick to be of positive upper density. A rather subtle criterion for the set $A = \{a_1 < a_2 < \cdots \}$ to be thick is to require that $\sum_{n=1}^{\infty} \frac{1}{a_n}$ diverges. A natural problem, due to Erdős, asks if this divergence alone guarantees that the set $A$ contains arbitrarily long arithmetic progressions. One special case of this is Szemerédi's theorem itself; another is the result by Green and Tao [127] that the set of primes contains arbitrarily long arithmetic progressions.

[65] (Page 178) For $k = 1$ this result is due to Birkhoff [32], and it may be viewed as a topological analogue of Poincaré recurrence. The case stated here is a special case of a result due to Furstenberg and Weiss [109]: if $T_1, \ldots, T_k$ are commuting homeomorphisms of a compact metric space $(X, \mathsf{d})$ then there is a point $x \in X$ which is simultaneously recurrent under all the maps in the sense that there is a sequence $n_j \to \infty$ with $\mathsf{d}(x, T_i^{n_j}(x)) \to 0$ as $j \to \infty$ for each $i, 1 \leqslant i \leqslant k$.

[66] (Page 180) Some condition on $p$ is needed for this result to hold. For example, if $p$ is a non-zero constant then the conclusion of the theorem clearly cannot hold; similarly if $p(n) = 2n^2 + 1$ and $E = 2\mathbb{Z}$ the conclusion cannot hold. A result of Kamae and Mendès France [178] gives the necessary and sufficient condition that $p(\mathbb{N}) \cap a\mathbb{Z}$ must be non-empty for all $a > 0$.

[67] (Page 180) This short proof is taken from Bergelson's notes [26, Th. 1.31]. Furstenberg's proof uses the Spectral Theorem (Theorem B.4), while the proof in Bergelson's notes uses softer Hilbert space methods, which are more amenable to generalization.

[68] (Page 192) The result that a subset of the integers of positive upper density contains a three-term arithmetic progression, equivalent to Theorem 7.14 by Furstenberg correspondence, was proved by Roth [318] using harmonic analysis.

[69] (Page 199) This is an example of a skew-product construction studied by Anzai [5] and von Neumann. More complicated examples of this sort arose in Sect. 4.4.3.

[70] (Page 226) The transfinite induction is not strictly needed. It may be shown that for averages related to $(k+1)$-term arithmetic progressions, only $k$ successive distal extensions are needed to construct a characteristic factor in the sense of Sect. 7.13.1 (this is shown in the original paper [102] of Furstenberg, and a convenient account may be found in a paper of Frantzikinakis [95, Th. 5.2]).

[71] (Page 227) The extension of Szemerédi's theorem to commuting transformations (the linear case of the result of Bergelson and Leibman) was carried out much earlier by Furstenberg and Katznelson [105]; this has now also been proved using combinatorial methods by Gowers [123] and Nagle, Rödl and Schacht [265]. In a different direction, the Hales–Jewett theorem [133] asserts that for every $r$ and $k$ there is an $n$ such that every $r$-coloring of $\{1, \ldots, k\}^n$ contains a *combinatorial line* (that is, a subset of $k$ elements in $\{1, \ldots, k\}^n$ obtained from some template in $(\{1, \ldots, k\} \cup \{*\})^n$ by replacing the symbol $*$ with $1, \ldots, k$ in turn). The Hales–Jewett theorem is a generalization of van der Waerden's theorem, and like the van der Waerden theorem it has a density version, proved by Furstenberg and Katznelson [106] using an extension of the ergodic techniques of Furstenberg in his proof of Szemerédi's theorem. Recent approaches include a paper of Polymath [291] giving an elementary proof of the result, and in particular a quantitative bound on $n(r, k)$ (a consequence is a different proof of Szemerédi's theorem) and work of Towsner [363] giving a proof of Szemerédi's theorem using ideas from model theory.

# Chapter 8
# Actions of Locally Compact Groups

The facet of ergodic theory coming from abstract mathematical models of dynamical systems evolving in time involves a single, iterated, measure-preserving transformation (action of $\mathbb{N}$ or of $\mathbb{Z}$) or a flow (action of the reals). For many reasons—including geometry, number theory, and the origins of ergodic theory in statistical mechanics—it is useful to study actions of groups more general than the integers or the reals. In this chapter we extend the definition of a measure-preserving system $(X, \mathscr{B}, \mu, T)$ to allow the possibility that $T$ is an action of a group $G$. This means that $T$ is a homomorphism $T : G \to \mathrm{MPT}(X, \mathscr{B}, \mu)$, where $\mathrm{MPT}(X, \mathscr{B}, \mu)$ denotes the group of invertible measure-preserving transformations of the probability space $(X, \mathscr{B}, \mu)$. We write $T_g$ or $x \mapsto g{\cdot}x$ for the measure-preserving transformation $T(g)$.

In addition to its many powerful applications, an attractive feature of the ergodic theory of group actions is the subtle interplay between algebraic and functional-analytic properties of the acting group $G$ on the one hand, and ergodic properties of its actions on the other. In keeping with our determination to remain in rather standard territory, acting groups are assumed to be locally compact and $\sigma$-compact (which we will abbreviate to $\sigma$-locally compact). Moreover, we will for simplicity restrict ourselves to continuous actions.

## 8.1 Ergodicity and Mixing

We start the discussion of general group actions with some basic definitions. A continuous $G$-action is—for our purposes[72]—given by a homomorphism from $G$ to the group of homeomorphisms of a $\sigma$-compact metric space; we write $x \mapsto g{\cdot}x$ for the action of $g \in G$. In particular, $g{\cdot}(h{\cdot}x) = (gh){\cdot}x$ for all $g, h \in G$ and $x \in X$. We also require continuity for the action as a whole (not just for the individual maps $g$) as below.

M. Einsiedler, T. Ward, *Ergodic Theory*, Graduate Texts in Mathematics 259, DOI 10.1007/978-0-85729-021-2_8, © Springer-Verlag London Limited 2011

**Definition 8.1.** An action of a $\sigma$-compact metric group $G$ on a $\sigma$-compact metric space $(X, \mathsf{d})$ is *continuous* if the map $G \times X \to X$ defined by

$$(g, x) \mapsto g{\cdot}x$$

is continuous. A measure $\mu \in \mathscr{M}(X)$ is *invariant under $G$* if $g_*\mu = \mu$ for all $g \in G$, where $(g_*\mu)(A) = \mu(g^{-1}{\cdot}A)$ for any Borel set $A \subseteq X$, and we write $\mathscr{M}^G(X)$ for the set of $G$-invariant measures.

*A priori* it is not clear whether invariant measures always exist in the setting of Definition 8.1. Example 8.2 shows that there are indeed groups some of whose continuous actions do not have invariant measures (Exercise 8.1.4 shows that $\mathrm{SL}_2(\mathbb{Z})$ has continuous actions with no invariant measures, for example). In Sect. 8.4 we will discuss *amenability*, which is a property of the acting group that ensures the existence of an invariant measure for any continuous action. Notice that, while Exercise 8.1.4 shows that $\mathrm{SL}_2(\mathbb{Z})$ is not amenable, its natural action by automorphisms of the 2-torus shows that it does have non-trivial actions with invariant measures. What is always true is that if an invariant probability measure exists, then an invariant ergodic probability measure exists, just as in the case of a single transformation (Theorem 4.4). We will address this result in Sect. 8.7, by establishing the ergodic decomposition for group actions.

*Example 8.2.* Let $X = \{z \in \mathbb{C} \mid |z - \mathrm{i}| = 1\}$ and let $T : X \to X$ be the North–South map defined in Example 4.3, so that $\mathscr{M}^T(X)$ consists only of the measures $p\delta_{2\mathrm{i}} + (1 - p)\delta_0$, $p \in [0, 1]$ that are supported on the two points $2\mathrm{i}$ and $0$. On the other hand, no invariant measure for the irrational rotation $R_\alpha$ defined by $z \mapsto \mathrm{e}^{2\pi\mathrm{i}\alpha}(z - \mathrm{i}) + \mathrm{i}$ with $\alpha \notin \mathbb{Q}$ could be atomic, since the orbit of every point is infinite (alternatively, recall that $R_\alpha$ is uniquely ergodic by Example 4.11, with $\mathscr{M}^{R_\alpha}(\mathbb{S}^1) = \{m\}$, the Lebesgue measure). Thus the continuous action of the group $\Gamma$ generated by $T$ and $R_\alpha$ on $X$ has no invariant measures.

The hierarchy of mixing properties discussed in Chap. 2 makes sense for actions of more general groups[73]. Since some of these will be used in several places, we collect them here in a rather general setting. Let $G$ act continuously on a compact metric space $(X, \mathsf{d})$, and let $\mu \in \mathscr{M}^G(X)$ be an invariant measure, so that $G$ acts by measure-preserving transformations of $(X, \mathscr{B}, \mu)$. For a sequence $(g_n)_{n \geqslant 1}$ of elements of $G$, we write $g_n \to \infty$ if, for any compact set $K \subseteq G$, there is an $N$ for which $g_n \notin K$ if $n \geqslant N$. The $G$-action is called

- *ergodic* if any set $A \in \mathscr{B}$ with $\mu\left(g^{-1}{\cdot}A \triangle A\right) = 0$ for all $g \in G$ has $\mu(A) = 0$ or 1;
- *weak-mixing* if the diagonal action $g \mapsto g \times g$ on the product space

$$(X \times X, \mathscr{B} \otimes \mathscr{B}, \mu \times \mu)$$

is ergodic;

- *mixing* if for any $A_0, A_1 \in \mathscr{B}$ and sequence $(g_n)$ with $g_n \to \infty$ as $n \to \infty$,

$$\mu \left( A_0 \cap g_n^{-1} \cdot A_1 \right) \longrightarrow \mu(A_0)\mu(A_1)$$

as $n \to \infty$;
- *mixing on $r$ sets* if for any $A_0, A_1, \ldots, A_{r-1} \in \mathscr{B}$ and list of $(r-1)$ sequences $(g_{j,n})$ with

$$g_{j,n} \to \infty, \qquad g_{i,n} g_{j,n}^{-1} \to \infty$$

for each $i \neq j$, $1 \leqslant i, j < r$ as $n \to \infty$,

$$\mu \left( A_0 \cap g_{1,n}^{-1} \cdot A_1 \cap \cdots \cap g_{r-1,n}^{-1} \cdot A_{r-1} \right) \to \mu(A_0) \cdots \mu(A_{r-1})$$

as $n \to \infty$;
- *mixing of all orders* if it is mixing on $r$ sets for each $r \geqslant 1$; and
- *rigid* if there is a sequence $(g_n)$ with $g_n \to \infty$ as $n \to \infty$ such that

$$\mu \left( A_0 \cap g_n^{-1} \cdot A_1 \right) \to \mu(A_0 \cap A_1)$$

as $n \to \infty$ for any sets $A_0, A_1 \in \mathscr{B}$.

Notice that for the $\mathbb{Z}$-action generated by a single invertible measure-preserving transformation, most of these definitions replicate those of Chap. 2.

Mixing (of any order) descends to the induced action obtained by restriction to a closed subgroup of $G$. Ergodicity does not descend in the same way. A trivial example illustrates this: if $(X, \mathscr{B}, \mu, T)$ is an ergodic measure-preserving system on a non-trivial space then the $\mathbb{Z}^2$-action defined by $(m, n) \longmapsto T^n$ is ergodic, but the action of the subgroup $\mathbb{Z} \times \{0\}$ is non-ergodic.

Proposition 2.14 generalizes to this setting: if the $G$-action is ergodic in the sense above, then any measurable $G$-invariant function is equal almost everywhere to a constant function.

As in the case of single transformations, it will be convenient to know that invariant sets modulo $\mu$ can be modified by a null set to become strictly invariant (see Proposition 2.14). This is a little more delicate in the setting of group actions, because the acting group may be uncountable. Proposition 8.3 will be proved at the end of Sect. 8.3, after we have assembled more information about Haar measure.

**Proposition 8.3.** *Let $G$ be a $\sigma$-compact metric group acting continuously on a compact metric space $X$ preserving a measure $\mu \in \mathscr{M}(X)$. Then for $B \in \mathscr{B}$ the following properties are equivalent:*

(1) *$B$ is invariant in the sense that $\mu(g \cdot B \triangle B) = 0$ for all $g \in G$;*
(2) *$B$ is invariant in the sense that there is a set $B' \in \mathscr{B}$ with $\mu(B \triangle B') = 0$ and with $g \cdot B' = B'$ for all $g \in G$.*

**Theorem 8.4.** *Let $G$ be a $\sigma$-locally compact metrizable group acting continuously on a $\sigma$-locally compact metrizable space $X$. Then the space $\mathscr{M}^G(X)$ of $G$-invariant measures is a closed convex subset of $\mathscr{M}(X)$. A measure $\mu$ in $\mathscr{M}^G(X)$ is extremal in $\mathscr{M}^G(X)$ if and only if $\mu$ is $G$-ergodic.*

PROOF. For $g \in G$ write $\mathscr{M}^g(X)$ for the space of $g$-invariant measures (that is, invariant measures for the transformation $x \mapsto g \cdot x$ for $x \in X$). Then we know that $\mathscr{M}^g(X)$ is a closed convex subset of $M(X)$ (since the induced map $g_*$ on $\mathscr{M}(X)$ is continuous and affine). As $\mathscr{M}^G(X) = \bigcap_{g \in G} \mathscr{M}^g(X)$, $\mathscr{M}^G(X)$ is also a closed convex set.

To prove the last claim, one can argue as in Theorem 4.4. If $\mu \in \mathscr{M}^G(X)$ is not $G$-ergodic, then there exists a measurable set $B \subseteq X$ with $0 < \mu(B) < 1$ for which $\mu\left(g^{-1} \cdot B \triangle B\right) = 0$ for all $g \in G$. Then

$$\mu = \mu(B) \left( \frac{1}{\mu(B)} \mu|_B \right) + \mu(X \smallsetminus B) \left( \frac{1}{\mu(X \smallsetminus B)} \mu|_{X \smallsetminus B} \right),$$

and one can quickly check that, by our assumption on $B$, the two normalized measures on the right are $G$-invariant. This shows that $\mu$ is not extremal.

For the converse, suppose that $\mu \in \mathscr{M}^G(X)$ is ergodic with

$$\mu = s\nu_1 + (1 - s)\nu_2$$

for some $s \in (0, 1)$ and $\nu_1, \nu_2 \in \mathscr{M}^G(X)$. Then $\nu_1 \ll \mu$, and we claim that the Radon–Nikodym derivative $f = \frac{d\nu_1}{d\mu}$ is $G$-invariant and measurable. This implies that $f \equiv 1$ almost everywhere, since $\mu$ is ergodic for the $G$-action, and so $\mu = \nu_1 = \nu_2$. To see the claim, fix $g \in G$ and a measurable set $B \subseteq X$. Then, by definition of $f$ and $G$-invariance of $\nu_1$ and $\mu$, we have

$$\int_B f \, d\mu = \nu_1(B) = \nu_1(g \cdot B) = \int_{g \cdot B} f \, d\mu = \int_{g \cdot B} f \circ g \circ g^{-1} \, d\mu = \int_B f \circ g \, d\mu, \tag{8.1}$$

(using Lemma 2.6 with $T(x) = g^{-1} \cdot x$), which implies that $f = f \circ g$ almost everywhere by uniqueness of the Radon–Nikodym derivative (see Theorem A.15). □

## Exercises for Sect. 8.1

**Exercise 8.1.1.** Let $(X, \mathscr{B}_X, \mu, T)$ and $(Y, \mathscr{B}_Y, \nu, S)$ be ergodic $\mathbb{Z}$-actions. Define a $\mathbb{Z}^2$-action on the product $(X \times Y, \mu \times \nu)$ by $(m, n) \mapsto T^m \times S^n$. Show that this action is ergodic, but has subgroups whose action is not ergodic.

**Exercise 8.1.2.** Construct an ergodic measure-preserving $\mathbb{Z}^2$-action $T$ with the property that each element $T^{\mathbf{n}}$ is not ergodic.

**Exercise 8.1.3.** Notice that the definition of mixing of a given order also makes sense for a measure-preserving action of the semigroup $\mathbb{N}^2$, simply by requiring that $g_{j,n} - g_{i,n} \to \infty$ as $n \to \infty$ when viewed as elements of $\mathbb{Z}^2$.

(1) Extend the construction of the invertible extension from Exercise 2.1.7 for the case of a single transformation to the case of an action of the semigroup $\mathbb{N}^2$.
(2) Find an example of a mixing $\mathbb{N}^2$-action whose invertible extension to a $\mathbb{Z}^2$-action is not mixing.
(3) Let $T$ be an action of $\mathbb{N}^2$ that is mixing on $k$ sets. Show that the invertible extension of $T$ is mixing on $(k-1)$ sets, but is not in general mixing on $k$ sets[74].

**Exercise 8.1.4.** Show that $\mathrm{SL}_2(\mathbb{Z})$ has continuous actions without invariant probability measures by studying the natural action

$$\begin{pmatrix} a & b \\ c & d \end{pmatrix} : [x, y] \longrightarrow [ax + by, cx + dy]$$

on the projective line $\mathbb{P}^1(\mathbb{R})$.

**Exercise 8.1.5.** Show that an ergodic rotation of a compact group is rigid.

**Exercise 8.1.6.** Show that a mixing transformation on a non-trivial probability space is not rigid[75].

## 8.2 Mixing for Commuting Automorphisms

As we have already seen, endomorphisms of compact groups are a class of measure-preserving transformations whose structure makes their ergodic properties particularly transparent. In this chapter we consider a natural class of measure-preserving systems which on the one hand have a transparent algebraic structure, while on the other already exhibit some distinctly higher-rank properties. These are the actions of $\mathbb{Z}^d$ by continuous automorphisms of compact abelian groups. Schmidt's monograph [332] gives a systematic treatment of these systems; we simply mention a few examples and then discuss some more recent work on rigidity properties. To simplify matters we focus for much of the chapter on two examples: the "$\times 2, \times 3$" system (see Sect. 8.2.2) and Ledrappier's "three-dots" example (see Sect. 8.2.1)[76].

Mixing, and mixing of higher orders, was defined for group actions in Sect. 8.1. As pointed out in Exercise 8.1.3, the definitions easily extend to actions of semigroups like $\mathbb{N}^2$, and we will use this in Sect. 8.2.2. As mentioned in Sect. 2.7, one of the outstanding open problems in classical ergodic theory is that of Rokhlin (see p. 50): does mixing imply mixing of all orders for a measure-preserving transformation? In Sect. 8.2.1 we describe a simple example of a type due to Ledrappier [221] that answers this negatively[77] for $\mathbb{Z}^d$-actions with $d > 1$. In Sect. 8.2.2 we consider the simple example of the $\mathbb{N}^2$-action generated by $x \mapsto 2x$ and $x \mapsto 3x$ modulo one on the circle, and show that it is mixing of all orders.

### 8.2.1 Ledrappier's "Three Dots" Example

Write $\mathbf{e}_1 = (1,0)$ and $\mathbf{e}_2 = (0,1)$ for the standard basis of $\mathbb{R}^2$, and define

$$X_{\bullet\bullet}^{\bullet} = \{x \in \{0,1\}^{\mathbb{Z}^2} \mid x_{\mathbf{n}+\mathbf{e}_1} + x_{\mathbf{n}+\mathbf{e}_2} + x_{\mathbf{n}} = 0 \text{ for all } \mathbf{n} \in \mathbb{Z}^2\},$$

where the addition is that of the compact group $\{0,1\}$. The notation $X_{\bullet\bullet}^{\bullet}$ corresponds to the shape of the condition defining the group (see Fig. 8.1). The conditions defining $X_{\bullet\bullet}^{\bullet}$ as a subset of the compact group $\{0,1\}^{\mathbb{Z}^2}$ are closed and homogeneous, so $X_{\bullet\bullet}^{\bullet}$ is a compact abelian group. Moreover, since the conditions are applied for all $\mathbf{n} \in \mathbb{Z}^2$, the group $X_{\bullet\bullet}^{\bullet}$ is invariant under the shift action $T$ of $\mathbb{Z}^2$ defined by

$$(x_{\mathbf{n}})_{\mathbf{n}} \xrightarrow{T_{\mathbf{m}}} (x_{\mathbf{n}+\mathbf{m}})_{\mathbf{n}}.$$

The Haar measure on $X_{\bullet\bullet}^{\bullet}$ is determined by its value on cylinder sets, and the measure of the cylinder set $C$ defined by specifying the coordinates in some finite set $A \subseteq \mathbb{Z}^2$ is given by

$$m(C) = \frac{1}{|\pi_A(X_{\bullet\bullet}^{\bullet})|}$$

if $C \cap X_{\bullet\bullet}^{\bullet} \neq \varnothing$, where $\pi_A : X_{\bullet\bullet}^{\bullet} \to \{0,1\}^A$ denotes the projection map obtained by restricting to the coordinates in $A$. To see this, notice that the cylinder set $\{x \in X_{\bullet\bullet}^{\bullet} \mid x_{\mathbf{a}} = 0 \text{ for all } \mathbf{a} \in A\}$ is a subgroup of index $|\pi_A(X_{\bullet\bullet}^{\bullet})|$, and $m$ is translation-invariant. Thus, for example,

$$m\left(\{x \in X_{\bullet\bullet}^{\bullet} \mid x_0 = x_{\mathbf{e}_1} = x_{\mathbf{e}_2} = x_{\mathbf{e}_1+\mathbf{e}_2} = 0\}\right) = \tfrac{1}{8}.$$

**Proposition 8.5.** *With respect to Haar measure* $m = m_{X_{\bullet\bullet}^{\bullet}}$, *the* $\mathbb{Z}^2$-*action* $T$ *is mixing but not mixing of all orders.*

PROOF. The proof that $T$ is mixing uses the same ideas as were used in the proof of Theorem 2.19. As in the case of a single transformation, mixing is equivalent to the property that

$$\int f_0(x) f_1(T_{\mathbf{n}} x)\, dm \longrightarrow \int f_0\, dm \int f_1\, dm \qquad (8.2)$$

as $\mathbf{n} \to \infty^*$, for any $f_0, f_1 \in L^2_m$. By approximating with trigonometric polynomials (finite linear combinations of characters on $X_{\bullet \bullet}$), it is sufficient to check (8.2) for individual characters. Thus $T$ is mixing if and only if for any characters $\chi_0, \chi_1$ on $X_{\bullet \bullet}$,

$$\int \chi_0(x) \chi_1(T_{\mathbf{n}} x)\, dm \longrightarrow \int \chi_0\, dm \int \chi_1\, dm$$

as $\mathbf{n} \to \infty$. The usual orthogonality relations for characters show that

$$\int \chi\, dm = \begin{cases} 1 & \text{if } \chi \equiv 1; \\ 0 & \text{if } \chi \text{ is non-trivial.} \end{cases}$$

This gives the following characterization: $T$ is mixing if and only if

$$\chi_0 \left( \chi_1 \circ T_{\mathbf{n}} \right) \equiv 1 \implies \mathbf{n} \text{ is bounded} \qquad (8.3)$$

for any characters $\chi_0$ and $\chi_1$, not both trivial.

From the relationship defining $X_{\bullet \bullet}$, any point $x \in X_{\bullet \bullet}$ is determined by the coordinates $x_{\mathbf{n}}$ for $\mathbf{n} \in F$ where

$$F = \{\mathbf{n} \mid n_2 = 0 \text{ or } n_1 = 0 \text{ and } n_2 < 0\}$$

(see Fig. 8.1). This means that $X_{\bullet \bullet}$ as a group is isomorphic to $\{0, 1\}^F$.

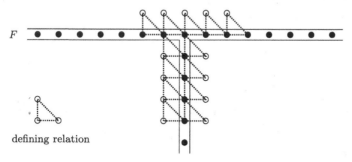

Fig. 8.1 The projection onto the coordinates in $F$ (marked $\bullet$) determines the other coordinates marked $\circ$ via the defining relation $x_{\mathbf{n}+\mathbf{e}_1} + x_{\mathbf{n}+\mathbf{e}_2} + x_{\mathbf{n}} = 0$

Moreover, the coordinates in any finite subset $A \subseteq F$ are determined by the coordinates in a set of the shape $E = \{\mathbf{m}, \mathbf{m} + \mathbf{e}_1, \ldots, \mathbf{m} + s\mathbf{e}_1\}$ (see

---

$^*$ Recall that this means $\mathbf{n}$ takes on a sequence of values $\mathbf{n}_1, \mathbf{n}_2, \ldots$ with the property that for any finite set $A \subseteq \mathbb{Z}^2$ there is an $R$ such that $r \geqslant R \implies \mathbf{n}_r \notin A$.

Fig. 8.2), and we may choose $m_1$ arbitrarily large negative, and any function
defined on the coordinates $A$ that does vary with the left- and right-most
coordinates will also vary with the left- and right-most coordinates in $E$ when
described in the coordinates $E$. Thus we may describe two given non-trivial
characters $\chi_0$ and $\chi_1$ by

$$\chi_i(x) = e^{\pi i \sum_{\mathbf{n} \in E_i} x_{\mathbf{n}}}$$

for sets $E_i \subseteq \{\mathbf{m}+s_i\mathbf{e}_1, \ldots, \mathbf{m}+t_i\mathbf{e}_i\}$ for $i = 1, 2$, with $\mathbf{m}+s_i\mathbf{e}_1, \mathbf{m}+t_i\mathbf{e}_i \in E_i$
for $i = 1, 2$. That is, each character depends on a line of coordinates with
fixed $y$ coordinate, and the lines are minimal with this property at the given
value $m_2$ of the $y$ coordinate.

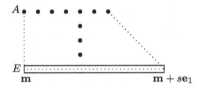

**Fig. 8.2** Any character is supported on a horizontal line of coordinates

Now given a character $\chi_0$ presented in this way by a horizontal line $E$, only
a character that can be described using the coordinates in the "shadow" $S(E)$
cast by $E$ (see Fig. 8.3) can *fail* to be orthogonal to $\chi_0$. Thus (8.3) is equivalent to the statement that $S(E_0)\triangle S(E_1+\mathbf{n}) = \varnothing$ requires that $\mathbf{n}$ be bounded
(since we can always describe $\chi_0$ and $\chi_1 \circ T_\mathbf{n}$ using a horizontal line of coordinates at the same level), which is clear. Thus $T$ is mixing with respect
to $m_{\bullet\,\bullet}$.

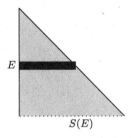

**Fig. 8.3** The shadow of a character defined on the line of coordinates $E$

It is easier to see that $T$ is not mixing on three sets. First, notice that the
condition $x_\mathbf{n} + x_{\mathbf{n}+\mathbf{e}_1} + x_{\mathbf{n}+\mathbf{e}_2} = 0$ implies that for any $k$

$$x_{2^k \mathbf{e}_2} = \sum_{j=0}^{2^k} \binom{2^k}{j} x_{j\mathbf{e}_1} = x_0 + x_{2^k \mathbf{e}_1} \pmod{2}. \tag{8.4}$$

This is clear for $k = 1$, and the general case follows by a simple induction. Now let $A = \{x \in X_{\bullet\bullet} \mid x_0 = 0\}$ and let $x_* \in X_{\bullet\bullet}$ be any element with $x_0 = 1$. Then $X_{\bullet\bullet}$ is the disjoint union of $A$ and $A + x_*$, so $m(A) = m(A + x_*) = \frac{1}{2}$. However, (8.4) shows that

$$x \in A \cap T_{-2^k \mathbf{e}_1} A \implies x \in T_{-2^k \mathbf{e}_2} A,$$

so

$$A \cap T_{-2^k \mathbf{e}_1} A \cap T_{-2^k \mathbf{e}_2}(A + x_*) = \varnothing$$

for all $k \geqslant 1$, which shows that $T$ cannot be mixing on three sets with respect to Haar measure $m$. $\qquad\square$

### 8.2.2 Mixing Properties of the ×2, ×3 System

Define an action $S$ of $\mathbb{N}^2$ on the circle $\mathbb{T}$ by writing

$$S_{\mathbf{n}} = S_2^{n_1} S_3^{n_2},$$

where $S_2 : x \mapsto 2x$ and $S_3 : x \mapsto 3x$ modulo 1. Notice that each of the maps $S_2$ and $S_3$ is measurably isomorphic to a one-sided full shift, so is mixing of all orders. This is related to a restricted kind of higher-order mixing for the whole system (see Exercise 8.2.4). The problem of higher-order mixing for the whole system is much more subtle, partly because there are many ways for a set of points to move apart in $\mathbb{N}^2$. In fact mixing of all orders for this system (more precisely, its generalization to algebraic dynamical systems on connected groups) is equivalent to a deep Diophantine result, and we begin by stating that result[78].

$S$-unit Theorem (van der Poorten and Schlickewei). Let $\mathbb{K}$ be a field of characteristic zero, and let $\Gamma \subseteq \mathbb{K}^\times = \mathbb{K}\backslash\{0\}$ be a finitely generated multiplicative subgroup. Then for any $n \geqslant 1$ and fixed coefficients $a_1, \ldots, a_n$ in $\mathbb{K}^\times$, the equation

$$a_1 x_1 + a_2 x_2 + \cdots + a_n x_n = 1 \tag{8.5}$$

has only finitely many solutions $(x_1, \ldots, x_n) \in \Gamma$ for which no proper subsum $\sum_{i \in I \subsetneq \{1, \ldots, n\}} a_i x_i$ vanishes.

Notice that it is clearly necessary to restrict solutions to avoid vanishing proper subsums: if there is such a subsum, then it generates an infinite family of solutions. The hypothesis that the field have characteristic zero is also needed: if the field has positive characteristic, then the Frobenius automorphism generates an infinite family of solutions. Indeed, the arithmetic reason behind the failure of three-fold mixing in Ledrappier's example comes from (8.4), which is equivalent to the infinite family of solutions

$$(1+t)^{2^k} + t^{2^k} = 1$$

to the equation $x_1 + x_2 = 1$ in the multiplicative subgroup $\ll 1+t, t \gg$ of the field $\mathbb{F}_2(1+t,t)$.

**Corollary: Mixing of all Orders.** The $\times 2, \times 3$ action of $\mathbb{N}^2$ is mixing of all orders with respect to Lebesgue measure $m_{\mathbb{T}}$.

Before embarking on the proof of mixing of all orders, we give an overview of how the argument works. We assume that the $\mathbb{N}^2$-action $S$ is not mixing on $r$ sets for some $r \geqslant 2$. Using Fourier analysis, this translates into an infinite family of solutions to an equation of the form (8.5). Thus, by the $S$-unit theorem, we must have a vanishing subsum. This in turn implies that the system is not mixing on $s$ sets for some $s < r$. The proof is completed by checking that $S$ is mixing (that is, is mixing on 2 sets). All the subtle work is hidden in the $S$-unit theorem.

Recall that the map sending $m \in \mathbb{Z}$ to the character $x \mapsto e^{2\pi i m x}$ is an isomorphism between $\mathbb{Z}$ and the character group $\widehat{\mathbb{T}}$ of $\mathbb{T}$.

PROOF OF MIXING ASSUMING THE $S$-UNIT THEOREM. Assume that the $\mathbb{N}^2$-action $S$ is not mixing on $r$ sets for some $r \geqslant 2$. Then there are sets $A_1, \ldots, A_r$ in $\mathscr{B}$ and a sequence

$$\left( \mathbf{n}_1^{(k)}, \ldots, \mathbf{n}_r^{(k)} \right)_{j \geqslant 1}$$

of $r$-tuples of elements of $\mathbb{Z}^2$ such that $n_1^{(k)} = 0^*$ for all $k \geqslant 1$ and

$$\mathbf{n}_i^{(k)} - \mathbf{n}_j^{(k)} \to \infty$$

as $k \to \infty$ for any $i \neq j$, $1 \leqslant i, j \leqslant r$, such that

$$m_{\mathbb{T}} \left( S^{-\mathbf{n}_1^{(k)}}(A_1) \cap \cdots \cap S^{-\mathbf{n}_r^{(k)}}(A_r) \right) \not\to \prod_{i=1}^{r} m_{\mathbb{T}}(A_i) \qquad (8.6)$$

as $k \to \infty$. By expanding each of the indicator functions $\chi_{A_i}$ as a Fourier series, we see that (8.6) is equivalent to the statement that there are characters $\chi_1, \ldots, \chi_r$ on $\mathbb{T}$, not all trivial, such that

$$\int_{\mathbb{T}} \chi_1(S^{\mathbf{n}_1^{(k)}} x) \cdots \chi_r(S^{\mathbf{n}_r^{(k)}} x) \, dm_{\mathbb{T}}(x) \not\to \prod_{i=1}^{r} \int_{\mathbb{T}} \chi_i(S^{\mathbf{n}_i^{(k)}} x) \, dm_{\mathbb{T}}(x) = 0 \quad (8.7)$$

---

* Notice that we may make this assumption after subtracting the first term from all the others, which means that the resulting sequence comprises $r$-tuples of elements of $\mathbb{Z}^2$ rather than $\mathbb{N}^2$.

since $\int_{\mathbb{T}} \chi \, dm_{\mathbb{T}} = 0$ for any non-trivial character (that is, $\int_0^1 e^{2\pi i m x} \, dx = 0$ for $m \in \mathbb{Z} \setminus \{0\}$).

Each character $\chi_i(x) = e^{2\pi i m_i x}$ for some $m_i \in \mathbb{Z}$,

$$S^{\mathbf{n}_i^{(k)}} x = 2^{n_i^{(k,1)}} 3^{n_i^{(k,2)}} x$$

(where we write $\mathbf{n}_i^{(k)} = (n_i^{(k,1)}, n_i^{(k,2)})$), and the function being integrated on the left-hand side of (8.7) is itself the character corresponding to the integer

$$2^{n_1^{(k,1)}} 3^{n_1^{(k,2)}} m_1 + \cdots + 2^{n_r^{(k,1)}} 3^{n_r^{(k,2)}} m_r.$$

It follows that $S$ is not mixing on $r$ sets if and only if there are non-zero integers $m_1, \ldots, m_r$, and a sequence $(\mathbf{n}_1^{(k)}, \ldots, \mathbf{n}_r^{(k)})_{j \geqslant 1}$ with $n_1^{(k)} = 0$ for all $k \geqslant 1$ and

$$\mathbf{n}_i^{(k)} - \mathbf{n}_j^{(k)} \to \infty$$

as $k \to \infty$ for any $i \neq j$, $1 \leqslant i, j \leqslant r$ such that

$$2^{n_1^{(k,1)}} 3^{n_1^{(k,2)}} m_1 + \cdots + 2^{n_r^{(k,1)}} 3^{n_r^{(k,2)}} m_r = 0 \qquad (8.8)$$

for infinitely many values of $k$. Rearranging (8.8), this is equivalent to

$$-\frac{m_2}{m_1} 2^{n_2^{(k,1)}} 3^{n_2^{(k,2)}} - \cdots - \frac{m_r}{m_1} 2^{n_r^{(k,1)}} 3^{n_r^{(k,2)}} = 1. \qquad (8.9)$$

Equation (8.9) gives an infinite family of solutions in the subgroup $\ll 2, 3 \gg$ to an $S$-unit equation in $\mathbb{Q}$, so by the $S$-unit theorem we must have some subset $I \subsetneq \{1, \ldots, r\}$ with the property that

$$\sum_{i \in I} 2^{n_i^{(k,1)}} 3^{n_i^{(k,2)}} m_i = 0 \qquad (8.10)$$

for infinitely many values of $k$ (after multiplying out $m_1$). Reversing the argument leading from $S$ not being mixing on $r$ sets to (8.8), we see that (8.10) says that $S$ is not mixing on $|I| < r$ sets.

Thus the $S$-unit theorem shows that if $S$ is not mixing on $r$ sets for some $r \geqslant 2$, then $S$ is not mixing on $s$ sets for some $s < r$. All that remains to complete the proof is to check that $S$ is mixing. To see this, we use (8.8) again: if $S$ is not mixing, then there are non-zero integers $m_1$ and $m_2$ and a sequence $(n_1^{(k)}, n_2^{(k)})$ going to infinity in $\mathbb{Z}^2$ with $m_1 + 2^{n_1^{(k)}} 3^{n_2^{(k)}} m_2 = 0$ for infinitely many values of $k$, which is impossible since $\gcd(2, 3) = 1$. $\qquad \square$

## Exercises for Sect. 8.2

**Exercise 8.2.1.** Prove that Ledrappier's example is mixing without using character theory, by showing that it is mixing on a generating algebra of cylinder sets using the geometry of the defining relation.

**Exercise 8.2.2.** Prove that Ledrappier's example is mixing using algebra (instead of the geometry of the defining relation) by the following steps.

(1) The group $\widehat{X}$ is isomorphic to the ring $R = \mathbb{F}_2[t^{\pm 1}, \frac{1}{1+t}]$, and under this isomorphism the generators of the action correspond to the commuting maps $r \mapsto tr$ and $r \mapsto (1+t)r$.

(2) It follows that the action is not mixing if and only if there is a sequence $\mathbf{n}_j \to \infty$ and elements $r_0, r_1 \in R$, not both zero, for which

$$r_0 + t^{n_{1,j}}(1+t)^{n_{2,j}} r_1 = 0 \tag{8.11}$$

for all $j \geqslant 1$.

(3) Embed the (8.11) into the field of fractions of $R$ and deduce that (8.11) can only hold for finitely many values of $j$.

**Exercise 8.2.3.** [79] Use the methods of Sect. 8.2.1 to show that a large class of algebraic dynamical systems on zero-dimensional groups fail to be mixing of all orders in a simple way as follows. For each prime ideal $\mathfrak{p}$ in $\mathbb{F}_p[u_1^{\pm 1}, \ldots, u_d^{\pm 1}]$ define a $\mathbb{Z}^d$-action by shifting on the compact group

$$X_{\mathfrak{p}} = \left\{ x \in \mathbb{F}_p^{\mathbb{Z}^d} \mid \sum_{\mathbf{n} \in \mathbb{Z}^d} f_{\mathbf{n}} x_{\mathbf{n}} = 0 \text{ for all } f(\mathbf{u}) = \sum_{\mathbf{n}} f_{\mathbf{n}} \mathbf{u}^{\mathbf{n}} \in \mathfrak{p} \right\},$$

where we write $\mathbf{u}^{\mathbf{n}}$ for the monomial $u_1^{n_1} \cdots u_d^{n_d}$. For example, Ledrappier's example may be obtained in this way with $d = 2$ and $\mathfrak{p} = \langle 1 + u_1 + u_2 \rangle$. Prove that $\mathfrak{p}$ contains a non-constant polynomial if and only if the resulting $\mathbb{Z}^d$-action is not mixing of all orders.

**Exercise 8.2.4.** [80] Prove directly that the $\mathbb{N}^2$-action generated by $S_2$ and $S_3$ on the circle is mixing of all orders "in positive cones": for any $r \geqslant 1$ and measurable sets $A_1, \ldots, A_r \in \mathscr{B}_{\mathbb{T}}$,

$$m_{\mathbb{T}}\left( S^{-\mathbf{n}_1^{(k)}}(A_1) \cap \cdots \cap S^{-\mathbf{n}_r^{(k)}}(A_r) \right) \to \prod_{i=1}^{r} m_{\mathbb{T}}(A_i)$$

as $k \to \infty$, under the assumption that $n_{i+1}^{(k)} - n_i^{(k)} \to \infty$ while remaining in $\mathbb{N}^2$ (rather than in $\mathbb{Z}^2$) as $k \to \infty$ for $1 \leqslant i < r$. Prove the same restricted mixing of all orders property for $X_{\bullet\bullet}$.

**Exercise 8.2.5.** Prove that the $\times 2$, $\times 3$ action of $\mathbb{Z}^2$ from Sect. 8.2.2 is mixing on three sets without using the $S$-unit theorem.

## 8.3 Haar Measure and Regular Representation

In this section we collect some information about the natural measure $m_G$ defined on the Borel $\sigma$-algebra $\mathscr{B}_G$ of a $\sigma$-locally compact metric group $G$, which will replace the counting measure on $\mathbb{N}$ used, for example, to form ergodic averages. Some of this material is also summarized in Appendix C on topological groups.

The natural measure is the left-invariant Haar measure $m_G$ on $G$, which satisfies the following properties:

(1) $m_G(gB) = m_G(B)$ for all $B \in \mathscr{B}_G$ and $g \in G$;
(2) $m_G(K) < \infty$ for any compact subset $K \subseteq G$; and
(3) $m_G(O) > 0$ for any open set $O \subseteq G$.

As we will show below, referring to "the" Haar measure is (almost) legitimate: this measure is unique up to a scalar multiple in the following sense. If $\mu_1$ and $\mu_2$ are two measures satisfying properties (1)–(3), then there is some $c > 0$ with $\mu_2 = c\mu_1$ (see Corollary 8.8).

The existence theorem below is standard and may be found in many books (see Appendix C for references).

**Theorem (Haar).** Let $G$ be a $\sigma$-locally compact metrizable group. Then there exists a left Haar measure $m_G$ satisfying properties (1)–(3) above.

We now sketch one argument for the existence of Haar measure; for complete details see Folland [94, Sec. 2.2] for example.

For any compact set $K \subseteq G$ and any set $L$ with non-empty interior, we define $[K : L]$ to be the minimal number of left translates $gL$ of $L$ needed to cover $K$. Let $V$ be a small neighborhood of the identity $e \in G$, and let $K_0$ be some fixed compact set with non-empty interior. Notice that if we allow $V$ to shrink to smaller and smaller neighborhoods (this is possible unless $G$ is discrete, in which case counting measure is a Haar measure) then we expect that $[K : V]$ will diverge to infinity. To take account of this we normalize relative to $K_0$ by defining

$$I_V(K) = \frac{[K : V]}{[K_0 : V]}$$

for any compact $K \subseteq G$. It is easy to check that $[gK : V] = [K : V]$ and

$$[K \cup K' : V] \leqslant [K : V] + [K' : V],$$

so $I_V(K)$ is a left-invariant, subadditive function defined on the compact subsets of $G$. Note, however, that if $V$ is the open ball on the left in Fig. 8.4 (which is drawn in $\mathbb{R}^2$, but the geometric argument explained below applies for any group $G$) and $K, K'$ are the two compact sets on the right, then

$$[K : V] = [K' : V] = 1 = [K \cup K' : V],$$

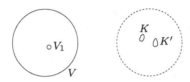

**Fig. 8.4** The sets $K$ and $K'$

so that $I_V$ is not additive, even though $K$ and $K'$ are disjoint.

Clearly this failure of additivity over disjoint sets occurs because the scaling device, namely the neighborhood $V$, is too coarse to discriminate between the two compact sets which are close together. If $V$ is replaced by $V_1$ as illustrated in Fig. 8.4, then

$$[K \cup K' : V_1] = [K : V_1] + [K' : V_1],$$

so $I_{V_1}(K \cup K') = I_{V_1}(K) + I_{V_1}(K')$. To see this, notice that any left translate of $V_1$ cannot intersect both $K$ and $K'$ if $V_1$ is, as illustrated, sufficiently small. Thus an optimal cover (that is, one for which the number of translates needed is minimal) can be split into an optimal cover for $K$ and an optimal cover for $K'$ once the open set is small enough. Clearly replacing $V$ by $V_1$ does not solve the problem completely, because other disjoint compact sets could be even closer together. To get around this, we must work with a sequence of neighborhoods $(V_n)$ which form a basis of the neighborhoods of $e \in G$. The functional

$$I(K) = \lim_{n \to \infty} I_{V_n}(K)$$

defined in this way would be additive on all disjoint compact sets, but in order to complete the argument the convergence would need to be shown. The following argument allows us to get round this problem.

By assumption, $K_0$ has non-empty interior, so $m = [K : K_0] < \infty$. Let

$$K \subseteq g_1 K_0 \cup g_2 K_0 \cup \cdots \cup g_m K_0$$

and

$$K_0 \subseteq h_1 V \cup h_2 V \cup \cdots \cup h_n V$$

be optimal covers (so, in particular, $n = [K_0 : V]$). Then

$$K \subseteq g_1 \left( h_1 V \cup h_2 V \cup \cdots \cup h_n V \right) \cup \cdots \cup g_m \left( h_1 V \cup h_2 V \cup \cdots \cup h_n V \right),$$

so that $[K : V] \leqslant [K : K_0][K_0 : V]$ or equivalently $I_V(K) \leqslant [K : K_0]$. By the same argument, $[K_0 : V] \leqslant [K_0 : K][K : V]$, and so $\frac{1}{[K_0:K]} \leqslant I_V(K)$, where we interpret $[K_0 : K]$ as infinite and $\frac{1}{[K_0:K]}$ as zero if $K_0$ cannot be covered by finitely many left translates of $K$. With this convention,

$$I_V \in \prod_{K \subseteq G} \left[ \tfrac{1}{[K_0:K]}, [K:K_0] \right] = \Re,$$

where the product is taken over all compact subsets of $G$, and the space $\Re$ is compact by Tychonoff's theorem (Theorem B.5; this space is not metric unless $G$ is countable). It follows that there must be an accumulation point $I$ of the set $\{I_{v_n} \mid n \in \mathbb{N}\}$, and this accumulation point is a function $I$, defined on all compact subsets of $G$, and with the property that

$$I(K) \in \left[ \tfrac{1}{[K_0:K]}, [K:K_0] \right],$$

such that for any neighborhood $U$ of $I$ in the product topology, and any $N$, there is some $n > N$ with $I_{v_n} \in U$. This implies that $I$ is a subadditive function, which is additive on disjoint compact sets by the following argument. If $K$ and $K'$ are disjoint compact sets then, for all $\varepsilon > 0$,

$$U = \big\{ J \in \Re \mid |J(K) - I(K)| < \varepsilon, |J(K') - I(K')| < \varepsilon,$$
$$\text{and } |J(K \cup K') - I(K \cup K')| < \varepsilon \big\}$$

is a neighborhood of $I$ in the product topology of $\Re$. By the discussion above, we have (for the given sets $K, K'$, and) for large enough $n$, that $I_{V_n}$ satisfies

$$I_{V_n}(K \cup K') = I_{V_n}(K) + I_{V_n}(K').$$

On the other hand, there are arbitrarily large values of $n$ with $I_{V_n} \in U$, so that

$$|I(K \cup K') - I(K) - I(K')| < 3\varepsilon$$

for all $\varepsilon > 0$.

From this additivity on compact sets one can use measure-theoretic arguments like those used in the proof of the Riesz representation theorem[81] to deduce that

$$m_G(O) = \sup\{I(K) \mid K \subseteq O, K \text{ compact}\}$$

and

$$m_G^*(B) = \inf\{m_G(O) \mid B \subseteq O, O \text{ open}\}$$

define a measure on open sets, and an outer measure on all sets, respectively. The latter, when restricted to $\mathscr{B}_G$, gives a left Haar measure on $G$.

### 8.3.1 Measure-Theoretic Transitivity and Uniqueness

The following elementary result will be useful in this and the following chapters. In the arguments of this section we will not use the uniqueness properties of Haar measure, instead it will emerge in Corollary 8.8.

**Proposition 8.6.** *Let $G$ be a $\sigma$-locally compact metrizable group, and let $m_G$ denote a left Haar measure on $G$. Then for any two Borel sets $B_1, B_2 \in \mathscr{B}_G$ with $m_G(B_1)m_G(B_2) > 0$, the set*

$$\{g \in G \mid m_G(gB_1 \cap B_2) > 0\}$$

*is open and non-empty. The same holds for $\{g \in G \mid m_G(B_1g \cap B_2) > 0\}$. Moreover, $m_G(B) > 0$ if and only if $m_G(B^{-1}) > 0$ for $B \in \mathscr{B}_G$.*

PROOF. Clearly

$$m_G(gB_1 \cap B_2) = \int \chi_{gB_1}(h)\chi_{B_2}(h)\,dm_G(h),$$

and notice that $h \in gB_1$ if and only if $g \in hB_1^{-1}$ so $\chi_{gB_1}(h) = \chi_{hB_1^{-1}}(g)$. Together, these facts imply by Fubini's theorem (Theorem A.13) that

$$\int m_G(gB_1 \cap B_2)\,dm_G(g) = \iint \chi_{gB_1}(h)\chi_{B_2}(h)\,dm_G(h)\,dm_G(g)$$

$$= \int \chi_{B_2}(h) \underbrace{\int \chi_{hB_1^{-1}}(g)\,dm_G(g)}_{= m_G(hB_1^{-1}) = m_G(B_1^{-1})}\,dm_G(h)$$

$$= \int_{B_2} m_G(B_1^{-1})\,dm_G(h)$$

$$= m_G(B_1^{-1})m_G(B_2).$$

In fact we have used Fubini's theorem for non-negative but not necessarily integrable functions, but this can easily be avoided by replacing $B_1$ and $B_2$ with sequences of subsets with compact closures. Setting $B_2 = G$, the above argument shows that

$$m_G(B_1)m_G(G) = m_G(B_1^{-1})m_G(G),$$

which shows the last statement of the proposition (as is usual in measure theory, $0 \cdot \infty = 0$ by convention).

Having established the last claim, and returning to a general Borel set $B_2$, we get

$$\int m_G(gB_1 \cap B_2)\,dm_G(g) = m_G(B_1^{-1})m_G(B_2) > 0,$$

which shows that the set $O = \{g \in G \mid m_G(gB_1 \cap B_2) > 0\}$ is non-empty.

We now prove that $O$ is open, completing the proof of the proposition. Assume that $g \in O$, so that $m_G(gB_1 \cap B_2) > 0$. Since $G$ is $\sigma$-compact we can write

$$B_1 = \bigcup_{n=1}^{\infty} A_n,$$

where each $A_n$ has compact closure. Since $m_G$ is a measure, we must have

$$\varepsilon = m_G(gA_n \cap B_2) > 0$$

for some $A_n$. Now

$$m_G(g_1 A_n \cap B) = \int \chi_{g_1 A_n} \chi_{B_2} \, dm_G = \int \chi_{A_n}(g_1^{-1}h)\chi_{B_2}(h) \, dm_G(h)$$

so that, with $f = \chi_{A_n}$, we have

$$\left| m_G(gA_n \cap B_2) - m_G(g_1 A_n \cap B_2) \right| \leqslant \left| \int \left( f(g^{-1}h) - f(g_1^{-1}h) \right) \chi_{B_2}(h) \, dm_G(h) \right|$$

$$= \| f(g^{-1}\cdot) - f(g_1^{-1}\cdot) \|_1,$$

and we claim that for $g_1$ sufficiently close to $g$ the last term is less than $\varepsilon$, so $O$ is open. This claim is shown in the following lemma, in a more general context.

For the second claim of the proposition, notice that we already established the last claim, so that $m_G(B_1^{-1}) > 0$ and $m_G(B_2^{-1}) > 0$ by assumption. Hence, again by the last claim,

$$\{ g \in G \mid m_G(B_1 g \cap B_2) > 0 \} = \{ h \in G \mid m_G(hB_1^{-1} \cap B_2^{-1}) > 0 \}^{-1}$$

is non-empty and open by the first claim of the proposition. $\qquad\square$

**Lemma 8.7.** *Let $G$ be a $\sigma$-locally compact metrizable group acting continuously on a locally compact, $\sigma$-compact, metrizable space $X$. Let $\mu$ be a locally finite measure on $X$ which is invariant under $G$. Then, for $p \in [1, \infty)$ and any $f \in L_\mu^p(X)$,*

$$(U_g f)(x) = f(g^{-1}\cdot x)$$

*defines an element $U_g f \in L_\mu^p(X)$ with $\| U_g f \|_p = \| f \|_p$ and, for any fixed function $f \in L_\mu^p(X)$, $g \mapsto U_g f$ is a continuous map from $G$ to $L_\mu^p(X)$ with respect to the norm $\| \cdot \|_p$.*

PROOF. First notice that for $f \in L_\mu^p$,

$$\int_X |f(g^{-1}\cdot x)|^p \, d\mu(x) = \int_X |f(x)|^p \, d\mu(x),$$

since by assumption $\mu$ is invariant under $g \in G$. To prove the second claim, recall that for every $\varepsilon > 0$ there exists some $f_0 \in C_c(X)$ such that

$$\| f - f_0 \|_p < \varepsilon. \tag{8.12}$$

Let $V_0 \subseteq G$ be a symmetric (that is, $V_0^{-1} = V_0$) compact neighborhood of $e \in G$, and let

$$K = V_0 \cdot \mathrm{Supp}(f_0) = \{v \cdot x \mid v \in V_0, x \in \mathrm{Supp}\, f_0\}$$

so that $K$ is also compact, by continuity of the $G$-action. By uniform continuity of $f_0$ on the compact set $K$, there exists some $\delta > 0$ such that

$$|f_0(x) - f_0(y)| < \varepsilon/\mu(K)^{1/p}$$

for $d_X(x,y) < \delta$. Moreover, again by uniform continuity of the map

$$V_0 \times K \ni (g,x) \mapsto g \cdot x,$$

there is a symmetric neighborhood $V \subseteq V_0$ such that $g \in V$ implies that $d_X(x, g \cdot x) < \delta$. Putting these facts together, we see that $g \in V$ implies that $|f_0(x) - f_0(g^{-1} \cdot x)| < \varepsilon/\mu(K)^{1/p}$, and so

$$\int_K |f_0(x) - f_0(g^{-1} \cdot x)|^p \, d\mu < \varepsilon^p.$$

However, for $x \notin K$ we also have $g^{-1} \cdot x \notin \mathrm{Supp}(f)$ and so

$$|f_0(x) - f_0(g^{-1} \cdot x)| = 0.$$

Therefore the bound above actually gives $\|f_0 - U_g f_0\|_p < \varepsilon$ for any $g \in V$. Together with (8.12), and its consequence $\|U_g f - U_g f_0\|_p < \varepsilon$, we get $\|f - U_g f\|_p < 3\varepsilon$ for $g \in V$. In general, $g_0 \in G$ and $g \in g_0 V$ implies that

$$\|U_{g_0} f_0 - U_g f_0\|_p = \|f_0 - U_{g_0^{-1}g} f_0\|_p < \varepsilon$$

as required.                                                                    $\square$

**Corollary 8.8.** *The Haar measure $m_G$ of a $\sigma$-locally compact metrizable group $G$ is unique up to a multiplicative factor. In particular, every continuous group automorphism $\phi : G \to G$ satisfies*

$$\phi_* m_G = \mathrm{mod}(\phi) m_G$$

*for some scalar $\mathrm{mod}(\phi) > 0$.*

See Appendix C for more discussion of the modular function $\mathrm{mod}(\phi)$. Recall that $G$ is called *unimodular* if the left Haar measure $m_G$ is also right invariant (equivalently, if mod applied to the inner automorphism $\phi_g$ gives $\mathrm{mod}(\phi_g) = 1$ for all $g \in G$, where $\phi_g(h) = ghg^{-1}$ for $h \in G$).

PROOF OF COROLLARY 8.8. Suppose $m_1$ and $m_2$ are two left-invariant Haar measures. We define $m = m_1 + m_2$, so that $m$ is also a left-invariant Haar measure, with the property that both $m_1$ and $m_2$ are absolutely continuous with respect to $m$. By Theorem A.15, it follows that there are measurable non-negative functions $f_1, f_2$ such that $\mathrm{d}m_j = f_j \, \mathrm{d}m$ for $j = 1, 2$. We claim that both $f_1 \equiv c_1$ and $f_2 \equiv c_2$ are constant, so that $m_1 = c_1 m$, $m_2 = c_2 m$ and $m_2 = \frac{c_1}{c_2} m_1$ as stated in the corollary.

Suppose the claim is false for $f_1$. Then there exist two measurable sets $B_1, B_2 \subseteq G$ such that $f_1(x_1) < f_1(x_2)$ for $x_1 \in B_1$, $x_2 \in B_2$ and both $B_1$ and $B_2$ have positive measure under $m$ (this follows from dividing $\mathbb{R}_{\geqslant 0}$ into intervals $[\frac{k}{n}, \frac{k+1}{n})$ and taking their pre-images under $f_1$; if $f_1$ is not constant then there exists some $n$ and $k_1 \neq k_2$ for which the pre-images both have positive measure). By Proposition 8.6 (applied to $m_G = m$) there exists some $g \in G$ such that $m_G(gB_1 \cap B_2) > 0$.

Now let $E \subseteq G$ be any measurable set. Then

$$\int_E f_1(x) \, \mathrm{d}m(x) = m_1(E) = m_1(g^{-1}E) = \int_{g^{-1}E} f_1 \, \mathrm{d}m = \int_E f_1(gx) \, \mathrm{d}m(x)$$

by definition of $f_1$, left-invariance of $m_1$, and left-invariance of $m$. Since this holds for all measurable $E \subseteq G$, the uniqueness of the Radon–Nikodym derivative (Theorem A.4) shows that $f_1(x) = f_1(gx)$ for $m$-almost every $x$. However, for $x \in B_1 \cap g^{-1}B_2$ we have, by construction of $B_1$ and $B_2$, that $f_1(x) < f_1(gx)$ and again by construction $m(B_1 \cap g^{-1}B_2) > 0$. This contradiction shows that $m_1 = c_1 m$ for some $c_1 > 0$, and by symmetry of $m_1$ and $m_2$ the first half of the corollary follows.

For the second half, notice that for a continuous automorphism $\phi : G \to G$ the measure $\phi_* m_G$ is also left-invariant for the following reason. For $B \subseteq G$ a measurable set and $g \in G$,

$$\phi_* m_G(gB) = m_G(\phi^{-1}(gB))$$
$$= m_G(\phi^{-1}(g)\phi^{-1}(B))$$
$$= m_G(\phi^{-1}(B))$$
$$= \phi_* m_G(B).$$

Also by assumption, $\phi^{-1}(K)$ is compact (resp. $\phi^{-1}(O)$ is open) whenever $K$ is compact (resp. $O$ is open), so $0 < \mathrm{mod}(\phi) < \infty$.  $\square$

The results on Haar measure from this section now allow Proposition 8.3 to be proved.

PROOF OF PROPOSITION 8.3. It is clear that (2) implies (1) whenever the action of $G$ preserves the measure $\mu$, so we need to show that (1) implies (2).

For a countable group $G$ the proof is similar to the case of a single transformation: Suppose that $B$ is as in (1) and define the measurable set $B' = \bigcap_{g \in G} g \cdot B$. Then $B'$ is invariant in the sense of (2), and equivalent to $B$ in the sense that

$$\mu(B \triangle B') = \mu(B \smallsetminus B') = \mu\left(\bigcup_{g \in G} B \smallsetminus g{\cdot}B\right) = 0$$

by the assumption on $B$ and countability of $G$.

So now let $G$ be a $\sigma$-compact metric group, and let $G'$ be a dense countable subgroup of $G$ (such a subgroup exists since $G$ is separable as a topological space, and a countable dense subset generates a countable dense subgroup). Let $B \in \mathscr{B}$ be a $G$-invariant set as in (1); by the first case we may assume without loss of generality that it is a strictly $G'$-invariant set as in (2).

As usual write $m_G$ for a left Haar measure on $G$; we will use $m_G$ to find an equivalent strictly $G$-invariant set $B'$. For this, we first need to analyze the subsets

$$B_x = \{g \in G \mid g{\cdot}x \in B\} \subseteq G$$

for each $x \in X$. By continuity of the map $g \mapsto g{\cdot}x$, we know that $B_x \subseteq G$ is measurable for each $x \in X$.

Notice that by strict $G'$-invariance of $B$ we have $hB_x = B_x$ for all $h \in G'$, and we claim that this implies $m_G(B_x) = 0$ or $m_G(G \smallsetminus B_x) = 0$. To prove this claim, suppose instead that $m_G(B_x) > 0$ and $m_G(G \smallsetminus B_x) > 0$. Then by Proposition 8.6 the set

$$O = \{g \in G \mid m_G(gB_x \smallsetminus B_x) > 0\}$$

is non-empty and open. Hence there exists some $h \in G' \cap O$ which contradicts the fact that $hB_x = B_x$.

Now define

$$B' = \{x \in X \mid m_G(B_x) > 0\} = \{x \in X \mid m_G(G \smallsetminus B_x) = 0\},$$

let $x \in B'$ and $h \in G$. Then

$$B_{hx} = \{g \in G \mid gh{\cdot}x \in B\} = B_x h^{-1}.$$

Since $m_G(B_x) > 0$ this implies that $m_G(B_{hx}) > 0$ by Proposition 8.6 and so $hx \in B'$; that is, $B'$ is strictly $G$-invariant as in (2).

All that remains is to show that $B'$ is measurable and that $\mu(B \triangle B') = 0$, and to do this we reformulate the definition of $B'$ as follows. Let $U \subseteq G$ be a measurable set of finite positive measure, and define

$$f(x) = \frac{1}{m_G(U)} \int_U \chi_B(g{\cdot}x) \, dm_G(g).$$

By the fact that $B_x$ is either a null set or a co-null set with respect to $m$, we see that $f = \chi_{B'}$, which shows that $B'$ is measurable by Fubini's theorem (Theorem A.13) applied to $U \times X$. Notice that $\chi_B(g{\cdot}x) = \chi_{g^{-1}{\cdot}B}(x)$. Combining this with Fubini's theorem again and our assumption, we have

$$\mu(B \triangle B') = \int (\chi_B + \chi_{B'} - 2\chi_B \chi_{B'}) \, d\mu$$

$$= \frac{1}{m_G(U)} \int_U \int (\chi_B + \chi_{g^{-1}B} - 2\chi_B \chi_{g^{-1}B}) \, d\mu \, dm_G(g)$$

$$= \frac{1}{m_G(U)} \int_U \mu \left( B \triangle g^{-1} B \right) \, d\mu = 0,$$

which proves the proposition.                                                      □

## 8.4 Amenable Groups

One of the fundamental ways in which measure-preserving transformations arise comes from the Kryloff–Bogoliouboff Theorem (Corollary 4.2), which in the language of this chapter says that any continuous $\mathbb{Z}$-action on a compact metric space must have an invariant probability measure. This motivates the definition of a class of groups all of whose continuous actions will turn out to have invariant measures. We have seen in Example 8.2 that not all groups have this property, so the definition of amenability below[82] is non-trivial. Amenability may be defined in several different ways; for the purposes of ergodic averaging (for example, as in the mean ergodic theorem), and in order to work with familiar groups, the next definition is a convenient one.

### 8.4.1 Definition of Amenability and Existence of Invariant Measures

**Definition 8.9.** A $\sigma$-locally compact group $G$ is *amenable* if, for any compact subset $K \subseteq G$ and $\varepsilon > 0$, there is a measurable set $F \subseteq G$ with compact closure such that $KF$ is a measurable* set with

$$m_G (F \triangle KF) < \varepsilon m_G(F), \tag{8.13}$$

where $m_G$ denotes a left Haar measure on $G$.

Such a set $F$ is said to be $(K, \varepsilon)$-invariant, and a sequence $(F_n)_{n \geqslant 1}$ of compact subsets of $G$ is called a *Følner sequence* if for every compact set $K$ and $\varepsilon > 0$, we have that $F_n$ is $(K, \varepsilon)$-invariant for all large enough $n$. Følner sequences in a group $G$ allow ergodic averages of $G$-actions to be formed. Notice that if $G$ is discrete, then the sets $F_n$ and $K$ are finite sets and the natural Haar measure is cardinality.

---

* This is clearly the case if, for example, $F$ is a countable union of compact sets, which one may always assume by regularity of the measure $m_G$.

A different characterization of amenability may be given in terms of the existence of invariant probability measures for continuous actions. This directly generalizes the Kryloff–Bogoliouboff Theorem [214] (Corollary 4.2). We shall only need one direction of this result, not the fact that it characterizes amenability.

**Theorem 8.10.** *If a locally compact group $G$ is amenable, then every continuous $G$-action $G \to \mathrm{Homeo}(X, \mathsf{d})$ on a compact metric space has an invariant probability measure.*

PROOF. Recall that we write $\mathscr{M}(X)$ for the space of probability measures on $(X, \mathsf{d})$ with the weak*-topology (see p. 97); in this argument we will make use of the theory of integration for functions taking values in the space of measures (see Chap. 5 and Sect. A.3).

A continuous action $G \to \mathrm{Homeo}(X)$, with the action of $g$ written $x \mapsto g \cdot x$, induces an action of $G$ on $\mathrm{Homeo}(\mathscr{M}(X))$ written $\nu \mapsto g_* \nu$. Let $\nu \in \mathscr{M}(X)$ be any measure, and let $(F_n)$ be a Følner sequence in $G$. The sequence of averaged measures defined by

$$\mu_n = \frac{1}{m_G(F_n)} \int_{F_n} g_* \nu \, \mathrm{d}m_G(g),$$

or more concretely by

$$\int f \, \mathrm{d}\mu_n = \frac{1}{m_G(F_n)} \int_{F_n} \int f(g \cdot x) \, \mathrm{d}\nu \, \mathrm{d}m_G(g)$$

for any $f \in C(X)$, for $n \geqslant 1$, has a weak*-convergent subsequence $\mu_{n_j} \to \mu$ as $j \to \infty$. Given any $f \in C(X)$ and $h \in G$, we have

$$\int f(h \cdot x) \, \mathrm{d}\mu_n(x) = \frac{1}{m_G(F_n)} \int_{F_n} \int f(hg \cdot x) \, \mathrm{d}\nu(x) \, \mathrm{d}m_G(g)$$

$$= \frac{1}{m_G(F_n)} \int_{hF_n} \int f(g \cdot x) \, \mathrm{d}\nu(x) \, \mathrm{d}m_G(g),$$

and so

$$\left| \int f(x) \, \mathrm{d}\mu_n(x) - \int f(h \cdot x) \, \mathrm{d}\mu_n(x) \right| \leqslant \frac{1}{m_G(F_n)} \int_{F_n \triangle hF_n} |f(g \cdot x)| \, \mathrm{d}\nu(x) \mathrm{d}m_G(g)$$

$$\leqslant \frac{m_G(F_n \triangle hF_n)}{m_G(F_n)} \|f\|_\infty \to 0$$

as $n \to \infty$. This implies that the weak*-limit $\mu$ is invariant under the action of $g \in G$. □

Abelian groups are amenable; since we do not need this fact itself we instead show that abelian groups share the invariant measure property of amenable groups.

**Theorem 8.11.** *If $G$ is an abelian locally compact group, then every continuous $G$-action $G \to \mathrm{Homeo}(X, d)$ on a compact metric space has an invariant probability measure.*

PROOF. The proof uses averaging and compactness very much like the proof of Theorem 4.1. For each $g \in G$ and $n \geqslant 0$ define a map $A_{n,g} : \mathscr{M}(X) \to \mathscr{M}(X)$ by

$$A_{n,g}(\mu) = \frac{1}{n} \sum_{j=0}^{n-1} g_*^j \mu;$$

this is well-defined since $\mathscr{M}(X)$ is convex. There is no reason for these maps to be invertible, so let $\mathscr{A}_G$ be the semigroup of maps $\mathscr{M}(X) \to \mathscr{M}(X)$ generated by the set $\{A_{n,g} \mid n \geqslant 0, g \in G\}$. Notice that the maps in $\mathscr{A}_G$ all commute, since $G$ is abelian. Since any map $A \in \mathscr{A}_G$ is continuous, the image $A(\mathscr{M}(X)) \subseteq \mathscr{M}(X)$ is compact.

Given finitely many maps $A_1, \dots, A_r \in \mathscr{A}_G$, let $A = A_1 \circ \cdots \circ A_r$. Then, since $\mathscr{A}_G$ is commutative, for any $i = 1, \dots, r$,

$$A(\mathscr{M}(X)) = A_i \left( \underbrace{A_1 \circ \cdots \circ A_r(\mathscr{M}(X))}_{\text{omitting } A_i} \right) \subseteq A_i(\mathscr{M}(X)).$$

It follows that

$$\bigcap_{i=1}^{r} A_i(\mathscr{M}(X)) \supseteq A(\mathscr{M}(X)) \neq \varnothing$$

for any finite collection $A_1, \dots, A_r$ of maps in $A_G$. By compactness, we deduce that

$$\bigcap_{A \in \mathscr{A}_G} A(\mathscr{M}(X)) \neq \varnothing.$$

Now let $\mu^*$ be a measure in $\bigcap_{A \in \mathscr{A}_G} A(\mathscr{M}(X))$. For $g \in G$ and any $n \geqslant 0$, there must be some $\mu$ for which $\mu^* = A_{n,g}\mu$, and therefore

$$\mu^* - g_*\mu^* = \frac{1}{n}(\mu - g_*^n \mu)$$

has operator norm $\|\mu^* - g_*\mu^*\| \leqslant \frac{2}{n}$ for all $n \geqslant 1$, so $\mu^* = g_*\mu^*$ for all $g \in G$. This shows that $\mu^*$ is $G$-invariant. $\square$

Compact groups are also amenable (they satisfy Definition 8.9); we may prove directly that their actions have invariant measures (see also Exercise 8.4.2).

**Lemma 8.12.** *If $G$ is compact, then any continuous action of $G$ has an invariant probability measure.*

PROOF. Let $G$ act continuously on a compact metric space $(X, d)$, and let $x$ be a point in $X$. Define $\phi : G \to X$ by $\phi(g) = g \cdot x$. Writing $m_G$ for the

Haar measure on $G$, we see that $\phi_*(m_G)$ is an invariant probability measure on $X$.                                                                          □

We also note that the statement in Exercise 8.4.2, is the reason why ergodic theory usually concerns itself with actions of non-compact groups.

As Example 8.2 shows, it is easy to exhibit non-amenable groups, though it has proved to be a difficult problem to characterize amenability group-theoretically.

## Exercises for Sect. 8.4

**Exercise 8.4.1.** Prove directly that a countable abelian group is amenable in the sense of Definition 8.9.

**Exercise 8.4.2.** Classify all ergodic invariant Borel probability measures for a compact group $G$ acting continuously on a locally compact metric space $X$ by showing that they all arise as push-forwards of the Haar measure on $G$ by the map $g \mapsto g \cdot x$ for some $x \in X$ (as in Lemma 8.12).

**Exercise 8.4.3.** Prove that the Heisenberg group

$$H = \left\{ \begin{pmatrix} 1 & x & z \\ & 1 & y \\ & & 1 \end{pmatrix} \mid x, y, z \in \mathbb{R} \right\}$$

is amenable and unimodular.

**Exercise 8.4.4.** Prove that the group

$$B = \left\{ \begin{pmatrix} a & b \\ & 1 \end{pmatrix} \mid a, b \in \mathbb{R}, a > 0 \right\},$$

which is also called the '$ax + b$' group to reflect its natural action on

$$\left\{ \begin{pmatrix} x \\ 1 \end{pmatrix} \mid x \in \mathbb{R} \right\},$$

is amenable but not unimodular.

## 8.5 Mean Ergodic Theorem for Amenable Groups

Følner sequences permit ergodic averages to be formed, and the mean and pointwise ergodic theorems hold under suitable conditions for measure-preserving actions of amenable groups.

In the theorem below we deal with integration of functions on the group taking values in the Hilbert space $L^2_\mu$ (see Sect. A.3).

The reader may find it a helpful exercise to specialize the proof to the case of countable discrete amenable groups, in which Haar measure is cardinality, the elements of a Følner sequence are finite sets, and the integrals appearing are simply finite sums.

We will also be using the *induced unitary representation*[*] of $G$ defined by

$$U_g(f)(x) = f(g^{-1} \cdot x)$$

for all $x \in X$. As every element of $G$ is assumed to preserve the measure $\mu$ in its action on $X$, we know that $U_g : L^2_\mu \to L^2_\mu$ is unitary. Moreover, if $g, h \in G$, then by definition

$$U_h(U_g(f))(x) = U_g(f)(h^{-1} \cdot x) = f(g^{-1}h^{-1} \cdot x) = f((hg)^{-1} \cdot x) = U_{hg}(f)(x),$$

which shows that $g \mapsto U_g$ for $g \in G$ defines an action of $G$ on $L^2_\mu$. (The inverse in the definition of $U_g$ is used to ensure that this is indeed an action.)

**Theorem 8.13.** *Let $G$ be a $\sigma$-locally compact amenable group with left Haar measure $m_G$ acting continuously on $X$, and let $\mu$ be a $G$-invariant Borel probability measure on $X$. Let $P_G$ be the orthogonal projection onto the closed subspace*

$$I = \{f \in L^2_\mu(X) \mid U_g f = f \text{ for all } g \in G\} \subseteq L^2_\mu(X).$$

*Then, for any Følner sequence $(F_n)$ and $f \in L^2_\mu(X)$,*

$$\frac{1}{m_G(F_n)} \int_{F_n} U_{g^{-1}} f \, dm_G(g) \xrightarrow[L^2_\mu]{} P_G f.$$

*Thus the action is ergodic if and only if*

$$\frac{1}{m_G(F_n)} \int_{F_n} U_{g^{-1}} f \, dm_G(g) \xrightarrow[L^2_\mu]{} \int_X f \, d\mu$$

*for all $f \in L^2_\mu(X)$ (or for all $f$ in a dense subset of $L^2_\mu(X)$).*

As will become clear, it is important that we average the expression

$$U_{g^{-1}} f(x) = f(g \cdot x)$$

instead of the expression $U_g f(x)$.

---

[*] For now we will just use the action of $U_g$ for $g \in G$ on $L^2_\mu(X)$ defined here; a more formal definition of the notion of unitary representation will be given in Sect. 11.3.

PROOF OF THEOREM 8.13. Let $u$ be a function of the form

$$u(x) = v(h{\cdot}x) - v(x)$$

for some $v \in L^2_\mu(X)$ and $h \in G$, that is $u = U_{h^{-1}}v - v$. Then

$$\int_{F_n} U_{g^{-1}}U_{h^{-1}}v \, dm_G(g) = \int_{F_n} U_{(hg)^{-1}}v \, dm_G(g) = \int_{hF_n} U_{g^{-1}}v \, dm_G(g),$$

and so

$$\left\| \frac{1}{m_G(F_n)} \int_{F_n} U_{g^{-1}}u \, dm_G(g) \right\|_2 = \left\| \frac{1}{m_G(F_n)} \left( \int_{hF_n} U_{g^{-1}}v \, dm_G(g) \right. \right.$$

$$\left. \left. - \int_{F_n} U_{g^{-1}}v \, dm_G(g) \right) \right\|_2$$

$$\leqslant \frac{1}{m_G(F_n)} \int_{F_n \triangle hF_n} \|U_{g^{-1}}v\| \, dm_G(g) \longrightarrow 0$$

as $n \to \infty$, by (8.13) and Sect. B.7. It follows that the same holds for any function in the $L^2_\mu$-closure $V$ of the space of finite linear combinations of such functions. Just as in the proof of the mean ergodic theorem (Theorem 2.21), if $u \perp V$ then, for every $v \in L^2_\mu$,

$$\langle U_g u, v \rangle = \langle u, U_{g^{-1}}v - v \rangle + \langle u, v \rangle = \langle u, v \rangle,$$

so $u \in I$. Thus $L^2_\mu = V \oplus I$, showing the first part of the theorem. As discussed on page 233, the $G$-action is ergodic if and only if $I = \mathbb{C}$, the constant functions, or equivalently if and only if $P_G f = \int_X f \, d\mu$, completing the proof.                                                                    $\square$

Just as the mean ergodic theorem (Theorem 2.21) readily implies a mean ergodic theorem in $L^1$ for single transformations (Corollary 2.22), Theorem 8.13 implies an $L^1$ theorem. See Sect. B.7 for an explanation of the meaning of the integral arising.

**Corollary 8.14.** *Let $G$ be a locally compact amenable group with left Haar measure $m_G$ acting continuously on $X$, and let $\mu$ be a $G$-invariant Borel probability measure on $X$. Then, for any Følner sequence $(F_n)$ and $f \in L^1_\mu(X)$,*

$$\frac{1}{m_G(F_n)} \int_{F_n} f \circ g \, dm_G(g) \longrightarrow E\left(f \,\middle|\, \mathscr{E}\right)$$

*in $L^1_\mu$, where $\mathscr{E}$ is the $\sigma$-algebra of $G$-invariant sets.*

## Exercises for Sect. 8.5

**Exercise 8.5.1.** Emulate the proof of Corollary 2.22 to deduce Corollary 8.14 from Theorem 8.13.

# 8.6 Pointwise Ergodic Theorems and Polynomial Growth

While the mean ergodic theorem (Theorem 2.21) extends easily to the setting of amenable group actions both in its statement and in its proof, extending the pointwise ergodic theorem (Theorem 2.30) is much more involved[(83)]. The general pointwise ergodic theorem for amenable groups is due to Lindenstrauss [233]; a condition is needed on the averaging Følner sequence used (but every amenable group has Følner sequences satisfying the condition). That some condition on the sequence is needed is already visible for single transformations ($\mathbb{Z}$-actions): del Junco and Rosenblatt [67] show that for any non-trivial measure-preserving system the pointwise ergodic theorem does not hold along the Følner sequence defined by $F_n = [n^2, n^2 + n) \cap \mathbb{Z}$.

### 8.6.1 Flows

We start by showing how Theorem 2.30 extends to the case of continuous time. As this is convenient, we state and prove the theorem in the measurable context. A *flow* is a family $\{T_t \mid t \in \mathbb{R}\}$ of measurable transformations of the probability space $(X, \mathscr{B}, \mu)$ satisfying the identity $T_s T_t = T_{s+t}$ for all $s, t \in \mathbb{R}$, and with $T_0 = I_X$. The flow is measure-preserving if $T_t$ preserves $\mu$ for each $t \in \mathbb{R}$, and is measurable (as a flow) if the map $(x, t) \mapsto T_t(x)$ is a measurable map from $(X \times \mathbb{R}, \mathscr{B} \otimes \mathscr{B}_{\mathbb{R}})$ to $(X, \mathscr{B})$. Similarly, a semi-flow is an action of the semigroup $\mathbb{R}_{\geqslant 0}$. The pointwise ergodic theorem for (semi-)flows is a direct corollary of Theorem 2.30.

**Corollary 8.15.** *Let $T$ be a measurable and measure-preserving (semi-)flow on the probability space $(X, \mathscr{B}, \mu)$. Then, for any $f \in L_\mu^1$, there is a measurable set of full measure on which*

$$\frac{1}{s} \int_0^s f(T_s x) \, \mathrm{d}s \to E\left(f \middle| \mathscr{E}\right)(x)$$

*converges everywhere and in $L_\mu^1$ to the expectation with respect to the $\sigma$-algebra $\mathscr{E} = \{B \in \mathscr{B} \mid \mu(B \triangle T_t^{-1}B) = 0 \text{ for all } t \in \mathbb{R}\}$.*

PROOF. The function $(x, s) \mapsto f(T_s(x))$ is integrable on $X \times [0, s]$ for any non-negative $s$ by Fubini's theorem (Theorem A.13). Thus the integral $\int_0^s f(T_t x)\, dt$ is well-defined for almost every $x \in X$. In particular, $F(x) = \int_0^1 f(T_t x)\, dt$ is well-defined for almost every $x$, and therefore defines a function in $L_\mu^1$. Now

$$\int_0^n f(T_t x)\, dt = \sum_{j=0}^{n-1} F(T_1^j x),$$

so, by Theorem 2.30, the averages

$$\frac{1}{N} \int_0^N f(T_t x)\, dt$$

converge almost everywhere as $N \to \infty$.

Moreover, Theorem 2.30 applied to $F_{\text{abs}} = \int_0^1 |f(T_s x)|\, ds$ also implies (by taking the difference between $\mathsf{A}_N(F_{\text{abs}})$ and $\frac{N+1}{N} \mathsf{A}_{N+1}(F_{\text{abs}})$, which converge to the same limit) that

$$\frac{1}{N} \int_0^1 |f(T_{t+N} x)|\, dt \to 0$$

as $N \to \infty$ almost everywhere. For any $s \in [N, N+1)$ we have

$$\left| \int_0^s f(T_t x)\, dt - \int_0^N f(T_t x)\, dt \right| \leqslant \int_0^1 |f(T_{t+N} x)|\, dt,$$

so the convergence almost everywhere follows.

To finish the proof we need to identify the limit $f^*$, and show convergence in $L_\mu^1$. By Theorem 6.1, the integral averages (from 0 to $N$) of $f \in L_\mu^1$ converge almost everywhere and in $L_\mu^1$ to the conditional expectation

$$f^* = E\left( F \big| \mathscr{E}^{T_1} \right)$$

with respect to the $\sigma$-algebra of sets invariant under $T_1$. To obtain convergence in $L_\mu^1$ of the integral from 0 to $s$, we can use the trick above for $s \in [N, N+1)$. Moreover, the same argument shows that $f^*$ is invariant under the flow $T_t$ (that is, $f^*$ is $\mathscr{E}$-measurable). Finally, using convergence in $L_\mu^1$, we see that for a measurable set $B \in \mathscr{E}$ we have

$$\int \chi_B(x) \frac{1}{N} \int_0^N f(T_t x)\, dt\, d\mu(x) = \frac{1}{N} \int_0^N \int \chi_B(x) f(x)\, d\mu(x)\, dt$$
$$\to \int_B f\, d\mu = \int_B f^*\, d\mu,$$

which proves that $f^* = E\left( f \big| \mathscr{E} \right)$.                                    $\square$

A quite different class of ergodic theorems for continuous time are the *local ergodic theorems*; we mention one simple instance. Local ergodic theorems were introduced by Wiener [382]; notice how closely they are related to the fundamental theorem of calculus, which states that

$$\lim_{\varepsilon \to 0} \frac{1}{\varepsilon} \int_s^{s+\varepsilon} f(t) \, dt = f(s)$$

for almost every $s \in \mathbb{R}$ if $f \in L^1_{m_{\mathbb{R}}}$ (see Royden [320, Ch. 5] or Theorem A.25).

**Theorem 8.16 (Wiener).** *Let $T$ be a measurable and measure-preserving flow on $(X, \mathscr{B}, \mu)$. Then, for any $f \in L^1_\mu$,*

$$\lim_{\varepsilon \to 0} \frac{1}{\varepsilon} \int_0^\varepsilon f(T_t x) \, dt = f(x)$$

*almost everywhere.*

Here we will write $m$ for Lebesgue measure $m_{\mathbb{R}}$ restricted to $[0, \infty)$.

PROOF OF THEOREM 8.16. Let

$$N = X \times [0, \infty) \smallsetminus \left\{ (x, t) \in X \times [0, \infty) \mid \lim_{\varepsilon \to 0} \tfrac{1}{\varepsilon} \int_0^\varepsilon f(T_{t+s}) \, ds = f(T_t x) \right\};$$

we wish to show that $N$ is a null set with respect to $\mu \times m$. Define

$$N^t = \{ x \in X \mid (x, t) \in N \}$$

for $t \geqslant 0$. Finally, let

$$N_x = \{ t \in [0, \infty) \mid (x, t) \in N \}.$$

Now for almost every $x$, $\int_0^s f(T_t x) \, dt$ is well-defined for all $s$, and $t \mapsto f(T_t x)$ is integrable on $[0, T]$ for any $T \geqslant 0$. Thus the fundamental theorem of calculus (Theorem A.25) implies that $m(N_x) = 0$ for almost every $x$. It follows that $(\mu \times m)(N) = 0$, and by Fubini's theorem (Theorem A.13), almost every $t$ has the property that $\mu(N^t) = 0$. On the other hand $N^t = T_t^{-1} N^0$, so $\mu(N^t)$ is independent of $t$, and therefore vanishes identically as required, since $T_t$ is measure-preserving. $\qquad \square$

### 8.6.2 Pointwise Ergodic Theorems for a Class of Groups

In this section we describe a more general version of the argument in Sect. 2.6.3 and the second proof in Sect. 2.6.5, using once again the finite Vitali covering lemma (Lemma 2.27). The pointwise ergodic theorem obtained here applies to a large collection of groups but is far from the most general pointwise theorems.

We assume here that $G$ is a locally compact unimodular group equipped with a right-invariant metric (that is, a metric d compatible with the topology and with $\mathsf{d}(g_1 h, g_2 h) = \mathsf{d}(g_1, g_2)$ for all $g_1, g_2, h \in G$) and with the following growth properties.

(P) For any $r > 0$ the metric $r$-ball $B_r^G = B_r^G(e)$ has compact closure (that is, the metric is *proper*), and $m_g(\overline{B_r^G} \setminus B_r^G) = 0$.

(D) The metric has a *doubling property*\*: there is a constant $C_G$ such that

$$m_G\left(B_{3r}^G\right) \leqslant C_G m_G\left(B_r^G\right).$$

(F) The metric balls form a *Følner sequence*: for any fixed $s \in \mathbb{R}$,

$$\frac{m_G\left(B_{r+s}^G\right)}{m_G\left(B_r^G\right)} \longrightarrow 1$$

as $r \to \infty$.

We will check that property (F) does indeed imply the amenability property in Definition 8.9 just after the statement of Theorem 8.19. It is clear that any metric induced by a norm on $\mathbb{R}^k \times \mathbb{Z}^\ell$ satisfies these properties. However, there are many other groups satisfying the properties above: for example, the Heisenberg group (defined in Exercise 8.4.3, with further properties described in Examples 9.13 and 9.15, and in Chap. 10) satisfies them.

We start as in Sect. 2.6 with a maximal inequality in $L^1(G)$, the analog of Lemma 2.29.

**Lemma 8.17.** *Let $G$ be a locally compact unimodular group with a right-invariant metric satisfying properties (P) and (D). Then for any $\phi \in L^1(G)$ and $\alpha > 0$ define the maximal function*

$$\phi^*(a) = \sup_{r>0} \frac{1}{m_G\left(B_r^G\right)} \int_{B_r^G} \phi(ga)\, \mathrm{d}m_G(g)$$

*and the set $E_\alpha^\phi = \{a \in G \mid \phi^*(a) > \alpha\}$. Then*

$$\alpha m_G\left(E_\alpha^\phi\right) \leqslant C_G \|\phi\|_1.$$

PROOF. By the assumption (P), $m_G(B_r^G)$ depends continuously on $r > 0$, so $\phi^*(a)$ can also be defined as the supremum over all rational $r > 0$. This shows that the map $a \mapsto \phi^*(a)$ is measurable. For any $a \in E_\alpha^\phi$ choose some $r(a)$ with

---

\* Strictly speaking this would be more naturally called a tripling condition, but it is equivalent to a condition on the measure of balls of doubled radius, and the term "doubling condition" is a standard one. The simplest example of a group without this property is given by the free group $F_2$ on two generators: in the natural word metric, $|B_{3r}^{F_2}| \approx |B_r^{F_2}|^3$.

$$\frac{1}{m_G(B^G_{r(a)})} \int_{B^G_{r(a)}} \phi(ga) \, dm_G(g) > \alpha. \tag{8.14}$$

Since $G$ is second countable (that is, has a countable basis for its topology), we can write the open set

$$O = \bigcup_{a \in E^\phi_\alpha} B^G_{r(a)} a$$

as a union of countably many sets,

$$O = \bigcup_{i=1}^{\infty} B^G_{r(a_i)} a_i$$

for some $\{a_1, a_2, \dots\} \subseteq E^\phi_\alpha$. For brevity, write $r_i = r(a_i)$. Notice that by the assumption of right-invariance for the metric, the right translate of a ball is also a ball: $B^G_r a = B^G_r(e)a = B^G_r(a)$ for any $r > 0$ and $a \in G$. Fix some $K \geq 1$ and apply the Vitali covering lemma (Lemma 2.27) to select indices $j(1), \dots, j(k) \in \{1, \dots, K\}$ such that the balls

$$B^G_{r_{j(1)}} a_{j(1)}, \dots, B^G_{r_{j(k)}} a_{j(k)} \tag{8.15}$$

are disjoint, and

$$B^G_{r_1} a_1 \cup \dots \cup B^G_{r_K} a_K \subseteq B^G_{3r_{j(1)}} a_{j(1)} \cup \dots \cup B^G_{3r_{j(k)}} a_{j(k)}. \tag{8.16}$$

By disjointness of the sets in (8.15),

$$\|\phi\|_1 \geq \sum_{i=1}^{k} \int_{B^G_{r_{j(i)}} a_{j(i)}} |\phi| \, dm_G$$

$$\geq \sum_{i=1}^{k} m_G(B^G_{r_{j(i)}} a_{j(i)}) \frac{1}{m_G(B^G_{r_{j(i)}})} \int_{B^G_{r_{j(i)}} a_{j(i)}} \phi(g a_{j(i)}) \, dm_G(g)$$

$$\geq \alpha \sum_{i=1}^{k} m_G(B^G_{r_{j(i)}} a_{j(i)}) \qquad \text{(by (8.14))}.$$

On the other hand, by (8.16), right-invariance of $m_G$, and property (D),

$$\sum_{j=1}^{K} m_G(B^G_{r_j} a_j) \leq \sum_{i=1}^{k} m_G(B^G_{3r_{j(i)}} a_{j(i)}) \leq C_G \sum_{i=1}^{k} m_G(B^G_{r_{j(i)}} a_{j(i)}).$$

Together with the bound on the latter sum, this gives

$$\alpha \sum_{j=1}^{K} m_G(B_{r_j}^G) \leqslant C_G \|\phi\|_1,$$

for any $K \geqslant 1$, which gives $\alpha m_G(O) \leqslant C_G \|\phi\|_1$ and hence the lemma, since $E_\alpha^\phi \subseteq O$. $\qquad\qquad\qquad\qquad\qquad\qquad\qquad\qquad\qquad\qquad\qquad\square$

We note that a right-invariant metric in a group satisfies the symmetry relation $\mathsf{d}(g,e) = \mathsf{d}(g^{-1},e)$ for $g \in G$ (which follows by right multiplication with $g^{-1}$) and the group triangle inequality $\mathsf{d}(gh,e) \leqslant \mathsf{d}(g,e) + \mathsf{d}(h,e)$ for $g, h \in G$. The latter follows by combining the right-invariance and the triangle inequality

$$\mathsf{d}(gh,e) = \mathsf{d}(g,h^{-1}) \leqslant \mathsf{d}(g,e) + \mathsf{d}(e,h^{-1}) = \mathsf{d}(g,e) + \mathsf{d}(h,e).$$

In the notation $B_r^G = B_r^G(e)$ for the balls of radius $r > 0$ around the identity $e \in G$ these properties may also be written as

$$(B_r^G)^{-1} = B_r^G$$

and

$$B_r^G B_s^G \subseteq B_{r+s}^G \qquad\qquad\qquad (8.17)$$

for any $r, s > 0$.

**Theorem 8.18 (Maximal Ergodic Theorem).** *Let $G$ be a unimodular locally compact group with a right-invariant metric satisfying properties* (P), (D) *and* (F). *Let $G$ act continuously on a locally compact $\sigma$-compact metric space $X$, preserving a Borel probability measure $\mu$ on $X$, and let $f \in L^1(X, \mathscr{B}, \mu)$. We define the maximal function*

$$f^*(x) = \sup_{r>0} \frac{1}{m_G(B_r^G)} \int_{B_r^G} f(g{\cdot}x) \,\mathrm{d}m_G(g),$$

*and for $\alpha > 0$ define*

$$E_\alpha^f = \{x \in X \mid f^*(x) > \alpha\}.$$

*Then $\alpha\mu(E_\alpha^f) \leqslant C_G \|f\|_1$.*

PROOF. In the proof below we will frequently use Fubini's theorem (Theorem A.13) applied to the functions

$$(g,x) \longmapsto f(g{\cdot}x)$$

and

$$(g,x) \longmapsto |f(g{\cdot}x)|,$$

for $(g, x) \in B_J^G \times X$ for some fixed $J > 1$. Notice that

$$\int |f(g{\cdot}x)|\,d\mu(x) = \|f\|_1$$

since $\mu$ is $G$-invariant, so

$$\int_{B_r^G} \int_X |f(g{\cdot}x)|\,d\mu(x)\,dm_G(g) = m_G(B_r^G)\|f\|_1,$$

and therefore these functions lie in $L^1(B_J^G \times X, m_G|_{B_J^G} \times \mu)$. In particular, for almost every $x \in X$,

$$\phi(g) = \begin{cases} f(g{\cdot}x) & \text{for } g \in B_J^G; \\ 0 & \text{for } g \notin B_J^G \end{cases}$$

defines a function $\phi \in L^1(G)$ (suppressing the dependence on $x \in X$ and on $J$).

Fix $\alpha > 0$. By Lemma 8.17,

$$\alpha m_G(E_\alpha^\phi) \leqslant C_G\|\phi\|_1. \tag{8.18}$$

Fix some $M > 0$ (later we will let $J \to \infty$ for $M$ fixed). Define the restricted maximal functions

$$\phi_M^*(a) = \sup_{0<r<M} \frac{1}{m_G(B_r^G)} \int_{B_r^G} \phi(ga)\,dm_G(g) \tag{8.19}$$

and

$$f_M^*(x) = \sup_{0<r<M} \frac{1}{m_G(B_r^G)} \int_{B_r^G} f(g{\cdot}x)\,dm_G(g), \tag{8.20}$$

and sets

$$E_{\alpha,M}^\phi = \{a \in G \mid \phi_M^*(a) > \alpha\}$$

and

$$E_{\alpha,M}^f = \{x \in X \mid f_M^*(x) > \alpha\}.$$

Fix $a \in B_{J-M}^G$ and $g \in B_M^G$. Then $ga \in B_J^G$ by (8.17), so

$$\phi(ga) = f(ga{\cdot}x)$$

by definition of $\phi$. Substituting this into (8.19) and (8.20) gives

$$\phi_M^*(a) = f_M^*(a{\cdot}x)$$

for $a \in B_{J-M}^G$. It follows that

$$\alpha m_G \left( \{ a \in B^G_{J-M} \mid a{\cdot}x \in E^f_{\alpha,M} \} \right) = \alpha m_G \left( B^G_{J-M} \cap E^\phi_{\alpha,M} \right)$$
$$\leqslant \alpha m_G \left( E^\phi_\alpha \right)$$
$$\leqslant C_G \|\phi\|_1, \tag{8.21}$$

where the last inequality is (8.18) again. We now integrate both sides of the inequality (8.21) over $X$ and use Fubini's theorem. More concretely, for the left-hand side of (8.21) we have

$$m_G(B^G_{J-M})\mu(E^f_{\alpha,M}) = \int_{B^G_{J-M}} \int_X \chi_{E^f_{\alpha,M}} (a{\cdot}x)\, d\mu(x)\, dm_G$$
$$= \int_X m_G \left( \{ a \in B^G_{J-M} \mid a{\cdot}x \in E^f_{\alpha,M} \} \right) d\mu(x) \tag{8.22}$$

by Fubini's theorem and the invariance of $\mu$ under the action of $G$. Similarly, for the right-hand side we may write

$$\int_X \|\phi\|_1 \, d\mu(x) = \int_X \int_{B^G_J} |f(g{\cdot}x)| \, dm_G(g)\, d\mu(x)$$
$$= \int_{B^G_J} \int_X |f(g{\cdot}x)| \, d\mu(x)\, dm_G(g) = m_G(B^G_J)\|f\|_1. \tag{8.23}$$

The identities (8.21)–(8.23) together give

$$\alpha m_G(B^G_{J-M})\mu(E^f_{\alpha,M}) \leqslant C_G m_G(B^G_J)\|f\|_1.$$

We divide by $m_G(B^G_J)$ and let $J \to \infty$, which gives

$$\alpha \mu(E^f_{\alpha,M}) \leqslant C_G\|f\|_1$$

by property (F). Finally, letting $M \to \infty$ gives the theorem. $\qquad\square$

It will be useful to extend the notation for ergodic averages from Sect. 2.5 to averages over balls for a group action, so we define

$$\mathsf{A}_r(f)(x) = \frac{1}{m_G(B^G_r)} \int_{B^G_r} f(g{\cdot}x)\, dm_G(g)$$

for a group action of $G$.

**Theorem 8.19.** *Let $G$ be a unimodular locally compact group with a right-invariant metric satisfying properties* (P), (D) *and* (F). *Let $G$ act continuously on a locally compact $\sigma$-compact metric space $X$, preserving a Borel probability measure $\mu$ on $X$, and let $f \in L^1(X, \mathscr{B}, \mu)$. Then*

$$\mathsf{A}_r(f)(x) \longrightarrow E\left(f \mid \mathscr{E}\right)(x)$$

*almost everywhere and in $L^1_\mu$, where as usual $\mathscr{E}$ denotes the $\sigma$-algebra of $G$-invariant sets.*

The proof of Theorem 8.19 will follow the pattern of the second proof in Sect. 2.6.5, where the pointwise ergodic theorem is deduced from the mean ergodic theorem using the maximal ergodic theorem. In order to use the mean ergodic theorem, we first verify that property (F) gives (8.13).

If $K \subseteq G$ is compact and non-empty, then $K \subseteq B^G_s$ for some $s > 0$, so

$$B^G_r \triangle K B^G_r = B^G_r \diagdown K B^G_r \cup K B^G_r \diagdown B^G_r \subseteq B^G_r \diagdown B^G_{r-s} \cup B^G_{r+s} \diagdown B^G_r$$

for all $r > s$, since $B^G_{r-s} \subseteq K B^G_r \subseteq B^G_{r+s}$ by (8.17). It follows that

$$m_G(B^G_r \triangle K B^G_r) \leqslant m_G(B^G_r \diagdown B^G_{r-s}) + m_G(B^G_{r+s} \diagdown B^G_r)$$
$$= m_G(B_{r+s}) - m_G(B_{r-s}),$$

so that $\frac{m_G(B_r \triangle K B^G_r)}{m_G(B^G_r)} \to 0$ as $r \to \infty$ by property (F).

PROOF OF THEOREM 8.19. Assume first that $f_0 \in \mathscr{L}^\infty$, so that

$$\mathsf{A}_n(f_0) \to F_0 = E\left(f_0 \mid \mathscr{E}\right)$$

as $n \to \infty$ in $L^1_\mu$ by the mean ergodic theorem for $L^1$ (Corollary 8.14). Pick some $M$ with $\|F_0 - \mathsf{A}_M(f_0)\|_1 < \varepsilon^2$. By the maximal ergodic theorem (Theorem 8.18) applied to $F_0 - \mathsf{A}_M(f_0)$ we have

$$\mu\left(\{x \in X \mid \sup_{r>0} \mathsf{A}_r\left(F_0 - \mathsf{A}_M(f_0)\right) > \varepsilon\}\right) < \varepsilon. \qquad (8.24)$$

By invariance of $F_0$ under the action, we have $\mathsf{A}_r(F_0) = F_0$. Just as in Sect. 2.6.5, we now estimate the difference between $\mathsf{A}_r(\mathsf{A}_M(f_0))$ and $\mathsf{A}_r(f_0)$ as $r \to \infty$. We have

$$\mathsf{A}_r\left(\mathsf{A}_M(f_0)\right)(x) = \frac{1}{m_G(B^G_r)} \int \mathsf{A}_M(f_0)(g{\cdot}x)\chi_{B^G_r}(g)\, dm_G(g)$$

$$= \frac{1}{m_G(B^G_r)m_G(B^G_M)} \iint \chi_{B^G_r}(g)\chi_{B^G_M}(h)f_0(hg{\cdot}x)\, dm_G(g)\, dm_G(h)$$

$$= \frac{1}{m_G(B^G_r)m_G(B^G_M)} \int \underbrace{\int \chi_{B^G_r}(h^{-1}g')\chi_{B^G_M}(h)\, dm_G(h)}_{\psi(g')}\, f_0(g'{\cdot}x)\, dm_G(g')$$

by Fubini's theorem (Theorem A.13) and left-invariance of $m_G$. Notice that for $g' \in B^G_{r-M}$ we have $h^{-1}g' \in B^G_r$ for all $h \in B^G_M$ by (8.17). It follows that $\psi(g') = m_G(B^G_M)$ if $g' \in B^G_{r-M}$. Similarly, if $g' \in B^G_{r+M}$ then $h^{-1}g' \notin B^G_r$ for any $h \in B^G_M$, so $\psi(g') = 0$. Finally, for $g' \in B^G_{r+M} \diagdown B^G_{r-M}$, we have

$$0 \leqslant \psi(g') \leqslant m_G(B_M^G).$$

Thus

$$|\mathsf{A}_r(\mathsf{A}_M(f_0)) - \mathsf{A}_r(f_0)| \leqslant \frac{1}{m_G(B_r^G)} \int \left| \frac{1}{m_G(B_M^G)} \psi(g) - \chi_{B_r^G}(g) \right| |f_0(g{\cdot}x)| \, \mathrm{d}m_G(g)$$

$$\leqslant \frac{m_G(B_{r+M}^G \setminus B_{r-M}^G)}{m_G(B_r^G)} \|f_0\|_\infty \to 0 \qquad (8.25)$$

as $r \to \infty$, by property (F). Together (8.24) and (8.25) show that

$$\mu(\{x \in X \mid \limsup_{r \to \infty} |F_0 - \mathsf{A}_r(f_0)| > \varepsilon\})$$

$$= \mu(\{x \in X \mid \limsup_{r \to \infty} |\mathsf{A}_r(F_0 - \mathsf{A}_M(f_0))| > \varepsilon\}) < \varepsilon,$$

which means that $\mathsf{A}_r(f_0) \to F_0$ pointwise almost everywhere as $r \to \infty$. The remainder of the proof follows the final part of the proof on p. 47. □

## Exercises for Sect. 8.6

**Exercise 8.6.1.** Fill in the details completing the proof of Theorem 8.19 as suggested there.

**Exercise 8.6.2.** Prove that the doubling property (D) from page 260 (for the left-invariant Haar measure and the right-invariant metric) implies that the group is unimodular.

## 8.7 Ergodic Decomposition for Group Actions

Just as for single transformations, the ergodic decomposition for a continuous measure-preserving action of a $\sigma$-compact metric group may be deduced from Choquet's theorem as in Theorem 4.8. The proof we will give in this section is related to the proof of Theorem 6.2, but a little more involved since we do not have a pointwise ergodic theorem for the action of $G$ on $X$. We will follow loosely work of Greschonig and Schmidt [129] where a similar but much stronger result is found using quasi-invariance (where the action is only assumed to take null sets to null sets).

**Theorem 8.20.** *Let $G$ be a $\sigma$-compact metric group acting continuously on a $\sigma$-compact metric space $(X, \mathsf{d})$. Let $\mu$ be a $G$-invariant probability measure* *on $X$, and let*

$$\mathscr{E} = \Big\{ B \subseteq X \mid B \text{ is measurable and } \mu\left(g{\cdot}B \triangle B\right) = 0 \text{ for all } g \in G \Big\}$$

*be the $\sigma$-algebra of $G$-invariant sets. Then*

$$\mu = \int \mu_x^{\mathscr{E}} \, d\mu(x)$$

*is the ergodic decomposition of $\mu$. That is, for $\mu$-almost every $x$ the conditional measure $\mu_x^{\mathscr{E}}$ is a $G$-invariant and ergodic probability measure on $X$.*

The next result will serve as a substitute for an ergodic theorem.

**Lemma 8.21.** *Let $P_1, P_2, \ldots$ be orthogonal projections, all defined on a separable Hilbert space $\mathscr{H}$. Define the operators*

$$Q_1 = P_1, \; Q_2 = Q_1 P_2 Q_1, \ldots, \; Q_{n+1} = Q_n P_{n+1} Q_n$$

*for any $n \geqslant 0$. Then the sequence $(Q_n v)$ converges in the norm topology to the orthogonal projection of $v$ onto the subspace $\bigcap_{m=1}^{\infty} \operatorname{Im} P_m$, for any $v \in \mathscr{H}$.*

As we will see in the proof, the peculiar structure of the definition of $Q_n$ will be used in two different ways. First, for any $v \in \mathscr{H}$ and any fixed $m$, the vector $Q_n v$ for $n \geqslant m$ is obtained from an element of $\operatorname{Im} P_m$ by a few (how many depending on $m$ but not on $n$) projections $P_j$ with $j < m$. Second, the symmetry of the definition will be used in the later part of the proof to miraculously convert weak\*-convergence into strong convergence.

To see how this can be helpful, notice the following peculiar geometrical fact. For a vector $v$ in a Hilbert space it is possible to find a finite number of projections $P_1, \ldots, P_n$ so that $v' = P_n \cdots P_1 v$ is orthogonal to $v$ but has almost the same length as $v$. A geometric picture of a composition of many projections that together behave like a rotation is illustrated in Fig. 8.5.

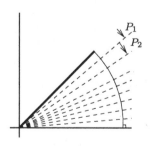

**Fig. 8.5** Many projections together acting like a rotation

The closer we require $\|v'\|$ to be to $\|v\|$, the more projections need to be taken. The bound on the number of projections that appear to the left of $P_m$ in the definition of $Q_n$ will in this way allow us to say that $Q_n v$ has not moved very far from $\operatorname{Im} P_m$, if indeed the length has not decreased much.

PROOF OF LEMMA 8.21. Fix some $v \in \mathscr{H}$, and let $w$ be a weak*-limit of a subsequence $(Q_{n_k} v)$. That is, $w \in \mathscr{H}$ satisfies

$$\langle Q_{n_k} v, \xi \rangle \to \langle w, \xi \rangle$$

as $k \to \infty$, for all $\xi \in \mathscr{H}$. Such a subsequence exists by Alaoglu's theorem (Theorem B.6) because $\|Q_n\| \leqslant 1$ for all $n$.

We wish to show that $w \in \bigcap_{m=1}^{\infty} \operatorname{Im} P_m$. We claim first that

$$P_1 Q_n = Q_n \qquad (8.26)$$

for all $n \geqslant 1$ by induction. For $n = 1$, $P_1 Q_1 = P_1^2 = P_1$ since $P_1$ is a projection, and if we assume that $P_1 Q_n = Q_n$ for some $n$, then

$$P_1 Q_{n+1} = P_1 Q_n P_{n+1} Q_n = Q_n P_{n+1} Q_n = Q_{n+1}.$$

Hence, if $\xi$ is an element of $(\operatorname{Im} P_1)^{\perp} = \ker P_1$, then

$$\langle Q_n v, \xi \rangle = \langle P_1 Q_n v, \xi \rangle = \langle Q_n v, P_1 \xi \rangle = 0$$

for all $n$, and so $\langle w, \xi \rangle = 0$ for all $\xi \in (\operatorname{Im} P_1)^{\perp}$ and $w \in \operatorname{Im} P_1$ as required.

To prove the same result for $m > 1$ we have to work a bit harder. To this end, define

$$c = \lim_{n \to \infty} \|Q_n v\|_2 = \inf_{n \to \infty} \|Q_n v\|_2,$$

where the equality holds since

$$\|Q_{n+1} v\| = \|Q_n P_{n+1} Q_n v\| \leqslant \|Q_n v\|,$$

as $Q_n P_{n+1}$ is a product of orthogonal projections.

We start by describing (in a very rough way) the case $m = 2$. Notice that when $\|Q_n v\|$ is close to $c$, then the same must be true of $\|P_2 S P_{n+1} Q_n v\|$, where $S$ is the product of projections $P_\ell$ with the property that $Q_n = P_1 P_2 S$. Therefore,

$$\|Q_{n+1} v\| = \|P_1 P_2 S P_{n+1} Q_n v\| \geqslant c$$

is close to $\|P_2 S P_{n+1} Q_n v\|$, which implies that $Q_{n+1} v$ is close to

$$P_2 S P_{n+1} Q_n v \in \operatorname{Im} P_2,$$

since the orthogonal projection $P_1$ either decreases the norm significantly, or does not move the vector much. It follows that $\langle Q_{n+1} v, \xi \rangle$ is close to zero for any $\xi \in (\operatorname{Im} P_2)^{\perp}$, which shows that $w \in \operatorname{Im} P_2$. This statement replaces (8.26) for the case $m = 2$.

We now give the formal argument for any $m \geqslant 1$. Fix $\varepsilon > 0$ and pick $n > m$ such that $\|Q_{n-1}v\| < c + \varepsilon$. Then $Q_n = Q_{m-1}P_m S P_n Q_{n-1}$ for some product $S$ of projections $P_\ell$ for various $\ell$ (this may be seen by a simple induction, starting with $n = m+1$ and using the particular inductive definition of $Q_n$). Let

$$v' = P_m S P_n Q_{n-1}v \in \operatorname{Im} P_m;$$

we wish to show that $Q_{m-1}v' = Q_n v$ is close to $\operatorname{Im} P_m$ by applying the projections $P_\ell$ appearing in $Q_{m-1}$ successively to $v'$. Let $P$ be the first projection appearing on the right in $Q_{m-1}$. Then

$$c \leqslant \|Pv'\| \leqslant \|v'\| < c + \varepsilon$$

and so

$$\|v' - Pv'\| = \sqrt{\|v'\|^2 - \|Pv'\|^2} \leqslant \sqrt{(c+\varepsilon)^2 - c^2} = \sqrt{2c\varepsilon + \varepsilon^2}.$$

Let $i(m)$ be the number of projections appearing in the definition of $Q_{m-1}$. By repeating the argument above, and using the triangle inequality, we get

$$\|v' - Q_n v\| = \|v' - Q_{m-1}v'\| \leqslant i(m)\sqrt{2c\varepsilon + \varepsilon^2}, \tag{8.27}$$

so $Q_n v$ is within $i(m)\sqrt{2c\varepsilon + \varepsilon^2}$ of $\operatorname{Im} P_m$. We will use this statement as a replacement for (8.26). If $\xi$ is an element of $(\operatorname{Im} P_m)^\perp$, then for large enough $k$ we have

$$|\langle Q_{n_k}v, \xi\rangle| = |\langle Q_{n_k}v, \xi\rangle - \langle v', \xi\rangle| \leqslant i(m)\sqrt{2c\varepsilon + \varepsilon^2}\|\xi\|$$

by (8.27) with $n = n_k$ and $v' \in \operatorname{Im} P_m$ depending on $n_k$. Therefore, $\langle w, \xi\rangle = 0$ for the weak*-limit $w$ of $Q_{n_k}v$ for all $\xi \in (\operatorname{Im} P_m)^\perp$ and all $m \geqslant 1$. The claim that $w \in \bigcap_{m=1}^\infty \operatorname{Im} P_m$ follows.

In particular $Q_{n_k}w = w$ and so

$$\langle Q_{n_k}(v - w), v - w\rangle = \langle Q_{n_k}(v), v - w\rangle - \langle w, v - w\rangle$$
$$\to \langle w, v - w\rangle - \langle w, v - w\rangle = 0$$

as $k \to \infty$. Since $Q_n^* = Q_n$ and $P_n^* = P_n = P_n^2$, we deduce that

$$\langle Q_{n_k}(v - w), v - w\rangle = \langle P_{n_k}Q_{n_k-1}(v - w), P_{n_k}Q_{n_k-1}(v - w)\rangle$$
$$= \|P_{n_k}Q_{n_k-1}(v - w)\|_2^2 \to 0$$

as $k \to \infty$. Hence $\|Q_n(v - w)\|_2 = \|Q_n v - w\|_2 \to 0$ as $n \to \infty$. That is, we have obtained norm convergence of the original sequence.

To see that $w$ is the projection of $v$ onto $\bigcap_{m \geqslant 1} \operatorname{Im} P_m$, notice that if $v'$ lies in $\bigcap_{m \geqslant 1} \operatorname{Im} P_m$ then

$$\langle w, v' \rangle = \lim_{n \to \infty} \langle Q_n v, v' \rangle = \lim_{n \to \infty} \langle v, \underbrace{Q_n v'}_{=v'} \rangle = \langle v, v' \rangle$$

as needed. □

We will use Lemma 8.21 together with the projections

$$P(f) = E_\mu \left( f \,\middle|\, \mathscr{B}^g \right) \tag{8.28}$$

for $f \in L^2(X, \mathscr{B}, \mu)$, where $\mathscr{B}^g = \{B \in \mathscr{B} \mid \mu(g \cdot B \triangle B) = 0\}$ is the $\sigma$-algebra of $g$-invariant sets for a given $g \in G$. By the pointwise ergodic theorem (Theorem 6.1) for the single transformation $g$, the operator $P(f)$ has a pointwise interpretation in the sense that

$$P(f)(x) = \lim_{m \to \infty} \frac{1}{m} \sum_{i=0}^{m-1} f \left( g^i \cdot x \right) \tag{8.29}$$

defines the projection $E_\mu \left( f \,\middle|\, \mathscr{B}^g \right)$, and the limit exists on the complement of a null set $N = N(f, g)$.

Assume first that $G$ is countable, and write $G = \{g_1, g_2, \dots\}$. The projection $P_n$ is then defined by (8.28) for $g = g_n$, and equivalently by (8.29). Moreover, in this case the operator of projection onto $\bigcap_{n=1}^\infty \operatorname{Im} P_n$ is precisely the map $f \mapsto E_\mu \left( f \,\middle|\, \mathscr{E} \right)$, where $\mathscr{E} = \bigcap_{n=1}^\infty \mathscr{B}^{g_n}$ is as in the statement of Theorem 8.20.

**Lemma 8.22.** *Let $G = \{g_1, g_2, \dots\}$ be a countable group acting continuously on a $\sigma$-compact metric space $(X, \mathrm{d})$, and let $\mu$ be a $G$-invariant probability measure on $X$. Then $\mu$ is ergodic if and only if, for any $f \in C_c(X)$ (or for $f$ in a countable dense subset of $C_c(X)$), we have*

$$(Q_{n_k} f)(x) \to \int f \, \mathrm{d}\mu$$

*as $k \to \infty$ for almost every $x$, for some subsequence $(n_k)$ which may depend on $f$.*

We note once more that for any $f \in L^2_\mu$ the function $(Pf)(x)$ in (8.29) is also in $L^2_\mu$, so that $Q_n f \in L^2_\mu$ for any $n \geqslant 1$ and $f \in C(X)$. Moreover, $Q_n f$ converges in $L^2_\mu$ to $E_\mu(f|\mathscr{E})$ by Lemma 8.21, so there always exists a subsequence $(n_k)$ such that $Q_{n_k} f$ converges almost everywhere to $E_\mu(f|\mathscr{E})$ by Corollary A.12.

PROOF OF LEMMA 8.22. Let $f \in C_c(X)$ and assume that $\mu$ is an ergodic probability measure. Then $Q_n f$ converges in $L^2_\mu$ to $\int f \, \mathrm{d}\mu$, so there exists a subsequence $(n_k)$ for which $Q_{n_k} f \to \int f \, \mathrm{d}\mu$ almost everywhere by Corollary A.12. On the other hand, if there exists a subsequence $(n_k)$ with $Q_{n_k} f \to \int f \, \mathrm{d}\mu$ almost everywhere, then the orthogonal projection of $f$ onto the subspace of $G$-invariant functions must give $\int f \, \mathrm{d}\mu$. Knowing this for all $f \in C_c(X)$

(or for all $f$ in a dense subset of $C_c(X)$) implies, by the density of $C_c(X)$ in $L^2_\mu$, that there are no almost everywhere $G$-invariant non-constant functions, so $\mu$ is ergodic. $\qquad\square$

PROOF OF THEOREM 8.20. Suppose first that $G = \{g_1, g_2, \dots\}$ is countable, let $D \subseteq C_c(X)$ be a dense countable subset, and let $f \in D$. We know that the limit $f^{(1)} = P_1(f)$ defined as in (8.29) (using $g = g_1$) converges on the complement of a null set. As $G$ is countable, we may assume that the null set is $G$-invariant. Similarly, $f^{(2)} = P_2(f^{(1)})$, $f^{(3)} = P_1(f^{(2)}) = Q_2(f), \dots$ (as in (8.29) using $g = g_2, g = g_1, \dots$) converges pointwise on the complement of a $G$-invariant null set. Here we use the group elements $g_n$ in the order in which the corresponding projections $P_n$ are used to define $Q_\ell$. Therefore, we obtain one $G$-invariant null set with the property that $Q_n(f)(x)$ is defined on its complement for all $n$—and here the definition of $Q_n(f)(x)$ is such that we do not use properties of the measure $\mu$, but rather only use the values of $f$ on the orbit $G \cdot x$. Taking once more a countable union of null sets we may assume that the above holds for all $f \in D$ and all $x \in X'$, for some $G$-invariant conull set $X'$.

Fix again some $f \in D$. By Lemma 8.21, $Q_n(f)$ converges to $E_\mu(f|\mathscr{E})$ in the $L^2_\mu$ norm. It follows that there is a sequence $(n_k = n_k(f))$ for which

$$Q_{n_k} f(x) \to E\left(f|\mathscr{E}\right) = \int f \, d\mu^{\mathscr{E}}_x \qquad (8.30)$$

almost everywhere as $k \to \infty$. We may assume that the convergence in (8.30) also holds for $x \in X'$, for all $f \in D$ and the respective subsequence $(n_k = n_k(f))$.

Since $\mathscr{E}$ is the $\sigma$-algebra of invariant sets (and so $\mathscr{E}$ is a $g$-invariant $\sigma$-algebra for each $g \in G$), we have

$$g_* \mu^{\mathscr{E}}_x = \mu^{\mathscr{E}}_x \qquad (8.31)$$

almost everywhere for any $g \in G$ by Corollary 5.24. Once more shrinking $X'$, we may assume that (8.31) holds for $x \in X'$, and finally that $\mu^{\mathscr{E}}_x(X') = 1$ for all $x \in X'$, without losing the property that $X'$ has full measure and is $G$-invariant.

We claim that $\mu^{\mathscr{E}}_x$ is a $G$-invariant and ergodic probability measure for any $x \in X'$. Invariance holds by (8.31) and the choice of $X'$. For ergodicity we will apply Lemma 8.22 with the measure $\mu^{\mathscr{E}}_x$. For this we need to show, for any $f \in D$, that $Q_{n_k}(f)(x)$ converges $\mu^{\mathscr{E}}_x$-almost everywhere to $\int f \, d\mu^{\mathscr{E}}_x$. For $x \in X'$ we have that $P_1(f)(x)$ defined as in (8.29) converges. Therefore, we obtain (using an ergodic theorem applied to $f$ and $\mu^E_x$) that $f^{(1)}$ can indeed be identified with the projection of $f$ onto the subspace of $g_1$-invariant functions in $L^2_{\mu^E_x}$. Similarly, we obtain that $f^{(\ell)}(x)$ as defined inductively above for $x \in X'$, can be identified, as an element of $L^2_{\mu^{\mathscr{E}}_x}$, with the outcome of the successive projections of $f$ onto the $g_1, g_2, g_1, \dots$-invariant subspaces of $L^2_{\mu^{\mathscr{E}}_x}$. In particular, as $Q_{n_k}(f)(x)$ converges for $x \in X'$ to $\int f d\mu^{\mathscr{E}}_x$, we obtain that $\mu^{\mathscr{E}}_x$ is $G$-ergodic by Lemma 8.22. Notice that for the transition from $\mu$ to $\mu^{\mathscr{E}}_x$ in this argument it is crucial that $Q_n(f)(x)$ is defined in a way which depends only on $x$ and not on the measure $\mu$ (resp. $\mu^{\mathscr{E}}_x$). This is ensured by (8.29) and the iterative definition of $Q_n$.

Now let $G$ be any $\sigma$-compact metric group. Then there exists a countable dense subgroup $G' \subseteq G$ for which the statement has already been proven. Notice that by continuity of the action and the density of $G'$ in $G$ we have that a measure $\mu \in \mathscr{M}(X)$ is $G$-invariant if and only if it is $G'$-invariant (this is proved by the argument used to prove that (3) $\implies$ (1) in Theorem 4.14, using functions in $C_c(X)$ instead of $C(X)$ to get around the problem that $X$ is only assumed to be $\sigma$-compact). Similarly, a measurable set $B$ is $G$-invariant modulo $\mu$ (resp. modulo $\mu^{\mathscr{E}}_x$) if and only if it is $G'$-invariant modulo $\mu$ (resp. modulo $\mu^{\mathscr{E}}_x$) by the argument in the proof of Proposition 8.3 on pages 249–251. This implies that the $\sigma$-algebra $\mathscr{E}$ can be defined either using $G$ or using $G'$. Applying the result proved for $G'$, we see that the $\mu^{\mathscr{E}}_x$ are $G'$-invariant and ergodic probability measures almost surely; that is, $\mu = \int \mu^{\mathscr{E}}_x \, d\mu$ is the ergodic decomposition for $\mu$ and the action of $G'$. However, by the remarks above, if $\mu^{\mathscr{E}}_x$ is $G'$-invariant and ergodic, it is also $G$-invariant and ergodic. The theorem follows. $\quad\square$

## Exercises for Sect. 8.7

**Exercise 8.7.1.** Give a different proof of the existence of the ergodic decomposition for a continuous action of a $\sigma$-compact metric group on a $\sigma$-compact metric space using Choquet's theorem.

## 8.8 Stationary Measures

As we have discussed in Sect. 8.1, there are continuous group actions on compact metric spaces that do not have any invariant probability measures. One way to overcome this fundamental obstacle is to restrict attention to actions of amenable groups. Another is to loosen the requirement of strict invariance for the probability measures considered.

Rather than requiring invariance under every element of the acting group, one can ask for invariance in some averaged sense, leading to the following definition[84].

**Definition 8.23.** Let $G$ be a $\sigma$-compact metrizable group equipped with a probability measure $\nu$. Suppose that $G$ acts continuously on a $\sigma$-compact metric space $X$. A probability measure $\mu \in \mathscr{M}(X)$ is $\nu$-*stationary* if

$$\mu = \int_G g_*\mu \, d\nu(g),$$

or equivalently if

$$\int_X f(x) \, d\mu(x) = \int_G \int_X f(g \cdot x) \, d\mu(x) \, d\nu(g)$$

for all $f \in C_c(X)$.

Informally, we may think of stationarity as invariance under the random walk on $X$ defined by the measure $\nu$. This random walk is defined as follows: Given some $x \in X$, choose at random (with respect to the measure $\nu$) an element $g \in G$, and then move $x$ to $g \cdot x$. This connection can be made more formal (see the proof of Proposition 8.24), and allows one to use the arguments we used for a single transformation to study general group actions.

**Proposition 8.24.** *Let $G$ be a $\sigma$-compact metrizable group, and let $\nu$ be a probability measure on $G$. Suppose that $G$ acts continuously on a compact metric space $X$. Then there exists a $\nu$-stationary measure on $X$.*

PROOF. Let $\mu_0 \in \mathscr{M}(X)$ be any probability measure on $X$, and define a new probability measure $\nu * \mu_0$ by

$$\nu * \mu_0 = \int_G g_*\mu_0 \, d\nu(g).$$

Also define the averages

$$\mu_N = \frac{1}{N} \sum_{n=0}^{N-1} \underbrace{\nu * (\nu * \cdots (\nu * \mu_0) \cdots)}_{n \text{ times}},$$

which define a sequence of measures in the compact space $\mathscr{M}(X)$. It follows that there is a subsequence $(N_k)$ for which $\mu_{N_k} \to \mu$ for some $\mu \in \mathscr{M}(X)$. Notice that

$$\nu * \mu_{N_k} - \mu_{N_k} = \frac{1}{N_k}(\nu * (\nu * \cdots (\nu * \mu_0) \cdots)) - \mu_0,$$

which shows that

$$\left| \int f \, d\nu * \mu_{N_k} - \int f \, d\mu_{N_k} \right| \leqslant \frac{2\|f\|_\infty}{N_k}$$

for $f \in C(X)$. Here

$$\int f \, d\nu * \mu_{N_k} = \iint f(g \cdot x) \, d\nu(g) \, d\mu_{N_k}(x) \qquad (8.32)$$

converges as $k \to \infty$ to

$$\iint f(g{\cdot}x)\,\mathrm{d}\nu(g)\,\mathrm{d}\mu(x) = \int f\,\mathrm{d}\nu * \mu. \qquad (8.33)$$

It follows that $\mu$ is $\nu$-stationary.                                    □

## Exercises for Sect. 8.8

**Exercise 8.8.1.** Check (8.32) and (8.33). Show that $\int f(g{\cdot}x)\,\mathrm{d}\nu(g)$ is a continuous function of $x \in X$, which is needed to deduce (8.33) from (8.32).

**Exercise 8.8.2.** Show how to deduce Proposition 8.24 from Theorem 4.1.

## Notes to Chap. 8

[72](Page 231) There is a rich theory of the topological dynamics of continuous group actions on compact topological spaces which is outside the topics covered here. The monograph of Gottschalk and Hedlund [121] is an influential early source; more recent sources include Auslander [9] (mainly dealing with structure theorems for topological dynamical systems) and Akin [3].

[73](Page 232) See also the survey of mixing properties of group actions, with particular emphasis on the natural action of $\mathrm{SL}_d(\mathbb{Z})$ on $\mathbb{T}^d$, by Bergelson and Gorodnik [27] and their paper [28]; further results on higher-order mixing in special cases may be found in papers of Bhattacharya [30], Kamiński [179] and Ward [375].

[74](Page 235) This exercise is taken from a note by Morris and Ward [261].

[75](Page 235) The relationship between mixing properties and rigidity for a measure-preserving transformation is rather involved. Motivated by a problem from the theory of transformations preserving an infinite measure, Furstenberg and Weiss [108] defined a measure-preserving transformation to be *mild-mixing* if it has no rigid factors, and showed that mild-mixing is strictly weaker than strong-mixing and strictly stronger than weak-mixing. Just as in note (27) on p. 50, there is a natural sense in which the typical measure-preserving transformation is both rigid and weak-mixing.

[76](Page 235) The mixing behavior of measure-preserving $\mathbb{Z}^d$-actions for $d > 1$ in general (that is, away from algebraic examples) is more involved, and in particular there are rigid examples with the property that every element is mixing (see [81]).

[77](Page 235) The higher-order mixing properties of $\mathbb{Z}^d$-actions by automorphisms of compact abelian groups are intimately connected with subtle Diophantine problems, as shown in Sect. 8.2.2; Schmidt [332, Chap. VIII] gives an overview. There is a sharp dichotomy between connected and zero-dimensional groups. For connected groups Schmidt and Ward [333] showed that mixing implies mixing of all orders; for zero-dimensional groups Masser [256] showed that the order of mixing is related to the algebraic property of mixing shapes. Actions of non-abelian groups by automorphisms of a compact group behave very differently; special cases are dealt with by Bergelson and Gorodnik [28] and Bhattacharya [30]. In a different direction, there are certain groups with the property that mixing implies mixing of all orders for any measure-preserving action (see Mozes [262] and Sect. 11.4, where a special case of Mozes' work is described).

(78)(Page 239) This result, a form of "$S$-unit theorem", is proved by van der Poorten and Schlickewei [295] and, independently, by Evertse [88]. In the case of a number field, explicit bounds on the number of solutions in terms of $n$, $[\mathbb{K} : \mathbb{Q}]$ and the rank of $\Gamma$ are found by Schlickewei [330].

(79)(Page 242) Much more is true: an action of $\mathbb{Z}^d$ by continuous automorphisms of a compact, zero-dimensional abelian group is mixing of all orders if and only if it has completely positive entropy (entropy theory extends to actions of amenable groups). This is shown by Kitchens and Schmidt [196] as part of a wider investigation of mixing and directional entropy invariants for such systems.

(80)(Page 242) In fact much more is true: there is a precise sense in which examples like that of Sect. 8.2.1 are mixing of all orders if a small subset of times are avoided. This is shown in certain cases by Arenas-Carmona, Berend and Bergelson [6].

(81)(Page 245) See, for example, Halmos [137, Sec. 56].

(82)(Page 251) Amenable groups were introduced and studied by von Neumann [266], who used them to explain aspects of the Banach–Tarski paradox; the word 'amenability' was introduced later by Day [65] (see Wagon [373] for an attractive account of the Banach–Tarski paradox). Significant monographs on the notion of amenability include those of Greenleaf [128], Pier [284] and Paterson [281]. In addition there are overviews of amenable groups in a survey paper of Day [66] and chapters in the books on harmonic analysis by Hewitt and Ross [151, Sect. 17] and by Reiter [307, Chap. 8]. The rich articulation between amenability, property (T), rigidity, and lattices in Lie groups is addressed in the monograph by Zimmer [394]. The ergodic theory of actions of amenable groups was given enormous impetus by the comprehensive work of Ornstein and Weiss [275] in which the entropy and isomorphism theory for such actions is developed. The pointwise ergodic theorem and relationship between entropy and mixing have been developed by Lindenstrauss [233] and Rudolph and Weiss [325] respectively. The theory of orbit equivalence and restricted orbit equivalences, interpolating between orbit equivalence and isomorphism, for amenable group actions has been developed by Kammeyer and Rudolph [180].

(83)(Page 257) Ergodic theorems for group actions have a long history, and we refer the interested reader to the extensive surveys by Nevo [271] and Gorodnik and Nevo [120], and the monograph of Tempel'man [358]. In particular, [271, Sect. 4] discusses the relationships between growth conditions on metric balls, unimodularity and asymptotic invariance in general, and [271, Sect. 5] describes pointwise ergodic theorems for ball averages. The use of Vitali covering lemmas and growth conditions goes back to Wiener [382] and Riesz [314], and was also used by Calderon [45]. These methods have been applied, for example, by Bourgain [41] to develop pointwise ergodic theorems along subsequences of arithmetic interest, and by Ornstein and Weiss [274, 275] in developing ergodic theory for amenable group actions.

(84)(Page 272) This notion was introduced by Fustenberg [104], and there is a large body of research related to it which we do not go into. In the setting of Proposition 8.24 the set of $\nu$-stationary measures is a compact convex subset of $\mathscr{M}(X)$. If in addition $G$ is abelian then for any measure $\nu$ with the property that $\nu(H) = 1$ for a closed subgroup $H \leqslant G$ implies that $H = G$, any $\nu$-stationary measure is invariant (in the terminology of [104], this means that any action of an abelian group is a *stiff* action). A striking recent result of Bourgain, Furman, Lindenstrauss, and Mozes [42] gives a dichotomy for $\nu$-stationary measures on $\mathbb{T}^2$ if the support of $\nu$ generates a sufficiently large subgroup of $\mathrm{SL}_2(\mathbb{Z})$. A more general (but less effective) stiffness result for homogeneous spaces has been found by Benoist and Quint [24].

# Chapter 9
# Geodesic Flow on Quotients of the Hyperbolic Plane

In this chapter we will start our analysis of actions on locally homogeneous spaces by studying the geodesic flow on hyperbolic surfaces. Since, throughout the book, we will not assume prior knowledge of Lie theory or differential geometry, the material needed will be introduced here. As an application, the geodesic flow will be used to give another proof of ergodicity for the Gauss measure.

The connection between the geodesic flow and continued fractions goes back to work of Artin [8][(85)].

## 9.1 The Hyperbolic Plane and the Isometric Action

Before we can discuss the geodesic flow we need to introduce the space on which it acts. Indeed, even the space on which it acts will be approached via a simpler space.

A convenient model for the hyperbolic plane is the upper half-plane

$$\mathbb{H} = \{x + iy \in \mathbb{C} \mid y > 0\}$$

with the hyperbolic metric. To define this metric, we need to introduce the tangent bundle* $T\mathbb{H} = \mathbb{H} \times \mathbb{C}$ comprising the disjoint union of the tangent planes $T_z\mathbb{H} = \{z\} \times \mathbb{C}$ for all $z \in \mathbb{H}$. One should think of $T_z\mathbb{H}$ as a plane touching $\mathbb{H}$ tangentially at $z$ and having no other intersection with $\mathbb{H}$. This suggests that $T_z\mathbb{H}$ is the natural space for derivatives in the following sense. If $\phi : [0,1] \to \mathbb{H}$ is differentiable at $t \in [0,1]$ with $\phi(t) = z$, then we define the derivative of $\phi$ at $t$ by

$$\mathrm{D}\,\phi(t) = (\phi(t), \phi'(t)) \in T_z\mathbb{H}.$$

---

* The tangent bundle can be defined abstractly on any manifold, but for our purposes we may think of it as the space in which derivatives live and use an *ad hoc* definition.

M. Einsiedler, T. Ward, *Ergodic Theory*, Graduate Texts in Mathematics 259,
DOI 10.1007/978-0-85729-021-2_9, © Springer-Verlag London Limited 2011

Here $\phi'(t)$ is the derivative of $\phi$ as a map into $\mathbb{C}$. We give $T_z\mathbb{H}$ the structure of a vector space inherited from the second component in $T_z\mathbb{H} = \{z\} \times \mathbb{C}$.

The *hyperbolic Riemannian metric* is defined as the collection of inner products[*]

$$\langle v, w \rangle_z = \frac{1}{y^2}(v \cdot w)$$

for $z = x + iy \in \mathbb{H}$ and $v, w \in T_z\mathbb{H}$. Here $(v \cdot w)$ denotes the usual inner product in $\mathbb{C}$ under the identification of $\mathbb{C}$ with $\mathbb{R}^2$ as real vector spaces.

The hyperbolic Riemannian metric induces the hyperbolic metric $\mathsf{d}(\cdot, \cdot)$. mentioned above as follows. If $\phi : [0, 1] \to \mathbb{H}$ is a continuous piecewise differentiable curve (we will refer to these as *paths*), then its length is defined by

$$L(\phi) = \int_0^1 \| \mathrm{D}\,\phi(t) \|_{\phi(t)} \, \mathrm{d}t$$

where $\| \mathrm{D}\,\phi(t) \|_{\phi(t)}$ denotes the length of the tangent vector

$$\mathrm{D}\,\phi(t) = (\phi(t), \phi'(t)) \in T_{\phi(t)}\mathbb{H}$$

with respect to the norm derived from $\langle \cdot, \cdot \rangle_{\phi(t)}$. We will refer to $\| \mathrm{D}\,\phi(t) \|_{\phi(t)}$ as the *speed* of the path $\phi$ at time $t$. The hyperbolic distance is now defined as

$$\mathsf{d}(z_0, z_1) = \inf_\phi L(\phi)$$

where the infimum is taken over all continuous piecewise differentiable curves $\phi$ with $\phi(0) = z_0$ and $\phi(1) = z_1$. It may be checked that this does indeed define a metric on $\mathbb{H}$.

Moreover, the hyperbolic metric on $\mathbb{H}$ induces the same topology on $\mathbb{H}$ as that induced by the Euclidean norm $\| \cdot \|_2$ on $\mathbb{C} \supseteq \mathbb{H}$ (see Exercise 9.1.1).

The real line $\mathbb{R} \subseteq \mathbb{C}$ together with a single point $\infty$ (from the one-point compactification of $\mathbb{C}$) is called the *boundary* $\partial\mathbb{H}$ of the hyperbolic plane. Notice that the hyperbolic distance from any point $z \in \mathbb{H}$ to any point $\alpha \in \partial\mathbb{H}$ is infinite, where the distance is defined as an infimum over paths

$$\phi : [0, 1] \to \mathbb{H} \cup \partial\mathbb{H}$$

with $\phi(t) \in \mathbb{H}$ for $t \in [0, 1)$, $\phi(0) = z$ and $\phi(1) = \alpha$.

We are going to introduce a group action on $\mathbb{H}$ by isometries of the hyperbolic metric. This will serve two purposes. First, it will help us to understand the hyperbolic metric; second, it will allow us to define important quotients of the space $\mathbb{H}$. Looking back at some of the natural examples of measure-preserving systems from Chap. 2 shows that very few arise naturally on spaces like $\mathbb{R}$ or $\mathbb{R}^2$, but instead live on the spaces $\mathbb{T} = \mathbb{R}/\mathbb{Z}$ or $\mathbb{T}^2 = \mathbb{R}^2/\mathbb{Z}^2$, which

---

[*] The scaling by $y^{-2}$ may seem arbitrary, but it is this exact scaling that gives a metric with respect to which the action of $SL_2(\mathbb{R})$ on $\mathbb{H}$ by Möbius transformations is isometric (see Lemma 9.1).

have finite volume with respect to a natural measure. In that setting, the subgroup $\mathbb{Z}$ or $\mathbb{Z}^2$ acts by translation, which is an isometry of the Euclidean metric on $\mathbb{R}$ or $\mathbb{R}^2$.

Recall that an action of a group $G$ on a set $X$ is said to be *transitive* if for any $x_1, x_2 \in X$ there is a $g \in G$ with $g \cdot x_1 = x_2$, and is *simply transitive* if there is a unique $g \in G$ with $g \cdot x_1 = x_2$.

Let $\mathrm{SL}_2(\mathbb{R})$ denote the *special linear group* consisting of $2 \times 2$ matrices with real entries and determinant one. The group $\mathrm{SL}_2(\mathbb{R})$ acts on $\mathbb{H}$ by the Möbius transformations

$$g = \begin{pmatrix} a & b \\ c & d \end{pmatrix} : z \mapsto \frac{az + b}{cz + d}. \tag{9.1}$$

Notice that $cz + d \neq 0$ for $z \in \mathbb{H}$ since $cz + d = 0$ requires $z = -\frac{d}{c} \in \mathbb{R}$ unless $c = 0$, in which case $d = 0$. If we identify $z \in \mathbb{C}$ with the point $\binom{z}{1}$ in projective space $\mathbb{P}^1(\mathbb{C})$, then this is simply the natural linear action

$$\begin{pmatrix} a & b \\ c & d \end{pmatrix} \begin{pmatrix} z \\ 1 \end{pmatrix} = \begin{pmatrix} az + b \\ cz + d \end{pmatrix} \sim \begin{pmatrix} \frac{az+b}{cz+d} \\ 1 \end{pmatrix}$$

viewed in affine coordinates. This implies that (9.1) defines an action once we have verified that $g(z) \in \mathbb{H}$ for all $z \in \mathbb{H}$ and $g \in \mathrm{SL}_2(\mathbb{R})$, and an easy calculation shows that

$$\Im\left(g(z)\right) = \frac{\Im(z)}{|cz + d|^2}, \tag{9.2}$$

which proves that $g(\mathbb{H}) \subseteq \mathbb{H}$ for all $g \in \mathrm{SL}_2(\mathbb{R})$. Notice that the matrix

$$-I_2 = \begin{pmatrix} -1 & 0 \\ 0 & -1 \end{pmatrix}$$

acts trivially on $\mathbb{H}$, so (9.1) defines an action of the *projective special linear group*

$$\mathrm{PSL}_2(\mathbb{R}) = \mathrm{SL}_2(\mathbb{R})/\{\pm I_2\}.$$

It will be convenient to continue to write $\left(\begin{smallmatrix} a & b \\ c & d \end{smallmatrix}\right)$ for the element $\pm\left(\begin{smallmatrix} a & b \\ c & d \end{smallmatrix}\right)$ of $\mathrm{PSL}_2(\mathbb{R})$.

**Lemma 9.1.** *The action of* $\mathrm{PSL}_2(\mathbb{R})$ *on* $\mathbb{H}$ *defined by (9.1) has the following properties.*

(1) *The action is isometric, meaning that*

$$\mathsf{d}\left(g(z_0), g(z_1)\right) = \mathsf{d}(z_0, z_1)$$

*for any* $z_0, z_1 \in \mathbb{H}$ *and* $g \in \mathrm{PSL}_2(\mathbb{R})$. *Moreover, the action of* $\mathrm{PSL}_2(\mathbb{R})$ *on* $\mathrm{T}\mathbb{H}$ *defined by the derivative* $\mathrm{D}\,g$ *of the action of* $g \in \mathrm{PSL}_2(\mathbb{R})$ *on* $\mathbb{H}$ *preserves the Riemannian metric.*

(2) *The action is transitive on* $\mathbb{H}$: *given any two points* $z_0, z_1 \in \mathbb{H}$ *there is a matrix* $g \in \mathrm{PSL}_2(\mathbb{R})$ *with* $g(z_0) = z_1$.

(3) *The stabilizer*

$$\mathrm{Stab}_{\mathrm{PSL}_2(\mathbb{R})}(\mathrm{i}) = \{g \in \mathrm{PSL}_2(\mathbb{R}) \mid g(\mathrm{i}) = \mathrm{i}\}$$

*of* $\mathrm{i} \in \mathbb{H}$ *is the projective special orthogonal group*

$$\mathrm{PSO}(2) = \mathrm{SO}(2)/\{\pm I_2\}$$

*where* $\mathrm{SO}(2) = \left\{ \begin{pmatrix} \cos\theta & -\sin\theta \\ \sin\theta & \cos\theta \end{pmatrix} \mid \theta \in \mathbb{R} \right\}$.

Notice that property (3) gives an identification

$$\mathbb{H} \cong \mathrm{PSL}_2(\mathbb{R})/\mathrm{PSO}(2),$$

and under the identification the coset $g\,\mathrm{PSO}(2)$ is sent to $g(\mathrm{i})$.

The derivative action $\mathrm{D}\,g$ of $g \in \mathrm{PSL}_2(\mathbb{R})$ used in (1) is defined as follows. Notice that $g$ when viewed as a map from $\mathbb{H}$ to $\mathbb{H}$ is complex differentiable, moreover

$$g'(z) = \frac{1}{(cz+d)^2}.$$

Now $\mathrm{D}\,g : \mathrm{T}\mathbb{H} \to \mathrm{T}\mathbb{H}$ is defined by

$$\mathrm{D}\,g(z,v) = (g(z), g'(z)v) = \left( \frac{az+b}{cz+d}, \frac{v}{(cz+d)^2} \right),$$

which for a fixed $z \in \mathbb{H}$ is a linear map

$$(\mathrm{D}\,g)_z : \mathrm{T}_z\mathbb{H} \to \mathrm{T}_{g(z)}\mathbb{H}.$$

The chain rule for differentiation shows that this really defines an action.

PROOF OF LEMMA 9.1 (1): Since the metric is defined in terms of the Riemannian metric, we need to start by proving the second claim. For $v, w$ in $\mathrm{T}_z\mathbb{H}$ we have $(\mathrm{D}\,g)_z v, (\mathrm{D}\,g)_z w \in \mathrm{T}_{g(z)}\mathbb{H}$ and (9.2) shows that

$$\langle (\mathrm{D}\,g)_z v, (\mathrm{D}\,g)_z w \rangle_{g(z)} = \left( \frac{y}{|cz+d|^2} \right)^{-2} \left( \frac{1}{(cz+d)^2} v, \frac{1}{(cz+d)^2} w \right)$$

$$= \frac{1}{y^2}(v,w) = \langle v, w \rangle_z, \tag{9.3}$$

where we have used the property that multiplication of $v$ and $w$ by a complex number $\lambda$ changes the Euclidean inner product by a factor of $|\lambda|^2$. In particular, (9.3) shows that $\mathrm{D}\,g$ does not change the length of a vector when the lengths are defined using the corresponding base points. This implies that for any continuous piecewise differentiable curve $\phi : [0,1] \to \mathbb{H}$ the action of $g$ does not change its length: $L(g \circ \phi) = L(\phi)$ (see Exercise 9.1.2). Finally it follows that

$$\mathsf{d}\left(g(z_0), g(z_1)\right) = \mathsf{d}(z_0, z_1)$$

for all $z_0, z_1 \in \mathbb{H}$.

(2): It is enough to show that for any $z = x + iy$ with $y > 0$ there is a matrix $g$ with $g(\mathrm{i}) = z$. The matrix $\left( \begin{smallmatrix} \sqrt{y} & x/\sqrt{y} \\ 0 & 1/\sqrt{y} \end{smallmatrix} \right)$ sends i to $x + iy$.

(3): If $g(\mathrm{i}) = \mathrm{i}$, then we must have $|c + \mathrm{i}d| = 1$ by (9.2). Thus there is a $\theta \in \mathbb{R}$ with $c = \sin\theta$, $d = \cos\theta$. Now $g(\mathrm{i}) = \mathrm{i}$, equivalently

$$\frac{a\mathrm{i} + b}{\sin\theta \mathrm{i} + \cos\theta} = \mathrm{i},$$

is equivalent to $a = \cos\theta$ and $b = -\sin\theta$. Thus the stabilizer of i is the subgroup $\mathrm{PSO}(2)$ of $\mathrm{PSL}_2(\mathbb{R})$. $\qquad\square$

By Lemma 9.1(1), the action of $\mathsf{D}\,g$ preserves the length of tangent vectors. Write

$$\mathrm{T}^1\mathbb{H} = \{(z,v) \in \mathrm{T}\mathbb{H} \mid \|v\|_z = 1\}$$

for the unit tangent bundle of $\mathbb{H}$ consisting of all unit vectors $v$ attached to all possible points $z \in \mathbb{H}$. The restriction of $\mathsf{D}\,g$ defines an action of $\mathrm{PSL}_2(\mathbb{R})$ on $\mathrm{T}^1\mathbb{H}$ (naturally extending the action on $\mathbb{H}$ itself).

**Lemma 9.2.** *The action of* $\mathrm{PSL}_2(\mathbb{R})$ *on* $\mathrm{T}^1\mathbb{H}$ *is simply transitive.*

Notice that this allows us to describe the unit tangent bundle to $\mathbb{H}$ as

$$\mathrm{T}^1\mathbb{H} \cong \mathrm{PSL}_2(\mathbb{R}). \tag{9.4}$$

In order to do this, we have to choose an arbitrary reference vector $(z_0, v_0)$ in $\mathrm{T}^1\mathbb{H}$ which corresponds to $I_2 \in \mathrm{PSL}_2(\mathbb{R})$; the identification is then given by $g \mapsto \mathsf{D}\,g(z_0, v_0)$. We will make the convenient choice $z_0 = \mathrm{i}$ and $v_0 = \mathrm{i}$. That is, the reference vector is the upward unit vector based at the imaginary unit $\mathrm{i} \in \mathbb{H}$. Under the resulting identification the action of $\mathrm{PSL}_2(\mathbb{R})$ on $\mathbb{H}$ is conjugated to the action of $\mathrm{PSL}_2(\mathbb{R})$ by left multiplication on $\mathrm{PSL}_2(\mathbb{R})$.

PROOF OF LEMMA 9.2. Since we already know that the action on $\mathbb{H}$ is transitive, it is enough to consider vectors $v \in \mathrm{T}_\mathrm{i}\mathbb{H}$ with base point i, and here we compute

$$\left(\mathsf{D}\,g\right)_\mathrm{i}(v) = \frac{1}{(\mathrm{i}\sin\theta + \cos\theta)^2} v = (\cos(2\theta) - \mathrm{i}\sin(2\theta))\, v.$$

So by varying $\theta$, $(\mathrm{D}\,g)_{\mathrm{i}}\,(v)$ could be any vector of modulus one. To see that the transitivity is simple, notice that $(\mathrm{D}\,g)_{\mathrm{i}}\,(v) = v$ implies that

$$2\theta \equiv 0 \quad (\mathrm{mod}\ 2\pi),$$

so $\theta \in \mathbb{Z}\pi$, and $g = \pm I_2$. $\qquad\qquad\qquad\qquad\qquad\qquad\qquad\qquad\qquad\Box$

## Exercises for Sect. 9.1

**Exercise 9.1.1.** Prove that the hyperbolic metric on $\mathbb{H}$ is indeed a metric, and that it induces the same topology as does the Euclidean norm $\|\cdot\|_2$ on $\mathbb{C} \supseteq \mathbb{H}$.

**Exercise 9.1.2.** Using (9.3), verify that $\mathrm{L}(g \circ \phi) = \mathrm{L}(\phi)$ for any continuous piecewise differentiable curve $\phi : [0,1] \to \mathbb{H}$ and $g \in \mathrm{PSL}_2(\mathbb{R})$.

## 9.2 The Geodesic Flow and the Horocycle Flow

In this section we describe some basic hyperbolic geometry, starting with distances, geodesics, and their parameterizations. We have defined the distance between two points $z_0, z_1 \in \mathbb{H}$ by an infimum over paths from $z_0$ to $z_1$; it is not clear *a priori* whether this infimum is attained by some minimizing path. The first step is to show that this is indeed the case, starting with a simple special case.

**Lemma 9.3.** *Let* $z_0 = y_0 \mathrm{i}$ *and* $z_1 = y_1 \mathrm{i}$ *with* $0 < y_0 < y_1$. *Then*

$$\mathsf{d}(z_0, z_1) = \log y_1 - \log y_0$$

*and*

$$\phi(t) = y_0 \left(\frac{y_1}{y_0}\right)^t \mathrm{i}$$

*for* $t \in [0,1]$ *defines a path in* $\mathbb{H}$ *from* $z_0$ *to* $z_1$ *with constant speed*

$$\log y_1 - \log y_0,$$

*so that*

$$\mathrm{L}(\phi) = \log y_1 - \log y_0.$$

*Moreover, the curve* $\phi$ *is uniquely determined: if* $\psi : [0,1] \to \mathbb{H}$ *is any path from* $z_0$ *to* $z_1$ *with* $\mathrm{L}(\psi) = \mathsf{d}(z_0, z_1)$ *then there is some increasing piecewise differentiable map* $f : [0,1] \to [0,1]$ *with* $\psi = \phi \circ f$.

Lemma 9.3 says that any two points on the vertical line $\{yi \mid y > 0\}$ have a unique path of minimal length joining them, and that minimizing path also lies in $\{yi \mid y > 0\}$. For that reason, the whole set $\{yi \mid y > 0\}$ will be called a geodesic curve for $\mathbb{H}$ or simply a *geodesic*, and the minimizing path $\phi$ will be called a *geodesic path*.

PROOF OF LEMMA 9.3. It is readily checked that the path $\phi$ defined in the lemma has constant speed equal to $\log y_1 - \log y_0 = L(\phi)$ as claimed. It follows that

$$\mathsf{d}(z_0, z_1) \leqslant \log y_1 - \log y_0.$$

Suppose now that $\eta : [0,1] \to \mathbb{H}$ is another path joining $z_0$ to $z_1$, and write $\eta(t) = \eta_x(t) + i\eta_y(t)$ with $\eta_x(t), \eta_y(t) \in \mathbb{R}$. Then

$$L(\eta) = \int_0^1 \frac{\|\eta'(t)\|_2}{\eta_y(t)}\, dt \geqslant \int_0^1 \frac{|\eta_y'(t)|}{\eta_y(t)}\, dt \geqslant \int_0^1 \frac{\eta_y'(t)}{\eta_y(t)}\, dt = \log y_1 - \log y_0.$$

Equality holds in the first inequality if and only if $\eta_x'(t) = \eta_x(t) = 0$ for all $t \in [0,1]$, and in the second if and only if $\eta_y'(t) \geqslant 0$ for all $t \in [0,1]$. This implies the remaining statement of the lemma. $\qquad\square$

From now on we will always parameterize geodesics so that they have constant speed equal to 1 (and therefore have a domain whose length is equal to the length of the path). Thus Lemma 9.3 may be thought of as saying that for $z_0 = iy_0$, $z_1 = iy_1$ there is a unique path

$$\phi : [0, \mathsf{d}(z_0, z_1)] \to \mathbb{H}$$

with unit speed and with $\phi(0) = z_0$ and $\phi(\mathsf{d}(z_0, z_1)) = z_1$, and that unique path is defined by

$$\phi(t) = y_0 e^t i.$$

It is clear that an isometry $g \in \mathrm{PSL}_2(\mathbb{R})$ sends a geodesic path (or curve) to another geodesic path (curve). The next result is a converse to this observation, and gives a description of all geodesics in $\mathbb{H}$.

**Proposition 9.4.** *For any two points* $z_0, z_1 \in \mathbb{H}$ *there is a unique path*

$$\phi : [0, \mathsf{d}(z_0, z_1)] \to \mathbb{H}$$

*of unit speed with* $\phi(0) = z_0$ *and* $\phi(\mathsf{d}(z_0, z_1)) = z_1$. *Moreover, there is a unique isometry* $g \in \mathrm{PSL}_2(\mathbb{R})$ *such that* $\phi(t) = g(e^t i)$.

PROOF. We first claim that there exists a $g \in \mathrm{PSL}_2(\mathbb{R})$ with $g^{-1}(z_0) = i$ and $g^{-1}(z_1) = iy_1$ for some $y_1 > 1$. By Lemma 9.1(2) we can certainly find some $\widetilde{g} \in \mathrm{PSL}_2(\mathbb{R})$ with $\widetilde{g}^{-1}(z_0) = i$, and we want to modify $\widetilde{g}$ to also satisfy the second condition. By Lemma 9.1(3) any element of $\mathrm{PSO}(2)$ fixes the point i. Hence we may suppose without loss of generality that $\Im\left(g^{-1}(z_1)\right)$ is maximal within $\Im\left(\mathrm{PSO}(2)\widetilde{g}^{-1}(z_1)\right)$. Let

$$h = \begin{pmatrix} \cos\theta & -\sin\theta \\ \sin\theta & \cos\theta \end{pmatrix}$$

and $\widetilde{g}^{-1}(z_1) = \widetilde{x}_1 + i\widetilde{y}_1$, so that

$$\Im\left(h(\widetilde{g}^{-1}(z_1))\right) = \frac{y_1}{|\sin\theta z_1 + \cos\theta|^2} \qquad \text{(by (9.2))}$$

has a maximum at $\theta = 0$. Choosing $g^{-1} = h\widetilde{g}^{-1}$ and taking the derivative shows that this implies that $\Re g^{-1}(z_1) = 0$.

Moreover, we must have $g^{-1}(z_1) = y_1 > 1$ since if $y_1 < 1$ then the map

$$k = \begin{pmatrix} 0 & -1 \\ 1 & 0 \end{pmatrix}$$

would increase the imaginary part, contradicting the maximality assumption. By Lemma 9.3, there is a unique geodesic path $\phi_0(t) = e^t i$ of unit speed from i to $iy_1$, so that $\phi(t) = g(e^t i)$ is the unique geodesic path of unit speed connecting $z_0$ and $z_1$.

Finally, we claim that not only is the geodesic unique, but so is the element $g \in \mathrm{PSL}_2(\mathbb{R})$. To see this, suppose that $t \mapsto g_1(e^t i)$ is any geodesic path from $z_0$ to $z_1$. Then $t \mapsto g^{-1} g_1(e^t i)$ is a geodesic path from i to $y_1 i$ of unit speed, and so must be equal to the path $t \mapsto e^t i$. Taking the derivative at $t = 0$ we see that $\mathrm{D}\left(g^{-1}g_1\right)(i,i) = (i,i)$, which shows that $g^{-1}g_1 = I_2$ by Lemma 9.2.                                                                                   □

Finally, we claim that a Möbius transformation $g$ maps the geodesic curve $\{yi \mid y > 0\}$ either to another vertical line (that is, a line normal to the real axis, with constant $x$ coordinate), or to the upper half of a circle

$$(x - f)^2 + y^2 = r^2$$

with center in the real axis (equivalently, meeting the real axis at right angles). Moreover, all of those curves do arise as images of $\{yi \mid y > 0\}$, so that this list of curves is precisely the list of geodesic curves in $\mathbb{H}$.

To see this, note first that a Möbius transformation of the form $z \mapsto az + b$ maps the vertical line $\{yi \mid y > 0\}$ to the vertical line $\Re(z) = b$ and maps the upper half of the circle $(x - f)^2 + y^2 = r^2$ to the upper half of the circle $(x - (af + b))^2 + y^2 = a^2 r^2$. Now the subgroup

$$U = \left\{ \begin{pmatrix} 1 & b \\ 0 & 1 \end{pmatrix} \mid b \in \mathbb{R} \right\} < \mathrm{SL}_2(\mathbb{R})$$

together with $w = \begin{pmatrix} 0 & 1 \\ -1 & 0 \end{pmatrix}$ generate $\mathrm{SL}_2(\mathbb{R})$, since

$$wUw^{-1} = \left\{ \begin{pmatrix} 1 & 0 \\ -b & 1 \end{pmatrix} \mid b \in \mathbb{R} \right\}.$$

So it remains to check that the Möbius transformation $z \mapsto -\frac{1}{z}$ corresponding to $w$ maps a vertical line or a semicircle with center in $\mathbb{R}$ to another vertical line or semicircle. In polar coordinates $(r, \phi)$ the transformation $z \mapsto -\frac{1}{z}$ is the transformation $(r, \phi) \mapsto (\frac{1}{r}, \pi - \phi)$. The claim follows since both vertical lines and circles with real centers can be defined by equations of the form

$$\alpha r^2 + \beta r \cos \phi + \gamma = 0$$

with $(\alpha, \beta, \gamma) \neq (0, 0, 0)$.

Historically, the hyperbolic plane $\mathbb{H}$ was important in solving a classical problem in geometry: the points and geodesics in $\mathbb{H}$ satisfy all the classical axioms of geometry apart from the parallel axiom, thus showing that the parallel axiom is not a consequence of the other axioms. Indeed, we have for instance that for any two different points in $\mathbb{H}$ there is a unique geodesic through them, and any two different geodesics intersect in at most one point. Angles and areas will be defined later, and they will behave well (if a little unusually). However, for every point $z \in \mathbb{H}$ and geodesic $\ell$ not containing $z$ there are uncountably many geodesics through $z$ that do not intersect $\ell$ and which are therefore "parallel" to $\ell$.

Proposition 9.4 shows that any two points $z_0, z_1 \in \mathbb{H}$ determine a unique geodesic $\ell$ passing through $z_0$ and $z_1$. Alternatively, a geodesic $\ell$ is also uniquely determined by a base point $z \in \mathbb{H}$ and a unit vector $\mathbf{v} \in \mathrm{T}_z^1 \mathbb{H}$ under the requirement that $\ell$ passes through $z$ in the direction of $\mathbf{v}$. The unique geodesic through i in the direction i is illustrated in Fig. 9.1, while a geodesic through $z$ in the direction $\mathbf{v}$ in general position is illustrated in Fig. 9.2.

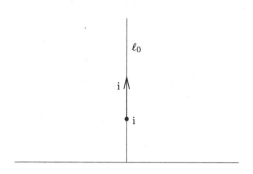

**Fig. 9.1** The geodesic $\ell_0$ through the pair $(\mathrm{i}, \mathrm{i})$

In fact there is a unique $g \in \mathrm{PSL}_2(\mathbb{R})$ with $\mathrm{D}\,g(\mathrm{i}, \mathrm{i}) = (z, \mathbf{v})$, and $\ell$ is the image of $\{\mathrm{i}y \mid y > 0\}$ under the Möbius transformation corresponding to $g$. Moreover, the vector $\mathbf{v}$ gives $\ell$ a direction and defines the unit speed parametrization $g(\mathrm{e}^t \mathrm{i})$ of the geodesic $\ell$ starting at $z$.

**Fig. 9.2** The unique geodesic $\ell$ defined by a pair $(z, \mathbf{v})$

The *geodesic flow* $g_t : \mathrm{T}^1\mathbb{H} \to \mathrm{T}^1\mathbb{H}$ is defined by following the uniquely defined parametrization of the geodesic $\ell$ determined by $(z, \mathbf{v})$ for time $t$. Here the direction of the image in $\mathrm{T}^1\mathbb{H}$ is the direction of the geodesic $\ell$ at time $t$. In the case of the reference vector $(\mathrm{i}, \mathrm{i})$ this means that

$$g_t\left((\mathrm{i}, \mathrm{i})\right) = (e^t\mathrm{i}, e^t\mathrm{i})$$

where we write $g_t\left((z, \mathbf{v})\right)$ for the action of the geodesic flow.

Notice that $(e^t\mathrm{i}, e^t\mathrm{i})$ is also the image of $(\mathrm{i}, \mathrm{i})$ under the derivative action of the matrix

$$\begin{pmatrix} e^{t/2} & 0 \\ 0 & e^{-t/2} \end{pmatrix},$$

that is

$$g_t\left((\mathrm{i}, \mathrm{i})\right) = \mathrm{D}\begin{pmatrix} e^{t/2} & 0 \\ 0 & e^{-t/2} \end{pmatrix}(\mathrm{i}, \mathrm{i}).$$

A word about notation: we will use $(z, \mathbf{v}) \mapsto g_t\left((z, \mathbf{v})\right)$ for the geodesic flow for time $t$ on $\mathrm{T}^1\mathbb{H}$; the corresponding action on the group $\mathrm{PSL}_2(\mathbb{R})$ under the identification in (9.4) will be written $h \mapsto R_{a_t}h$. We now explicitly describe this flow in terms of matrices.

For an arbitrary point $(z, \mathbf{v}) = g(\mathrm{i}, \mathrm{i})$ the isometric Möbius transformation defined by $g$ maps the unit speed parametrization of the geodesic defined by the reference vector to the one defined by $(z, \mathbf{v})$. It follows that

$$g_t\left((z, \mathbf{v})\right) = \mathrm{D}\,g\left(g_t\left((\mathrm{i}, \mathrm{i})\right)\right) = \mathrm{D}\,g\left(\mathrm{D}\begin{pmatrix} e^{t/2} & 0 \\ 0 & e^{-t/2} \end{pmatrix}(\mathrm{i}, \mathrm{i})\right) = \mathrm{D}\left(ga_t^{-1}\right)(\mathrm{i}, \mathrm{i}),$$

where we write

$$a_t = \begin{pmatrix} e^{-t/2} & 0 \\ 0 & e^{t/2} \end{pmatrix}.$$

This describes the geodesic flow on $\mathrm{T}^1\mathbb{H} \cong \mathrm{PSL}_2(\mathbb{R})$ as *right multiplication by the inverse\* of the matrix* $a_t$, which we write as

---

\* The inverse in the multiplication on the right is the only way to define an *action* of the full group $\mathrm{PSL}_2(\mathbb{R})$, which we will consider later.

$$R_{a_t}(g) = g a_t^{-1}$$

for $g \in \mathrm{PSL}_2(\mathbb{R})$. Recall that the derivative action of $\mathrm{PSL}_2(\mathbb{R})$ on

$$\mathrm{T}^1 \mathbb{H} \cong \mathrm{PSL}_2(\mathbb{R})$$

corresponds to *left* multiplication.

The second flow on $\mathrm{T}^1 \mathbb{H} \cong \mathrm{PSL}_2(\mathbb{R})$ that we will study is the *horocycle flow*. As we will see, the horocycle flow is intimately linked to the geodesic flow, and it is in this context that we introduce it.

Notice first that the geodesic for our reference vector $(\mathrm{i}, \mathrm{i})$ and for a second vector $(x + \mathrm{i}, \mathrm{i})$ for any $x \in \mathbb{R}$ are both vertical lines. The respective geodesic trajectories move along parallel to each other, with

$$g_t\left((\mathrm{i}, \mathrm{i})\right) = (\mathrm{e}^t \mathrm{i}, \mathrm{e}^t \mathrm{i});$$
$$g_t\left((x + \mathrm{i}, \mathrm{i})\right) = (x + \mathrm{e}^t \mathrm{i}, \mathrm{e}^t \mathrm{i})$$

for $t \in \mathbb{R}$, and so the hyperbolic distance between their base points does not exceed $\frac{|x|}{\mathrm{e}^t}$ (since this is the length of a horizontal path from $\mathrm{e}^t \mathrm{i}$ to $x + \mathrm{e}^t \mathrm{i}$). This suggests that points $g_t\left((\mathrm{i}, \mathrm{i})\right)$ and $g_t\left((x + \mathrm{i}, \mathrm{i})\right)$ on the orbits of the two points under the geodesic flow approach each other as $t \to \infty$. On the other hand if $(z, \mathbf{v})$ has the property that the distance between the base point of $g_t\left((\mathrm{i}, \mathrm{i})\right)$ and that of $g_t\left((z, \mathbf{v})\right)$ converges to zero, then $\Im(z) = 1$ and $\mathbf{v} = \mathrm{i}$ (see Exercise 9.2.1). Combining these two statements, we expect that the set of upward-pointing vectors on the horizontal line through $\mathrm{i}$ will form the *stable manifold* for the geodesic flow through the point $(\mathrm{i}, \mathrm{i})$. That is, these are precisely the set of points $(z, \mathbf{v})$ with the property that the distance between $g_t\left((\mathrm{i}, \mathrm{i})\right)$ and $g_t\left((z, \mathbf{v})\right)$ converges to zero as $t \to \infty$. We will be able to verify this once we have defined a metric on $\mathrm{T}^1 \mathbb{H}$. The orbit of $(\mathrm{i}, \mathrm{i})$ under the subgroup

$$U^- = \left\{ \begin{pmatrix} 1 & s \\ & 1 \end{pmatrix} \mid s \in \mathbb{R} \right\} \tag{9.5}$$

is precisely the set described (see Fig. 9.3).

More generally, for any $(z, \mathbf{v}) = \mathrm{D}\, g(\mathrm{i}, \mathrm{i})$ we similarly define the *(stable) horocycle flow* on $\mathrm{T}^1 \mathbb{H}$ by

$$u^-(s)\left((z, \mathbf{v})\right) = \mathrm{D}\left(g \begin{pmatrix} 1 & -s \\ & 1 \end{pmatrix}\right)(\mathrm{i}, \mathrm{i}),$$

with the corresponding description on $\mathrm{PSL}_2(\mathbb{R})$ by

$$R_{u^-(s)}(h) = h u^-(-s)$$

for $h \in \mathrm{PSL}_2(\mathbb{R})$ and $u^-(s) = \begin{pmatrix} 1 & s \\ 0 & 1 \end{pmatrix} \in U^-$. Similarly, one may check that the subgroup

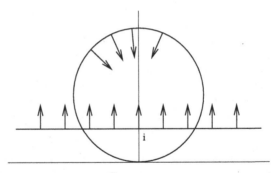

**Fig. 9.3** Two particular orbits of the stable horocycle $U^-$ on $T^1\mathbb{H}$

$$U^+ = \left\{ \begin{pmatrix} 1 & \\ s & 1 \end{pmatrix} \mid s \in \mathbb{R} \right\}$$

gives rise to the *unstable manifolds*, and the corresponding flow is the *unstable horocycle flow.*

In order to make the discussion above meaningful, we need to define a metric on $T^1\mathbb{H} \cong \mathrm{PSL}_2(\mathbb{R})$ with respect to which we can determine whether or not a pair of sequences with no limits in the space may nonetheless be converging to each other. This metric will be constructed in a more general setting in the next section.

## Exercises for Sect. 9.2

**Exercise 9.2.1.** Show directly that if $(z, \mathbf{v})$ has the property that the distance between the base points of $g_t\,((\mathrm{i},\mathrm{i}))$ and that of $g_t\,((z,\mathbf{v}))$ converges to zero as $t \to \infty$, then $\Im(z)' = 1$ and $\mathbf{v} = \mathrm{i}$.

**Exercise 9.2.2.** Show that the action of $\mathrm{PSL}_2(\mathbb{R})$ on $\mathbb{H}$ (and hence also on $T^1\mathbb{H}$) is *proper.* That is, for any compact set $P \subseteq \mathbb{H}$ there exists a compact set $L \subseteq \mathrm{PSL}_2(\mathbb{R})$ such that $z, g(z) \in P$ for some $g \in \mathrm{PSL}_2(\mathbb{R})$ implies $g \in L$. Deduce that the bijection between $T^1\mathbb{H}$ and $\mathrm{PSL}_2(\mathbb{R})$ is a homeomorphism.

## 9.3 Closed Linear Groups and Left Invariant Riemannian Metric

The general linear group $\mathrm{GL}_d(\mathbb{R})$, comprising invertible $d \times d$ matrices over the reals, has a natural topology when viewed as a subset of $\mathrm{Mat}_{dd}(\mathbb{R}) \cong \mathbb{R}^{d^2}$, and in this topology it is an open subset. A *closed linear group* $G$ is any group which can be embedded into $\mathrm{GL}_d(\mathbb{R})$ for some $d \geqslant 1$ such that the image is a closed set in $\mathrm{GL}_d(\mathbb{R})$. What we are going to describe in this section also works for the larger class of *Lie groups*, but we will not need this level of abstraction. Since the map $\det : \mathrm{GL}_d(\mathbb{R}) \to \mathbb{R}$ is continuous, a simple example of a closed linear group is $\mathrm{SL}_d(\mathbb{R})$, the group of real matrices with determinant 1.

The group $\mathrm{PSL}_2(\mathbb{R})$, which plays such a big role in this chapter, is also a closed linear group. To see this, define

$$(\phi(g))\,(m) = gmg^{-1}$$

for $g \in \mathrm{SL}_2(\mathbb{R})$, $m \in \mathrm{Mat}_{22}(\mathbb{R})$, and notice that $\phi(-I) = \phi(I)$ is the identity map. This shows that $\phi$ descends to a homomorphism

$$\widehat{\phi} : \mathrm{PSL}_2(\mathbb{R}) \to \mathrm{GL}(\mathrm{Mat}_{22}(\mathbb{R})) \cong \mathrm{GL}_4(\mathbb{R}),$$

where here and below we write $\mathrm{GL}(V)$ for the group of invertible linear automorphisms of a vector space $V$. We claim that $\widehat{\phi}$ is injective. Moreover, $\widehat{\phi}$ (or, equivalently, $\phi$) is a *proper map*: that is, $\phi^{-1}(K)$ is compact for any compact set $K \subseteq \mathrm{GL}(\mathrm{Mat}_{22}(\mathbb{R}))$. To see this, consider first the basis vector $m = \left(\begin{smallmatrix} 0 & 1 \\ 0 & 0 \end{smallmatrix}\right)$ of $\mathrm{Mat}_{22}(\mathbb{R})$. If the image

$$\begin{pmatrix} a & b \\ c & d \end{pmatrix} \begin{pmatrix} 0 & 1 \\ 0 & 0 \end{pmatrix} \begin{pmatrix} d & -b \\ -c & a \end{pmatrix} = \begin{pmatrix} -ac & a^2 \\ -c^2 & ac \end{pmatrix}$$

is $m$, then $a = \pm 1$ and $c = 0$. Moreover, if this image is bounded then both $a$ and $c$ are bounded. A similar argument using $m = \left(\begin{smallmatrix} 0 & 0 \\ 1 & 0 \end{smallmatrix}\right)$ shows injectivity and that if the image is bounded then $b$ and $d$ are bounded. It follows that $\widehat{\phi}$ is proper with a closed injective image in $\mathrm{GL}(\mathrm{Mat}_{22}(\mathbb{R}))$.

### 9.3.1 The Exponential Map and the Lie Algebra of a Closed Linear Group

Our goal is to define a left-invariant Riemannian metric on any closed linear group $G$. For this we need to identify the tangent bundle $TG$, and we begin this by analyzing the tangent space of $G$ at the identity using the exponential and logarithm maps. The *exponential map* $\exp : \mathrm{Mat}_{dd}(\mathbb{C}) \to \mathrm{Mat}_{dd}(\mathbb{C})$ is defined by the absolutely convergent power series

$$\exp(v) = \sum_{n=0}^{\infty} \frac{1}{n!} v^n \tag{9.6}$$

for $v \in \mathrm{Mat}_{dd}(\mathbb{C})$. We will mainly use this for real matrices, but the properties are most easily seen if we allow complex entries. The following basic properties of the exponential map are readily checked.

- If $v = \begin{pmatrix} \alpha_1 & v_{12} & \cdots & v_{1d} \\ & \ddots & & \vdots \\ & & & \alpha_d \end{pmatrix} \in \mathrm{Mat}_{dd}(\mathbb{C})$ is upper-triangular, then so is $\exp(v)$,
  and the diagonal entries of $\exp(v)$ are $\mathrm{e}^{\alpha_1}, \ldots, \mathrm{e}^{\alpha_d}$.
- If $v \in \mathrm{Mat}_{dd}(\mathbb{C})$ and $g \in \mathrm{GL}_d(\mathbb{C})$, then
$$\exp(gvg^{-1}) = g\exp(v)g^{-1}.$$

- It follows that
$$\exp\left(\mathrm{Mat}_{dd}(\mathbb{R})\right) \subseteq \mathrm{GL}_d(\mathbb{R})$$
  and
$$\det\left(\exp(v)\right) = \mathrm{e}^{\mathrm{tr}(v)}, \tag{9.7}$$
  where the trace $\mathrm{tr}(v)$ is the sum of the diagonal entries in $v$.

If $v$ is an upper-triangular matrix then (9.7) follows from the first property of the exponential map. More generally, for any matrix $v$ there is some $g \in \mathrm{GL}_d(\mathbb{C})$ for which $gvg^{-1}$ is upper-triangular, and hence (using the complex case of the first two properties)

$$\begin{aligned} \det(\exp(v)) &= \det(g)\det(\exp(v))\det(g^{-1}) \\ &= \det(g\exp(v)g^{-1}) \\ &= \det(\exp(gvg^{-1})) \\ &= \mathrm{e}^{\mathrm{tr}(gvg^{-1})} = \mathrm{e}^{\mathrm{tr}(v)}, \end{aligned}$$

in which the last equality follows from the easily checked property

$$\mathrm{tr}(vw) = \mathrm{tr}(wv)$$

for $v, w \in \mathrm{Mat}_{dd}(\mathbb{C})$.

- If $v, w \in \mathrm{Mat}_{dd}(\mathbb{C})$ commute, then
$$\exp(v + w) = \exp(v)\exp(w) = \exp(w)\exp(v).$$

This can be shown directly using the power series. In particular,

$$\exp(nv) = (\exp(v))^n$$

for any $v \in \mathrm{Mat}_{dd}(\mathbb{C})$ and $n \in \mathbb{Z}$.
- For any $v \in \mathrm{Mat}_{dd}(\mathbb{R})$ the map $t \mapsto \exp(tv)$ from $\mathbb{R}$ to $\mathrm{GL}_d(\mathbb{R})$ is a homomorphism; the image is called a *one-parameter subgroup*.
- Taking the derivative of such a homomorphism $t \mapsto \exp(tv)$ with respect to $t$ gives

$$\frac{d}{dt}\exp(tv) = \frac{d}{dt}\sum_{n=0}^{\infty}\frac{1}{n!}(tv)^n = \sum_{n=0}^{\infty}\frac{n}{n!}t^{n-1}v^n = v\exp(tv) = \exp(tv)v.$$

- The exponential map $\exp : \mathrm{Mat}_{dd}(\mathbb{R}) \to \mathrm{GL}_d(\mathbb{R})$ is *locally invertible* at 0. Its inverse is given by the *logarithm map*

$$\log g = \sum_{n=1}^{\infty}\frac{(-1)^{n+1}}{n}(g-I)^n,$$

which is convergent when $g$ is close enough to the identity (for then the absolute value of the entries may be bounded by a convergent geometric series).
- The derivative of $\exp(\cdot)$ at $0 \in \mathrm{Mat}_{dd}(\mathbb{R})$, and the derivative of $\log(\cdot)$ at $\exp(0) = I \in \mathrm{GL}_d(\mathbb{R})$, are the identity map $I \in \mathrm{GL}(\mathrm{Mat}_{dd}(\mathbb{R}))$.

The next result is one of the main reasons why the exponential and logarithm maps are useful in studying closed linear groups.

**Proposition 9.5.** *For any closed linear group $G \subseteq \mathrm{GL}_d(\mathbb{R})$ there is a neighborhood $B$ of $I \in G$ with the property that $\log B \subseteq \mathrm{Mat}_{dd}(\mathbb{R})$ is a neighborhood of 0 inside a linear subspace $\mathfrak{g} \subseteq \mathrm{Mat}_{dd}(\mathbb{R})$. Here $\mathfrak{g}$ is characterized by either of the properties*

- $\mathfrak{g}$ *is the subspace consisting of all derivatives $\phi'(t)$ of paths*

$$\phi : [a,b] \to G \subseteq \mathrm{GL}_d(\mathbb{R})$$

*at points $t \in [a,b]$ with $\phi(t) = I$; or*
- $\mathfrak{g}$ *is the maximal linear subspace of $\mathrm{Mat}_{dd}(\mathbb{R})$ with $\exp(\mathfrak{g}) \subseteq G$.*

Thus the exponential and logarithm maps give a canonical linear coordinate system to a neighborhood of the identity in $G$. The following lemma will be useful for the proof of Proposition 9.5.

**Lemma 9.6.** *There is a neighborhood $B$ of $0 \in \mathrm{Mat}_{dd}(\mathbb{R})$ such that for any $v \in B$, and any sequence $(v_m)$ with $v_m \to v$ as $m \to \infty$, we have*

$$\left(I + \frac{1}{m}v_m\right)^m \longrightarrow \exp(v)$$

*as $m \to \infty$.*

PROOF. If $v$ is sufficiently small, then for large $m$ we may use the geometric series to get the estimate

$$m\log\left(I + \frac{1}{m}v_m\right) = m\left(\frac{1}{m}v_m - \frac{1}{2m^2}v_m^2 + \cdots\right)$$
$$= v_m + \mathrm{O}\left(1/m\right).$$

This implies that $m \log \left( I + \frac{1}{m} v_m \right)$ is still small, and that it converges to $v$ as $m \to \infty$. Thus

$$\exp \left( m \log \left( I + \frac{1}{m} v_m \right) \right) = \left( I + \frac{1}{m} v_m \right)^m \longrightarrow \exp(v).$$

□

The argument below can be described roughly as follows. We study the group multiplication of elements $\exp(v)$ and $\exp(w)$ for small $v, w \in \mathfrak{g}$ up to an error of smaller order, and then Lemma 9.6 will be used as a magnifying device.

PROOF OF PROPOSITION 9.5. Define

$$\mathfrak{g} = \{ v \in \mathrm{Mat}_{dd}(\mathbb{R}) \mid \exp(tv) \in G \text{ for all } t \in \mathbb{R} \}.$$

We claim that $\mathfrak{g}$ is a linear subspace of $\mathrm{Mat}_{dd}(\mathbb{R})$. It is clear that $\mathbb{R}\mathfrak{g} = \mathfrak{g}$, so it remains to show that $\mathfrak{g}$ is closed under addition. Suppose that $v, w \in \mathfrak{g}$ are given, and choose $t > 0$ so that $t(v + w) \in B$ (where $B$ is chosen as in Lemma 9.6). Define a sequence $(g_n)$ in $G$ by

$$g_n = \left( \exp \left( \tfrac{t}{n} v \right) \exp \left( \tfrac{t}{n} w \right) \right)^n$$

for all $n \geqslant 1$. The approximation

$$\exp \left( \tfrac{t}{n} u \right) = I + \tfrac{t}{n} u + \mathrm{O} \left( 1/n^2 \right)$$

shows that

$$g_n = \left( I + \tfrac{1}{n} \left( t(v + w) + \mathrm{O} \left( 1/n \right) \right) \right)^n.$$

By Lemma 9.6,

$$g_n \longrightarrow \exp \left( t(v + w) \right) \in G$$

since $G$ is a closed subgroup of $\mathrm{GL}_d(\mathbb{R})$. In other words, for every $v, w \in \mathfrak{g}$ and every sufficiently small $t \in \mathbb{R}$ we have $\exp \left( t(v + w) \right) \in G$. This implies that $v + w \in \mathfrak{g}$ by the definition of $\mathfrak{g}$, as required.

Let $V \subseteq \mathrm{Mat}_{dd}(\mathbb{R})$ be a linear complement of the subspace $\mathfrak{g}$ in $\mathrm{Mat}_{dd}(\mathbb{R})$ and consider the map

$$\psi : \mathfrak{g} \times V \longrightarrow \mathrm{GL}_d(\mathbb{R}) \subseteq \mathrm{Mat}_{dd}(\mathbb{R})$$

$$(u, v) \longmapsto (\exp(u)) (\exp(v)).$$

The derivative of $\psi$ is the embedding $\mathfrak{g} \times V \to \mathrm{Mat}_{dd}(\mathbb{R})$ obtained by adding the components. By the choice of $V$ as a linear complement to $\mathfrak{g}$, this map is invertible. It follows that $\psi$ is locally invertible, and thus every $g$ in some neighborhood $B_1$ of $I \in \mathrm{GL}_d(\mathbb{R})$ can be written in the form $g = \exp(u)\exp(v)$ with $u \in \mathfrak{g}$ and $v \in V$. Notice that since $G$ is a group, $\exp(v) \in G$ if $g \in G$.

The main statement of the proposition is that there is some neighborhood $B \subseteq B_1$ of $I$ for which

$$\log(B \cap G) \subseteq \mathfrak{g}.$$

If this were not the case then by the argument above there would exist a sequence $(v_m)$ in $V \smallsetminus \{0\}$ with $v_m \to 0$ as $m \to \infty$ with $\exp(v_m) \in G$. Now by compactness of the unit ball in $V$ and after passing to a subsequence, we may assume that

$$\frac{v_m}{\|v_m\|} \longrightarrow w \in V.$$

Moreover, $\exp(\mathbb{Z}v_m) \leqslant G$. Thus the sequence of discrete subgroups $\mathbb{Z}v_m \leqslant V$ (whose generators become arbitrarily small) converge to the subspace $\mathbb{R}w$ in $V$. This implies that $\exp(\mathbb{R}w) \subseteq G$, and hence $w \in \mathfrak{g}$ which contradicts the definition of $V$. More formally, for any $t \in \mathbb{R}$ there exists a sequence $(m_n)$ in $\mathbb{Z}$ with $m_n\|v_n\| \to t$ and

$$\exp(tw) = \lim_{n \to \infty} \exp\left(m_n\|v_n\|\frac{v_n}{\|v_n\|}\right) = \lim_{n \to \infty} (\exp(v_n))^{m_n} \in G.$$

The equivalence of the two characterizations of $\mathfrak{g}$ given in the proposition now follow easily.                                                                □

Although we will not need this, an argument similar to the proof of Proposition 9.5 shows that $\mathfrak{g}$ is closed under another operation, namely the *Lie bracket* (see Exercise 9.3.1). For $v, w \in \mathrm{Mat}_{dd}(\mathbb{R})$ we define the Lie bracket of $v$ and $w$ by

$$[v, w] = vw - wv.$$

Then the *Lie algebra* $\mathfrak{g}$ is the linear subspace of Proposition 9.5 together with the restriction of the Lie bracket $[\cdot, \cdot]$ to $\mathfrak{g}$. It may readily be checked that the Lie bracket satisfies the properties

- $(v, w) \longmapsto [v, w]$ is bilinear;
- $[v, w] + [w, v] = 0$ for all $v, w \in \mathfrak{g}$; and
- $[u, [v, w]] + [v, [w, u]] + [w, [u, v]] = 0$ for all $u, v, w \in \mathfrak{g}$.

The third of these is the *Jacobi identity*. Part of the general theory of Lie groups[86] develops the precise relationship between abstract Lie algebras (vector spaces with a map $[\cdot, \cdot]$ satisfying these three properties) and abstract Lie groups (which generalize closed linear groups).

Recall that in a topological space $X$ the connected component of a point $x \in X$ is the largest connected subset of $X$ containing $x$. Also note that we use the ambient vector space $\mathrm{Mat}_{dd}(\mathbb{R}) \geqslant G$ to determine whether a function $f : O \to G$ defined on an open subset $O \subseteq \mathbb{R}^m$ is differentiable.

**Corollary 9.7.** *For any closed linear group $G \subseteq \mathrm{GL}_d(\mathbb{R})$ the Lie algebra $\mathfrak{g}$ uniquely determines the connected component $G^0$ of the identity in $G$. Indeed, $G^0$ is the group generated by $\exp(\mathfrak{g})$. Moreover, $G^0$ is an open, closed, normal, subgroup of $G$, and is path-connected via smooth curves.*

PROOF. Define, for $n \geqslant 1$,

$$(\exp(\mathfrak{g}))^n = \{\exp(v_1)\exp(v_2)\cdots\exp(v_n) \mid v_1,\ldots,v_n \in \mathfrak{g}\}$$

and

$$H = \bigcup_{n=1}^{\infty} (\exp(\mathfrak{g}))^n.$$

By definition, $H$ is closed under multiplication, and under taking inverses since $\exp(v)^{-1} = \exp(-v)$ for $v \in \mathfrak{g}$. It follows that $H \leqslant G$ is a subgroup.

By Proposition 9.5, $\exp(\mathfrak{g})$ is a neighborhood of $I$ inside $G$. Therefore, for every $g \in (\exp(\mathfrak{g}))^n$ the set $g\exp(\mathfrak{g}) \subseteq \exp(\mathfrak{g})^{n+1}$ is a neighborhood of $g$ inside $G$ which shows that $H$ is open in $G$. This implies that any coset $Hg$ is open for $g \in G$, so the complement $G \smallsetminus H$ of $H$ in $G$, which is a union of cosets of $H$, is open and therefore $H$ is closed. It follows that $G^0 \subseteq H$.

To prove that $H \subseteq G^0$ it is enough to show the last claim of the corollary, namely that any two points $g_1, g_2 \in H$ can be connected by a differentiable curve. Since translation by a group element is a homeomorphism, it is sufficient to consider the identity $I$ and an arbitrary

$$g = \exp(v_1)\cdots\exp(v_n) \in H.$$

In this case the path defined by

$$\phi(t) = \exp(tv_1)\cdots\exp(tv_n)$$

for $t \in [0,1]$ is smooth, and has $\phi(0) = I$ and $\phi(1) = g$.

It remains to check that $H \leqslant G$ is a normal subgroup. For $g \in G$ the conjugation map $h \mapsto ghg^{-1}$ sends $G$ to $G$, and for some neighborhood $B'$ of $I$ we also have

$$gB'g^{-1} \subseteq B$$

where $B$ is chosen as in Proposition 9.5. This implies that

$$g\left(\log(B')\right)g^{-1} = \log\left(gB'g^{-1}\right) \subseteq \mathfrak{g}$$

and so $g\mathfrak{g}g^{-1} \subseteq \mathfrak{g}$ by linearity. Taking the exponential shows that

$$g\exp(\mathfrak{g})g^{-1} \subseteq \exp(\mathfrak{g})$$

and therefore $gHg^{-1} \subseteq H$. □

We will see more examples of the relationship between closed linear groups and their Lie algebras later; for now we return to the important example of $\mathrm{SL}_d(\mathbb{R})$.

*Example 9.8.* By definition

$$\mathrm{SL}_d(\mathbb{R}) = \{g \in \mathrm{Mat}_{dd}(\mathbb{R}) \mid \det(g) = 1\}$$

is the closed set of solutions to the polynomial equation $\det(g) = 1$; by the basic properties of the determinant it is a closed subgroup of $\mathrm{GL}_d(\mathbb{R})$. By (9.7) we have for $v \in \mathrm{Mat}_{dd}(\mathbb{R})$ that

$$\exp(v) \in \mathrm{SL}_d(\mathbb{R}) \Leftrightarrow v \in \mathfrak{sl}_d(\mathbb{R}) = \{v \in \mathrm{Mat}_{dd}(\mathbb{R}) \mid \mathrm{tr}(v) = 0\}$$

which shows that $\mathfrak{sl}_d(\mathbb{R})$ is the Lie algebra of $\mathrm{SL}_d(\mathbb{R})$. We claim that $\mathrm{SL}_d(\mathbb{R})$ is connected, so that $\mathrm{SL}_d(\mathbb{R})$ is uniquely determined by $\mathfrak{sl}_d(\mathbb{R})$ in the sense of Corollary 9.7. Let $E_{ij}$ denote the matrix with a single non-zero entry 1 in the $i$th row and $j$th column. In order to prove the claim, it will be enough to show that $\mathrm{SL}_d(\mathbb{R})$ is generated as a group by the subgroups

$$U_{ij} = \{I + tE_{ij} \mid t \in \mathbb{R}\}$$

with $i \neq j$. Equivalently, it is enough to note that any $g \in \mathrm{SL}_d(\mathbb{R})$ can be reduced to $I$ by a finite sequence of row operations consisting of adding a multiple of the $j$th row to the $i$th for any $i \neq j$.

### 9.3.2 The Left-Invariant Riemannian Metric

As indicated earlier, our main goal in this section is to define a left-invariant Riemannian metric and to derive a left-invariant metric from it. For this we need to identify the tangent bundle $\mathrm{T}G$. We have already described the tangent space at the identity in Sect. 9.3.1, and using the group structure again allows us to define $\mathrm{T}G$ as $G \times \mathfrak{g}$. There are two ways to make this precise, depending on a choice of left or right multiplication; we will use left translation as follows.

**Definition 9.9.** The tangent bundle $\mathrm{T}G$ to the closed linear group $G$ is defined to be $G \times \mathfrak{g}$ with the understanding that for a path $\phi : [0, 1] \to G$ which is differentiable at $t_0 \in [0, 1]$ the corresponding tangent vector is

$$\mathrm{D}\,\phi(t_0) = \big(\phi(t_0), \phi(t_0)^{-1}\phi'(t_0)\big).$$

For $g \in G$ we define the tangent space of $G$ at $g$ by $\mathrm{T}_g G = \{g\} \times \mathfrak{g}$.

Note that the derivative $\phi'(t_0)$ is meant in the usual sense:

$$\phi : [0,1] \to G \subseteq \mathrm{Mat}_{dd}(\mathbb{R})$$

is viewed as a vector-valued function of dimension $d^2$. In order to be sure this definition makes sense, we need to check that

$$\mathrm{D}\,\phi(t_0) \in \mathrm{T}G;$$

that is, that $\phi(t_0)^{-1}\phi'(t_0) \in \mathfrak{g}$. We know that the curve defined by

$$\eta(t) = \phi(t_0)^{-1}\phi(t)$$

has values in $G$ and has $\eta(t_0) = I$. By Proposition 9.5,

$$\eta'(t_0) = \phi(t_0)^{-1}\phi'(t_0) \in \mathfrak{g}$$

as required.

**Proposition 9.10.** *Let $G$ be a closed linear group, let $\phi : [0,1] \to G$ be a continuous curve which is differentiable at $t_0 \in [0,1]$, and let $g \in G$. Then the curves $(g\phi)(t) = g\phi(t)$ and $(\phi g^{-1})(t) = \phi(t)g^{-1}$ are also differentiable at $t_0$, and if $\mathrm{D}\,\phi(t_0) = (\phi(t_0), v)$ then*

$$\mathrm{D}(g\phi)(t_0) = (g\phi(t_0), v);$$
$$\mathrm{D}(\phi g^{-1})(t_0) = (\phi(t_0)g^{-1}, gvg^{-1}).$$

PROOF. By definition,

$$\mathrm{D}(g\phi)(t_0) = \big(g\phi(t_0), (g\phi(t_0))^{-1}g\phi'(t_0)\big) = \big(g\phi(t_0), \phi(t_0)^{-1}\phi'(t_0)\big)$$

which shows the first claim since the $\mathfrak{g}$ component is unchanged by $g$. Similarly,

$$\mathrm{D}(\phi g^{-1})(t_0) = \big(\phi(t_0)g^{-1}, (\phi(t_0)g^{-1})^{-1}\phi'(t_0)g^{-1}\big)$$
$$= \big(\phi(t_0)g^{-1}, g(\phi(t_0)^{-1}\phi'(t_0))g^{-1}\big).$$

$\square$

By reading the chain rule for differentiation backwards, Proposition 9.10 may be interpreted as follows. The derivative of left translation

$$L_g : G \to G$$
$$h \longmapsto gh$$

at a point $h$ is the map

$$(\mathrm{D}\,L_g)_h : \mathrm{T}_h G \to \mathrm{T}_{gh} G$$

sending $(h, v)$ to $(gh, v)$. Thus the derivative of left translation moves the base point but acts as the identity in the fiber. Similarly, the derivative of right translation

$$R_g : G \longrightarrow G$$
$$h \longmapsto hg^{-1}$$

at a point $h$ is the map

$$(\mathrm{D}\, R_g)_h : \mathrm{T}_h G \to \mathrm{T}_{hg^{-1}} G$$

sending $(h, v)$ to $(hg^{-1}, gvg^{-1})$.

As obtained in the proof of Corollary 9.7, for $g \in G$ and $v \in \mathfrak{g}$ we have $gvg^{-1} \in \mathfrak{g}$. This action of $G$ on $\mathfrak{g}$ is called the *adjoint representation*, and is denoted by

$$\mathrm{Ad}_g(v) = gvg^{-1}$$

for $v \in \mathfrak{g}, g \in G$. Notice that the linear map $\mathrm{Ad}_g$ describes conjugation by $g$ on $G$ within the Lie algebra $\mathfrak{g}$ in the sense that

$$\exp\,(\mathrm{Ad}_g v) = g \exp(v) g^{-1}.$$

In particular, $\mathrm{Ad}_g : \mathfrak{g} \to \mathfrak{g}$ is the derivative of the conjugation map $h \mapsto ghg^{-1}$ at the identity.

Now choose an inner product $\langle \cdot, \cdot \rangle$ on $\mathfrak{g}$, and define a Riemannian metric by letting

$$\langle u, v \rangle_g = \langle u, v \rangle \qquad (9.8)$$

for $u, v \in \mathrm{T}_g G$; one might equally write $\langle (g, u), (g, v) \rangle$ for this, but as the base point is the same it is moved to a subscript. Equation (9.8) means that for any two vectors in the same tangent space, we define their inner product using our initially chosen inner product $\langle \cdot, \cdot \rangle$. This construction is similar to the case of the hyperbolic plane on p. 278, where the hyperbolic metric was constructed, and (9.8) above is analogous to (9.3) on p. 280*.

---

* More correctly, in the definition of the hyperbolic metric we used $\frac{1}{y^2}$ as a correcting factor for the inner product, and later showed that with this factor the Riemannian metric is invariant under the action of $\mathrm{SL}_2(\mathbb{R})$. In (9.8) we do not have a normalizing factor which might be surprising. The explanation lies in the way we have chosen to realize the tangent bundles as concrete objects: In the case of $\mathbb{H}$ we defined $\mathrm{T}\mathbb{H}$ as $\mathbb{H} \times \mathbb{C}$ with the understanding that when we take the derivative of a curve in $\mathbb{H}$ we consider it as a curve in $\mathbb{C}$. In the case of $G$ we defined $\mathrm{T}G$ in Definition 9.9 as the product of $G$ with the *tangent plane* $\mathfrak{g}$ *of* $G$ *at the identity*, and then used left translation to define the derivatives of curves passing through other points in $G$. This has the effect of making the derivative $\mathrm{D}\, L_g$ of the left translation map $L_g$ look very simple: it is the map that moves $(h, v) \in \mathrm{T}G$ to $(gh, v)$. Recall that the derivative of the action of $\mathrm{SL}_2(\mathbb{R})$ was not that simple. We respond to this by making a simpler definition in (9.8), which gives us the left-invariant metric that we are after.

**Corollary 9.11.** *For any closed linear group $G$, the Riemannian metric defined by equation (9.8) defines a left-invariant metric on $G^0$. That is, if we define the length of a piecewise smooth curve $\phi : [0,1] \to G$ by*

$$\mathrm{L}(\phi) = \int_0^1 \| \mathrm{D} \, \phi(t) \|_{\phi(t)} \, dt,$$

*and use this to define a metric*

$$\mathsf{d}_G(g_0, g_1) = \inf_\phi \mathrm{L}(\phi),$$

*where the infimum is taken over all such curves with $\phi(0) = g_0$ and $\phi(1) = g_1$, then*

$$\mathsf{d}_G(hg_0, hg_1) = \mathsf{d}_G(g_0, g_1)$$

*for all $h, g_0, g_1 \in G^0$.*

The first statement in Corollary 9.11 is an immediate consequence of Proposition 9.10 and the definitions, and the rest follows just as in our earlier discussion of the hyperbolic plane $\mathbb{H}$. The only difference lies in the fact that we need to restrict to $g_0, g_1 \in G^0$ in order that the definition of the metric makes sense. More generally, it is enough to require only that $g_0, g_1$ lie in the same coset of $G^0$. We may extend the metric by initially defining $\mathsf{d}_G(g_0, g_1) = \infty$ if $g_0 G^0 \neq g_1 G^0$ and then defining a metric $\overline{\mathsf{d}}_G$ in the usual sense by

$$\overline{\mathsf{d}}_G(g_0, g_1) = \begin{cases} \frac{\mathsf{d}_G(g_0,g_1)}{1+\mathsf{d}_G(g_0,g_1)} & \text{if } g_0 G^0 = g_1 G^0, \\ 1 & \text{otherwise.} \end{cases}$$

**Lemma 9.12.** *For any closed linear group $G$ the topology induced by $\mathsf{d}_G$ (or $\overline{\mathsf{d}}_G$) is the subspace topology inherited from $G$ (or, equivalently, from the space $\mathrm{Mat}_{dd}(\mathbb{R})$). Moreover, for any $g \in G$ there is a neighborhood on which $\mathsf{d}_G$ and the metric derived from any norm on $\mathrm{Mat}_{dd}(\mathbb{R})$ are Lipschitz equivalent.*

PROOF. We may assume that the norm on $\mathrm{Mat}_{dd}(\mathbb{R})$ and the inner product on $\mathfrak{g}$ are both induced from an inner product on $\mathrm{Mat}_{dd}(\mathbb{R})$. In what follows all topological statements are meant (unless stated otherwise) with respect to the topology induced from $\mathrm{Mat}_{dd}(\mathbb{R})$. To prove the lemma it is sufficient to find, for any $g \in G$, a neighborhood $B$ which is also a neighborhood with respect to $\mathsf{d}_G$ such that $\mathsf{d}_G$ and the norm on $\mathrm{Mat}_{dd}(\mathbb{R})$ are Lipschitz equivalent if restricted to that neighborhood. Left multiplication by $g$ and by $g^{-1}$ are both Lipschitz maps with respect to the norm on $\mathrm{Mat}_{dd}(\mathbb{R})$ and are isometries with respect to $\mathsf{d}_G$, so it is enough to consider the case $g = e$. Let $B$ be a neighborhood of $I \in G$ with the property that $\log B \subseteq \mathfrak{g}$ is a neighborhood of $0 \in \mathfrak{g}$ as in Proposition 9.5. We may assume that $B$ is compact and connected, indeed we may assume that $\log B$ is a closed metric ball around $0 \in \mathfrak{g}$.

Now fix a compact neighborhood $B'$ contained in the interior $B^o$ of $B$ (see Fig. 9.4). Since for a piecewise smooth curve $\phi : [0,1] \to B$ left multiplication by $\phi(t)^{-1}$ in the definition of $\mathrm{D}\,\phi$ cannot change the norm of $\phi'(t)$ too much, there is some $c > 0$ with

$$c\|\phi(0) - \phi(1)\| \leqslant \mathrm{L}(\phi).$$

Note that the restriction to curves that stay in $B$ is necessary to make this uniform statement.

This implies that curves which start in $B'$ and leave $B$ must have a certain positive length, that is there is some $\ell_0 > 0$ such that any piecewise smooth curve $\phi$ with $\phi(0) \in B'$ and $\phi(t) \notin B$ for some $t \in (0,1]$ has length at least $\ell_0$. Hence, for some constant $c'$,

$$c'\|g_0 - g_1\| \leqslant \mathrm{d}_G(g_0, g_1)$$

for any $g_0, g_1 \in B'$ (the constant $c$ needs to be modified to accommodate paths that leave $B$; since there is a positive lower bound on the lengths of such paths, $c'$ is still positive).

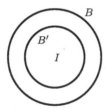

**Fig. 9.4** The neighborhoods $B$ and $B'$

For the reverse inequality, note that for $g_0, g_1 \in B$ the path

$$\phi(t) = \exp\left(\log g_0 + t(\log g_1 - \log g_0)\right)$$

stays in $B$ and that both the exponential and logarithm maps are Lipschitz on small neighborhoods. This gives

$$\mathrm{L}(\phi) \leqslant C\|g_0 - g_1\|$$

for some constant $C$, which implies that

$$\mathrm{d}_G(g_0, g_1) \leqslant C\|g_0 - g_1\|$$

for all $g_0, g_1 \in B$. Thus $\mathsf{d}_G$ and the metric induced from the norm on $B'$ are Lipschitz equivalent (which implies that the respective topologies agree). $\square$

Notice that by definition, if $H \leqslant G$ is a closed subgroup of a closed linear group then $H$ is also a closed linear group. Naturally the Lie algebra $\mathfrak{h}$ of $H$ is a subspace of the Lie algebra $\mathfrak{g}$ of $G$. Given an inner product on $\mathfrak{g}$ we get the restriction to $\mathfrak{h}$, and thence Riemannian metrics on $G$ and $H$. As discussed above, this results in a left-invariant metric $\mathsf{d}_G$ on $G$ and a left-invariant metric $\mathsf{d}_H$ on $H$. However, it is not automatically the case that $\mathsf{d}_H$ is the restriction of $\mathsf{d}_G$ to $H$: in the restriction of $\mathsf{d}_G$ to $H$ we use paths $\phi : [0, 1] \to G$ connecting points in $H$, and among those we may find shorter paths than any path entirely in $H$ connecting the same points. Clearly any path in $H$ is also a path in $G$, so all we can say in general is $\mathsf{d}_G(h_0, h_1) \leqslant \mathsf{d}_H(h_0, h_1)$ for $h_0, h_1$ in $H$. Before we show in an example that an inequality between $\mathsf{d}_G$ and $\mathsf{d}_H$ may really arise, notice that locally $\mathsf{d}_G$ and $\mathsf{d}_H$ are Lipschitz equivalent by Lemma 9.12.

*Example 9.13.* The Heisenberg group

$$G = \left\{ \begin{pmatrix} 1 & x & z \\ & 1 & y \\ & & 1 \end{pmatrix} \mid x, y, z \in \mathbb{R} \right\}$$

has the Lie algebra

$$\mathfrak{g} = \left\{ \begin{pmatrix} 0 & u & w \\ 0 & 0 & v \\ 0 & 0 & 0 \end{pmatrix} \mid u, v, w \in \mathbb{R} \right\}.$$

The closed subgroup

$$H = \left\{ \begin{pmatrix} 1 & 0 & z \\ & 1 & 0 \\ & & 1 \end{pmatrix} \mid z \in \mathbb{R} \right\}$$

has Lie algebra

$$\mathfrak{h} = \left\{ \begin{pmatrix} 0 & 0 & w \\ 0 & 0 & 0 \\ 0 & 0 & 0 \end{pmatrix} \mid w \in \mathbb{R} \right\}.$$

Notice that we simply omit matrix entries if they are always zero, which makes it easier to focus attention on the structure of the groups in question; for the Lie algebras we retain the zeros because there are no non-zero diagonal entries to orient positions in the matrices. Pick an inner product on $\mathfrak{g}$ for which the vectors corresponding to the variables $u, v, w$ are orthornormal. The restriction to $\mathfrak{h}$ gives rise to $\mathsf{d}_H$, which is the usual metric on $H \cong \mathbb{R}$. However, it is clear that

$$d_G\left(\begin{pmatrix}1&x&0\\&1&0\\&&1\end{pmatrix},I\right)\leqslant|x|,\quad d_G\left(\begin{pmatrix}1&0&0\\&1&y\\&&1\end{pmatrix},I\right)\leqslant|y|,$$

and so

$$d_G\left(\begin{pmatrix}1&x&0\\&1&0\\&&1\end{pmatrix}\begin{pmatrix}1&0&0\\&1&y\\&&1\end{pmatrix}\begin{pmatrix}1&-x&0\\&1&0\\&&1\end{pmatrix}\begin{pmatrix}1&0&0\\&1&-y\\&&1\end{pmatrix},I\right)\leqslant2|x|+2|y|$$

by left invariance and (8.17)*. Thus if we choose

$$h=\begin{pmatrix}1&0&z\\&1&0\\&&1\end{pmatrix}=\begin{pmatrix}1&x&0\\&1&0\\&&1\end{pmatrix}\begin{pmatrix}1&0&0\\&1&y\\&&1\end{pmatrix}\begin{pmatrix}1&-x&0\\&1&0\\&&1\end{pmatrix}\begin{pmatrix}1&0&0\\&1&-y\\&&1\end{pmatrix}$$

with $z>0$ and $x=y=\sqrt{z}$, then $d_G(h,I)\leqslant4\sqrt{z}$. For large enough $z$, this shows that $d_H$ is not the restriction of $d_G$ to $H$.

Notice that Proposition 9.10 also shows that $L(\phi g^{-1})\leqslant L(\phi)\|\operatorname{Ad}_g\|$ where $\|\operatorname{Ad}_g\|$ is the operator norm of the adjoint representation with respect to the inner product on $\mathfrak{g}$. In particular, the map $R_g:G\to G$ is Lipschitz.

### 9.3.3 Discrete Subgroups of Closed Linear Groups

As we will see in the rest of this chapter, an important[87] role is played by discrete subgroups in closed linear groups; the example to have in mind is the discrete subgroup $SL_2(\mathbb{Z})$ of $SL_2(\mathbb{R})$.

Recall that a subset $D$ of a metric space is *discrete* if every point $x\in D$ has a neighborhood intersecting $D$ only in the point $x$. Moreover, if a subgroup $\Gamma\subseteq G$ of a closed linear group is discrete, then it is automatically uniformly discrete in the following sense: there exists some $\eta>0$ such that $d_G(\gamma,\gamma')>\eta$ for all $\gamma,\gamma'\in\Gamma$ with $\gamma\neq\gamma'$. To see this, notice that if there is a sequence of pairs of points $\gamma_n,\gamma'_n\in\Gamma$ with $d_G(\gamma'_n,\gamma_n)\to0$ as $n\to\infty$, then we also have $d_G(\gamma_n^{-1}\gamma'_n,e)\to0$ which gives points in $\Gamma$ arbitrarily close to the identity.

We define a metric on the space $X=\Gamma\backslash G$ of right cosets $\Gamma g$ of $\Gamma$ in $G$ by

$$d_X(\Gamma g_1,\Gamma g_2)=\inf_{\gamma_1,\gamma_2\in\Gamma}d_G(\gamma_1 g_1,\gamma_2 g_2)=\inf_{\gamma\in\Gamma}d_G(g_1,\gamma g_2).\qquad(9.9)$$

---

* The formula (8.17) was derived for a right-invariant metric, but the same argument applies to a left-invariant metric.

The equality between the two expressions in (9.9) for $d_X$ follows from the left invariance of $d_G$. The triangle inequality for $d_X$ follows by a simple argument from the first expression. The second expression is useful for the following reason. As we have seen, $R_{g_2}$ is a Lipschitz map, so $\Gamma g_2$ (like $\Gamma$) is uniformly discrete, and hence

$$\inf_{\gamma \in \Gamma} d_G(g_1, \gamma g_2) > 0$$

unless $g_1 = \gamma g_2$ for some $\gamma \in \Gamma$, showing that $d_X$ is a metric on $\Gamma \backslash G$.

Notice that the right translation map $R_g : X \to X$ sending $\Gamma h$ to $\Gamma h g^{-1}$ is well-defined on $X$ (in general, the left translation $L_g$ is not well-defined on $X$).

**Proposition 9.14.** *Let $G$ be a closed linear group and $\Gamma \leqslant G$ a discrete subgroup. Then for any $x \in X = \Gamma \backslash G$ there exists some $r > 0$ such that the map from*

$$B_r^G = \{g \in G \mid d_G(g, e) < r\}$$

*to*

$$B_r^X(x) = \{y \in X \mid d_X(x, y) < r\}$$

*defined by $g \mapsto xg$ is an isometry. For a compact subset $K \subseteq X$ we can choose $r > 0$ so that the above property holds for all $x \in K$.*

The number $r$ arising in Proposition 9.14 is called an *injectivity radius* at $x$. The proposition shows that locally $X$ still looks like $G$, but the radius of the domain to which that isometric similarity extends potentially varies with $x$.

PROOF OF PROPOSITION 9.14. Let $x = \Gamma h$ and fix some $r > 0$. Then, for $g_1, g_2 \in B_r^G$,

$$d_X(\Gamma h g_1, \Gamma h g_2) = \inf_{\gamma \in \Gamma} d_G(h g_1, \gamma h g_2) = \inf_{\gamma \in \Gamma} d_G(g_1, h^{-1} \gamma h g_2).$$

We wish to show that (with a suitable choice of $r > 0$) the infimum is achieved for $\gamma = e$. Suppose that $\gamma \in \Gamma$ has

$$d_G(g_1, h^{-1} \gamma h g_2) \leqslant d_G(g_1, g_2) < 2r$$

then

$$d_G(h^{-1} \gamma h g_2, e) < 3r$$

and

$$d_G(h^{-1} \gamma h, e) < 4r.$$

Since $h^{-1} \Gamma h$ is also a discrete subgroup, for small enough $r > 0$ this implies that $\gamma = e$ and therefore proves the first claim in the proposition.

The last claim follows by compactness and since for $x$ and $r$ as above it is easily checked that any $y \in B^X_{r/2}(x)$ satisfies the first claim of the proposition with $r$ replaced by $r/2$.          □

Notice that by definition of the metric $\mathsf{d}_X$, the canonical projection map

$$\pi : G \longrightarrow X = \Gamma \backslash G$$
$$g \longmapsto \Gamma g$$

satisfies

$$\mathsf{d}_X\left(\pi(g_1), \pi(g_2)\right) \leqslant \mathsf{d}_G\left(g_1, g_2\right)$$

for $g_1, g_2 \in G$, and so is continuous.

Of course the simplest example of such a space is the torus $\mathbb{T}^d = \mathbb{Z}^d \backslash \mathbb{R}^d$, which is itself both a compact metric space and a compact abelian group. We end this section with an example in which the coset space $X$ is not a group.

*Example 9.15.* Let

$$G = \left\{ \begin{pmatrix} 1 & x & z \\ & 1 & y \\ & & 1 \end{pmatrix} \mid x, y, z \in \mathbb{R} \right\}$$

be the Heisenberg group of Example 9.13, and define the discrete subgroup

$$\Gamma = G \cap \mathrm{Mat}_{33}(\mathbb{Z}).$$

We claim that the quotient space $X = \Gamma \backslash G$ is compact but is not a group with respect to the canonical multiplication of coset representative inherited from the group structure on $G$. The statement that $X$ is not a group is simply the statement that $\Gamma$ is not a normal subgroup, which is easily seen:

$$\begin{pmatrix} 1 & x & 0 \\ & 1 & 0 \\ & & 1 \end{pmatrix} \begin{pmatrix} 1 & 0 & 0 \\ & 1 & 1 \\ & & 1 \end{pmatrix} \begin{pmatrix} 1 & -x & 0 \\ & 1 & 0 \\ & & 1 \end{pmatrix} = \begin{pmatrix} 1 & 0 & x \\ & 1 & 1 \\ & & 1 \end{pmatrix} \notin \Gamma$$

if $x \notin \mathbb{Z}$.

To see that $X$ is compact, it is enough to find a compact subset $K \subseteq G$ with the property that the canonical quotient map

$$\pi : G \to X$$
$$g \mapsto \Gamma g$$

restricted to $K$ is onto. Let

$$K = \left\{ \begin{pmatrix} 1 & x & z \\ & 1 & y \\ & & 1 \end{pmatrix} \mid 0 \leqslant x, y, z \leqslant 1 \right\}$$

and

$$g = \begin{pmatrix} 1 & x & z \\ & 1 & y \\ & & 1 \end{pmatrix} \in G.$$

Then (recall that we write $\{\cdot\}$ for the fractional part and $\lfloor \cdot \rfloor$ for the integer part)

$$\begin{pmatrix} 1 & -\lfloor x \rfloor & 0 \\ & 1 & -\lfloor y \rfloor \\ & & 1 \end{pmatrix} g = \begin{pmatrix} 1 & \{x\} & z' \\ & 1 & \{y\} \\ & & 1 \end{pmatrix} = g'$$

already has two entries in $[0,1]$, and

$$\begin{pmatrix} 1 & 0 & -\lfloor z' \rfloor \\ & 1 & 0 \\ & & 1 \end{pmatrix} g' = \begin{pmatrix} 1 & \{x\} & \{z'\} \\ & 1 & \{y\} \\ & & 1 \end{pmatrix} \in K.$$

This shows that $\pi(K) = X$. Notice that while $X$ is obtained from $K \equiv [0,1]^3$ by gluing parallel faces, the resulting space is not $\mathbb{T}^3$ since some of the sides are twisted in the gluing process*. We will study dynamics on this quotient space in Chap. 10.

## Exercises for Sect. 9.3

**Exercise 9.3.1.** Prove that the Lie algebra $\mathfrak{g}$ of a closed linear group (as in Proposition 9.5) is closed under the Lie bracket $[v,w] = vw - wv$ in two different ways, as follows.
(a) Do this by taking the limit of

$$\left(\exp(\tfrac{v}{n})\exp(\tfrac{w}{n})\exp(-\tfrac{v}{n})\exp(-\tfrac{w}{n})\right)^{n^2}$$

as $n \to \infty$.
(b) Do this by taking the derivative of

$$\mathrm{Ad}_{\exp(tv)}(w) = \log\left(\exp(tv)\exp(w)\exp(-tv)\right) \in \mathfrak{g}$$

with respect to $t$.

**Exercise 9.3.2.** Show that the series in (9.6) defining the exponential of a matrix converges absolutely, and use this to check the claimed basic properties of the exponential function.

---

* The group $G$ is simply connected (topologically it is $\mathbb{R}^3$) and the map $\pi : G \to X$ is a covering map, so the fundamental group of $X$ is isomorphic to $\Gamma$, whereas the fundamental group of $\mathbb{T}^3$ is $\mathbb{Z}^3$.

**Exercise 9.3.3.** Describe the gluing of the faces in Example 9.15 to give an explicit description of the space $\Gamma\backslash G$.

**Exercise 9.3.4.** Let $X = \Gamma\backslash G$ be as in Example 9.15, and define

$$T : X \to X$$

to be the map defined by multiplication on the right by the matrix $\begin{pmatrix} 1 & a & c \\ & 1 & b \\ & & 1 \end{pmatrix}$.

(a) Show that $T$ preserves the measure $m$ induced on $X$ by Haar measure on $G$.

(b) Prove that if $T$ is ergodic with respect to $m$, then $1, a$ and $b$ are linearly independent over $\mathbb{Q}$.

(c) Can you prove that if $1, a, b$ are linearly independent over $\mathbb{Q}$, then $T$ is uniquely ergodic? (The methods to prove this are slightly beyond the material in this chapter, and will be discussed in Chap. 10.)

# 9.4 Dynamics on Quotients

Using the last section we will now introduce a metric on $\mathrm{PSL}_2(\mathbb{R}) \cong \mathrm{T}^1\mathbb{H}$. For any inner product on the Lie algebra of $\mathrm{PSL}_2(\mathbb{R})$, we can associate to it a left-invariant metric on $\mathrm{PSL}_2(\mathbb{R})$ using Corollary 9.11.

Using this metric we can verify the claim made on p. 287 that the orbit of a point $g_0 \in \mathrm{PSL}_2(\mathbb{R})$ under the right action of $U^-$ gives the stable manifold in the following sense ($U^-$ is defined in (9.5)). A point $g_1 \in \mathrm{PSL}_2(\mathbb{R})$ is in the orbit $g_0 U^-$ if and only if

$$\mathsf{d}\left(R_{a_t}(g_0), R_{a_t}(g_1)\right) \longrightarrow 0 \tag{9.10}$$

as $t \to \infty$. To see this, note first that by using left invariance we have

$$\mathsf{d}\left(R_{a_t}(g_0), R_{a_t}(g_1)\right) = \mathsf{d}\left(g_0 a_t^{-1}, g_1 a_t^{-1}\right) = \mathsf{d}\left(I_2, a_t g_0^{-1} g_1 a_t^{-1}\right).$$

Now if $\begin{pmatrix} a & b \\ c & d \end{pmatrix} = g_0^{-1} g_1$, then

$$a_t \begin{pmatrix} a & b \\ c & d \end{pmatrix} a_t^{-1} = \begin{pmatrix} a & be^{-t} \\ ce^t & d \end{pmatrix}$$

and it follows that the convergence (9.10) holds if and only if $g_0^{-1} g_1 = \begin{pmatrix} 1 & b \\ 0 & 1 \end{pmatrix}$ for some $b \in \mathbb{R}$, or equivalently $g_1 \in g_0 U^-$.

In fact, this argument shows that $g_i$ belongs to $g_0 U^-$ if and only if there exists $C > 0$ such that

$$\mathsf{d}\left(R_{a_t}(g_0), R_{a_t}(g_1)\right) \leqslant Ce^{-t}.$$

Even though the discussion above gives a satisfactory picture of the local dynamics of the geodesic flow in terms of stable and unstable directions, the global dynamics of the geodesic flow on $T^1\mathbb{H}$ is not very interesting: Given any $g \in \mathrm{PSL}_2(\mathbb{R})$ the orbit eventually leaves any compact set, so the dynamics exhibits no recurrence. As indicated in Sect. 9.3, in order to obtain interesting dynamics we need[88] to look at quotients of $\mathrm{PSL}_2(\mathbb{R})$ by a discrete subgroup $\Gamma$. In a sense we will be most interested in the case of discrete subgroups where the quotient $\Gamma\backslash\mathrm{PSL}_2(\mathbb{R})$ is small. It turns out to be too restrictive to ask for a compact quotient; the most useful requirement is that the quotient space has finite volume in the following sense.

### 9.4.1 Hyperbolic Area and Fuchsian Groups

**Lemma 9.16.** *The* hyperbolic area form $\mathrm{d}A = \frac{1}{y^2}\,\mathrm{d}x\,\mathrm{d}y$ *on* $\mathbb{H}$, *and the hyperbolic volume form*

$$\mathrm{d}m = \frac{1}{y^2}\,\mathrm{d}x\,\mathrm{d}y\,\mathrm{d}\theta$$

*on* $T^1\mathbb{H}$, *where* $\theta$ *gives the angle of the unit tangent vector at* $z = x + \mathrm{i}y$, *are both invariant under the respective actions of* $\mathrm{PSL}_2(\mathbb{R})$.

PROOF. Recall that the complex derivative of $z \mapsto g(z) = \frac{az+b}{cz+d}$ is $\frac{1}{(cz+d)^2}$, so the Jacobian is $\frac{1}{|cz+d|^4}$. Therefore, for any continuous function $f : \mathbb{H} \to \mathbb{R}$ with compact support we may apply a substitution to obtain (using (9.2) once more)

$$\int_{\mathbb{H}} f \circ g\,\mathrm{d}A = \int_{\mathbb{H}} f\left(g(z)\right)\frac{1}{y(z)^2}\,\mathrm{d}x\,\mathrm{d}y$$

$$= \int_{\mathbb{H}} f\left(g(z)\right)\frac{\frac{1}{|cz+d|^4}}{\frac{y(z)^2}{|cz+d|^4}}\,\mathrm{d}x\,\mathrm{d}y$$

$$= \int_{\mathbb{H}} f(z')\frac{1}{(y(z'))^2}\,\mathrm{d}x'\,\mathrm{d}y' = \int_{\mathbb{H}} f\,\mathrm{d}A$$

where $x = x + \mathrm{i}y$ and $z' = g(z) = x(z') + \mathrm{i}y(z')$. The proof for $T^1\mathbb{H}$ is identical once we have calculated the Jacobian for the derivative action $\mathrm{D}\,g$ of $g \in \mathrm{PSL}_2(\mathbb{R})$ on $T^1\mathbb{H}$ in the $(x, y, \theta)$ coordinate system. We claim that in these coordinates, the derivative $\mathrm{D}\,(\mathrm{D}\,g)$ of $\mathrm{D}\,g$ takes the form

$$\left(\begin{array}{c|c} \mathrm{D}\,g & 0 \\ \hline * & 1 \end{array}\right). \tag{9.11}$$

Here $\mathrm{D}\,g$ stands as before for the derivative of $g$ acting on $\mathbb{H}$ and describes the derivatives of the $z$-coordinate along the $z$-coordinate which we see since the action of $\mathrm{D}\,g$ on $\mathrm{T}^1\mathbb{H}$ extends the action of $g$ on $\mathbb{H}$. The last column of the matrix in (9.11) stands for the derivatives of the $z$-coordinate and the $\theta$-coordinate along the $\theta$-coordinate. Again because the action of $\mathrm{D}\,g$ on $\mathrm{T}^1\mathbb{H}$ extends the action of $g$ on $\mathbb{H}$ the former must be zero. Moreover, we know that $g$ is complex-differentiable and so $\mathrm{D}\,g$ only rotates for a given $z$ the angle $\theta$—that is, the derivative of $\mathrm{D}\,g$ along $\theta$ (9.11). Thus the Jacobian of $\mathrm{D}\,g$ (the determinant of the matrix in (9.11)) is equal to the Jacobian of $g$. $\square$

**Definition 9.17.** A *Fuchsian group* is a discrete subgroup $\Gamma \leqslant \mathrm{PSL}_2(\mathbb{R})$. A *lattice* in $\mathrm{PSL}_2(\mathbb{R})$ is a discrete subgroup $\Gamma \leqslant \mathrm{PSL}_2(\mathbb{R})$ such that a fundamental domain for the quotient space $\Gamma\backslash\mathrm{PSL}_2(\mathbb{R})$ has finite measure with respect to $m$. A lattice is *uniform* if the quotient space is compact.

Here a *fundamental domain* $F$ for $\Gamma\backslash\mathrm{PSL}_2(\mathbb{R})$ is a measurable subset of $\mathrm{PSL}_2(\mathbb{R})$ with the property that for every $g \in \mathrm{PSL}_2(\mathbb{R})$ we have

$$|F \cap \Gamma g| = 1.$$

We will also use a slightly more relaxed definition of fundamental domain (in which the intersection can be larger or the union smaller, but only by null sets). Where the distinction becomes important, we will refer to a *strict* fundamental domain.

A particularly important non-uniform lattice is the *modular group*

$$\mathrm{PSL}_2(\mathbb{Z}) = \mathrm{SL}_2(\mathbb{Z})/\{\pm I_2\}.$$

The next result is well-known and a proof may be found in any number theory book dealing with the modular group; this argument is taken from the book of Serre [338].

**Proposition 9.18.** *The set $E = \{z \in \mathbb{H} \mid |z| \geqslant 1, |\Re(z)| \leqslant \frac{1}{2}\}$ illustrated in Fig. 9.5 is a fundamental domain for the action of $\mathrm{PSL}_2(\mathbb{Z})$ on $\mathbb{H}$ in the following sense:*

$$A(\gamma E \cap E) = 0 \tag{9.12}$$

*for $\gamma \in \mathrm{PSL}_2(\mathbb{Z})\backslash\{I_2\}$, and*

$$\mathbb{H} = \bigcup_{\gamma \in \mathrm{PSL}_2(\mathbb{Z})} \gamma E. \tag{9.13}$$

*In particular, $\mathrm{PSL}_2(\mathbb{Z})$ is a lattice in $\mathrm{PSL}_2(\mathbb{R})$.*

Sets $F$ and $\gamma F$ for $\gamma \in \mathrm{PSL}_2(\mathbb{Z})$ with the property in (9.12) are called *almost disjoint*. This property should be understood as a replacement for the more restrictive requirement that

$$\gamma E \cap E = \varnothing$$

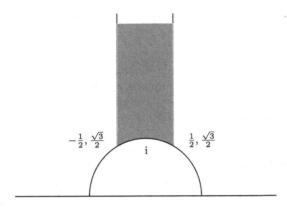

$-\frac{1}{2}, \frac{\sqrt{3}}{2}$                              $\frac{1}{2}, \frac{\sqrt{3}}{2}$

i

**Fig. 9.5** A fundamental region for $\mathrm{PSL}_2(\mathbb{Z})$ acting on $\mathbb{H}$

for all $\gamma \in \mathrm{PSL}_2(\mathbb{Z}) \smallsetminus \{I_2\}$ mentioned above. Notice that it is easy to obtain a measurable subset $E'$ of $E$ which is a fundamental domain in the strict sense by removing parts of the boundary of $E$. Moreover, the set $F = \{g \in \mathrm{PSL}_2(\mathbb{R}) \mid g(\mathrm{i}) \in E'\}$ is then a strict fundamental domain for $\mathrm{PSL}_2(\mathbb{Z}) \backslash \mathrm{PSL}_2(\mathbb{R})$.

In order to start to understand the action of $\mathrm{PSL}_2(\mathbb{Z})$ on $\mathbb{H}$, consider the action of the elements $\tau = \left( \begin{smallmatrix} 1 & 1 \\ 0 & 1 \end{smallmatrix} \right)$ and $\sigma = \left( \begin{smallmatrix} 0 & -1 \\ 1 & 0 \end{smallmatrix} \right)$ on the set $E$. Notice that $\sigma z = -\frac{1}{z}$, $\tau z = z + 1$, and

$$\sigma^2 = (\sigma \tau)^3 = I_2, \tag{9.14}$$

the identity in $\mathrm{PSL}_2(\mathbb{R})$.

The images of $E$ under a few elements of $\mathrm{PSL}_2(\mathbb{Z})$ are shown in Fig. 9.6. To see how this picture is explained, notice that the boundary of $E$ is made out of three pieces of geodesics, and that a Möbius transformation $\gamma \in \mathrm{PSL}_2(\mathbb{Z})$ will map a geodesic to a geodesic. To determine the image geodesic it is enough to consider the images of the two limit points of the original geodesic on $\partial\mathbb{H}$.

PROOF OF PROPOSITION 9.18. Let $z \in \mathbb{H}$. We first show that there is some element $\gamma \in \mathrm{PSL}_2(\mathbb{Z})$ with $\gamma z \in E$, proving (9.13). Recall that for $\gamma = \left( \begin{smallmatrix} a & b \\ c & d \end{smallmatrix} \right)$,

$$\Im(\gamma z) = \frac{\Im(z)}{|cz + d|^2}. \tag{9.15}$$

Since $c$ and $d$ are integers, there must be a matrix $\gamma \in \mathrm{PSL}_2(\mathbb{Z})$ with

$$\Im(\gamma z) = \max\{\Im(\eta z) \mid \eta \in \mathrm{PSL}_2(\mathbb{Z})\}. \tag{9.16}$$

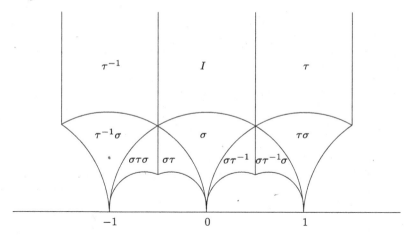

**Fig. 9.6** The action of $\sigma$ and $\tau$ on $E$

Choose $k \in \mathbb{Z}$ so that $\left| \Re \left( \tau^k \gamma z \right) \right| \leqslant \frac{1}{2}$. We claim that $w = \tau^k \gamma z \in E$: if $|w| < 1$ then $\Im(-\frac{1}{w}) > \Im(z)$, contradicting (9.16). So $|w| \geqslant 1$ and $w \in E$ as required.

Now let $z, w \in E$ have the property that $\gamma z = w$ for some $\gamma \in \mathrm{PSL}_2(\mathbb{Z})$. We claim that either $|\Re(z)| = \frac{1}{2}$ (and $z = w \pm 1$), or $|z| = 1$ (and $w = -1/z$). This shows (9.12). Let $\gamma$ be given by the matrix $\left( \begin{smallmatrix} a & b \\ c & d \end{smallmatrix} \right)$. If $\Im(\gamma z) < \Im(z)$ replace the pair $(z, \gamma)$ by $(\gamma(z), \gamma^{-1})$ so that we may assume without loss of generality that $\Im(\gamma z) \geqslant \Im(z)$. This gives $|cz + d| \leqslant 1$ by (9.15). Since $z \in E$ and $d \in \mathbb{Z}$, this requires that $|c| < 2$, so $c = 0, \pm 1$.

If $c = 0$, then $d = \pm 1$ and the map $\gamma$ is translation by $\pm b$. By assumption, $|\Re(z)| \leqslant \frac{1}{2}$ and $|\Re(\gamma z)| \leqslant \frac{1}{2}$ so this implies that $b = 0$ and $\gamma = I_2$ or that $b = \pm 1$ and $\{\Re(z), \Re(\gamma z)\} = \{\frac{1}{2}, -\frac{1}{2}\}$.

Now write $\kappa = -\frac{1}{2} + \mathrm{i}\frac{\sqrt{3}}{2}$. If $c = 1$, the condition $z \in E$ and $|z + d| \leqslant 1$ implies that $d = 0$ unless $z = \kappa$ or $z = -\overline{\kappa}$. Taking $d = 0$ forces $|z| \leqslant 1$ and so $|z| = 1$. If $c = -1$ then replace $\left( \begin{smallmatrix} a & b \\ c & d \end{smallmatrix} \right)$ by $\left( \begin{smallmatrix} -a & -b \\ -c & -d \end{smallmatrix} \right)$, which defines the same element of $\mathrm{PSL}_2(\mathbb{Z})$, and apply the argument above.

This shows that $E$ is a fundamental domain in the sense given.

Finally, to estimate the volume of the fundamental domain $E$, notice that any $z \in E$ has $\Im(z) \geqslant \sqrt{3}/2$, so

$$\mathrm{volume}(E) = \int_{z \in E} \mathrm{d}A \leqslant \int_{\sqrt{3}/2}^{\infty} \int_{-1/2}^{1/2} \frac{\mathrm{d}x \, \mathrm{d}y}{y^2}$$

$$= \int_{\sqrt{3}/2}^{\infty} \frac{1}{y^2} \, \mathrm{d}y = \frac{2}{\sqrt{3}} < \infty.$$

$\square$

As mentioned above, with a little more work all the overlaps between $E$ and $\gamma E$ for $\gamma \in \mathrm{PSL}_2(\mathbb{R})$ can be described, and moreover it may be shown that the elements $\sigma$ and $\tau$ generate $\mathrm{PSL}_2(\mathbb{Z})$. Indeed there are no relations other than those in (9.14). That is,

$$\mathrm{PSL}_2(\mathbb{Z}) = \langle \sigma, \tau \mid \sigma^2 = (\sigma\tau)^3 = I_2 \rangle$$

is a presentation of $\mathrm{PSL}_2(\mathbb{Z})$ as a free product of a cyclic group of order 2 generated by $\sigma$ and a cyclic group of order 3 generated by $\sigma\tau$ (see Exercise 9.4.4 or Kurosh [216] for a complete proof).

The fact that the fundamental domain $E$ of $\mathrm{PSL}_2(\mathbb{Z})$ is a rather concrete geometrical object is not a coincidence. In fact, for every Fuchsian group one can define a hyperbolic polygon which gives a fundamental domain (see Sect. 11.1).

### 9.4.2 Dynamics on $\Gamma\backslash \mathrm{PSL}_2(\mathbb{R})$

Let $\Gamma \leqslant \mathrm{PSL}_2(\mathbb{R})$ be a lattice, for example $\Gamma = \mathrm{PSL}_2(\mathbb{Z})$ as in Proposition 9.18. Let $X = \Gamma\backslash \mathrm{PSL}_2(\mathbb{R})$ be the quotient viewed as a metric space. The geodesic flow on $X$ is still defined via right multiplication

$$R_{a_t}(x) = x a_t^{-1}$$

for any $x = \Gamma g \in X$ and $t \in \mathbb{R}$. The name geodesic is still appropriate since $\Gamma$ can also be used to define a hyperbolic surface $\Gamma\backslash\mathbb{H}$ whose unit tangent bundle (up to some minor technicalities[89]) can be identified with $X$, and the flow can be understood as following the arrow for the given amount of time. In the specific case $\Gamma = \mathrm{PSL}_2(\mathbb{Z})$ we can also use the fundamental domain $E$ and its induced set $F \subseteq \mathrm{PSL}_2(\mathbb{R})$ from Proposition 9.18 to view this action, as in Fig. 9.7.

Suppose that $x = \Gamma g \in X = \mathrm{PSL}_2(\mathbb{Z})\backslash \mathrm{PSL}_2(\mathbb{R})$ satisfies $g \in F$ and represent $x$ by the corresponding arrow based at the point $g(\mathrm{i}) \in E$. Then the geodesic flow $R_{a_t}(g)$ follows the geodesic determined by the arrow until it hits the boundary of $E$ (if it ever does) with the arrow pointing outwards from $E$, and at that point we apply $\tau^{\pm 1}$ or $\sigma^{\pm 1}$ so that the image is again in the boundary of $E$ with the arrow pointing inwards to $E$. Then the geodesic flow follows the geodesic determined by the new arrow until the next time the boundary of $E$ is hit (see Fig. 9.7).

Note that every $g \in \mathrm{PSL}_2(\mathbb{R})$ still acts on $X$ by the right translation

$$R_g(x) = x g^{-1}$$

which defines an action of $\mathrm{PSL}_2(\mathbb{R})$ on $X$. In particular, the action of the subgroups $U^-, U^+ \leqslant \mathrm{PSL}_2(\mathbb{R})$ discussed earlier is still defined and we will still

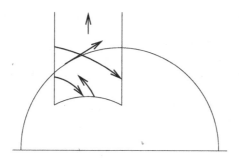

**Fig. 9.7** Geodesic trajectories viewed on $X = \mathrm{PSL}_2(\mathbb{Z})\backslash\mathrm{PSL}_2(\mathbb{R})$ via pieces of hyperbolic geodesics in $E$

refer to them as the stable and unstable horocycle flows (see Exercise 9.4.1). Thus whenever $\Gamma$ is a discrete subgroup we have many one-parameter subgroups acting on $X$. The reader probably already expects that in the case of a lattice, we should have a finite (and after normalization, a probability) measure $m_X$ on $X$ which is invariant under the right action of $\mathrm{PSL}_2(\mathbb{R})$. This will be the start of ergodic theory on $X$.

**Proposition 9.19.** *If $\Gamma \leqslant \mathrm{PSL}_2(\mathbb{R})$ is a lattice, the hyperbolic measure defined by the volume form $\mathrm{d}m = \frac{1}{y^2}\,\mathrm{d}x\,\mathrm{d}y\,\mathrm{d}\theta$ in Lemma 9.16 induces a finite $\mathrm{PSL}_2(\mathbb{R})$-invariant measure $m_X$ on $X = \Gamma\backslash\mathrm{PSL}_2(\mathbb{R})$. In fact if*

$$\pi : \mathrm{PSL}_2(\mathbb{R}) \to X$$

*is the canonical quotient map $\pi(g) = \Gamma g$ for $g \in \mathrm{PSL}_2(\mathbb{R})$ and $F$ is a finite volume fundamental domain, then*

$$m_X(B) = m\left(F \cap \pi^{-1}B\right)$$

*for $B \subseteq X$ measurable defines the $\mathrm{PSL}_2(\mathbb{R})$-invariant measure on $X$.*

### 9.4.3 Lattices in Closed Linear Groups

Rather than prove Proposition 9.19 in isolation, we give some general comments which will lead to a natural generalization (Proposition 9.20). Notice that the measure $m$ on $\mathrm{PSL}_2(\mathbb{R})$ is invariant under the left action of $\mathrm{PSL}_2(\mathbb{R})$ on $\mathrm{PSL}_2(\mathbb{R})$ by Lemma 9.16, that is $m$ is a Haar measure on $\mathrm{PSL}_2(\mathbb{R})$.

Recall from p. 248 that the left Haar measure $m_G$ of a locally compact metric group $G$ is unique up to scalar multiples (see Sect. C.2), and note that right multiplication by elements of $G$ sends $m_G$ to another Haar measure, since

$$(R_g)_* \, m_G(hB) = m_G(hBg) = m_G(Bg) = (R_g)_* \, m_G(B)$$

for all measurable sets $B \subseteq G$. Since Haar measures are unique up to scalars, this defines a continuous modular homomorphism $\mathrm{mod} : G \to \mathbb{R}_{>0}$ into the multiplicative group of positive reals (such a homomorphism is also called a *character*) by

$$(R_g)_* \, (m_G) = \mathrm{mod}(g) m_G.$$

A group $G$ is unimodular if $\mathrm{mod}(G) = \{1\}$, that is if $m_G$ is both a left and a right Haar measure on $G$. Part of the proof of the following more general statement consists of showing that the group $G$ appearing is unimodular.

**Proposition 9.20.** *Let $G$ be a closed linear group, and let $\Gamma \leqslant G$ be a lattice in the sense that $\Gamma$ is discrete and that there is a fundamental domain $F$ for $X = \Gamma \backslash G$ with finite left Haar measure. Then any fundamental domain has the same measure as $F$, $G$ is unimodular, and the Haar measure $m_G$ induces a finite measure $m_X$ on $X$ via*

$$m_X(B) = m_G \left( \pi^{-1}(B) \cap F \right)$$

*for all measurable $B \subseteq X$. Moreover, the right $G$-action $R_g(x) = xg^{-1}$ for $x \in X$ and $g \in G$ leaves the measure $m_X$ invariant.*

Despite the fact that in general $X$ is not a group, we will nonetheless refer to the measure $m_X$ on $X$ as the Haar measure on $X$.

PROOF OF PROPOSITION 9.20. We first show that any two fundamental domains $F, F' \subseteq G$ for $\Gamma \backslash G$ have the same volume. In fact we claim that if $B, B' \subseteq G$ are measurable sets with the property that $\pi|_B$ and $\pi|_{B'}$ are injective, and $\pi(B) = \pi(B')$, then $m_G(B) = m_G(B')$.

By assumption, for every $g \in B$ there is a unique $\gamma \in \Gamma$ with $g \in \gamma B'$, so

$$B = \bigsqcup_{\gamma \in \Gamma} B \cap \gamma B'$$

and similarly

$$B' = \bigsqcup_{\gamma' \in \Gamma} B' \cap \gamma' B.$$

However, these two decompositions are equivalent in the sense that one can be used to derive the other: given $\gamma \in \Gamma$ and a chosen set $B \cap \gamma B'$ we get

$$\gamma^{-1}(B \cap \gamma B') = B' \cap \gamma^{-1} B.$$

For the left Haar measure $m_G$ we therefore have

$$m_G(B) = \sum_{\gamma \in \Gamma} m_G(B \cap \gamma B') = \sum_{\gamma \in \Gamma} m_G(B' \cap \gamma^{-1} B) = m_G(B').$$

This proves the claim, and in particular $m_G(F) = m_G(F')$ for any two fundamental domains $F$ and $F'$.

Now notice that for any $g \in G$ the set $F' = Fg$ is another fundamental domain whose measure satisfies

$$m_G(F) = m_G(F') = \text{mod}(g)m_G(F).$$

Since our assumption is that $m_G(F) < \infty$, and $m_G(F) > 0$ since $\Gamma$ is discrete, we deduce that $\text{mod}(G) = \{1\}$ and that $G$ is unimodular.

Now let $B \subseteq X$ be a measurable set. We define $m_X(B) = m_G(\pi^{-1}(B) \cap F)$ and note that this definition is independent of the choice of fundamental domain $F$ by the claim above. Now write $C = \pi^{-1}(B) \cap F$ and note that

$$Cg = \pi^{-1}(Bg) \cap F' \subseteq F' = Fg.$$

Then by the above $m_G(C) = m_G(Cg)$ and

$$m_X(Bg) = m_G(Cg) = m_G(C) = m_X(B)$$

so $m_X(B) = m_X(R_g^{-1}(B))$ as claimed. $\qquad\qquad\square$

# Exercises for Sect. 9.4

**Exercise 9.4.1.** Let $\Gamma \subseteq \text{PSL}_2(\mathbb{R})$ be a uniform lattice and fix a point $x$ in $X = \Gamma \backslash \text{PSL}_2(\mathbb{R})$. Show that $xU^-$ consists precisely of all points $y \in X$ for which

$$\text{d}\left(R_{a_t}(x), R_{a_t}(y)\right) \longrightarrow 0$$

as $t \to \infty$.

**Exercise 9.4.2.** Show that the closed linear group

$$T = \left\{ \begin{pmatrix} e^{t/2} & s \\ & e^{-t/2} \end{pmatrix} \mid s, t \in \mathbb{R} \right\}$$

does not contain a lattice. That is, $T$ does not contain a discrete subgroup with a fundamental domain of finite left Haar measure.

**Exercise 9.4.3.** Show that $[\text{SL}_d(\mathbb{R}), \text{SL}_d(\mathbb{R})] = \text{SL}_d(\mathbb{R})$ where

$$[g, h] = g^{-1}h^{-1}gh$$

for $g, h \in G$ denotes the *commutator* of $g$ and $h$ in a group $G$, and $[G, G]$ denotes the *commutator subgroup* generated by all the commutators in $G$. Deduce that $\text{SL}_d(\mathbb{R})$ is unimodular for all $d \geqslant 2$.

**Exercise 9.4.4.** Prove that $\mathrm{PSL}_2(\mathbb{Z})$ is a free product of an element of order 2 and an element of order 3.

**Exercise 9.4.5.** Extend the arguments of Proposition 9.18 to show that the subgroup $\mathrm{PSL}_2(\mathbb{Z})$ is a non-uniform lattice in $\mathrm{PSL}_2(\mathbb{R})$.

## 9.5 Hopf's Argument for Ergodicity of the Geodesic Flow

The fundamental result about the geodesic flow on a quotient by a lattice, proved in greater generality than we need by Hopf [156] (see also his later paper [157]), is that it is ergodic.

**Theorem 9.21.** *Let $\Gamma \leqslant \mathrm{PSL}_2(\mathbb{R})$ be a lattice. Then any non-trivial element of the geodesic flow (that is, the map $R_{a_t}$ for some $t \neq 0$) is an ergodic transformation on $X = \Gamma \backslash \mathrm{PSL}_2(\mathbb{R})$ with respect to $m_X$.*

In the proof we will use the following basic idea: If a uniformly continuous function $f : X \to \mathbb{R}$ is invariant under $R_{a_t}$, then it is also invariant under $U^-$ and $U^+$, and is therefore constant.

To see this, we will consider the points $x, y = xu^- \in X$ and will show that $R_{a_t}^n(y) = R_{a_t}^n(x) a_t^n u^- a_t^{-n}$ and $R_{a_t}^n(x)$ are very close together for large enough $n$, and so by invariance and uniform continuity of $f$,

$$f(x) = f\left(R_{a_t}^n(x)\right) \approx f\left(R_{a_t}^n(y)\right) = f(y) \qquad (9.17)$$

are close together for large $n$, which shows that $f(x) = f(y)$ as claimed. Essentially the same idea will be used in the proof for a measurable invariant function, which is what is needed to prove ergodicity. In this outline, we could use a large $n$, but when working with a measurable function (as we must to establish ergodicity) we will need to be more careful in the choice of the variable $n$.

For the proof we will make use of Proposition 8.6, which gives a kind of "ergodicity" for the right action of a locally compact group on itself.

PROOF OF THEOREM 9.21. Normalize the Haar measure $m_X$ to ensure that $m_X(X) = 1$ and let $f : X \to \mathbb{R}$ be a measurable $R_{a_t}$-invariant function for some $t \neq 0$. Fix $\varepsilon > 0$ and choose a compact set $K \subseteq X$ of measure $m(K) > 1 - \varepsilon$ with the property that $f|_K$ is continuous (this is possible by Lusin's Theorem, Theorem A.20).

We claim that

$$B = \left\{ x \mid \lim_{n \to \infty} \frac{1}{n} \sum_{\ell=0}^{n-1} \chi_K \left(R_{a_t}^\ell x\right) > \frac{1}{2} \right\} \qquad (9.18)$$

has measure $m_X(B) \geq 1 - 2\varepsilon$. (Roughly speaking, simultaneous times $\ell \geq 0$ with $R^\ell_{a_t} x \in K$ and $R^\ell_{a_t} y \in K$ will be used below much like the argument outlined in (9.17), and knowing that more than $\frac{1}{2}$ of the future belongs to $K$ will allow us to find similar times for $x$ and $y$.) In fact

$$g^*(x) = \lim_{n \to \infty} \frac{1}{n} \sum_{\ell=0}^{n-1} \chi_K \left( R^\ell_{a_t} x \right) \in [0, 1]$$

exists almost everywhere, and $\int g^* \, dm_X = m_X(K) \geq 1 - \varepsilon$ by the ergodic theorem (Theorem 2.30). So

$$1 - \varepsilon \leq \int_B g^* \, dm_X + \int_{X \smallsetminus B} g^* \, dm_X \leq m_X(B) + \tfrac{1}{2} m_X(X \smallsetminus B)$$

$$= m_X(B) + \tfrac{1}{2} \left( 1 - m_X(B) \right)$$

$$= \tfrac{1}{2} m_X(B) + \tfrac{1}{2},$$

and thus $m_X(B) \geq 1 - 2\varepsilon$ as claimed. This argument should be compared with the discussion in Sect. 37 motivating the maximal ergodic theorem (Theorem 2.24).

Suppose now that $x$ and $y = R_{u^-(s)} x$ for some $s \in \mathbb{R}$ both belong to the set $B$. Then

$$f(x) = f(R^\ell_{a_t}(x)), f(y) = f(R^\ell_{a_t}(y)) \tag{9.19}$$

for all $\ell \geq 1$, by $R_{a_t}$-invariance of $f$, and as discussed earlier,

$$\mathsf{d}_X \left( R^\ell_{a_t}(x), R^\ell_{a_t}(y) \right) = \mathsf{d}_X \left( x a_t^{-\ell}, x u^-(-s) a_t^{-\ell} \right)$$

$$\leq \mathsf{d}_{\mathrm{PSL}_2(\mathbb{R})} \left( I_2, a_t^\ell u^-(-s) a_t^{-\ell} \right) \to 0$$

as $\ell \to \infty$.

Since asymptotically both $x$ and $y$ spend more than half of their future in $K$, there is a common sequence $\ell_n \to \infty$ for $n \to \infty$ of these close returns with

$$R^{\ell_n}_{a_t}(x), R^{\ell_n}_{a_t}(y) \in K.$$

On the set $K$, $f$ is uniformly continuous by compactness and so $f \left( R^{\ell_n}_{a_t}(x) \right)$ and $f \left( R^{\ell_n}_{a_t}(y) \right)$ are closer and closer along an unbounded sequence of $n$. Together with (9.19), this implies that

$$f(x) = f(R_{u^-(s)}(x))$$

whenever $x$ and $R_{u^-(s)}(x)$ are in $B$.

If $\varepsilon_1 < \varepsilon$ then we can choose a compact subset $K_1 \subseteq X$ with

$$m_X(K_1) > 1 - \varepsilon_1$$

316 9 Geodesic Flow on Quotients of the Hyperbolic Plane

for which $f|_{K_1}$ is continuous. We may assume that $K \subseteq K_1$ since continuity of $f|_K$ and $f|_{K_1}$ implies continuity of $f|_{K \cup K_1}$. Hence the set $B_1$ defined from $K_1$ just as in (9.18) satisfies $B_1 \supseteq B$. Since $\varepsilon$ was arbitrary we conclude that there is a set $X'$ with $m_X(X') = 1$ with the property that for $x, y = R_{u^-(s)}(x) \in X'$ we have $f(x) = f(y)$.

The same argument applied to $R_{a_t}^{-1}$ gives the same conclusion for points $x$ and $y = R_{u^+(s)}(x)$ in $X''$ on some other set of full measure $X''$. Taking the intersection $X_1 = X' \cap X''$ we get both conclusions on $X_1$. Therefore, if (for example)

$$g = u^+(s_4)u^-(s_3)u^+(s_2)u^-(s_1) \qquad (9.20)$$

then the set

$$X_g = X_1 \cap R_{u^-(s_1)}^{-1}(X_1) \cap R_{u^+(s_2)u^-(s_1)}^{-1}(X_1)$$
$$\cap R_{u^-(s_3)u^+(s_2)u^-(s_1)}^{-1}(X_1) \cap R_g^{-1}(X'')$$

has full measure, and we claim it satisfies

$$f(x) = f(R_g(x))$$

for all $x \in X_g$. This follows from the argument above:

$$x \in X_1 \cap R_{u^-(s_1)}^{-1}(X_1) \implies f(x) = f(R_{u^-(s_1)}(x)),$$

$$y = R_{u^-(s_1)}(x) \in X_1 \cap R_{u^+(s_2)}^{-1}(X_1) \implies f(x) = f(y) = f(R_{u^+(s_2)}(y)),$$

and continuing in this way implies the claim.

Row operations (only involving adding one row to another) show that the subgroups $U^+$ and $U^-$ generate $\mathrm{SL}_2(\mathbb{R})$ (cf. Exercise 9.5.1). Indeed, every element $g \in \mathrm{SL}_2(\mathbb{R})$ can be written as a product of four elements as in (9.20).

Assume now that $f : X \to \mathbb{R}$ is not constant almost everywhere with respect to $m_X$. Then there exist disjoint intervals $I_1, I_2 \subseteq \mathbb{R}$ for which

$$C_j = \{h \in \mathrm{PSL}_2(\mathbb{R}) \mid f(\Gamma h) \in I_j\}$$

for $j = 1, 2$, are neither null nor conull sets with respect to $m_{\mathrm{PSL}_2(\mathbb{R})}$. By Proposition 8.6, it follows that there is some $g \in G$ with

$$m_{\mathrm{PSL}_2(\mathbb{R})}(C_1 \cap C_2 g) > 0.$$

However, we then have that the set

$$D_g = \{h \in \mathrm{PSL}_2(\mathbb{R}) \mid \Gamma h \in X_g\}$$

is a conull set with respect to $m_{\mathrm{PSL}_2(\mathbb{R})}$, and so there is some element

$$h \in C_1 \cap C_2 g \cap D_g.$$

This gives a contradiction since

$$f(\Gamma h) = f(\Gamma h g^{-1})$$

by definition of $D_g$ and $X_g$, $f(\Gamma h) \in I_1$ and $f(\Gamma h g^{-1}) \in I_2$ by definition of $C_1$ and $C_2$. $\qquad\qquad\qquad\qquad\qquad\qquad\qquad\qquad\qquad\qquad\qquad\quad\square$

A shorter and more abstract proof of Theorem 9.21 will be given in Sect. 11.3.

## Exercises for Sect. 9.5

**Exercise 9.5.1.** Prove that the subgroups $U^+$ and $U^-$ generate $\mathrm{SL}_2(\mathbb{R})$.

## 9.6 Ergodicity of the Gauss Map

In this section we use the ergodicity of the geodesic flow established above to prove ergodicity of the Gauss map. Arguably the most direct proof is the one given in Chap. 3, but the connection between the Gauss map and the geodesic flow is important in itself.

We write $X_2 = \mathrm{PSL}_2(\mathbb{Z}) \backslash \mathrm{PSL}_2(\mathbb{R})$. Since $-I_2$ is an element of $\mathrm{SL}_2(\mathbb{Z})$, we can also think of $X_2$ as $\mathrm{SL}_2(\mathbb{Z}) \backslash \mathrm{SL}_2(\mathbb{R})$, which will be useful later. Recall that there is a unique geodesic through any pair $(z, \mathbf{v})$ in $T^1\mathbb{H}$, as illustrated in Fig. 9.2. For any geodesic for which $\mathbf{v} \neq \pm i$ (that is, for a geodesic that looks like a semi-circle in $\mathbb{H}$) there are two uniquely associated real numbers; the first is the limit point in $\partial\mathbb{H}$ for the past of the geodesic, the second is the limit in $\partial\mathbb{H}$ for the future of the geodesic. As in Figs. 9.8 and 9.9 we will write $-\tilde{y}$ and $y$ (respectively $\tilde{y}$ and $-y$) for these points according to the orientation of the geodesic. Most of the arguments in this section will take place in $\mathbb{H}$ using some simple geometry. We will be studying the geodesic flow in a similar way to the discussion in Sect. 9.4.2. When referring to a geodesic line in $\mathbb{H}$ we will simply say "geodesic", while for the orbit in $X_2$ under the geodesic flow we will use "geodesic flow".

Define subsets of $T^1\mathbb{H}$ by

$$C_+ = \left\{ (ib, \mathbf{v}) \mid \Re\mathbf{v} > 0, \Im\mathbf{v} < 0, y \in [0,1), \tilde{y} \geqslant 1 \text{ as in Fig. 9.8} \right\}$$

and

$$C_- = \left\{ (ib, \mathbf{v}) \mid \Re\mathbf{v} < 0, \Im\mathbf{v} < 0, y \in [0,1), \tilde{y} \geqslant 1 \right\}.$$

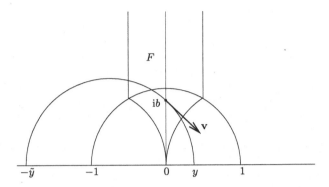

**Fig. 9.8** The reals $y$ and $\tilde{y}$ associated to a geodesic; the point $(ib, \mathbf{v}) \in C_+$

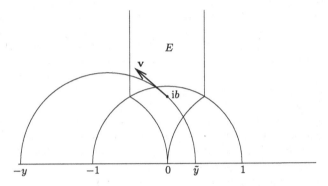

**Fig. 9.9** The reals $y$ and $\tilde{y}$ associated to a geodesic; the point $(ib, \mathbf{v}) \notin C_-$

In Fig. 9.9 we see a vector that does not belong to $C_-$ since its forward limit point $-y \in \partial \mathbb{H}$ of the geodesic flow is to the left of $-1$. Write $C = C_+ \cup C_-$. There are natural coordinates for $C_+$ and $C_-$ given by $(y, z)$ where

$$z = \frac{1}{y + \tilde{y}} \leqslant \frac{1}{1 + y}$$

is the reciprocal of the diameter of the geodesic.

The connection between the Gauss map and the geodesic flow is described in the following lemma. Recall that $\pi$ denotes the quotient map from $\mathrm{PSL}_2(\mathbb{R})$ to $X = \mathrm{PSL}_2(\mathbb{Z}) \backslash \mathrm{PSL}_2(\mathbb{R})$. Using our usual identification between $\mathrm{T}^1 \mathbb{H}$ and $\mathrm{PSL}_2(\mathbb{R})$ we will also use the same letter for the induced map from $\mathrm{T}^1 \mathbb{H}$ to $X$.

**Lemma 9.22.** *Let $x = (ib, \mathbf{v})$ be in $C_+$ with natural coordinates $(y, z)$. The next visit, if there is one, of the geodesic flow to the set $\pi(C)$ occurs in $\pi(C_-)$ and has coordinates*

$$\overline{T}(y,z) = \left(\left\{\tfrac{1}{y}\right\}, y(1 - yz)\right).$$

*A similar property holds for points in* $C_-$.

PROOF. The isometry $z \mapsto -\tfrac{1}{z}$ sends the usual fundamental domain $E$ to another fundamental domain illustrated in Fig. 9.10; this figure also shows a geodesic that never returns to them.

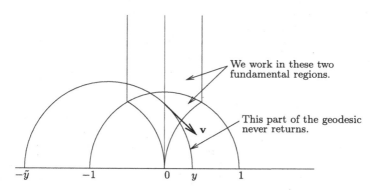

We work in these two fundamental regions.

This part of the geodesic never returns.

$-\tilde{y}$  $-1$  $0$  $y$  $1$

**Fig. 9.10** A geodesic that never returns to $E \cup \sigma E$

Write $D = \{(z, \mathbf{v}) \in T^1 \mathbb{H} \mid \Re(z) = 0\}$ for the collection of arrows attached to the vertical geodesic through $i \in \mathbb{H}$. The image of the geodesic in Fig. 9.10 under the isometry $\sigma : z \mapsto -\tfrac{1}{z}$ is shown in Fig. 9.11. Notice that the map $\sigma$ always reverses the orientation of the geodesic, so the new geodesic goes from right to left. The next return of the geodesic flow to $\pi(D)$, which is marked as $(1)$, is not in $\pi(C)$ because the corresponding limit point for the future does not satisfy the property required for $C_+$ nor for $C_-$. In this case, we continue the geodesic flow to the next return to $\pi(D)$ and repeat this until the return is to $\pi(C)$. Indeed, if we apply the isometry $\tau^n$ for that $n$ uniquely determined by the property that $-\tfrac{1}{y} + n \in (-1, 0]$, as illustrated in Fig. 9.12, then the intersection of the image of the geodesic with $\{z \mid \Re(z) = 0\}$ describes this first return to $\pi(C)$, which is to $\pi(C_-)$ since the orientation was reversed once.

The new coordinates

$$\left(\frac{1}{y} - n, \frac{1}{\frac{1}{\tilde{y}} + \frac{1}{y}}\right)$$

are as claimed, since

$$\frac{1}{\frac{1}{\tilde{y}} + \frac{1}{y}} = \frac{y\tilde{y}}{y + \tilde{y}} = y\left(1 - \frac{y}{y + \tilde{y}}\right) = y(1 - yz)$$

as required. $\qquad \square$

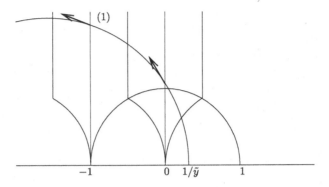

**Fig. 9.11** The first visit (1) to $\pi(D)$ is not in $\pi(C)$

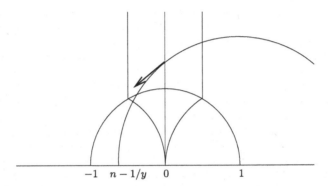

**Fig. 9.12** The return to $\pi(C)$

For $x \in \pi(C)$ the function

$$r_C(x) = \min\{t \mid t > 0, R_{a_t} x \in \pi(C)\} \tag{9.21}$$

is called the *return time function* for the geodesic. Note that the function $r_C$ gives the hyperbolic length of a geodesic between two intersections with vertical geodesics at integer coordinates. From this it follows that $r_C$ is smooth on the sets where this integer does not change; in the natural coordinate system this is the case for $(y, z)$ with $\frac{1}{n+1} < y < \frac{1}{n}$ for some $n \geqslant 1$. Also notice that for large values of $n$ the return time $r_C$ is large (that is, given $n \geqslant 1$, the set of points on which $r_C$ takes the value $n$ is non-empty), so that $r_C$ is unbounded. Lemma 9.23 will show how the unboundedness of $r_C$ is connected to the very complicated dynamics of the geodesic flow. We now show how the geodesic flow can be reconstructed in a measurable way from the Gauss map and this function $r_C$.

Following Chap. 8, we say that a homomorphism $\phi : \mathbb{R} \to \mathrm{MPT}(Y, \mu)$ into the group of invertible measure-preserving transformations of a measure space $(Y, \mu)$ is a *flow*. We will also consider flows of the positive reals defined by a map $\phi : \mathbb{R}^+ \to \mathrm{MPT}(Y, \mu)$ with $\phi_s \circ \phi_t = \phi_{s+t}$ for all $s, t > 0$. In either case, the flow is *ergodic* if any measurable function $f : Y \to \mathbb{R}$ with $f = f \circ \phi_t$ almost everywhere for all (positive) $t$ is equal to a constant almost everywhere (see Sect. 8.1).

**Lemma 9.23.** *Let $T : (Y, \mu) \to (Y, \mu)$ be an invertible measure-preserving map on a probability space, and let $r : Y \to \mathbb{R}^+$ be a measurable function. Then a flow may be defined on the space*

$$X_r = \{(y, s) \mid y \in Y, 0 \leqslant s < r(y)\},$$

*(with measure $m_r$ defined by the restriction of $\mu \times m_{\mathbb{R}}$ to $X_r$) by*

$$T_t(y, s) = \begin{cases} (y, s+t) & \text{if } 0 \leqslant s+t < r(y), \\ (Ty, s+t-r(y)) & \text{if } 0 \leqslant s+t-r(y) < r(Ty), \\ (T^2 y, s+t-r(y)-r(Ty)) & \text{if } 0 \leqslant s+t-r(y)-r(Ty) < r(T^2 y), \\ \vdots \end{cases}$$

*(see Fig. 9.13).*

PROOF. The proof is a measurable analog of the construction of the measure $m_X$ in the proof of Proposition 9.20. Define the map

$$\widetilde{T}(y, s) = (Ty, s - r(y))$$

on the space $Y \times \mathbb{R}$, which we equip with the infinite measure $\mu \times m_{\mathbb{R}}$. Notice that $\widetilde{T}$ preserves $\mu \times m_{\mathbb{R}}$. The inverse of $\widetilde{T}$ is

$$\widetilde{T}^{-1}(y, s) = (T^{-1} y, s + r(T^{-1} y))$$

so $\widetilde{T}$ defines a $\mathbb{Z}$-action on $Y \times \mathbb{R}$ (which plays the role of the isometries in $\Gamma$). The set $X_r$ defines a fundamental domain for this action in the sense that

- $\widetilde{T}^n X_r \cap X_r = \varnothing$ for all $n \neq 0$;
- for $\mu \times m_{\mathbb{R}}$ a.e. $x = (y, s) \in Y \times \mathbb{R}$ there exists a unique $n \in \mathbb{Z}$ with the property that $\widetilde{T}^n x \in X_r$.

The first property is easily seen from the definition. To see the second property let $\mathscr{E} \subseteq \mathscr{B}_Y$ be the $\sigma$-algebra of $T$-invariant sets, so that

$$E\left(r \mid \mathscr{E}\right)(y) = \lim_{n \to \infty} \frac{1}{n} \sum_{k=0}^{n-1} r(T^k y) = \int r \, \mathrm{d}\mu_y^{\mathscr{E}} > 0$$

almost everywhere, by Theorem 6.1 and Theorem 5.14. Therefore, for almost every $y \in Y$ we have $\sum_{k=0}^{n-1} r(T^k y) \nearrow \infty$ (which is trivial if $r$ is bounded from below by a positive constant, which will be the case considered later). So if $(y, s) \in Y \times \mathbb{R}$ has $s > 0$ then there exists some $n$ for which

$$s - \sum_{k=0}^{n} r(T^k y) < 0$$

but

$$s - \sum_{k=0}^{n-1} r(T^k y) \geqslant 0.$$

Then

$$\widetilde{T}^{n-1}(y, s) = \left( T^{n-1}s, s - \sum_{k=0}^{n-1} r(T^k y) \right) \in X_r.$$

The case $s < 0$ is similar. Clearly $\widetilde{T}$ commutes with the flow

$$T_t(y, s) = (y, s + t)$$

defined on $Y \times \mathbb{R}$. Therefore, the image $T_t(X_r)$ is another fundamental domain which can be decomposed into pieces for which a power of $\widetilde{T}$ brings them back to $X_r$. The definition of $T_t$ on $X_r$ precisely describes this process. Since both $T_t$ and $\widetilde{T}$ preserve the infinite measure $\mu \times m_{\mathbb{R}}$ on $Y \times \mathbb{R}$, the action on $X_r$ preserves the restriction of the measure.     □

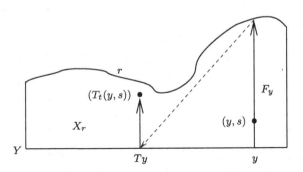

**Fig. 9.13** The flow built under the ceiling function $r$

The flow constructed in Lemma 9.23 is called the *special flow* for $T$ built under the ceiling function $r$.

**Lemma 9.24.** *For the special flow in Lemma 9.23,*

(1) *the measure $m_r$ is finite if and only if $r$ is integrable, and*
(2) *assuming $m_r$ is finite, the map $T$ is ergodic if and only if the flow $\{T_t\}$ is ergodic.*

In the space $X_r = \{(y, s) \mid y \in Y, 0 \leqslant s < r(y)\}$, we call a set of the form

$$F_y = \{(y, s) \mid 0 \leqslant s < r(y)\}$$

a (vertical) fiber (see Fig. 9.13).

PROOF OF LEMMA 9.24. The first statement is clear, since

$$m_r(X_r) = \int_{X_r} \mathrm{d}m_r = \int_Y r(y)\,\mathrm{d}\mu(y).$$

Assume first that the map $T$ is ergodic, and let $f$ be a bounded function on $X_r$ with $f \circ T_t = f$ almost everywhere for all $t \in \mathbb{R}$. We claim that $f$ can be modified on a null set to ensure that $f \circ T_t = f$ everywhere for all $t \in \mathbb{R}$ (just as in Proposition 8.3). Indeed, we may assume that $f \circ T_t = f$ everywhere for all $t \in \mathbb{Q}$ simply by taking a union of countably many null sets (and redefining $f$ to be 0 on the resulting null set). It follows that for every $x \in X$ the function $f_x : \mathbb{R} \to \mathbb{R}$ defined by $f_x(t) = f(T_t(x))$ is translation invariant under $\mathbb{Q}$. In particular,

$$\int_0^1 f_x(s)\,\mathrm{d}s = \int_t^{t+1} f_x(s)\,\mathrm{d}s$$

agrees with $f_x(t)$ for all $t$ (first for $t \in \mathbb{Q}$, and then by continuity of the integral for $t \in \mathbb{R}$), and so by Fubini's theorem we also have

$$f(x) = \int_0^1 f(T_s(x))\,\mathrm{d}s$$

almost everywhere, which proves the claim. In particular, $\overline{f}(y) = f((y,0))$ defines a $T$-invariant function on $Y$ which must be constant $\mu$-almost everywhere by ergodicity of $T$. This implies that $f$ is $m_r$-almost everywhere constant, and so the flow is ergodic also.

The reverse direction will be more important for us. Assume that the flow $T_t$ is ergodic with respect to $m_r$, and let $f = f \circ T$ be a measurable function on $Y$ that is strictly $T$-invariant. We define $\overline{f}(y,s) = f(y)$ and obtain a measurable function on $X_r$ that is strictly invariant under the flow (that is, $\overline{f} = \overline{f} \circ T_t$ everywhere and for all $t \in \mathbb{R}$). By ergodicity, this implies that $\overline{f}$ is constant almost everywhere, which implies that $f$ is constant almost everywhere and hence that $T$ is ergodic. □

**Proposition 9.25.** *The Gauss map* $T(y) = \{\frac{1}{y}\}$ *on* $[0,1] = Y$ *is ergodic with respect to the Gauss measure* $\mathrm{d}\mu = \frac{1}{\log 2}\frac{1}{1+x}\,\mathrm{d}x$. *The return time function* $r_C$ *is integrable.*

PROOF. We will prove ergodicity for the invertible extension of the Gauss map discussed in Sect. 3.4. Recall that this system is the map $\overline{T} : \overline{Y} \to \overline{Y}$ given by

$$\overline{T}(y,z) = (Ty, y(1-yz))$$

on the set

$$\overline{Y} = \{(y, z) \in [0, 1)^2 \mid 0 \leqslant z \leqslant \tfrac{1}{1+y}\}$$

(the set $\overline{Y}$ is illustrated in Fig. 3.2). This implies the ergodicity of the Gauss map easily: if $f$ is an invariant function for $T$ on $Y = [0, 1]$, then $\overline{f}(y, z) = f(y)$ is an invariant function for $\overline{T}$ on $\overline{Y}$. To take care of the alternating sign in the cross-section $C$ we define

$$\widetilde{Y} = \overline{Y} \times \{\pm 1\}$$

and the transformation $\widetilde{T} : \widetilde{Y} \to \widetilde{Y}$ by

$$\widetilde{T}(y, z, \varepsilon) = (\overline{T}(y, z), -\varepsilon).$$

We will write $r_C(y, z)$ for the return time of a point in $\pi(C)$ with natural coordinates $(y, z, \varepsilon)$; because of the symmetry between $C_+$ and $C_-$ the function $r_C$ is independent of $\varepsilon$.

We claim that the special flow $X_{r_C}$ of $\widetilde{T}$ under $r_C$ can be embedded into

$$X_2 = \mathrm{PSL}_2(\mathbb{Z}) \backslash \mathrm{PSL}_2(\mathbb{R})$$

in the sense described below. Indeed, for $(y, z, \varepsilon, s) \in \overline{Y} \times \{\pm 1\} \times \mathbb{R}$ satisfying $0 \leqslant s < r_C(y, z)$ we define

$$\phi(y, z, \varepsilon, s) = R_{a_s}(x(y, z, \varepsilon)) \tag{9.22}$$

where $x(y, z, \varepsilon) \in \pi(C)$ is the point with natural coordinates $(y, z, \varepsilon)$. Notice that $\phi(y, z, \varepsilon, s) \notin \pi(C)$ unless $s = 0$.

More generally, we claim that $\phi : X_{r_C} \to X_2$ is injective. Suppose therefore that

$$x = \phi(y_1, z_1, \varepsilon_1, s_1) = \phi(y_2, z_2, \varepsilon_2, s_2).$$

Then $s_1 = \min\{s \mid R_{a_{-s}}(x) \in \pi(C)\} = s_2$ as $r_C$ is defined to be the time of the first return from $\pi(C)$ to $\pi(C)$. It follows that $(y_1, z_1, \varepsilon_1) = (y_2, z_2, \varepsilon_2)$ are the natural coordinates of the point $R_{a_{-s_1}}(x) \in \pi(C)$.

We next claim that $\phi(\widetilde{T}_t(y, z, \varepsilon, s)) = R_{a_t}(\phi(y, z, \varepsilon, s))$. This means that the special flow under $r_C$ corresponds under the map $\phi$ to the geodesic flow. After the discussion below concerning the measures, we will know that the geodesic flow is actually conjugate to the special flow. The claim follows by considering the various cases used in Lemma 9.23 to define the special flow $T_t$. If $0 \leqslant s + t < r_C(y, z)$, then, by (9.22),

$$\phi(T_t(y, z, \varepsilon, s)) = \phi(y, z, \varepsilon, s + t) = R_{a_{s+t}}(x(y, z, \varepsilon)) = R_{a_t}(\phi(y, z, \varepsilon, s)).$$

If $r_C(y, z) \leqslant s + t < r_C(\overline{T}(y, z))$, then

$$\phi(T_t(y, z, \varepsilon, s)) = \phi(\overline{T}(y, z), -\varepsilon, s+t-r_c(y, z)) = R_{a_{s+t-r_C(y,z)}}(x(\overline{T}(y, z), -\varepsilon)).$$

However, by the properties of the first return to $\pi(C)$ discussed above and the definition of $r_C$ as the time of first return, we also have

$$R_{a_{r_C(y,z)}}\left(x(y,z,\varepsilon)\right) = x\left(\overline{T}(y,z),-\varepsilon\right).$$

This gives

$$\phi(T_t(y,z,\varepsilon,s)) = R_{a_{s+t}}\left(x(y,z,\varepsilon)\right) = R_{a_t}\left(\phi(y,z,\varepsilon,s)\right)$$

in this case. The other cases follow in the same way, or by applying the argument above with a sufficiently small value of $t$, sufficiently often.

We will describe $\phi_*(m_{r_C})$ (see p. 97), where $m_{r_C}$ is the measure in Lemma 9.23 constructed for the special flow for $\widetilde{T}$ on $\widetilde{Y}$ with respect to $r_C$ and $\phi$ is as defined above. This gives a measure on $X_2$, which at this stage is not known to be finite. Our claim in regard to this image measure is that $\phi_*(m_{r_C}) = cm_{X_2}$ for some finite $c > 0$.

Let us assume the claim for now. Then it follows that $m_{r_C}$ is a finite measure (and so $r_C$ is integrable), which may be normalized to be a probability measure. Moreover, in this case $\phi$ is a conjugacy between the special flow of $\widetilde{T}$ under $r_C$ and the geodesic flow (measurability of $\phi^{-1}$ is automatic since $\phi$ is defined by a piecewise differentiable invertible map). As the geodesic flow is ergodic with respect to $m_{X_2}$ by Theorem 9.21, so is the special flow of $\widetilde{T}$ under $r_C$. By Lemma 9.24 this gives ergodicity of $\widetilde{T}$, which in turn gives ergodicity of the invertible extension $\overline{T}$ of the Gauss map, since an invariant function for $\overline{T}$ immediately gives an invariant function for $\widetilde{T}$.

Therefore, the claim that $\phi_*(m_{r_C}) = cm_{X_2}$ for some finite $c > 0$ implies all the statements of the proposition. The main idea of the proof is that the smoothness of $\phi$ gives absolute continuity of $\phi_* m_{r_C}$ with respect to the (smooth) measure $m_{X_2}$, and then the ergodicity of the geodesic flow shows that the Radon–Nikodym derivative $\dfrac{\mathrm{d}\phi_*(m_{r_C})}{\mathrm{d}m_{X_2}} = c$ is constant almost everywhere.

To see the absolute continuity of $\phi_* m_{r_C}$ with respect to $m_{X_2}$, we apply the substitution rule in the form of Lemma A.26. For this we have to check that $\phi$ is injective (which we already know) and smooth with non-vanishing Jacobian. Instead of computing the Jacobian, we will study the map step-by-step, and will work in $\mathrm{T}^1\mathbb{H}$ instead of $X_2$. Without loss of generality we only consider points $(y,z,+1) \in \widetilde{Y}$ corresponding to points in $\pi(C_+)$. Notice that we may assume that $y > 0$ and $z > 0$, which leads to the endpoints $y$ and $-\widetilde{y} = -\frac{1}{z} + y$ of the geodesic that defines the point $x(y,z,+1) = (ib, \exp(i\theta)) \in C_+$. One can check easily that the map

$$\begin{pmatrix} y \\ z \end{pmatrix} \longmapsto \begin{pmatrix} y \\ \widetilde{y} \end{pmatrix} \longmapsto \begin{pmatrix} b \\ 0 \end{pmatrix} \in \mathbb{R}^+ \times \left(-\frac{\pi}{2},0\right)$$

is smooth with non-vanishing Jacobian for $y > 0, z > 0$. Now consider the Jacobian of

$$\phi : \begin{pmatrix} y \\ z \\ s \end{pmatrix} \longmapsto \begin{pmatrix} a' + ib' \\ \theta' \end{pmatrix} \in \mathbb{H} \times \left( -\tfrac{\pi}{2}, \tfrac{\pi}{2} \right) \tag{9.23}$$

for some $y > 0, z > 0$ and $s = 0$. As the application of $R_{a_s}$ is smooth, this Jacobian exists. From the structure of the map, the derivative of (9.23) at $s = 0$ is of the form

$$\begin{pmatrix} 0 & 0 & * \\ * & * & * \\ * & * & * \end{pmatrix},$$

where the rows correspond to $a', b', \theta'$ and the columns to $y, z, s$. However, as the partial derivative of $a'$ with respect to $s$ is non-zero (since $\theta \neq -\tfrac{\pi}{2}$), it follows that the Jacobian is non-zero for $s = 0$. Moreover, application of $R_{a_s}$ is smooth on $\mathrm{T}^1 \mathbb{H}$ with smooth inverse $R_{a_{-s}}$, which shows that the same holds for any $(y, z, s)$. Therefore, by Lemma A.26, we have that $\phi_* m_{r_C}$ is absolutely continuous with respect to $\mathrm{d}a\,\mathrm{d}b\,\mathrm{d}\theta$, and hence with respect to $\mathrm{d}m_{X_2}$ by Lemma 9.16. It follows that

$$\mathrm{d}\left( \phi_* m_{r_C} \right) = F \,\mathrm{d}m_{X_2} \tag{9.24}$$

where $F \geqslant 0$ is the Radon–Nikodym derivative. However, as $m_{r_C}$ is invariant under the special flow by construction, and $\phi$ intertwines the special flow and the geodesic flow (that is, $R_{a_t} \circ \phi = \phi \circ T_t$) we get

$$(R_{a_t})_* \phi_* m_{r_C} = (R_{a_t} \phi)_* m_{r_C} = (\phi T_t)_* m_{r_C} = \phi_* m_{r_C},$$

showing invariance of $\phi_* m_{r_C}$ under the geodesic flow. Applying this to (9.24), together with invariance of $m_{X_2}$, gives

$$\int f \circ R_{a_t} F \,\mathrm{d}m_{X_2} = \int f F \,\mathrm{d}m_{X_2} = \int f F \circ R_{a_t} \,\mathrm{d}m_{X_2}$$

for any measurable $f \geqslant 0$. This implies that $F = F \circ R_{a_t}$ almost everywhere for all $t \in \mathbb{R}$, which by ergodicity shows that $F = c$ almost everywhere for some constant $c$. It remains to check that $c > 0$ is finite. This follows from smoothness, which makes it easy to find an open set of finite measure in the special flow (viewed as a subset of $\mathbb{R}^3$) that is sent onto an open set in $X_2$. $\square$

# Exercises for Sect. 9.6

**Exercise 9.6.1.** Show that real numbers $x$ and $y$ in $[0, 1]$, with continued fraction digits $(a_n(x))_{n \geqslant 1}$ and $(a_n(y))_{n \geqslant 1}$, have the property that the digits eventually agree (there is some $N$ and $k$ with $a_n(x) = a_{n+k}(y)$ for all $n \geqslant N$) if and only if $x = \frac{ay+b}{cy+d}$ for some $\left( \begin{smallmatrix} a & b \\ c & d \end{smallmatrix} \right) \in \mathrm{GL}_2(\mathbb{Z})$. Here $\mathrm{GL}_2(\mathbb{Z})$ denotes the $2 \times 2$ integer matrices with determinant $\pm 1$.

**Exercise 9.6.2.** Describe the forward and backward orbit closure of the geodesic through the point

$$z = \rho + \mathrm{i}$$

in the direction i, where $\rho$ is the golden ratio $\frac{1+\sqrt{5}}{2}$.

**Exercise 9.6.3.** [90] Describe the orbits of the geodesic flow corresponding to the real numbers constructed in Exercise 3.3.1.

**Exercise 9.6.4.** Give a different proof that

$$\sum_{k=0}^{\infty} r(T^k y) = \sum_{k=0}^{\infty} r(T^{-k} y) = \infty$$

almost everywhere as follows. Let $Y_n = \{y \in Y \mid r(y) \geqslant \frac{1}{n}\}$. Show that

$$Y = \bigcup_{n=1}^{\infty} Y_n,$$

and deduce the result by applying Poincaré recurrence to the set $Y_n$.

## 9.7 Invariant Measures and the Structure of Orbits

In this section we show that there are many invariant measures for the geodesic flow. The real thrust of the conclusion is that there are invariant measures for the geodesic flow that cannot be described algebraically. Because this is a negative result (it is paucity of invariant measures rather than abundance that has powerful consequences) the arguments will only be outlined.

### 9.7.1 Symbolic Coding

In Sect. 9.6, we constructed a return-time function $r_C : Y \to \mathbb{R}$ and in the proof of Proposition 9.25 a map $\phi$ from the flow associated to $\widetilde{T}$ and the function $r_C$ into $X_2$. Moreover, this proof shows that every invariant probability measure for $\widetilde{T}$ induces an invariant measure (possibly an infinite one) for the special flow, and hence for the geodesic flow.

**Lemma.** After removing a countable union of line segments and curves, there is a bijection from $\overline{Y}$ to $\mathbb{N}^{\mathbb{Z}}$ that intertwines the map $\overline{T} : \overline{Y} \to \overline{Y}$ and the shift $\sigma : \mathbb{N}^{\mathbb{Z}} \to \mathbb{N}^{\mathbb{Z}}$.

The Gauss map $T : [0,1) \to [0,1)$ is conjugated, after removing $[0,1) \cap \mathbb{Q}$, to the one-sided shift

$$\sigma : \mathbb{N}^{\mathbb{N}} \to \mathbb{N}^{\mathbb{N}}$$

via the continued fraction expansion itself (cf. p. 79). The lemma extends this to the invertible extension of both maps.

Notice that the size of the digits in the continued fraction expansion can be used to bound the circumference of the corresponding geodesic. This shows that the height $b$ of the point $(ib, \mathbf{v}) \in C$ in the construction is similarly bounded (by a function of the digits). After applying $\sigma$ we also get a lower bound for $b$. Together this implies that $r_C$ is bounded by a function depending on the first two digits in the continued fraction expansion.

**Corollary.** Any rapidly-decaying probability vector $\mathbf{p} \in [0,1]^{\mathbb{N}}$ induces an invariant measure on $X_2$ via the probability measure $\mathbf{p}^{\mathbb{Z}}$ on $\mathbb{N}^{\mathbb{Z}}$.

The corollary follows from the above construction: the rapid decay assumption ensures integrability of $r_C$ so that we obtain a finite measure.

There are many more measures than those described in the corollary. For any stochastic matrix $P = (p_{ij})$ there is an associated measure $\mu$ on $\{1,2\}^{\mathbb{Z}}$ defined by

$$\mu\left([a_0, a_1, \dots, a_n]\right) = v_{a_0} p_{a_0 a_1} \cdots p_{a_{n-1} a_n}$$

where $\mathbf{v}$ is a normalized left eigenvector for $P$. All these measures are supported on closed invariant subsets, of which there are also many.

All of these examples are still concrete and relatively well-behaved in many aspects. However, there are many more invariant probability measures on $\{1,2\}^{\mathbb{Z}}$ with unusual behavior.

### 9.7.2 Measures Coming from Orbits

There are other means of showing results of this kind. For example, one can construct orbits of various behaviors using the shadowing lemma of hyperbolic dynamics (see Exercise 9.7.2 and Katok and Hasselblatt [182, Sect. 18.1]). These orbits can then be used to construct measures by the method of Theorem 4.1. Similarly, weak*-limits of orbit measures give invariant measures. For any point $x$,

$$\mu_{x,T} = \frac{1}{T} \int_0^T (R_{a_t})_* \delta_x \, \mathrm{d}t$$

is a measure supported on the part of the geodesic starting at $x$ and ending at $R_{a_T} x$. Any weak*-limit of $\mu_{x,T}$ along a sequence $T_j \to \infty$ is an invariant measure (though there is no guarantee this will be a probability measure). To get some idea of what may arise in this fashion, note the following.

- Any ergodic invariant measure can arise, simply by taking $x$ to be a generic point for that measure.
- More generally, for any set $E \in \mathscr{E}^{R_{a_t}}$ of ergodic measures there exists an initial point $x$ with the property that the set of weak*-limits obtained from $x$ contains $E$. Again this kind of result is easy to see for the shift on $\mathbb{N}^{\mathbb{Z}}$. Here it is enough to consider a countable set $E$ since the set of limit points is closed and the space of probability measures is separable, and a suitable point can be constructed by concatenating longer and longer orbit pieces approximating each measure in turn, using either the symbolic description from Sect. 9.7.1 or orbit shadowing (see Exercise 9.7.2).
- Non-ergodic measures may also arise as weak*-limits. For example, let $\mu_2$ denote the Bernoulli measure on $\mathbb{N}^{\mathbb{Z}}$ corresponding to the probability vector $(\frac{1}{2}, \frac{1}{2}, 0, \dots)$ and let $\mu_3$ denote the Bernoulli measure corresponding to the probability vector $(\frac{1}{3}, \frac{1}{3}, \frac{1}{3}, 0, \dots)$. Let $(x_n)$ be a sequence in $\mathbb{N}^{\mathbb{Z}}$ that is equidistributed with respect to $\mu_2$, and let $(y_n)$ be equidistributed with respect to $\mu_3$. Then the point

$$z = (\dots, 1, 1, x_1, y_1, x_1, x_2, y_1, y_2, x_1, x_2, x_3, y_1, y_2, y_3, \dots)$$

in which the blocks chosen from $(x_n)$ and from $(y_n)$ keep growing linearly in length, has the property that the weak*-limit exists (both on $\mathbb{N}^{\mathbb{Z}}$ and on $X_2$) and coincides with the non-ergodic convex combination $\frac{1}{2}\mu_2 + \frac{1}{2}\mu_3$
- Mass can be lost: If $x$ corresponds to the reference vector $(i, i)$ whose geodesic tends to infinity, then $\mu_{x,T}$ converges to the zero measure. This again can be combined with other possible behaviors to produce a partial loss of mass in the weak*-limit.

# Exercises for Sect. 9.7

**Exercise 9.7.1.** Describe the measure on the shift $\sigma : \mathbb{N}^{\mathbb{Z}} \to \mathbb{N}^{\mathbb{Z}}$ corresponding to the Gauss measure for the continued fraction map via the symbolic coding (see Sect. 9.7.1).

**Exercise 9.7.2.** (a) Show the *shadowing lemma* for the geodesic flow on the space $X = \Gamma \backslash \mathrm{PSL}_2(\mathbb{R})$. That is, show that for any two nearby points $x$ and $y$ there exists another point $z$ with the property that $R_{a_t}(x)$ and $R_{a_t}(z)$ are close for all $t \geqslant 0$ while $R_{a_t}(y)$ and $R_{a_t}(z)$ are close for all $t \leqslant 0$.
(b) Given two nearby points $x, y$ and times $s, t \geqslant 1$ such that $R_{a_s}(x)$ and $R_{\dot{a}_t}(y)$ are also close to $x$ and $y$, find a point $z$ with the property that

the weak*-limit of the orbit measure for $z$ exists, and is supported on some neighborhood of the set

$$\{R_{a_r}(x) \mid 0 \leqslant r \leqslant s\} \cup \{R_{a_r}(y) \mid 0 \leqslant r \leqslant t\}.$$

If these two pieces of the orbits of $x$ and of $y$ are significantly different, ensure that the weak*-limit is not supported on periodic orbits.

# Notes to Chap. 9

[85] (Page 277) Artin's work [8] showed how the continued fraction relates to the geodesic flow on the modular surface. Hedlund [144] proved ergodicity for the continued fraction map and deduced ergodicity for the geodesic flow; we reverse the direction of this argument and use the geometry of the geodesic flow to show ergodicity for the continued fraction map. Hedlund earlier showed ergodicity for the flow defined by a specific Fuchsian group with compact quotient [143], and Martin [254] showed ergodicity (meaning that an invariant measurable set of positive measure has a complement of zero measure) for the action of the modular group on the real axis, also using properties of continued fractions. The papers of Manning [242] and Series [336, 337] are accessible sources for this material.

[86] (Page 293) We will not pursue the general theory of Lie groups and Lie algebras here; an extensive treatment may be found in Knapp [204], with the theory of closed linear groups in Chap. 0. Lie theory is developed in the context of ergodic theory in the book of Feres [90].

[87] (Page 301) The monograph [248] by Margulis provides an extensive treatment of discrete subgroups in Lie groups and their importance in ergodic theory.

[88] (Page 306) This is analogous to the action of a hyperbolic matrix in $\mathrm{SL}_d(\mathbb{Z})$ on $\mathbb{R}^d$: every point apart from the fixed point moves to infinity, converging to the expanding subspace. More interesting dynamics is found after projecting the action onto the quotient $\mathbb{R}^d/\mathbb{Z}^d$.

[89] (Page 310) The technicalities mentioned arise in the following way. For the surface $\mathrm{PSL}_2(\mathbb{Z})\backslash\mathbb{H}$ the points corresponding to i and $\frac{1}{2} + \mathrm{i}\frac{\sqrt{3}}{2}$ are special, because their open neighborhoods are not injective images of neighborhoods of i and $\kappa$ in $\mathbb{H}$ respectively. These special points can be very useful (see Serre [338]) but we will ignore them and work in $\Gamma\backslash\mathrm{PSL}_2(\mathbb{R})$.

[90] (Page 327) See the paper of McMullen [259]; this gives bounded closed geodesics of arbitrarily long length.

# Chapter 10
# Nilrotation

In earlier chapters we have seen how ergodic circle rotations and their associated properties (which include unique ergodicity in Example 4.11, absence of mixing in Example 2.40, equidistribution in Example 4.18) provide an important example of the Kronecker systems studied in Sect. 6.4. In this chapter we introduce the wider class of rotations on quotients of nilpotent groups by studying an important example: the continuous Heisenberg group.

## 10.1 Rotations on the Quotient of the Heisenberg Group

We begin by recalling from Example 9.13 the continuous Heisenberg group

$$G = \left\{ \begin{pmatrix} 1 & x & z \\ & 1 & y \\ & & 1 \end{pmatrix} \mid x, y, z \in \mathbb{R} \right\}$$

and the lattice

$$\Gamma = \left\{ \begin{pmatrix} 1 & \ell & n \\ & 1 & m \\ & & 1 \end{pmatrix} \mid \ell, m, n \in \mathbb{Z} \right\}$$

from Example 9.15. By the argument in Example 9.15, the quotient space $\Gamma \backslash G$ is compact and the set

$$F = \left\{ \begin{pmatrix} 1 & a & c \\ & 1 & b \\ & & 1 \end{pmatrix} \mid 0 \leqslant a, b, c < 1 \right\} .$$

is a fundamental domain for $\Gamma$ in $G$.

Even though $G$ is non-abelian, it is nonetheless very close to being abelian in the following sense. Recall from Exercise 9.4.3 that the *commutator* of the

M. Einsiedler, T. Ward, *Ergodic Theory*, Graduate Texts in Mathematics 259, 331
DOI 10.1007/978-0-85729-021-2_10, © Springer-Verlag London Limited 2011

elements $g, h \in G$ is the element

$$[g, h] = g^{-1}h^{-1}gh,$$

so that the commutator is a measure of the extent to which $g$ and $h$ fail to commute (since $[g, h] = I$, the identity, if and only if $g$ and $h$ commute). For

$$g = \begin{pmatrix} 1 & x & z \\ & 1 & y \\ & & 1 \end{pmatrix}, \qquad h = \begin{pmatrix} 1 & a & c \\ & 1 & b \\ & & 1 \end{pmatrix}$$

we find

$$[g, h] = \begin{pmatrix} 1 & -x & xy - z \\ & 1 & -y \\ & & 1 \end{pmatrix} \begin{pmatrix} 1 & -a & ab - c \\ & 1 & -b \\ & & 1 \end{pmatrix} \begin{pmatrix} 1 & x & z \\ & 1 & y \\ & & 1 \end{pmatrix} \begin{pmatrix} 1 & a & c \\ & 1 & b \\ & & 1 \end{pmatrix}$$

$$= \begin{pmatrix} 1 & 0 & bx - ay \\ & 1 & 0 \\ & & 1 \end{pmatrix}, \tag{10.1}$$

so the *commutator subgroup* is

$$[G, G] = \Big\langle [g, h] \mid g, h \in G \Big\rangle = \left\{ \begin{pmatrix} 1 & 0 & z \\ & 1 & 0 \\ & & 1 \end{pmatrix} \mid z \in \mathbb{R} \right\}.$$

The calculation in (10.1) also shows that an element $g \in G$ commutes with all elements $h \in G$ if and only if $g \in [G, G]$, so the *center* of $G$ is

$$C_G = \big\{ g \in G \mid gh = hg \text{ for all } h \in G \big\} = \left\{ \begin{pmatrix} 1 & 0 & z \\ & 1 & 0 \\ & & 1 \end{pmatrix} \mid z \in \mathbb{R} \right\}.$$

We write $C = C_G = [G, G]$ for the center of the Heisenberg group. The map

$$\phi_G : \begin{pmatrix} 1 & a & c \\ & 1 & b \\ & & 1 \end{pmatrix} \longmapsto (a, b)$$

is a homomorphism $\phi_G : G \to \mathbb{R}^2$ with kernel $C$, so $\phi_G$ induces an isomorphism $G/C \cong \mathbb{R}^2$. Thus $G$ is close to being abelian in this sense: the center of $G$ is an abelian normal subgroup $C$ and the quotient $G/C$ is abelian. A group with this property is called a *2-step nilpotent group* (see Sect. 10.6 for the general definition).

The normal subgroup $C \lhd G$ is also useful in the discussion of the quotient space $X = \Gamma \backslash G$. The homomorphism $\phi_G$ sends $\Gamma$ to $\mathbb{Z}^2 \subseteq \mathbb{R}^2$, so that

$$\Gamma C/C^{\backslash G/C} \cong \mathbb{T}^2.$$

Equivalently, the group $C/\Gamma \cap C \cong \mathbb{T}$ acts on $X$ by right multiplication, since

$$R_\gamma x = x \gamma^{-1} = \Gamma g \gamma^{-1} = \Gamma \gamma^{-1} g = x$$

if $x = \Gamma g$ and $\gamma \in \Gamma \cap C$. Finally, the map

$$\Gamma \begin{pmatrix} 1 & a & c \\ & 1 & b \\ & & 1 \end{pmatrix} \longmapsto (a, b) \pmod{\mathbb{Z}^2}$$

should be thought of as the quotient map

$$\phi_X : X = \Gamma \backslash G \longrightarrow \Gamma C/C^{\backslash G/C} \cong \mathbb{T}^2 \tag{10.2}$$

by the action of $C/C \cap \Gamma \cong \mathbb{T}$. In more geometrical language, $X$ is a *bundle* over $\mathbb{T}^2$ with *fibers* equal to $\mathbb{T}$.

## 10.2 The Nilrotation

We fix some $\tau = \begin{pmatrix} 1 & \alpha & \delta \\ & 1 & \beta \\ & & 1 \end{pmatrix} \in G$, and define $S(x) = x\tau$. Notice that the factor map $\phi_X$ in (10.2) is really a factor map between the continuous transformation $S : X \to X$ and the rotation map

$$T : \mathbb{T}^2 \longrightarrow \mathbb{T}^2$$

$$(a, b) \longmapsto (a, b) + (\alpha, \beta) \pmod{\mathbb{Z}^2}.$$

We recall from Corollary 4.15 that the factor $T$ is ergodic with respect to Lebesgue measure $m_{\mathbb{T}^2}$ if and only if $T$ is uniquely ergodic if and only if $1, \alpha, \beta$ are linearly independent over $\mathbb{Q}$.

It is easy to see that the 3-dimensional Lebesgue measure on $\mathbb{R}^3$ is sent to the (simultaneously left- and right-invariant) Haar measure $m_G$ on $G$ under the map

$$\mathbb{R}^3 \ni (a, b, c) \longmapsto \begin{pmatrix} 1 & a & c \\ & 1 & b \\ & & 1 \end{pmatrix} \in G.$$

It follows that the Haar measure $m_X$ on $X$ (more precisely, the measure induced on $X$ by Haar measure on $G$, which we will also refer to as Haar

measure) is just the 3-dimensional Lebesgue measure on $F$, which may be
thought of as $[0,1)^3$ using the same coordinates as above. It follows that the
measure-preserving system $(\mathbb{T}^2, \mathscr{B}_{\mathbb{T}^2}, m_{\mathbb{T}^2}, T)$ is a factor of $(X, \mathscr{B}_X, m_X, S)$.
This observation gives the downward implications of the following theorem*;
we only need to note that ergodicity of $S$ with respect to $m_X$ implies ergodicity of $T$ with respect to $m_{\mathbb{T}^2}$.

**Theorem 10.1.** *For the transformation* $S(x) = x\tau$ *on* $X$, *the following are
equivalent:*

- $S$ *is uniquely ergodic.*
- $S$ *is ergodic with respect to* $m_X$.
- $\tau = \begin{pmatrix} 1 & \alpha & \delta \\ & 1 & \beta \\ & & 1 \end{pmatrix}$ *and* $1, \alpha, \beta$ *are linearly independent over* $\mathbb{Q}$.

## 10.3 First Proof of Theorem 10.1

Assume that $1, \alpha, \beta$ are linearly independent over $\mathbb{Q}$ (in the notation of Theorem 10.1). Even though Theorem 4.21 does not apply directly, since $X$ is
not topologically the direct product $\mathbb{T}^2 \times \mathbb{T}$, the argument used to prove that
theorem does generalize to give Lemma 10.2 (see Exercise 10.3.1; we will not
give a proof here). Instead of the product, we have a factor map $\phi_X : X \to \mathbb{T}^2$
with the pre-image of each point being given by orbits of $C/\Gamma \cap C \cong \mathbb{T}$.

**Lemma 10.2.** *If* $S$ *is ergodic (and hence* $T$ *is uniquely ergodic) with respect
to* $m_X$, *then* $S$ *is uniquely ergodic.*

Thus what is left in order to prove Theorem 10.1 is the implication that
linear independence of $1, \alpha, \beta$ over $\mathbb{Q}$ implies that $S$ is ergodic. In this first
proof we will use *matrix coefficients*; this argument is a simple instance of
a more general principle that will be discussed in Sect. 11.3. The method in
this case uses invariance of a function under $\tau$ to deduce invariance under all
of $C$, reducing the problem to the 2-torus.

FIRST PROOF OF THEOREM 10.1. Assume that $1, \alpha, \beta$ are linearly independent over $\mathbb{Q}$, and let $f \in L^2(X)$ be an $S$-invariant function. We associate to
the function $f$ the *matrix coefficients*

$$m(g) = \langle U_g f, f \rangle \, ;$$

the function $m$ defined in this way is a continuous function on $G$ by
Lemma 8.7. Moreover,

---

* Notice that $\delta$ does not play a role in the theorem.

$$m(\tau^n g \tau^{-n}) = \int f(x\tau^n g \tau^{-n})\overline{f(x)}\,dm_X$$

$$= \int f(yg\tau^{-n})\overline{f(y\tau^{-n})}\,dm_X \quad \text{(since } m_X \text{ is } \tau\text{-invariant)}$$

$$= \int f(yg)\overline{f(y)}\,dm_X \quad \text{(since } f \text{ is } S\text{-invariant)}$$

$$= m(g) \tag{10.3}$$

for all $n \geqslant 1$. Now let

$$g = \begin{pmatrix} 1 & \varepsilon & 0 \\ & 1 & 0 \\ & & 1 \end{pmatrix}$$

and compute

$$\tau^n g \tau^{-n} = \begin{pmatrix} 1 & n\alpha & \binom{n}{2}\alpha\beta + n\gamma \\ & 1 & n\beta \\ & & 1 \end{pmatrix} \begin{pmatrix} 1 & \varepsilon & 0 \\ & 1 & 0 \\ & & 1 \end{pmatrix} \begin{pmatrix} 1 & -n\alpha & \binom{n+1}{2}\alpha\beta - n\gamma \\ & 1 & -n\beta \\ & & 1 \end{pmatrix}$$

$$= \begin{pmatrix} 1 & \varepsilon & -n\varepsilon\beta \\ & 1 & 0 \\ & & 1 \end{pmatrix}$$

for $n \geqslant 1$. For any fixed $t \in \mathbb{R}$ we may now choose $n_\varepsilon = \lfloor \frac{t}{\varepsilon\beta} \rfloor$ so that $n_\varepsilon\varepsilon\beta \to t$ as $\varepsilon \to 0$. Taking the limit in (10.3) as $\varepsilon \to 0$ shows that

$$m\left(\begin{pmatrix} 1 & 0 & -t \\ & 1 & 0 \\ & & 1 \end{pmatrix}\right) = m\left(\begin{pmatrix} 1 & & \\ & 1 & \\ & & 1 \end{pmatrix}\right) = \|f\|_2^2.$$

Thus $\langle U_h f, f \rangle = \langle f, f \rangle$ for any $h \in C$. This equality in the Cauchy–Schwartz inequality can only occur if $U_h f = f$, so we deduce that $f$ is invariant under $C$. Hence, up to a set of measure zero, $f = F \circ \phi_X$ for some function $F$ on $\mathbb{T}^2$. Since $f$ is $S$-invariant, $F$ is $T$-invariant, and so must be constant $m_{\mathbb{T}^2}$-almost everywhere since $T$ is ergodic. This shows that $f$ is constant almost everywhere, so that $S$ is ergodic with respect to $m_X$. By Lemma 10.2, $S$ is also uniquely ergodic. $\qquad\square$

# Exercises for Sect. 10.3

**Exercise 10.3.1.** Give a more general formulation of Theorem 4.21 that includes both that theorem and Lemma 10.2, and check whether the method of proof of Theorem 4.21 also applies to the generalization.

**Exercise 10.3.2.** Find the eigenfunctions of the transformation $S : X \to X$ (assuming ergodicity).

# 10.4 Second Proof of Theorem 10.1

In this section we give an independent proof* of the difficult direction in Theorem 10.1. We assume again that $1, \alpha, \beta$ are linearly independent over $\mathbb{Q}$, and that $\mu$ is some $S$-invariant ergodic probability measure on $X$. We will conclude by showing that $\mu = m_X$[(91)].

### 10.4.1 A Commutative Lemma; The Set $K$

**Lemma 10.3.** *Let $S : X \to X$ be a continuous map on a compact metric space equipped with an $S$-invariant and ergodic Borel probability measure $\mu$, and let $R : X \to X$ be another continuous map that commutes with $S$. If there exists a point $x \in X$ that is $\mu$-generic with respect to $S$ such that $R(x)$ is also $\mu$-generic, then $R$ preserves $\mu$.*

PROOF. The proof proceeds quite directly from the definitions. Let $f \in C(X)$; it is enough to show that

$$\int f \, dR_* \mu = \int f \circ R \, d\mu = \int f \, d\mu.$$

Notice that by continuity of $R$, we also have $f \circ R \in C(X)$. Since $x$ is generic with respect to $\mu$ and $S$, we have

$$\int f \circ R \, d\mu = \lim_{N \to \infty} \frac{1}{N} \sum_{n=0}^{N-1} f(RS^n x) \quad \text{(since } f \circ R \in C(X))$$

$$= \lim_{N \to \infty} \frac{1}{N} \sum_{n=0}^{N-1} f(S^n Rx) \quad \text{(since } RS = SR)$$

$$= \int f \, d\mu \quad \text{(since } Rx \text{ is generic)},$$

which proves the lemma.  $\square$

---

* We will be using essentially the same calculation in $G$, but will argue quite differently, in particular without applying Theorem 4.21. It may be argued that the two proofs are the same at a deeper level, but both arguments are worth presenting as they generalize in inequivalent ways.

Clearly in the setting of Theorem 10.1 we may use Lemma 10.3 with the map $R(x) = xc$ for any $c \in C$, since $C$ is the center of $G$. However, as we don't know anything about $\mu$ initially, we do not know whether the assumption regarding the existence of generic points $x$, $xc$ can be satisfied. We will construct many such tuples below by a limiting argument; in order to ensure that the limit points are indeed generic points we restrict ourselves in part of the argument to a compact subset consisting entirely of generic points. Recall that $\mu$ is ergodic for $S$, so by Corollary 4.20 the set $X'$ of $\mu$-generic points for $S$ has full $\mu$-measure, and therefore there exists a compact set $K \subseteq X'$ with $\mu(K) > 0.99$ say.

### 10.4.2 Studying Divergence; The Set $X_1$

Let $x, x' \in X$ be two nearby points. Then there is some $g \in G$ close to the identity with $x' = xg$; we refer to this $g$ as the *displacement* between $x$ and $x'$. We now analyze how $x$ and $x'$ move apart under iterates of $S$ (if they do), in terms of $g = \begin{pmatrix} 1 & a & c \\ & 1 & b \\ & & 1 \end{pmatrix}$. This is quite straightforward: Replacing the pair $x, x'$ by $Sx = x\tau, Sx' = x'\tau$ replaces the displacement $g$ by $\tau^{-1}g\tau$ since

$$(x\tau)(\tau^{-1}g\tau) = xg\tau = x'\tau.$$

This can be iterated so that

$$S^n x' = (S^n x)(\tau^{-n} g \tau^n),$$

but (as we will see) for large $n$ the element $\tau^{-n} g \tau^n$ may become large, and so may not represent the shortest displacement between $S^n x$ and $S^n x'$. We calculate

$$g_n = \tau^{-n} g \tau^n = \begin{pmatrix} 1 & -n\alpha & \binom{n+1}{2}\alpha\beta - n\delta \\ & 1 & -n\beta \\ & & 1 \end{pmatrix} \begin{pmatrix} 1 & a & c \\ & 1 & b \\ & & 1 \end{pmatrix} \begin{pmatrix} 1 & n\alpha & \binom{n}{2} + n\delta \\ & 1 & n\beta \\ & & 1 \end{pmatrix}$$

$$= \begin{pmatrix} 1 & a & c + n(a\beta - b\alpha) \\ & 1 & b \\ & & 1 \end{pmatrix} \quad\quad (10.4)$$

for $n \geqslant 1$, and see that there are two possibilities. Either $(a\beta - b\alpha) = 0$, in which case the two points have parallel orbits (that is, the displacement between the points $g_n = g$ is constant), or $(a\beta - b\alpha) \neq 0$, and the two points drift apart at linear speed and in the direction of $C$, as illustrated in Fig. 10.1.

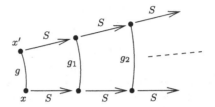

**Fig. 10.1** Linear growth in displacement in the Heisenberg group

Assume that we are in the latter situation, and that $g$ is very close to the identity. Then we can choose $n$ so that $g_n$ is close to

$$\begin{pmatrix} 1 & 0 & t \\ & 1 & 0 \\ & & 1 \end{pmatrix} \in C$$

for some fixed $t \in \mathbb{R}$. If the original pair of points $x, x'$ are generic, then so are $S^n x, S^n x'$ (this is easy to check), and these two points differ approximately by the fixed element in $C$. If we could take the limit of the tuple of points $S^n x, S^n x'$ along a sequence which constantly improves the strength of the statement that they differ approximately by

$$\begin{pmatrix} 1 & 0 & t \\ & 1 & 0 \\ & & 1 \end{pmatrix}$$

and ensures that the points remain generic points, then Lemma 10.3 could be applied. However, we can only do this if $S^n x, S^n x'$ lie in $K$, since in general the set of generic points is not closed*. This in turn restricts the possible values of $n$ we may use. Happily the maximal ergodic theorem (Theorem 2.24; also see Example 2.25) applied to the set $B = X \smallsetminus K$ shows that the set

$$X_1 = \left\{ x \in X \mid \frac{1}{N} \sum_{n=0}^{N-1} \chi_K(S^n x) \geqslant 0.9 \text{ for all } N \right\} \qquad (10.5)$$

has $\mu(X_1) > 0.9$. Thus if we choose our initial points $x, x'$ both from $X_1$, then the restriction mentioned above is a mild one: Regardless of the value of $N$, for most (at least 80%) of the integers $n \in [0, N-1]$ *both* $S^n x$ and $S^n x'$ lie in $K$.

---

* Of course once we have established unique ergodicity we know that the set of generic points is closed.

### 10.4.3 Combining Linear Divergence and the Maximal Ergodic Theorem

After the motivational discussion above we are now ready to start the formal argument.

SECOND PROOF OF THEOREM 10.1. From the discussion above, we will need the following ingredients:

- the compact set $K$ of generic points with $\mu(K) > 0.99$.
- the calculation of the displacement $g_n$ between $S^n x$ and $S^n x'$ from (10.4); and
- the set $X_1$ defined in (10.5).

Suppose for now that we can find, for any $\varepsilon_\ell = \frac{1}{\ell} > 0$ two points $x_\ell, x'_\ell \in X_1$, with displacement

$$g^{(\ell)} = \begin{pmatrix} 1 & a_\ell & c_\ell \\ & 1 & b_\ell \\ & & 1 \end{pmatrix}$$

satisfying $|a_\ell|, |b_\ell|, |c_\ell| < \frac{1}{\ell}$ and $a_\ell \beta - b_\ell \alpha \neq 0$. Fix $t > 0$, and define

$$N_\ell = \left\lfloor \frac{t}{|a_\ell \beta - b_\ell \alpha|} \right\rfloor.$$

Then, as discussed above, for at least 80% of all $n_\ell \in [0, N_\ell - 1]$ we have

$$S^{n_\ell} x, S^{n_\ell} x' \in K.$$

Choose some $n_\ell \in \left[ \frac{N_\ell - 1}{2}, N_\ell - 1 \right]$ with this property (notice that at least 30% of all $n \in [0, N_\ell]$ satisfy all three conditions); it follows that

$$\frac{t}{3} < |a_\ell \beta - b_\ell \alpha| n_\ell \leqslant t$$

for large enough $\ell$. Thus we may choose a sequence $\ell_i \to \infty$ for which

$$S^{n_{\ell_i}} x \to z \in K,$$

$$S^{n_{\ell_i}} x' \to z' \in K,$$

and

$$g^{(\ell_i)}_{n_{\ell_i}} = \tau^{-n_{\ell_i}} g^{(\ell_i)} \tau^{n_{\ell_i}} \cdot \to c = \begin{pmatrix} 1 & 0 & t' \\ & 1 & 0 \\ & & 1 \end{pmatrix}$$

as $i \to \infty$, for some $t'$ with $|t'| \in [\frac{1}{3}t, t]$. By Lemma 10.3, $\mu$ must be preserved by $c$. Since $t > 0$ was arbitrary, it follows that $\mu$ is invariant under a dense subgroup of $C$, and so $\mu$ must be invariant* under all of $C$ (or equivalently, under $C/\Gamma \cap C \cong \mathbb{T}$). As $(\phi_X)_* \mu$ is a $T$-invariant measure on $\mathbb{T}^2$ and $T$ is uniquely ergodic by the hypothesis on $1, \alpha, \beta$, we must have $(\phi_X)_* \mu = m_{\mathbb{T}^2}$. It follows that, for any $f \in C(X)$,

$$\int f \, \mathrm{d}\mu = \int_{\mathbb{T}} \int f(xc) \, \mathrm{d}\mu(x) \, \mathrm{d}m_{\mathbb{T}}(c)$$

$$= \underbrace{\iint f(xc) \, \mathrm{d}m_{\mathbb{T}}(c) \mathrm{d}\mu}_{\phi_X^{-1} \mathscr{B}_{\mathbb{T}^2}\text{-measurable}}$$

$$= \iint f(xc) \, \mathrm{d}m_{\mathbb{T}}(c) \, \mathrm{d}m_X \quad (\text{since } (\phi_X)_* \mu = (\phi_X)_* m_X)$$

$$= \int f \, \mathrm{d}m_X,$$

so $\mu = m_X$ as desired. In order to reach this conclusion, we relied on the assumption that for any $\varepsilon_\ell$ we can find points $x_\ell, x_\ell' \in X_1$ with special properties for this chosen small displacement. Below we will dispense with this assumption, thereby completing the proof.

**Lemma 10.4.** *Fix $\ell \geqslant 1$. If $V$ is a sufficiently small neighborhood of $e \in G$ and $z \in \operatorname{Supp} \mu|_{X_1}$, then for $\mu$-almost every $x \in zV \cap X_1$ there exists an $m \geqslant 1$ such that $x' = S^m x \in zV \cap X_1$, and the displacement*

$$g = \begin{pmatrix} 1 & a & c \\ & 1 & b \\ & & 1 \end{pmatrix}$$

*between $x$ and $x' = xg$ satisfies $|a|, |b|, |c| < \frac{1}{\ell}$ and $a\beta - b\alpha \neq 0$.*

PROOF. The first claim regarding the existence of such an $m$ is just Poincaré recurrence (Theorem 2.11) since $\mu(zV \cap X_1) > 0$ by assumption. Also, the displacement $g$ lies in $V^{-1}V$ (since we may choose $h, h' \in V$ with $x = zh$ and $x' = zh'$, and then $x' = xh^{-1}h'$). If $V$ is sufficiently small this will guarantee that $|a|, |b|, |c| < \frac{1}{\ell}$. Now write $x = \Gamma h$, $x' = S^m x = \Gamma h \tau^m = \Gamma h g$, which shows that $h\tau^m = \gamma h g$. Taking this modulo $C$ (that is, taking the image under $\phi_G$) gives

$$(m\alpha, m\beta) = (a + i, b + j)$$

for some $i, j \in \mathbb{Z}$. If $a\beta - b\alpha = \det \begin{pmatrix} a & b \\ \alpha & \beta \end{pmatrix} = 0$ then there is some $s \in \mathbb{R}$ with $a = s\alpha$ and $b = s\beta$ and hence

---

* This follows from the argument used in the proof on p. 108 that (3) $\implies$ (1) in Theorem 4.14.

$$\frac{\alpha}{\beta} = \frac{(m-s)\alpha}{(m-s)\beta} = \frac{m\alpha - a}{m\beta - b} = \frac{i}{j} \in \mathbb{Q},$$

contradicting the assumption that $1, \alpha, \beta$ are linearly independent over $\mathbb{Q}$. $\square$

This completes the second proof of Theorem 10.1. $\square$

## 10.5  A Non-ergodic Nilrotation

In this section we describe a case where the equivalent conditions of Theorem 10.1 fail. For simplicity, we only consider the case

$$\tau = \begin{pmatrix} 1 & 0 & 0 \\ & 1 & \beta \\ & & 1 \end{pmatrix},$$

which gives the map $S$ the form

$$S\left( \Gamma \begin{pmatrix} 1 & a & c \\ & 1 & b \\ & & 1 \end{pmatrix} \right) = \Gamma \begin{pmatrix} 1 & a & c \\ & 1 & b \\ & & 1 \end{pmatrix} \begin{pmatrix} 1 & 0 & 0 \\ & 1 & \beta \\ & & 1 \end{pmatrix} = \Gamma \begin{pmatrix} 1 & a & c+a\beta \\ & 1 & b+\beta \\ & & 1 \end{pmatrix}.$$

More concretely, assume that $\begin{pmatrix} 1 & a & c \\ & 1 & b \\ & & 1 \end{pmatrix} \in F$, so $0 \leqslant a, b, c < 1$. Then

$$\begin{pmatrix} 1 & 0 & -\lfloor c+a\beta \rfloor \\ & 1 & -\lfloor b+\beta \rfloor \\ & & 1 \end{pmatrix} \begin{pmatrix} 1 & a & c+a\beta \\ & 1 & b+\beta \\ & & 1 \end{pmatrix} = \begin{pmatrix} 1 & a & \{c+a\beta\} \\ & 1 & \{b+\beta\} \\ & & 1 \end{pmatrix}$$

is the coset representative in $F$ of the image of $\begin{pmatrix} 1 & a & c \\ & 1 & b \\ & & 1 \end{pmatrix}$ under $S$. This description shows that for any $a \in [0,1)$ the set

$$X_a = \Gamma \left\{ \begin{pmatrix} 1 & a & c \\ & 1 & b \\ & & 1 \end{pmatrix} \mid b, c \in [0,1) \right\}$$

is invariant under $S$, and that $S$ restricted to $X_a$ is the rotation map

$$(b, c) \longmapsto (b+\beta, c+a\beta) \pmod{\mathbb{Z}^2}$$

on the torus $\mathbb{T}^2$. Depending on whether or not $1, \beta, a\beta$ are linearly independent over $\mathbb{Q}$, this restriction to $X_a$ may or may not be (uniquely) ergodic. To give this observation more structure, notice that

$$X_a = \Gamma \begin{pmatrix} 1 & a & 0 \\ & 1 & 0 \\ & & 1 \end{pmatrix} H$$

is an orbit of the group

$$H = \left\{ \begin{pmatrix} 1 & 0 & c \\ & 1 & b \\ & & 1 \end{pmatrix} \mid b, c \in \mathbb{R} \right\}.$$

These remarks make the following result straightforward to prove (see Exercise 10.5.1).

**Proposition 10.5.** *Let $X = \Gamma \backslash G$, and let*

$$S : x \longmapsto x \begin{pmatrix} 1 & 0 & 0 \\ & 1 & \beta \\ & & 1 \end{pmatrix}$$

*for $x \in X$ be the rotation defined above, with $\beta$ irrational. Then for any $x$ the forward orbit $\overline{\{S^n x \mid n \geqslant 1\}}$ is $xL$ for some closed subgroup $L \subseteq H$ containing*

$$\begin{pmatrix} 1 & 0 & 0 \\ & 1 & \beta \\ & & 1 \end{pmatrix},$$

*the map $S$ restricted to $xL$ is uniquely ergodic, and in particular the point $x$ is generic for the unique $L$-invariant probability measure on $xL$.*

## Exercises for Sect. 10.5

**Exercise 10.5.1.** Prove Proposition 10.5.

**Exercise 10.5.2.** State and prove an analog of Proposition 10.5 for the map

$$S : x \longmapsto x \begin{pmatrix} 1 & \frac{1}{2} & 0 \\ & 1 & \beta \\ & & 1 \end{pmatrix}$$

with $\beta$ irrational.

## 10.6 The General Nilrotation

In this section we discuss generalizations of the Heisenberg group and their corresponding nilsystems. The simplest generalization of the Heisenberg group to higher dimensions is the $(2n + 1)$-dimensional group

$$H_n = \left\{ \begin{pmatrix} 1 & x_1 & x_2 & \cdots & x_n & z \\ & 1 & 0 & \cdots & 0 & y_1 \\ & & \ddots & \ddots & \vdots & \vdots \\ & & & 1 & 0 & y_{n-1} \\ & & & & 1 & y_n \\ & & & & & 1 \end{pmatrix} \,\middle|\, x_1, \ldots, x_n, y_1, \ldots, y_n, z \in \mathbb{R} \right\}.$$

The group $H_n$, which is also called a Heisenberg group, shares many properties with the group $G = H_1$ considered in Sect. 10.1. For example,

$$C = [H_n, H_n] = C_{H_n} = \left\{ \begin{pmatrix} 1 & 0 & \cdots & 0 & z \\ & 1 & 0 & \cdots & 0 \\ & & \ddots & \ddots & \vdots \\ & & & 1 & 0 \\ & & & & 1 \end{pmatrix} \,\middle|\, z \in \mathbb{R} \right\}$$

is again the commutator subgroup and the center of $H_n$, and $H_n$ is still a 2-step nilpotent group.

A different generalization of the Heisenberg group is the (full) upper triangular group

$$U = \left\{ \begin{pmatrix} 1 & a_{12} & \cdots & \cdots & a_{1n} \\ & 1 & a_{23} & \cdots & a_{2n} \\ & & \ddots & & \\ & & & 1 & a_{n-1,n} \\ & & & & 1 \end{pmatrix} \,\middle|\, a_{ij} \in \mathbb{R} \text{ for } j - i \geqslant 1 \right\}.$$

The group $U$ is an $(n-1)$-*step nilpotent group* in the following sense. The *lower central series* or *descending central series* of any group $G$ is the sequence of subgroups

$$G = G_1 \rhd G_2 \rhd \cdots \rhd G_n \rhd \cdots$$

in which $G_{n+1} = [G_n, G]$, the subgroup of $G$ generated by all commutators $[v, u]$ with $v \in G_n$ and $u \in G$, and $G$ is a $k$-step nilpotent group if the lower central series of $G$ terminates in the trivial group in $k$ steps. For $U$, we find that

$$U_2 = [U, U] = \{e + (a_{ij})_{ij} \mid a_{ij} \in \mathbb{R} \text{ and } a_{ij} = 0 \text{ for } j - i < 2\},$$

$$U_3 = [U_2, U] = \{e + (a_{ij})_{ij} \mid a_{ij} \in \mathbb{R} \text{ and } a_{ij} = 0 \text{ for } j - i < 3\},$$

and so on, ending with $U_{(n-1)} = \{e\}$, showing that $U$ is an $(n-1)$-step nilpotent group.

Let

$$\Gamma = \{u = e + (a_{ij})_{ij} \mid a_{ij} \in \mathbb{Z} \text{ but } a_{ij} = 0 \text{ for } j - i < 1\};$$

then $\Gamma$ is a lattice in $U$, and

$$F = \{u = e + (a_{ij})_{ij} \mid a_{ij} \in [0,1) \text{ but } a_{ij} = 0 \text{ for } j - i < 1\}$$

is a fundamental domain for $\Gamma$ in $U$.

Now let $G$ be a $k$-step nilpotent closed linear group, with $\Gamma$ a lattice in $G$, and let $\tau \in G$. We can define the transformation $Sx = x\tau$ for $x \in X = \Gamma\backslash G$, and the topological dynamical system $(X, S)$ is called a *nilsystem*. It turns out that the quotient map

$$\phi_G : G \to G/[G,G] = G^{(\mathrm{ab})}$$

from $G$ onto the abelianization $G^{(\mathrm{ab})}$ of $G$ sends $\Gamma$ into a lattice in the abelian group $G/[G,G]$, and $G^{(\mathrm{ab})}$ takes the form $\mathbb{R}^{k_1} \times \mathbb{Z}^{k_2}$ for some $k_1, k_2 \in \mathbb{N}_0$. Hence it makes sense to ask whether the rotation map $T$ induced by $\phi_G(\tau)$ on $G^{(\mathrm{ab})}/\phi_G(\Gamma) = X^{(\mathrm{ab})}$ is ergodic, and this is characterized by Theorem 4.14.

The next two theorems are not needed later and will not be proved here[92].

**Theorem.** In the notation above, the following are equivalent:

- $T$ is uniquely ergodic;
- $T$ is ergodic with respect to $m_{X^{(\mathrm{ab})}}$, the measure induced by Haar measure on $G^{(\mathrm{ab})}$;
- $S$ is uniquely ergodic; and
- $S$ is ergodic with respect to $m_X$, the measure induced by Haar measure on $G$.

Just as in Sect. 10.5, the non-ergodic case is also highly structured.

**Theorem.** In the notation above, let $x \in X = \Gamma\backslash G$. Then there is a closed subgroup $L \leqslant G$ such that the forward orbit closure

$$\overline{\{S^n x \mid n \geqslant 1\}} = xL$$

is the closed $L$-orbit of $x$, and $x$ is generic with respect to the $S$-ergodic $L$-invariant Haar measure $m_{xL}$ on the orbit $xL$.

# Exercises for Sect. 10.6

**Exercise 10.6.1.** Generalize Theorem 10.1 to the group $H_n$.

**Exercise 10.6.2.** Prove Theorem 10.1 for nilsystems on $\Gamma\backslash U$.

# Notes to Chap. 10

[91] (Page 336) The argument we present is a variation of the *H-principle* introduced by Ratner in a different context, see [299, 300]; see also a paper of Witte Morris [386].

[92] (Page 344) The first of these may be found in the monograph of Auslander, Green and Hahn [10] (which also contains several earlier results in this chapter). The second theorem follows from more general work of Ratner [305]. More recent results in this direction have been shown by Lesigne [226], Leibman [223, 224], and others.

# Chapter 11
# More Dynamics on Quotients of the Hyperbolic Plane

In addition to the geodesic flow, whose study we started in Chap. 9, we introduced the natural action of $\mathrm{PSL}_2(\mathbb{R})$ on $X = \Gamma \backslash \mathrm{PSL}_2(\mathbb{R})$. In this chapter we will study this natural action in more detail. We will show ergodicity[93] of the horocycle flow, mixing of $\mathrm{PSL}_2(\mathbb{R})$, and go on to deduce from the mixing property of the geodesic flow an "almost unique ergodicity" property for the horocycle flow, which we will refer to as an instance of *rigidity of invariant measures*. Finally, we shall use this together with the ergodic decomposition to establish equidistribution for individual orbits of the horocycle flow.

In many ways the horocycle flow is complementary to the geodesic flow considered in Chap. 9. It has already featured in the proof of ergodicity for the geodesic flow, and this link between the two flows will become stronger in this chapter, where we will use them alternately to prove stronger and stronger statements about both flows. We will see that despite this close linkage, the two flows have fundamentally different dynamical behaviors—indeed they are in many senses opposite extremes. For instance, as discussed in Sect. 9.7, the geodesic flow has an abundance of invariant measures, while (as already mentioned) we will see that the horocycle flow exhibits rigidity of invariant measures.

In this chapter we will switch back and forth between a geometric and an algebraic point of view. For the latter we will consider the slightly more general setting of quotients of $\mathrm{SL}_2(\mathbb{R})$ instead of quotients of $\mathrm{PSL}_2(\mathbb{R})$.

## 11.1 Dirichlet Regions

In Chap. 9 we constructed a standard fundamental domain for the discrete group $\mathrm{PSL}_2(\mathbb{Z})$ in the group $\mathrm{PSL}_2(\mathbb{R})$, and used its geometry to understand the relationship between the geodesic flow and the Gauss map. Here we present a generalization[94] of Proposition 9.18 which will give a description of the geometry of fundamental domains for other Fuchsian groups. This de-

M. Einsiedler, T. Ward, *Ergodic Theory*, Graduate Texts in Mathematics 259,
DOI 10.1007/978-0-85729-021-2_11, © Springer-Verlag London Limited 2011

scription will be needed later in the discussion of dynamics on more general quotients.

**Definition 11.1.** Let $Z$ be a locally compact metric space carrying an action of a countable group $\Gamma$ by homeomorphisms. The action is said to be *properly discontinuous* if for any compact set $P \subseteq Z$ the set $\{\gamma \in \Gamma \mid \gamma P \cap P \neq \varnothing\}$ is finite. A measurable set $F \subseteq Z$ is a *fundamental domain* if $|\Gamma z \cap F| = 1$ for all $z \in Z$. An open set $F \subseteq Z$ is called an *open fundamental domain* for the action if

(1) if $g_1 \neq g_2$ then $g_1 F \cap g_2 F = \varnothing$, and
(2) $\bigcup_{g \in G} g\overline{F} = Z$.

Thus, for example, the interior of the set $F$ in Proposition 9.18 is an open fundamental domain* for the natural action of $\mathrm{PSL}_2(\mathbb{Z})$ on $\mathbb{H}$.

Recall from Definition 9.17 that a Fuchsian group is a discrete subgroup $\Gamma \subseteq \mathrm{PSL}_2(\mathbb{R})$. Write $d$ for the hyperbolic metric as in Sect. 9.1.

**Lemma 11.2.** *An infinite subgroup $\Gamma \subseteq \mathrm{PSL}_2(\mathbb{R})$ is a Fuchsian group if and only if its action on $\mathbb{H}$ is properly discontinuous.*

PROOF. If $\Gamma$ is not discrete then we may choose a sequence of elements $(g_n)$ with $g_n \neq e$ for all $n \geqslant 1$ and $g_n \to e$ as $n \to \infty$. If $P$ is a compact set containing an open set, then $g_n P \cap P \neq \varnothing$ for all large $n$, showing that the action of $\Gamma$ is not properly discontinuous.

Conversely, assume that $\Gamma$ is discrete. Then $\{g \in \Gamma \mid gP \cap P \neq \varnothing\}$ will be finite for any compact $P$ if the set $B = \{g \in \mathrm{SL}_2(\mathbb{R}) \mid gP \cap P \neq \varnothing\}$ is compact. (This follows easily from Exercise 9.2.2, but we give a concrete argument here which effectively solves it.) Since the set $P$ is compact, $B$ is certainly closed, so it is enough to show that $B$ is a bounded set in $\mathrm{SL}_2(\mathbb{R})$ when viewed as a subset of $\mathbb{R}^4$.

By compactness, there are constants $R, \varepsilon > 0$ such that every $w \in P$ has $|w| \leqslant R$ and $\Im(w) \geqslant \varepsilon$. It follows that if $g = \begin{pmatrix} a & b \\ c & d \end{pmatrix} \in B$ (and so $gz \in P$ for some $z \in P$), then $\left| \frac{az+b}{cz+d} \right| \leqslant R$ and $\Im\left(\frac{az+b}{cz+d}\right) = \frac{\Im(z)}{|cz+d|^2} \geqslant \varepsilon$ by (9.15). Thus

$$|cz + d|^2 \leqslant \frac{1}{\varepsilon}\Im(z) \leqslant \frac{R}{\varepsilon}$$

and

$$|az + b|^2 \leqslant R^2|cz + d|^2 \leqslant \frac{R^3}{\varepsilon}.$$

Since $z$ belongs to some fixed compact subset of $\mathbb{H}$, this readily implies that the coefficients in the matrices of $B$ lie in a bounded subset of $\mathbb{R}^4$.    □

---

* Just as in the discussion after Proposition 9.18, we will be interested in cases where $F$ does not differ from a fundamental domain $F'$ too much. In the case we will consider we will only need to take the union of $F$ with some subset of the lower-dimensional boundary $\partial F$ to obtain $F'$.

**Definition 11.3.** Let $\Gamma$ be an infinite Fuchsian group, and let $p \in \mathbb{H}$ be a point not fixed by any element of $\Gamma$ other than the identity. Then the set

$$D = D(p) = \{z \in \mathbb{H} \mid \mathsf{d}(z,p) < \mathsf{d}(z,\gamma p) \text{ for all } \gamma \in \Gamma \smallsetminus \{e\}\}$$

is called a *Dirichlet region* for $\Gamma$.

Notice that Dirichlet regions always exist; since a Fuchsian group is countable and any non-trivial element can only fix at most two points in $\mathbb{C}$ (indeed, at most one in $\mathbb{H}$), there must be points in $\mathbb{H}$ not fixed by any non-identity element. Moreover, a Dirichlet region is the intersection of the hyperbolic half planes

$$\{z \in \mathbb{H} \mid \mathsf{d}(z,p) < \mathsf{d}(z,\gamma p)\}$$

for each $\gamma \in \Gamma \smallsetminus \{e\}$ (see Lemma 11.4 for a justification of the terminology).

**Lemma 11.4.** *For any $\gamma \in \mathrm{PSL}_2(\mathbb{R})$ the open set*

$$D_\gamma = \{z \in \mathbb{H} \mid \mathsf{d}(z,p) < \mathsf{d}(z,\gamma p)\}$$

*is the connected component of $\mathbb{H} \smallsetminus L_\gamma$ containing $p$, where $L_\gamma$ is the geodesic in $\mathbb{H}$ defined by the equation*

$$\mathsf{d}(z,p) = \mathsf{d}(z,\gamma p).$$

*It follows that a Dirichlet region is connected and convex in the sense that the hyperbolic geodesic path joining any two points in $D$ lies entirely inside $D$.*

PROOF. To see that $L_\gamma$ is a geodesic and the description of $D_\gamma$ is valid, notice that both depend only on the points $p$ and $\gamma p$. We may apply an isometry $g \in \mathrm{PSL}_2(\mathbb{R})$ to map those two points to $-r+\mathrm{i}$ and $r+\mathrm{i}$ respectively; choosing $r$ suitably we can ensure that $\mathsf{d}(-r+\mathrm{i}, r+\mathrm{i}) = \mathsf{d}(p, \gamma p)$, and then the existence of $g$ follows from Proposition 9.4. However, the set of points in $\mathbb{H}$ equidistant from the points $-r+\mathrm{i}$ and $r+\mathrm{i}$ is precisely the upper half of the $y$-axis. Clearly, $D_\gamma$ is convex (which is again easily seen in the case $p = -r+\mathrm{i}$ and $\gamma p = r+\mathrm{i}$) and the intersection of convex sets is again convex. Therefore $D$ is convex as claimed. $\square$

**Lemma 11.5.** *Any Dirichlet region for an infinite Fuchsian group $\Gamma$ is an open fundamental domain for the action of $\Gamma$ on $\mathbb{H}$. The boundary of a Dirichlet region is made up of geodesic segments contained in geodesics defined by*

$$L_\gamma = \{z \in \mathbb{H} \mid \mathsf{d}(z,p) = \mathsf{d}(z,\gamma p)\}$$

*for $\gamma \in \Gamma \smallsetminus \{e\}$.*

PROOF. Let $D = D(p)$ be a Dirichlet region. Since the action of $\Gamma$ is properly discontinuous by Lemma 11.2, we have that for any $z \in \mathbb{H}$ there are only finitely many $\gamma \in \Gamma$ with

$$d(\gamma z, p) \leqslant d(z, p) + 1,$$

say $\gamma_1, \ldots, \gamma_n$. Note that for $z' \in B_{1/2}(z)$, this list of elements will include those $\gamma \in \Gamma$ for which $d(\gamma z', p) \leqslant d(z', p)$. In particular, there is some $w \in \Gamma z$ with

$$d(w, p) \leqslant d(\gamma z, p) = d(z, \gamma^{-1} p)$$

for all $\gamma \in \Gamma$. Without loss of generality, assume that $z = w$. If $z \in D$ then a point $z'$ close to $z$ belongs to $D_{\gamma_1} \cap \cdots \cap D_{\gamma_n}$ by Lemma 11.4, so that $z' \in D$ and hence $D$ is open. If $z \notin D$ then $z$ belongs to some of the boundaries of the sets $D_{\gamma_1}, \ldots, D_{\gamma_n}$—assume that

$$z \in L_{\gamma_1} \cap \cdots \cap L_{\gamma_m} \cap D_{\gamma_{m+1}} \cap \cdots \cap D_{\gamma_n}.$$

Then by Lemma 11.4 the geodesic path $\phi$ joining $z$ to $p$ will have $\phi(0) = z \notin D$ but $\phi(t) \in D$ for $t \in (0, 1]$, so $z \in \overline{D}$.

To see property (1) of Definition 11.1, assume that points $z$ and $w$ in $D$ have $z = \gamma w$ for some $\gamma \in \Gamma \smallsetminus \{e\}$. Then

$$d(w, p) < d(w, \gamma^{-1} p) = d(z, p)$$

and

$$d(z, p) < d(z, \gamma p) = d(w, p),$$

which is a contradiction. $\square$

As before, we write

$$\overline{\mathbb{H}} = \mathbb{H} \cup \partial \mathbb{H}$$

for the union of $\mathbb{H}$ with its boundary $\partial \mathbb{H} = \mathbb{R} \cup \{\infty\}$.

Let $D$ be a subset of $\mathbb{H}$. We will also sometimes write $\partial^{\overline{\mathbb{H}}} D$ and $\overline{D}^{\overline{\mathbb{H}}}$ for boundaries and closures taken in the set $\overline{\mathbb{H}}$. Given a finite set of points in $\overline{\mathbb{H}}$, we can define a convex polygon by successively taking points on geodesic paths connecting two vertices or points obtained earlier. The smallest set of points that can be used to define a given polygon is the set of vertices. Alternatively, given a unit-speed parametrization of a subset of $\partial D$, we will refer to the points where the parametrization is not locally along a geodesic as *vertices* of $D$. Moreover, any point of $\overline{D}^{\overline{\mathbb{H}}} \cap \partial \mathbb{H}$ will be called a vertex also.

**Lemma 11.6.** *The boundary of a Dirichlet region $\partial D$ is the union of at most countably many connected components. Each connected component of $\partial D$ is the image of a piecewise geodesic path $\phi : \mathbb{R} \to \mathbb{H}$. Either $D$ is a convex polygon and this path periodically traverses $\partial D$, or any such path connects two (not necessarily distinct) points of (vertices of $D$ in) $\partial \mathbb{H}$.*

PROOF. Fix some $R > 1$ and write $B_R$ for the hyperbolic ball of radius $R$ around $p$, the chosen point defining $D$. Then we may find finitely many elements $\gamma_0 = I, \gamma_1, \ldots, \gamma_n \in \Gamma$ such that if $z \in B_R$ and $\gamma(z) \in B_R$ for some $\gamma \in \Gamma$ then $\gamma = \gamma_i$ for some $i, 0 \leqslant i \leqslant n$ (since the action of $\Gamma$ is properly discontinuous). This implies that

$$D \cap B_R = \bigcap_{i=1}^{n} D_{\gamma_i} \cap B_R$$

and

$$(\partial D) \cap B_R = \left( \partial \bigcap_{i=1}^{n} D_{\gamma_i} \right) \cap B_R.$$

It follows by induction on $n$ that $\partial \bigcap_{i=1}^{n} D_{\gamma_i}$ can be described as in the lemma. If for some $R > 0$ we have $D \subseteq B_R$ then $D = \bigcap_{i=1}^{n} D_{\gamma_i}$ is a convex polygon.

If not, then letting $R \to \infty$ (increasing $n$ as needed) we see that for every $R$ the boundary $(\partial D) \cap B_R$ is as claimed. Notice that as $R$ increases, the number of connected components of $(\partial D) \cap B_R$ can increase (when a new $D_{\gamma_i}$ is needed to describe $D \cap B_R$ and its boundary does not connect to the previous boundary pieces) or decrease (when the boundary pieces contained in some $L_{\gamma_i}$ and some $L_{\gamma_j}$ connect).

Fix some $z \in \partial D$ and let $\phi_R$ be a piecewise geodesic parametrization of the boundary component of $\bigcap_{i=1}^{n} D_{\gamma_i}$ normalized to have $\phi_R(0) = z$. As $R \to \infty$ the paths $\phi_R$ converge to some path $\phi$—indeed for fixed $t$, $\phi_R(t) = \phi(t)$ when $R$ is sufficiently large. After applying some isometry we may assume without loss of generality that $z = i$, part of the boundary of $D$ is a segment in the imaginary axis, and $D \subseteq \{z \mid \Re(z) > 0\}$ (see Fig. 11.1). Then it follows that $\Re(\phi(t))$ is eventually monotone, both for $t \to \infty$ and for $t \to -\infty$. Therefore, the limits $\lim_{t \to \infty} \phi(t)$ and $\lim_{t \to -\infty} \phi(t)$ taken in $\overline{\mathbb{H}}$ exist and belong to $\partial \mathbb{H}$.                                                                □

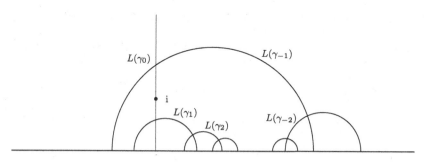

**Fig. 11.1** A possible (and representative) scenario in the proof of Lemma 11.6

As mentioned above, the sides of a Dirichlet region are segments of geodesics. A much less obvious result is that a Dirichlet region for a lattice in $\mathrm{PSL}_2(\mathbb{R})$ is in fact a hyperbolic polygon.

**Theorem 11.7.** *A Dirichlet region for a lattice in* $\mathrm{PSL}_2(\mathbb{R})$ *has finitely many sides. That is, it is a convex hyperbolic polygon.*

Before proving this result, we show how the hyperbolic area form from Lemma 9.16 gives a simple classical formula for the hyperbolic area of a polygon (a region bounded by finitely many geodesics).

**Proposition 11.8.** [GAUSS–BONNET FORMULA] *Let* $P$ *be a hyperbolic* $n$-*sided convex polygon in* $\mathbb{H}$ *with* $n \geqslant 3$ *vertices in* $\overline{\mathbb{H}}$, *with angles* $\alpha_1, \ldots, \alpha_n$ *at the* $n$ *vertices. Then the hyperbolic area of* $P$ *is*

$$(n-2)\pi - (\alpha_1 + \cdots + \alpha_n).$$

Here we measure angles between geodesics intersecting in $\mathbb{H}$ using the inner product at the intersection point; equivalently this is the angle in $\mathbb{C}$ between the circles (one of which might be a line) at the intersection point. For a vertex in $\partial\mathbb{H}$ we set the angle to be zero since the circles are tangential there.

PROOF OF PROPOSITION 11.8. Assume for the purposes of an induction that the formula holds for all polygons with no more than $(n-1)$ sides, and let $P$ be a polygon with $n$ sides as in the statement of the lemma. By cutting off one triangle $T$ (see Fig. 11.2) to leave an $(n-1)$-gon $Q$ we see that

$$
\begin{aligned}
\mathrm{Area}(P) &= \mathrm{Area}(Q) + \mathrm{Area}(T) \\
&= (n-3)\pi - (\alpha_1 + \cdots + \alpha_{n-3} + (\alpha_{n-2} - \beta_2) + (\alpha_{n-1} - \beta_1)) \\
&\quad + \pi - (\beta_1 + \beta_2 + \alpha_n) \\
&= (n-2)\pi - (\alpha_1 + \cdots + \alpha_n),
\end{aligned}
$$

showing that the result reduces to the case of a triangle.

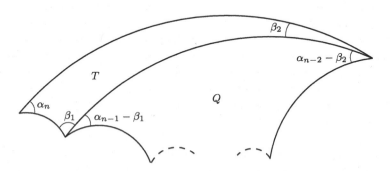

**Fig. 11.2** Decomposing a hyperbolic $n$-gon into a triangle and an $(n-1)$-gon

It is therefore enough to show that the formula holds when $n = 3$, so let $T$ be a triangle with angles $\alpha_1, \alpha_2, \alpha_3$. If $T$ has a vertex $z_1$ at infinity (this means the corresponding angle is zero), then we may apply a suitable transformation from $\mathrm{PSL}_2(\mathbb{R})$ to place the other two vertices on the unit circle (Lemma 9.16 shows that such a transformation preserves area) obtaining the triangle shown in Fig. 11.3. Then

$$\mathrm{Area}(T) = \int_{\cos(\pi-\alpha_2)}^{\cos(\alpha_3)} \left( \int_{\sqrt{1-x^2}}^{\infty} \frac{dy}{y^2} \right) dx = \pi - (\alpha_2 + \alpha_3).$$

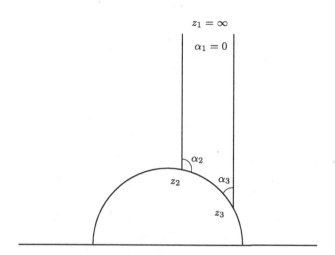

**Fig. 11.3**   A hyperbolic triangle with one vertex at infinity

Now assume that all three vertices lie in $\mathbb{H}$ and assume without loss of generality that $z_1$ and $z_2$ do not have the same real part. Continue the geodesic through $z_1$ and $z_2$ to meet the real axis at $w$ (a point at infinity), as shown in Fig. 11.4. Then the triangle $z_1 z_3 w$ has interior angles $\pi - \alpha_1, \beta, 0$ while the triangle $z_2 z_3 w$ has interior angles $\alpha_2, \alpha_3 + \beta, 0$.
  Thus

$$\mathrm{Area}(z_1 z_2 z_3) = \mathrm{Area}(z_2 z_3 w) - \mathrm{Area}(z_1 z_3 w)$$
$$= \pi - (\alpha_2 + \alpha_3 + \beta) - (\pi - (\pi - \alpha_1 + \beta))$$
$$= \pi - (\alpha_1 + \alpha_2 + \alpha_3),$$

showing the formula holds for a triangle.                                    □

PROOF OF THEOREM 11.7. Recall that the Haar measure on $\mathrm{PSL}_2(\mathbb{R})$ is (under the isomorphism to $\mathrm{T}^1\mathbb{H}$) given by

$$dm = \frac{1}{y^2}\, dx\, dy\, d\theta,$$

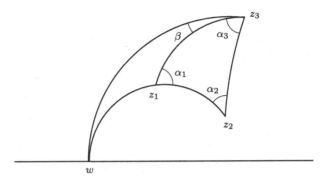

**Fig. 11.4** A hyperbolic triangle with all vertices in $\mathbb{H}$

where $\theta \in [0, 2\pi)$ corresponds to the angle of the unit vector, and

$$\mathrm{d}A = \frac{1}{y^2}\,\mathrm{d}x\,\mathrm{d}y$$

is the hyperbolic area form (see Lemma 9.16 and Proposition 9.19). This shows that a Dirichlet region $D = D(p)$ (which is a fundamental domain by Lemma 11.5) for a lattice $\Gamma$ must have finite hyperbolic area.

Recall that the interior angle of a vertex at infinity is zero. We claim that the Gauss–Bonnet formula (Proposition 11.8) shows that the number of vertices at infinity cannot exceed $\frac{1}{\pi}\mathrm{Area}(D) + 2$, and in particular is finite. For if we had more points on the closure of $D$ (taken in $\overline{\mathbb{H}}$) that lie on $\partial\mathbb{H}$ we could take such points $z_1, \dots, z_n$ with $n > \frac{1}{\pi}\mathrm{Area}(D) + 2$ and consider the convex polygon $P$ generated by them. By the Gauss–Bonnet formula we have $\mathrm{Area}(P) = (n-2)\pi > \mathrm{Area}(D)$. To obtain a contradiction, approximate each $z_i$ by some $w_i \in D$, so that the resulting polygon $Q$ generated by the $w_i$ will satisfy $Q \subseteq D$ by convexity of $D$ and have $\mathrm{Area}(Q) > \mathrm{Area}(D)$, which is impossible. Below we will ignore the boundaries for arguments like this one, and simply say that $P$ is essentially contained in $D$.

We proceed by again using the Gauss–Bonnet formula to describe the boundary $\partial D$ of $D$ that is contained in $\mathbb{H}$. For example, it follows that the boundary of $D$ lying in $\mathbb{H}$ can only have finitely many connected components. Let $C$ be one of the finitely many connected components of $\partial D$. Pick an arbitrary point in $C$ and a direction. Let $x_1, x_2, \dots$ be the vertices in $C$ along that chosen direction, and let $\omega_1, \omega_2, \dots$ be the corresponding internal angles. Of course we would like to show that there can be only finitely many vertices. As a first step towards this, we claim that all but finitely many of the angles must be close to $\pi$. That is, $\partial D$ is almost straight at most of its vertices.

Write $T_k$ for the triangle with vertices $x_k, x_{k+1}, p$ and write $\alpha_k, \beta_k, \varepsilon_k$ for the respective internal angles as shown in Fig. 11.5. Thus $\omega_k = \beta_{k-1} + \alpha_k$

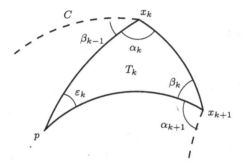

**Fig. 11.5**  A connected component $C$ of $\partial(D)$

for $1 \leqslant k$. Since $D$ is convex, for any $n$ the triangles $T_1, \ldots, T_{n-1}$ all lie essentially in $D$ and are essentially disjoint, so

$$\sum_{k=1}^{n-1} \text{Area}(T_k) = \sum_{k=1}^{n-1} \pi - (\alpha_k + \beta_k + \varepsilon_k)$$

$$= \pi - \alpha_1 - \beta_{n-1} + \sum_{k=2}^{n-1} (\pi - \omega_k) - \sum_{k=1}^{n-1} \varepsilon_k$$

$$\leqslant \text{Area}(D) < \infty.$$

Now the angles $\varepsilon_k$ are those for disjoint arcs at $p$, and so $\sum_{k=1}^{n-1} \varepsilon_k \leqslant 2\pi$. Therefore

$$\sum_{k=1}^{n-1} (\pi - \omega_k) \leqslant \alpha_1 + \beta_{n-1} + \pi + \text{Area}(D)$$

is uniformly bounded. This establishes our claim. In particular, it follows that we can only have

$$(\pi - \omega_k) > \tfrac{\pi}{4}$$

for finitely many $k$, so

$$\left| \{ k \mid \omega_k \leqslant \tfrac{3\pi}{4} \} \right| < \infty. \tag{11.1}$$

Since we have already established that there are only finitely many connected components of $\partial D$ this holds across all vertices of $\partial D$. We will now use this claim twice to establish that there are only finitely many vertices.

We show next that any vertex of $D$ has only finitely many other vertices of $D$ in its $\Gamma$-orbit. Choose a vertex $x_k$ of $D$ in $\mathbb{H}$, let $\gamma_1 x_k = x_k, \gamma_2 x_k, \ldots$ (with $\gamma_1 = I$, $\gamma_i \in \Gamma$, and $\gamma_i x_k \neq \gamma_j x_k$ for $i \neq j$) be the vertices of $D$ that lie on the orbit of $x_k$ under the action of $\Gamma$, and let the internal angles of $D$ at these vertices be $\delta_1, \delta_2, \ldots$.

For any $j \geqslant 1$, $x_k$ is a vertex of $\gamma_j^{-1} D$ with internal angle $\delta_j$. The sets

$$D = \gamma_1^{-1} D, \gamma_2^{-1} D, \ldots$$

are pairwise disjoint since the Dirichlet region $D$ is an open fundamental domain by Lemma 11.5. Hence

$$\delta_1 + \delta_2 + \cdots \leqslant 2\pi. \tag{11.2}$$

It follows from (11.1) that the number of vertices

$$x_k = \gamma_1 x_k, \gamma_2 x_k, \ldots, \gamma_n x_k$$

on the $\Gamma$-orbit of $x_k$ must be finite.

We finish the proof by showing that every class of these finite subsets of orbits described above has to contain at least one vertex whose internal angle is smaller than $\frac{2\pi}{3} < \frac{3\pi}{4}$. Together with the bound (11.1) this then implies that there are only finitely many vertices of $D$.

To do this, choose again a vertex $x_k$, let $\gamma_1, \ldots, \gamma_n$ be as above, and let

$$\Gamma_k = \{ \gamma \in \Gamma \mid \gamma x_k = x_k \}$$

be the stabilizer of $x_k$ in $\Gamma$. The group $\Gamma_k$ is conjugate to a discrete subgroup of the compact group $\mathrm{PSO}(2)$ and so must be finite. Using $\Gamma_k$ we now refine (11.2) as follows. The images of $D$ under the action of $\Gamma$ which have $x_k$ as a vertex are precisely those of the form $\gamma \gamma_i^{-1} D$ for all $\gamma \in \Gamma_k$ and $1 \leqslant i \leqslant n$. Since $D$ is a fundamental domain, these are all disjoint and together they essentially cover a neighborhood of $x_k$, so

$$|\Gamma_k|(\delta_1 + \cdots + \delta_n) = 2\pi.$$

Since $0 < \delta_j < \pi$ for $1 \leqslant j \leqslant n$, we must have $n \geqslant 3$, and hence $\delta_j \leqslant \frac{2\pi}{3}$ for some $j$. As explained above, this implies the desired statement. □

The geometry of a Dirichlet region for a lattice detects whether or not the lattice is uniform.

**Lemma 11.9.** *A lattice* $\Gamma \subseteq \mathrm{PSL}_2(\mathbb{R})$ *is uniform (that is,* $\Gamma \backslash \mathrm{PSL}_2(\mathbb{R})$ *is compact) if and only if every vertex of any Dirichlet region for* $\Gamma$ *lies in* $\mathbb{H}$ *(that is, has compact closure).*

PROOF. Let $D$ be a Dirichlet region for $\Gamma$. If the boundary of $D$ lies in $\mathbb{H}$, then the closure of $D$ is a compact subset of $\mathbb{H}$. The compact subset

$$F = \{ g \in \mathrm{PSL}_2(\mathbb{R}) \mid g(\mathrm{i}) \in \overline{D} \}$$

then maps continuously onto $\Gamma \backslash \mathrm{PSL}_2(\mathbb{R})$, so $\Gamma$ is uniform.

Conversely, assume that $\Gamma$ is uniform. By Proposition 9.14 there is, for every $x \in X$, a neighborhood $B_r^X(x)$ which is the isometric image of the neighborhood $B_r^G(g)$, where $x = \Gamma g$. By compactness of $X = \Gamma \backslash \mathrm{PSL}_2(\mathbb{R})$ there is a finite subcover of the open cover defined by the set $B_{r=r(x)}^X(x)$, and so there exists some bounded subset

$$B \subseteq \mathrm{PSL}_2(\mathbb{R})$$

which maps onto $X$. Let $p \in \mathbb{H}$ be not fixed by any non-trivial element of $\Gamma$, and let $D = D(p)$ be the Dirichlet region defined by $p$. Then for any $z \in \mathbb{H}$ there is some $g \in \mathrm{PSL}_2(\mathbb{R})$ with $g(p) = z$, some $\gamma \in \Gamma$ with $\gamma g \in B$, and hence $\gamma(z) \in B(p)$. By the argument in the proof of Lemma 11.2 (that is, the properness of the $\mathrm{PSL}_2(\mathbb{R})$-action on $\mathbb{H}$), the set $B(p) \subseteq \mathbb{H}$ has compact closure. It follows that the vertices of $D$ lie in $\mathbb{H}$.                    □

Let $D$ be a Dirichlet domain for a lattice $\Gamma$ in $\mathrm{PSL}_2(\mathbb{R})$. Points of the closure of $D$ taken in $\overline{\mathbb{H}}$ that belong to $\partial \mathbb{H}$ (that is, those that are elements of $\overline{D}^{\overline{\mathbb{H}}} \cap \partial \mathbb{H}$) are called *cusps*. Thus, for example, if $\Gamma = \mathrm{PSL}_2(\mathbb{Z})$ then $\infty \in \partial \mathbb{H}$ is a cusp. More precisely, if there are several points in $\overline{D}^{\overline{\mathbb{H}}} \cap \partial \mathbb{H}$ then we identify points on the same $\Gamma$-orbit (for the natural action of $\Gamma \subseteq \mathrm{PSL}_2(\mathbb{R})$ on $\partial \mathbb{H} = \overline{\mathbb{R}}$)—cusps are then the equivalence classes. We will see an example where this distinction is important in the next section.

# Exercises for Sect. 11.1

**Exercise 11.1.1.** Show that a lattice in $\mathrm{PSL}_2(\mathbb{R})$ is finitely generated. Hint: Choose a Dirichlet region $D$ for the lattice, use Theorem 11.7 to show that there are only finitely many elements $\gamma$ with $\overline{D} \cap \gamma \overline{D} \neq \varnothing$, and finally show that these elements must generate the lattice.

**Exercise 11.1.2.** Show that the fundamental domain constructed in Sect. 9.4 for $\mathrm{PSL}_2(\mathbb{Z})$ in $\mathrm{PSL}_2(\mathbb{R})$ is a Dirichlet domain.

# 11.2 Examples of Lattices

In order to avoid the impression that $\mathrm{PSL}_2(\mathbb{Z})$ is the only interesting lattice in $\mathrm{PSL}_2(\mathbb{R})$, in this section we will discuss some other lattices.

Notice first that the canonical map $\mathrm{SL}_2(\mathbb{R}) \to \mathrm{PSL}_2(\mathbb{R}) = \mathrm{SL}_2(\mathbb{R})/\{\pm I_2\}$ is two-to-one and the push-forward of the Haar measure on $\mathrm{SL}_2(\mathbb{R})$ under this map gives the Haar measure on $\mathrm{PSL}_2(\mathbb{R})$. As before, we may identify the Haar measure on $\mathrm{PSL}_2(\mathbb{R})$ with the measure $m$ described in Sect. 9.4.1 (which we use to normalize the measure on $\mathrm{SL}_2(\mathbb{R})$). It follows that we can determine whether a discrete subgroup of $\mathrm{SL}_2(\mathbb{R})$ is a lattice by analyzing a fundamental region for its action on $\mathbb{H}$. In particular, it follows that any lattice in $\mathrm{PSL}_2(\mathbb{R})$ gives a lattice in $\mathrm{SL}_2(\mathbb{R})$ by taking the pre-image under the map $\mathrm{SL}_2(\mathbb{R}) \to \mathrm{PSL}_2(\mathbb{R})$.

### 11.2.1 Arithmetic and Congruence Lattices in $\mathrm{SL}_2(\mathbb{R})$

Notice that any discrete subgroup of $\mathrm{SL}_2(\mathbb{R})$ which contains a lattice is itself a lattice. Moreover, a finite-index subgroup $\Lambda$ of a lattice $\Gamma$ is also a lattice, since in this case the union of finitely many copies of a fundamental domain for $\Gamma$ will form a fundamental domain for $\Lambda$. A *principal congruence lattice* of $\mathrm{SL}_2(\mathbb{R})$ is a discrete subgroup of $\mathrm{SL}_2(\mathbb{R})$ of the form

$$\Gamma(N) = \left\{ \begin{pmatrix} a & b \\ c & d \end{pmatrix} \in \mathrm{SL}_2(\mathbb{Z}) \mid a \equiv d \equiv 1, c \equiv b \equiv 0 \quad (\mathrm{mod}\ N) \right\}$$

for some $N \geqslant 1$. A *congruence lattice* is a lattice that contains a principal congruence lattice. A discrete subgroup $\Lambda$ with the property that $\Lambda \cap \mathrm{SL}_2(\mathbb{Z})$ has finite index in both $\Lambda$ and in $\mathrm{SL}_2(\mathbb{Z})$ is called an *arithmetic lattice* of $\mathrm{SL}_2(\mathbb{R})$ (equivalently, a lattice is called arithmetic if it has the property that $\Lambda \cap \mathrm{SL}_2(\mathbb{Z})$ is also a lattice).

We note that there are other arithmetic and congruence lattices that are not constructed from $\mathrm{SL}_2(\mathbb{Z})$ but from other types of integer lattices (see also Exercise 11.6.3).

*Example 11.10.* The subgroup $\Gamma(2)$ described above has index 6 in $\mathrm{SL}_2(\mathbb{Z})$ since $\mathrm{SL}_2(\mathbb{Z})/\Gamma(2) \cong \mathrm{SL}_2(\mathbb{F}_2)$ has order 6. The subgroup

$$\Gamma_0(2) = \left\{ \begin{pmatrix} a & b \\ c & d \end{pmatrix} \in \mathrm{SL}_2(\mathbb{Z}) \mid c \equiv 0 \quad (\mathrm{mod}\ 2) \right\}$$

is a congruence lattice of index 3 in $\mathrm{SL}_2(\mathbb{Z})$. The index may be seen by applying the orbit-stabilizer theorem to the natural linear action of $\mathrm{SL}_2(\mathbb{Z})$ on the vector space $\mathbb{F}_2^2$; $\Gamma_0(2)$ is the stabilizer of $\left(\begin{smallmatrix} 1 \\ 0 \end{smallmatrix}\right) \in \mathbb{F}_2^2$, which has an orbit of size 3.

### 11.2.2 A Concrete Principal Congruence Lattice of $\mathrm{SL}_2(\mathbb{R})$

Let $D$ be the convex 4-gon spanned by the points $-1, 0, 1, \infty \in \partial\mathbb{H}$ as in Fig. 11.6.

**Lemma 11.11.** *The image of the lattice $\Gamma(2)$ (cf. Example 11.10) in $\mathrm{PSL}_2(\mathbb{R})$ is freely generated by $\left(\begin{smallmatrix} 1 & 2 \\ 0 & 1 \end{smallmatrix}\right)$ and $\left(\begin{smallmatrix} 1 & 0 \\ 2 & 1 \end{smallmatrix}\right)$, and $D$ (as in Fig. 11.6) is its Dirichlet domain for the point $p = i$.*

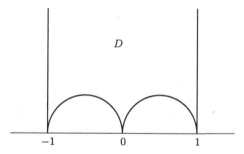

**Fig. 11.6** The convex polygon $D$

The proof that $\Gamma(2)/\{\pm I\}$ is a free group is based on a simple version of the so-called *ping-pong lemma*. Notice that the fact that $\Gamma(2)$ is a lattice with $D$ as its fundamental region in $\mathbb{H}$ can also be deduced from Proposition 9.18 by analyzing $\Gamma(2)\backslash \mathrm{SL}_2(\mathbb{R})$. We will give an independent proof here.

PROOF OF LEMMA 11.11. We begin by analyzing the action of $A = \begin{pmatrix} 1 & 2 \\ 0 & 1 \end{pmatrix}$ and $B = \begin{pmatrix} 1 & 0 \\ 2 & 1 \end{pmatrix}$ on $\mathbb{H}$ with respect to the partition of $\mathbb{H}$ into $D$ and the four connected components of $\mathbb{H}\backslash D$, labeled as in Fig. 11.7.

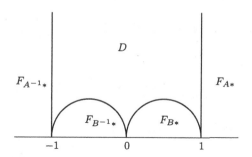

**Fig. 11.7** Connected components of $\mathbb{H}\smallsetminus D$

A calculation shows that

$$A\left(\mathbb{H}\smallsetminus F_{A^{-1}*}\right) \subseteq F_{A*}, \quad A^{-1}\left(\mathbb{H}\smallsetminus F_{A*}\right) \subseteq F_{A^{-1}*} \tag{11.3}$$

since the action of $A$ on $\mathbb{H}$ is given by $A(z) = z + 2$. Similarly,

$$B\left(\mathbb{H}\smallsetminus F_{B^{-1}*}\right) \subseteq F_{B*}, \quad B^{-1}\left(\mathbb{H}\smallsetminus F_{B*}\right) \subseteq F_{B^{-1}*}. \tag{11.4}$$

This can be checked either by calculating the images of the geodesics (which may be done by finding the images of the end points of the geodesics in $\partial\mathbb{H}$)

under $B$ and $B^{-1}$, or by conjugating with $\left(\begin{smallmatrix} 0 & -1 \\ 1 & 0 \end{smallmatrix}\right)$, thereby reducing (11.4) to the case of (11.3).

Now let $w = A^{i_0} B^{j_1} A^{i_1} B^{j_2} \cdots A^{i_m} B^\ell$ be any reduced word in the generators $A$ and $B$ of $\Gamma_{r,s}$. We claim that the image $w(D)$ of $D$ uniquely determines the exponents $i_0, j_1, i_1, j_2, \ldots, i_m, \ell \in \mathbb{Z}$, which implies that the group $\langle A, B \rangle$ is freely generated. To prove the claim it is sufficient to show that $w(D) \subseteq F_{A^{\pm 1}{}_*}$ if and only if the reduced word begins on the left with a positive power of $A^{\pm 1}$ respectively, and similarly for $B$. For words of length one (that is, for the generators $A, B$ and their inverses), this follows from (11.3) and (11.4). Using these equations again in an induction argument then proves the claim.

By the Gauss–Bonnet formula (Proposition 11.8) the four-gon $D$ has finite volume. We claim that $D$ is a Dirichlet domain in $\mathbb{H}$ for the action of $\langle A, B \rangle$. Note that the vertical line $x = -1$ (or $x = 1$) coincides with the line $L_\gamma$ in Lemma 11.5 where $\gamma = A^{-1}$ (or $\gamma = A$ respectively) for the point $p = $ i. Similarly, one may check that the other two geodesic boundaries of $D$ are of the form $L_\gamma$ for suitable $\gamma$. It follows that $D$ is contained in the Dirichlet domain for $\langle A, B \rangle$ and $p = $ i. Moreover, the argument above shows that no two interior points of $D$ are images of one another under the action of $\langle A, B \rangle$, so $D$ is the Dirichlet region as required.

It remains to show that $A$, $B$ and $-I$ together generate $\Gamma(2)$. This may be seen by the following version of the Euclidean algorithm. Let

$$\gamma = \begin{pmatrix} a & b \\ c & d \end{pmatrix} \in \Gamma(2).$$

Notice that $a$ is odd, $c$ is even,

$$\begin{pmatrix} 1 & 2 \\ 0 & 1 \end{pmatrix} \begin{pmatrix} a & b \\ c & d \end{pmatrix} = \begin{pmatrix} a + 2c & b + 2d \\ c & d \end{pmatrix}, \tag{11.5}$$

and

$$\begin{pmatrix} 1 & 0 \\ 2 & 1 \end{pmatrix} \begin{pmatrix} a & b \\ c & d \end{pmatrix} = \begin{pmatrix} a & b \\ c + 2a & d + 2b \end{pmatrix}. \tag{11.6}$$

If $|a| > |c|$ then (11.5) can be used repeatedly to find $A^n \gamma = \left(\begin{smallmatrix} a' & b' \\ c & d \end{smallmatrix}\right)$ with $|a'| < |c|$. If $|c| > |a|$ then (11.6) can be used repeatedly to find $B^n \gamma = \left(\begin{smallmatrix} a & b \\ c' & d' \end{smallmatrix}\right)$ with $|c'| < |a|$. Iterating these two cases shows that there is an element $C \in \langle A, B \rangle$ such that $C\gamma = \left(\begin{smallmatrix} a & b \\ 0 & d \end{smallmatrix}\right)$ with $a = \pm 1$ and $b$ even. Clearly $d = a$, and so $C\gamma = A^{b/2}(aI)$ as required. $\qquad\square$

### 11.2.3 Uniform Lattices

In this section we outline two constructions of uniform lattices in $\mathrm{PSL}_2(\mathbb{R})$.

**Lemma 11.12.** *There is a uniform lattice in* $\mathrm{PSL}_2(\mathbb{R})$.

In the proof a convex $n$-gon will be called *regular* if all of its interior angles are equal and all its sides have the same length.

SKETCH PROOF OF LEMMA 11.12. First notice that there is a regular four-gon $D$ for which all of the internal angles are equal to $\frac{\pi}{3}$. To see this, draw two geodesics intersecting at i in a normal angle, and consider the four points on this pair of geodesics at distance $t$ from i. For small values of $t$, the area of the regular four-gon spanned by these points is small, and the internal angles are close to $\frac{\pi}{2}$ (these are equivalent statements by the Gauss–Bonnet formula (Proposition 11.8)). As $t \to \infty$ these angles converge to zero, so for some $t$ they must be equal to $\frac{\pi}{3}$ (it is clear that the angles vary continuously in $t$). In Fig. 11.8 this is visualized in the *disc model*[95] of $\mathbb{H}$, which is obtained from $\mathbb{H}$ by applying an inversion with respect to a circle outside $\mathbb{H}$ but tangent to $\partial\mathbb{H}$ (for example, the map $\mathbb{H} \ni z \mapsto \frac{1}{z+i} \in \mathbb{C}$).

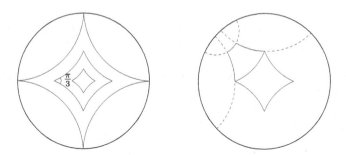

**Fig. 11.8** A regular polyhedron and a tiling in the ball model for $\mathbb{H}$

Now notice that 6 copies of $D$ that are isometric under the action of $\mathrm{PSL}_2(\mathbb{R})$ can be put together edge-to-edge to cover a neighborhood of one of their vertices. By iterating this, we obtain a tiling of $\mathbb{H}$ by tiles isometric to $D$. That is, we end up with countably many isometric images of $D$ with the property that any two of the images intersect either in one of the sides, or in one corner, or not at all. Notice that the same tiling can be constructed from any of the copies of $D$ in the tiling by placing new copies of $D$ edge-to-edge to already existing copies.

Now consider the group $\Gamma$ of all matrices in $\mathrm{PSL}_2(\mathbb{R})$ that map the tiling onto itself. It is clear that $\Gamma$ is discrete since the set of vertices of the copies of $D$ in the tiling forms a discrete subset of $\mathbb{H}$ and must be mapped onto itself by $\Gamma$. Finally, any copy of $D$ in the tiling can be mapped to any other by some element of $\Gamma$. It follows that $D$ contains a fundamental region and so $\Gamma$ is a uniform lattice.                                    $\square$

Another construction of co-compact lattices $\Gamma \subseteq \mathrm{SL}_2(\mathbb{R})$ comes from the uniformization theorem[96] which states that any connected Riemannian surface is a quotient of $\mathbb{H}$ or of $\mathbb{C}$, or it has a bijective holomorphic map to the Riemann sphere $\overline{\mathbb{C}}$. Moreover, if the surface has genus two or more, it must be a quotient $\Gamma \backslash \mathbb{H}$. Thus it is enough to construct some Riemann surface of genus two, and one way to do this is illustrated in Fig. 11.9.

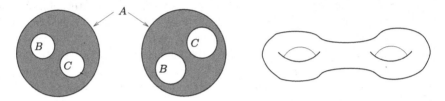

**Fig. 11.9** A Riemann surface of genus two

The Riemann surface is constructed by gluing the big circles labeled $A$ together via the holomorphic inversion on the circle followed by the correct translation, and similarly for the smaller circles $B$ and $C$. Strictly speaking, the manifold is defined by two charts which are small neighborhoods of the regions between the big and small circles and the chart maps. This produces a surface that topologically can be viewed as the surface of a three-dimensional body in the shape of a figure eight.

Another arithmetic construction of uniform lattices in $\mathrm{SL}_2(\mathbb{R})$ will be discussed in Exercise 11.6.3.

## Exercises for Sect. 11.2

**Exercise 11.2.1.** Show rigorously that the tiling used in the sketch proof of Lemma 11.12 exists.

**Exercise 11.2.2.** Show that the *Hecke triangle group* $G_n$ for $n \geqslant 3$, generated by

$$S = \begin{pmatrix} 0 & -1 \\ 1 & 0 \end{pmatrix}$$

and

$$T_n = \begin{pmatrix} 1 & 2\cos\frac{\pi}{n} \\ 0 & 1 \end{pmatrix}$$

is a non-uniform lattice in $\mathrm{SL}_2(\mathbb{R})$. Find the associated Dirichlet region for the point $p = 2i$.

**Exercise 11.2.3.** This exercise shows how to construct uncountably many non-conjugate lattices. Fix a parameter $x \in (-1, 1)$, and let

$$V_x = \begin{pmatrix} 1 & x \\ 0 & 1 \end{pmatrix} \begin{pmatrix} 1 & 0 \\ t_x & 1 \end{pmatrix} \begin{pmatrix} 1 & -x \\ 0 & 1 \end{pmatrix}.$$

The matrix $V_x$ acts on $\mathbb{C}$ via Möbius transformations as in (9.1); verify that $V_x(x) = x$ for any $t_x$. Now add the requirement that $V_x(-1) = 1$, and check that this uniquely determines $t_x$ in terms of $x$.
(a) Prove that $\Gamma_x$, the group generated by $\left(\begin{smallmatrix} 1 & 2 \\ 0 & 1 \end{smallmatrix}\right)$ and $V_x$, is a lattice in $\mathrm{SL}_2(\mathbb{R})$, freely generated by $\left(\begin{smallmatrix} 1 & 2 \\ 0 & 1 \end{smallmatrix}\right)$ and $V_x$, and show that the domain illustrated in Fig. 11.10 is a fundamental domain for $\Gamma_x$.

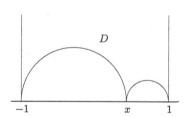

**Fig. 11.10** A fundamental domain for $\Gamma_x$

(b) Calculate the trace of the element

$$\gamma = V_x \begin{pmatrix} 1 & 2 \\ 0 & 1 \end{pmatrix} \in \Gamma_x,$$

and deduce that $\Gamma_x \backslash \mathrm{SL}_2(\mathbb{R})$ contains a periodic orbit, say of $\Gamma_x g$, for the geodesic flow associated to $\gamma$ in the sense that $\gamma g = ga$ for some diagonal element $a$. Express the length of the period as a function of $x$.
(c) Show that for any lattice $\Gamma \subseteq \mathrm{SL}_2(\mathbb{R})$ there are only countably many periodic orbits for the geodesic flow on $\Gamma \backslash \mathrm{SL}_2(\mathbb{R})$.
(d) Show that if $\Gamma^g = g\Gamma g^{-1}$ is the conjugate of a lattice $\Gamma \subseteq \mathrm{SL}_2(\mathbb{R})$ by some element $g \in \mathrm{SL}_2(\mathbb{R})$, then the sets of lengths of the periodic orbits on $\Gamma \backslash \mathrm{SL}_2(\mathbb{R})$ and on $\Gamma^g \backslash \mathrm{SL}_2(\mathbb{R})$ agree.
(e) Deduce that $\mathrm{SL}_2(\mathbb{R})$ has uncountably many non-conjugate lattices.

# 11.3 Unitary Representations, Mautner Phenomenon, and Ergodicity

The "Mautner phenomenon" has its origins in his study of geodesic flows [257]. The phenomenon refers to situations where invariance of an observable (a function) along orbits of one flow implies invariance along the orbits of certain transverse flows[97]. An instance of this has already been seen in the proof of Theorem 9.21.

### 11.3.1 Three Types of Actions

For any discrete subgroup $\Gamma \leqslant \mathrm{SL}_2(\mathbb{R})$, the geodesic flow is defined on the space $X = \Gamma \backslash \mathrm{SL}_2(\mathbb{R})$ by

$$R_{a_t}(x) = x a_t^{-1} = x \begin{pmatrix} e^{t/2} & 0 \\ 0 & e^{-t/2} \end{pmatrix}$$

for $x \in X$ and $t \in \mathbb{R}$. The stable (and analogously unstable) horocycle flow is defined by

$$R_{u^-(s)}(x) = x u^-(-s) = x \begin{pmatrix} 1 & -s \\ 0 & 1 \end{pmatrix}$$

for $x \in X$ and $s \in \mathbb{R}$. This represents a unit-speed parametrization of the stable manifold of the geodesic flow $g_t$.

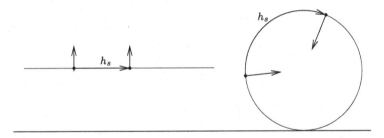

**Fig. 11.11** The action of the horocycle flow

Taken together, the subgroups

$$A = \left\{ \begin{pmatrix} e^{-t/2} & \\ & e^{t/2} \end{pmatrix} \mid t \in \mathbb{R} \right\}$$

and

$$U^- = \left\{ \begin{pmatrix} 1 & s \\ & 1 \end{pmatrix} \mid s \in \mathbb{R} \right\}$$

contain representatives of all but one type of element of $SL_2(\mathbb{R})$ in the following sense. Recall that group elements $g_1, g_2 \in G$ are said to be *conjugate* if there is some $h \in G$ with $g_1 = hg_2h^{-1}$, and recall also the group $SO_2(\mathbb{R})$ defined in Lemma 9.1.

**Lemma 11.13.** *Every* $g \in SL_2(\mathbb{R})$ *is conjugate to an element of* $\pm A$, $\pm U^-$, *or* $SO_2(\mathbb{R})$.

PROOF. If $g \in SL_2(\mathbb{R})$ is diagonalizable over $\mathbb{R}$, then it is conjugate to an element of $A$ or $-A$; if it is diagonalizable over $\mathbb{C}$ but not over $\mathbb{R}$ then it is conjugate to an element of $SO_2(\mathbb{R})$. In fact the two eigenvalues must be $\lambda, \lambda^{-1}$ for some $\lambda = e^{i\theta}$, and so in the right basis the matrix is a rotation by $\theta$. If $g$ is not diagonalizable, then it can only have one eigenvalue $\lambda$, which must satisfy $\det(g) = \lambda^2 = 1$. It follows that the Jordan normal form of either $g$ or $-g$ belongs to $U^-$.                                                      □

Lemma 11.13 is useful because of the following result.

**Lemma 11.14.** *Let* $\Gamma \leqslant G$ *be a discrete subgroup of a closed linear group. Suppose that* $g_2 = hg_1h^{-1}$ *for* $g_1, g_2, h \in G$. *Then the maps* $R_{g_1}$ *and* $R_{g_2}$ *on* $X = \Gamma \backslash G$ *are conjugate via* $R_h$. *In particular, if* $\Gamma$ *is a lattice, then the measure-preserving systems* $(X, \mathscr{B}_X, m_X, R_{g_1})$ *and* $(X, \mathscr{B}_X, m_X, R_{g_2})$ *are conjugate (measurably isomorphic).*

PROOF. This is clear since $R_{g_2} = R_h R_{g_1} R_h^{-1}$, and if $\Gamma$ is a lattice then $R_h$ preserves the finite measure $m_X$.                                                      □

Thus the study of the dynamics of elements of $SL_2(\mathbb{R})$ on $X$ reduces to three cases (ignoring the possibility of a negative sign), namely:

- diagonalizable elements (that is, those elements with a conjugate in $A$), which are called *hyperbolic*;
- elements conjugate to an element of $U^-$, called *parabolic*; and
- elements that are conjugate to an element of $SO(2)$, called *elliptic*.

Since the latter group is compact it exhibits little* interesting dynamics. For example, the ergodic measures for the $SO(2)$-action are precisely images of the Haar measure $m_{SO(2)}$. More precisely, consider a point $x \in X$ and the map

---

* There are nonetheless interesting dynamical properties of compact orbits (for example, one can ask about the behavior of compact orbits under translation by other group elements), but they are not relevant to the discussion in this chapter.

$$\phi_x(g) = R_g(x) = xg^{-1}$$

for $g \in \mathrm{SO}(2)$. Then $(\phi_x)_* \left(m_{\mathrm{SO}(2)}\right)$ is an ergodic measure for the $\mathrm{SO}(2)$-action for which the resulting measure-preserving system is a factor of the action of $\mathrm{SO}(2)$ on itself. These are all of the ergodic measures (see Exercise 11.3.1). This shows that we have to exclude elements of $\mathrm{SO}(2)$ in the discussion of ergodicity.

### 11.3.2 Ergodicity

**Theorem 11.15.** *Let $\Gamma \leqslant \mathrm{SL}_2(\mathbb{R})$ be a lattice, and write $X = \Gamma \backslash \mathrm{SL}_2(\mathbb{R})$. Let $g \in \mathrm{SL}_2(\mathbb{R})$ be an element that is not conjugate to an element of $\mathrm{SO}(2)$. Then $R_g$ acts ergodically on $(X, \mathscr{B}_X, m_X)$.*

As discussed above, we have to consider two cases, namely (after replacing $g$ by $g^2$ if necessary) elements of $A$ and elements of $U^-$. Even though we already dealt with the former, we will give here a different proof covering both cases using the language of unitary representations.

**Definition 11.16.** Let $G$ be a metrizable group and let $\mathscr{H}$ be a Hilbert space. An action $G \times \mathscr{H} \to \mathscr{H}$ of $G$ on $\mathscr{H}$ is a *unitary representation* if every $g \in G$ acts unitarily on $\mathscr{H}$ and for every $v \in \mathscr{H}$ the map $g \longmapsto g(v)$ is continuous with respect to the metric on $G$ and the norm on $\mathscr{H}$.

**Lemma 11.17.** *Let $X$ be a locally compact metric space, and let $\mu$ be a probability measure on $X$. Assume that $G$ is a metrizable group that acts continuously on $X$ (see p. 231) and preserves the measure $\mu$. Then the action of $G$ on $L_\mu^2(X)$ defined by $g : f \mapsto f \circ g^{-1}$ is a unitary representation.*

In particular, this lemma may be applied to the right action on $X = \Gamma \backslash G$ where $\Gamma$ is a lattice in a closed linear group $G$ and $\mu = m_X$ is the Haar measure.

The induced action on $L_\mu^2(X)$ may also be written $(g(f))(x) = f(xg)$; we avoid the usual notation for the unitary operator associated to a measure-preserving system to avoid confusion with the distinguished subgroups $U^\pm$.

PROOF OF LEMMA 11.17. Since every $g \in G$ preserves $\mu$, we already know that $f \mapsto g(f)$ is unitary on $L_\mu^2(X)$. All that remains is to check the continuity requirement, and this follows from the more general result in Lemma 8.7. $\square$

Since ergodicity of an action is characterized by the absence of invariant functions in $L_0^2$ (the space of square-integrable functions with integral zero) (see Chap. 2 and Exercise 2.3.5 in particular), the following proposition (Proposition 11.18) will become useful in our non-commutative setting. In a metric group $G$ with a left-invariant metric $\mathsf{d}_G$, write

$$B_\delta^G = B_\delta^G(I) = \{g \in G \mid \mathsf{d}_G(g, I) < \delta\}$$

for the metric open ball of radius $\delta > 0$ around the identity, and

$$B_\delta^G(h) = hB_\delta^G$$

for the metric open ball of radius $\delta$ around $h \in G$. The following simple but powerful argument is due to Margulis [249].

**Proposition 11.18.** *Let $\mathscr{H}$ be a Hilbert space carrying a unitary representation of a metric group $G$. Suppose that $v_0 \in \mathscr{H}$ is fixed by some subgroup $L \subseteq G$. Then $v_0$ is also fixed by any other element $h \in G$ with the property that*

$$B_\delta^G(h) \cap LB_\delta^G L \neq \varnothing$$

*for every $\delta > 0$.*

In particular, this proposition applies in general for an element $g \in G$ (with $L = g^{\mathbb{Z}}$) and elements $h$ contained in its *stable horospherical subgroup*

$$U_g^- = \{h \in G \mid g^n h g^{-n} \to I \text{ as } n \to \infty\}$$

or its *unstable horospherical subgroup*

$$U_g^+ = \{h \in G \mid g^n h g^{-n} \to I \text{ as } n \to -\infty\}.$$

As mentioned just after Theorem 11.15, this gives a simple (and less geometric) proof of Theorem 9.21. If $f \in L_{m_X}^2(X)$ is $R_g$-invariant, then by Proposition 11.18 it is also $R_{U_g^-}$ and $R_{U_g^+}$-invariant. In the case

$$g = a_t \in G = \mathrm{SL}_2(\mathbb{R}),$$

the subgroups $U_g^- = U^-$ and $U_g^+ = U^+$ generate all of $\mathrm{SL}_2(\mathbb{R})$, so the function $f$ is $\mathrm{SL}_2(\mathbb{R})$-invariant and therefore equal to a constant almost everywhere. Note that here the invariance is always understood in $L^2(X)$, while in the proof of Theorem 9.21 on p. 314 we worked with concrete points.

Summarizing the discussion above gives the following more general result.

**Corollary 11.19.** *Let $\Gamma \leqslant G$ be a lattice in a closed linear group and let $X$ be the homogeneous space $\Gamma\backslash G$. If $g \in G$ has the property that $G$ is generated by $U_g^-$ and $U_g^+$ then $R_g$ is ergodic with respect to the Haar measure $m_X$.*

PROOF OF PROPOSITION 11.18. Without loss of generality we may assume that $\|v_0\| = 1$. We define the auxiliary function (a so-called *matrix coefficient*)

$$p(h) = \langle h(v_0), v_0 \rangle$$

for $h \in G$. Notice that by the continuity requirement in Definition 11.16, $p(h)$ depends continuously on $h$. Moreover, for $g_1, g_2 \in L$ and $h \in G$,

$$p(g_1 h g_2) = \langle g_1 h(g_2(v_0)), v_0 \rangle = \langle h(v_0), g_1^{-1}(v_0) \rangle = p(h) \tag{11.7}$$

(since $v_0$ is fixed by $g_1, g_2 \in L$). Now $h \in G$ acts unitarily, so $\|h(v_0)\| = 1$. We claim that $p(h) = 1$ implies $h(v_0) = v_0$. This may be seen as a consequence of the fact that equality in the Cauchy–Schwartz inequality $|\langle v, w\rangle| \leqslant \|v\|\|w\|$ only occurs if $v$ and $w$ are linearly dependent.

Now let $h \in G$ be as in the statement of the proposition, and choose sequences $g_n \to e$ (the identity in $G$), $(\ell_n)$ and $(\ell'_n)$ in $L$ with

$$\ell_n g_n \ell'_n \to h$$

as $n \to \infty$. Then, on the one hand, by (11.7) we have

$$p(\ell_n g_n \ell'_n) = p(g_n) \to p(e) = \|v_0\|^2 = 1$$

as $n \to \infty$, while on the other

$$p(\ell_n g_n \ell'_n) \to p(h)$$

as $n \to \infty$ by continuity. It follows that $p(h) = 1$, and so $h(v_0) = v_0$ by the claim above.

$\square$

The proof of Theorem 11.15 would be easier if we were only interested in proving ergodicity of the horocycle flow, rather than of a single element. This distinction becomes important in (11.8), where we would in the former case be free to make the top right-hand entry in the final matrix vanish. In fact the case of the full horocycle flow is the only case that we will need later.

PROOF OF THEOREM 11.15. By Lemmas 11.13 and 11.14 it is enough to consider the cases $g = a_t$ for some $t \neq 0$ and $g = u_s^-$ for some $s \neq 0$. The case $g = a_t$ is covered by Corollary 11.19. So let $g = u_s^-$ (notice that in this case Corollary 11.19 tells us nothing since $U_{u_s^-}^- = U_{u_s^-}^+ = \{I_2\}$). Suppose that $f \in L^2_{m_X}(X)$ is invariant under $g$. We are going to apply Proposition 11.18. For $m, n \in \mathbb{Z}$,

$$h_\varepsilon = g^n u_\varepsilon^+ g^m = \begin{pmatrix} 1 & ns \\ 0 & 1 \end{pmatrix} \begin{pmatrix} 1 & 0 \\ \varepsilon & 1 \end{pmatrix} \begin{pmatrix} 1 & ms \\ 0 & 1 \end{pmatrix}$$
$$= \begin{pmatrix} 1 + ns\varepsilon & (1 + ns\varepsilon)ms + ns \\ \varepsilon & 1 + ms\varepsilon \end{pmatrix}. \tag{11.8}$$

For small $\varepsilon$ we may choose $n$ so that $1 + ns\varepsilon$ is close to 2, specifically so that

$$|(1 + ns\varepsilon) - 2| < s\varepsilon.$$

Now choose $m$ with $\left|ms + \frac{ns}{1+ns\varepsilon}\right| < 1$, so that

$$|(1 + ns\varepsilon)ms + ns| < 3.$$

Let $\varepsilon \to 0$ and choose a subsequence for which the matrix $h_\varepsilon$ converges to

$$h = \begin{pmatrix} 2 & r \\ 0 & \frac{1}{2} \end{pmatrix} \tag{11.9}$$

for some $r$, $|r| \leqslant 3$. For any $\delta > 0$ we can choose $\varepsilon > 0$ so that

$$h_\varepsilon \in B_\delta^G(h) \cap g^{\mathbb{Z}} B_\delta^G g^{\mathbb{Z}},$$

so by Proposition 11.18 we see that $f$ is invariant under $R_h$. Since $h$ is conjugate to an element of $A$, the theorem follows from the previous case. $\square$

### 11.3.3 Mautner Phenomenon for $\mathrm{SL}_2(\mathbb{R})$

Examining the abstract arguments above suggests a general principle. This principle, first discovered and exploited in a geometric context by Mautner [257], was generalized by Moore [260] to the unitary setting.

**Proposition 11.20.** *Let $\mathscr{H}$ be a Hilbert space carrying a unitary representation of $\mathrm{SL}_2(\mathbb{R})$. Let $g \in \mathrm{SL}_2(\mathbb{R})$ be an element that is not conjugate to an element of $\mathrm{SO}(2)$. Then any vector $v_0 \in \mathscr{H}$ that is fixed by $g$ is also fixed by all of $\mathrm{SL}_2(\mathbb{R})$.*

In ergodic theory this theorem gives a surprising corollary regarding the hereditary behavior of ergodicity. Notice that whenever a group $G$ acts by measure-preserving transformations on a probability space, the same holds for any subgroup $H \subseteq G$. However, if the $G$-action is in addition ergodic, there is no reason to expect the $H$-action to be ergodic, since we expect more functions to be invariant under the action of the smaller group (see Exercise 8.1.2 for an abelian example of this). Nonetheless, by combining Proposition 11.20 with Lemma 11.17 we get the following corollary for $\mathrm{SL}_2(\mathbb{R})$.

**Corollary 11.21.** *Let $X$ be a locally compact metric space with a Borel probability measure $\mu$, and suppose that $\mu$ is ergodic for a measure-preserving action of $\mathrm{SL}_2(\mathbb{R})$. Then any element of $\mathrm{SL}_2(\mathbb{R})$ that is not conjugate to an element of $\mathrm{SO}(2)$ acts ergodically.*

PROOF OF PROPOSITION 11.20. The proof is virtually the same as the proof of Theorem 11.15. An element $g \in \mathrm{SL}_2(\mathbb{R})$ that is not conjugate to an element of $\mathrm{SO}(2)$ is either conjugate to an element of $A$ or to an element of $U^-$. In the former case we can apply Proposition 11.18 for $g$, any $h \in U^-$ and any $h \in U^+$ to conclude that $v_0$ is fixed under $\langle U^-, U^+ \rangle = \mathrm{SL}_2(\mathbb{R})$. In the latter case we proceed as before, using the matrix calculation in (11.8) and

Proposition 11.18 to see that $v_0$ is fixed by some element $h$ as in (11.9), placing us back in the first case.    $\square$

## Exercises for Sect. 11.3

**Exercise 11.3.1.** For any point $x \in X = \Gamma \backslash \mathrm{SL}_2(\mathbb{R})$, define a map

$$\phi_x : \mathrm{SO}(2) \to X$$

by $\phi_x(g) = R_g(x) = xg^{-1}$ for $g \in \mathrm{SO}(2)$. Show that $(\phi_x)_* (m_{\mathrm{SO}(2)})$ is an ergodic measure for the $\mathrm{SO}(2)$-action on $X$, for which the resulting measure-preserving system is a factor of the action of $\mathrm{SO}(2)$ on itself. Show that any ergodic measure for the action of $\mathrm{SO}(2)$ is of this form.

**Exercise 11.3.2.** State and prove a generalization of Proposition 11.20 to unitary representations of $\mathrm{SL}_3(\mathbb{R})$ (and, more generally, to $\mathrm{SL}_d(\mathbb{R})$ for $d \geqslant 2$).

## 11.4 Mixing and the Howe–Moore Theorem

Using ergodicity of the horocycle flow we can now improve our understanding of the dynamical properties of the action of $\mathrm{SL}_2(\mathbb{R})$.

We say that a square matrix $M \in \mathrm{GL}_k(\mathbb{R})$ is *unipotent* if $(M - I)^k = 0$, that is if $M - I$ is *nilpotent*. In $\mathrm{SL}_2(\mathbb{R})$ a unipotent matrix is one which is conjugate to an element of $U^-$.

**Theorem 11.22.** *Let $\Gamma \leqslant \mathrm{SL}_2(\mathbb{R})$ be a lattice. Then the action of $\mathrm{SL}_2(\mathbb{R})$ on $X = \Gamma \backslash \mathrm{SL}_2(\mathbb{R})$ is mixing.*

### 11.4.1 First Proof of Theorem 11.22

In this section we will prove a stronger and more general result in the language of unitary representations, which will be related to Proposition 11.18 and which has Theorem 11.22 as a consequence. For this, the following notation will be useful. Let $G$ be a locally compact group, and let $\alpha = (a_n)$ be a sequence of elements of $G$. Define

$$S(\alpha) = \left\{ g \in G \mid e \in \overline{\{a_n^{-1} g a_n \mid n \in \mathbb{N}\}} \right\}$$

where $e$ is the identity element in $G$. The reader may think of the case $a_{t_n} \in A$ for a sequence $t_n \to \infty$, in which case it is easy to describe the set $S(\alpha)$. In fact, this is the only case that we will need later (in Sect. 11.5).

**Proposition 11.23.** *Let $G$ be a locally compact group and let $\mathscr{H}$ be a Hilbert space carrying a unitary representation of $G$. Let $\alpha = (a_n) \in G^{\mathbb{N}}$ be a sequence in $G$, and suppose for some $v \in \mathscr{H}$ the sequence $a_n(v)$ converges in the weak\*-topology to $v_0 \in \mathscr{H}$. Then $gv_0 = v_0$ for all $g$ in the closure of the subgroup generated by the set $S(\alpha)$.*

PROOF. Clearly it is sufficient to show that $v_0$ is fixed by each $g \in S(\alpha)$. So suppose that $g \in S(\alpha)$ and let $a_{n_k}$ be a subsequence such that

$$\lim_{k \to \infty} a_{n_k}^{-1} g a_{n_k} = e. \tag{11.10}$$

Then for any $w \in \mathscr{H}$ we have (by definition of weak\*-convergence)

$$\langle gv_0, w \rangle = \langle v_0, g^{-1}w \rangle = \lim_{k \to \infty} \langle a_{n_k} v, g^{-1} w \rangle = \lim_{k \to \infty} \langle g a_{n_k} v, w \rangle$$

and similarly

$$\langle v_0, w \rangle = \lim_{k \to \infty} \langle a_{n_k} v, w \rangle .$$

By the unitary property and the continuity of the representation, it follows that

$$|\langle gv_0, w \rangle - \langle v_0, w \rangle| = \lim_{k \to \infty} \left| \langle (a_{n_k}^{-1} g a_{n_k}) v, a_{n_k}^{-1} w \rangle - \langle v, a_{n_k}^{-1} w \rangle \right|$$

$$\leqslant \lim_{k \to \infty} \left\| (a_{n_k}^{-1} g a_{n_k}) v - v \right\| \|w\| = 0.$$

Since this holds for all $w \in \mathscr{H}$ we get $gv_0 = v_0$ as claimed. $\qquad \square$

To be able to apply this we need to be able to find non-trivial elements of $S(\alpha)$. We do this now for $G = \mathrm{SL}_2(\mathbb{R})$ in preparation for the proof of Theorem 11.22.

**Lemma 11.24.** *Let $\alpha = (g_n)$ be a sequence in $\mathrm{SL}_2(\mathbb{R})$ converging to $\infty$ (that is, such that for any compact subset $K \subseteq \mathrm{SL}_2(\mathbb{R})$ there are only finitely many $n$ with $g_n \in K$). Then the set $S(\alpha)$ contains a non-trivial unipotent element.*

PROOF. Recall from the discussion of $\mathrm{SL}_2(\mathbb{R})$ as a closed linear group on p. 289 that the homomorphism $\phi : \mathrm{SL}_2(\mathbb{R}) \to \mathrm{GL}\,(\mathrm{Mat}_{22}(\mathbb{R}))$ defined by $(\phi(g))\,(v) = gvg^{-1}$ is a proper map. That is, the norm of $\phi(g_n)$ goes to infinity when $(g_n)$ is a sequence that leaves compact subsets. Notice that $\mathrm{Mat}_{22}(\mathbb{R}) = \mathbb{R}I_2 \oplus \mathfrak{sl}_2(\mathbb{R})$ splits into a sum of two invariant subspaces on the first of which the action is trivial. Thus the norm of $\mathrm{Ad}_{g_n}$ goes to infinity; that is we can choose a sequence of vectors $(v_n)$ in $\mathfrak{sl}_2(\mathbb{R})$ for which $\|v_n\| \to 0$

while $\| \text{Ad}_{g_n}(v_n) \| = c > 0$ (where $c$ is some fixed small constant chosen so that the exponential map is injective on the ball of radius $2c$). By exponentiating this sequence, and choosing an appropriate subsequence we get $h_n = \exp(v_n) \to I_2$ and $g_n h_n g_n^{-1} \to u \neq I_2$. Since for large $n$ the element $h_n$ is close to the identity, its eigenvalues are close to 1. The conjugated element $g_n h_n g_n^{-1}$ has the same eigenvalues, so the limit element has 1 as its only eigenvalue—thus $u$ is unipotent and non-trivial. $\qquad\square$

PROOF OF THEOREM 11.22. Let $(a_n)$ be a sequence in $\text{SL}_2(\mathbb{R})$ converging to $\infty$. Let $f \in L^2(X)$ (with $X$ as in the statement of the theorem); we wish to show that $a_n(f)$ converges in the weak*-topology to $\int f \, dm_X$. Since

$$\| a_n(f) \|_2 = \| f \|_2,$$

we know that any subsequence of $(a_n(f))$ has a weak*-convergent subsequence (since the closed ball of radius $\| f \|_2$ is weak*-compact by the Alaoglu–Tychonoff theorem (Theorem B.6)).

So let $(a_{n_k}(f))$ be a sequence converging weak* to $f_0$. By Lemma 11.24, the set $S(\alpha)$ contains a non-trivial unipotent element of $\text{SL}_2(\mathbb{R})$ where $\alpha = (a_{n_k})$. It follows that $f_0$ is invariant under a unipotent element by Proposition 11.23 and is therefore constant by Theorem 11.15. This implies that $f_0 = \int f \, dm_X$ as claimed, proving the theorem. $\qquad\square$

## 11.4.2 Vanishing of Matrix Coefficients for $\text{PSL}_2(\mathbb{R})$

The proof in Sect. 11.4.1 also gives the following theorem due to Howe and Moore [160].

**Theorem 11.25.** *Let $\mathscr{H}$ be a Hilbert space carrying a unitary representation of $\text{SL}_2(\mathbb{R})$ without any invariant vectors. Then for any $v, w \in \mathscr{H}$ the matrix coefficients $\langle gv, w \rangle$ for $g \in \text{SL}_2(\mathbb{R})$ vanish at $\infty$:*

$$\langle g_n v, w \rangle \longrightarrow 0$$

*as $g_n \to \infty$.*

## 11.4.3 Second Proof of Theorem 11.22; Mixing of All Orders

In the section we present a different proof for mixing, which also leads to a proof of mixing of all orders[98]. The result and approach presented here is due to Mozes [262] and it holds more generally than the case considered here.

**Theorem 11.26.** *Let $X$ be a $\sigma$-compact metric space equipped with a continuous $\mathrm{SL}_2(\mathbb{R})$-action. Let $\mu$ be an $\mathrm{SL}_2(\mathbb{R})$-invariant ergodic probability measure on $X$. Then the $\mathrm{SL}_2(\mathbb{R})$-action is mixing of all orders with respect to $\mu$. That is, for any $r \geqslant 1$, functions $f_0, f_1, \ldots, f_{r-1} \in L^\infty(X)$ and $g^{(1)}, \ldots, g^{(r-1)}$ in $\mathrm{SL}_2(\mathbb{R})$, we have*

$$\int f_0(x) f_1(g^{(1)} {\cdot} x) f_2(g^{(2)} {\cdot} x) \cdots f_{r-1}(g^{(r-1)} {\cdot} x) \, \mathrm{d}\mu(x) \to \int f_0 \, \mathrm{d}\mu \cdots \int f_{r-1} \, \mathrm{d}\mu$$

*as $g^{(i)} \to \infty$ and $g^{(i)}(g^{(j)})^{-1} \to \infty$ for $i \neq j$, $1 \leqslant i, j \leqslant r$.*

We start with the special case $r = 2$, which is essentially the statement of Theorem 11.22.

SECOND PROOF OF THEOREM 11.22. Suppose that $g_n = g_n^{(1)} \in \mathrm{SL}_2(\mathbb{R})$ eventually leaves any compact subset of $\mathrm{SL}_2(\mathbb{R})$. We wish to show that for $f_1, f_2 \in C_c(X)$ we have

$$\int f_1(x) f_2(g_n {\cdot} x) \, \mathrm{d}\mu \longrightarrow \int f_1 \, \mathrm{d}\mu \int f_2 \, \mathrm{d}\mu. \qquad (11.11)$$

This then extends by approximation to functions $f_1, f_2 \in L^2_\mu(X)$, showing the mixing property (we will say more about this approximation argument in the proof of Theorem 11.26).

Consider the diagonal measure $\Delta$ on $X \times X$ defined by the relation

$$\int_{X \times X} F(x_1, x_2) \, \mathrm{d}\Delta(x_1, x_2) = \int_X F(x, x) \, \mathrm{d}m_X(x)$$

for all $F \in C_c(X \times X)$, and the push-forward

$$\mu_n = (I, g_n)_* \Delta,$$

where the action is simply the product action

$$(h_1, h_2) {\cdot} (x_1, x_2) = (h_1 {\cdot} x_1, h_2 {\cdot} x_2)$$

for $h_1, h_2 \in \mathrm{SL}_2(\mathbb{R})$. Then

$$\int_{X \times X} f_1(x_1) f_2(x_2) \, \mathrm{d}\mu_n(x_1, x_2) = \int_{X \times X} f_1(x_1) f_2(g_n {\cdot} x_2) \, \mathrm{d}\Delta(x_1, x_2)$$

$$= \int_X f_1(x) f_2(g_n {\cdot} x) \, \mathrm{d}\mu(x)$$

gives precisely the left-hand side of (11.11). Therefore we wish to show that $\mu_n$ converges weakly to $\mu \times \mu$.

Note that $\mu_n$ projects to $\mu$ in both coordinates: that is, $(\pi_j)_*(\mu_n) = \mu$ for $j = 1, 2$, where $\pi_j$ is as usual the projection $(x_1, x_2) \mapsto x_j$ onto the $j$th coordinate. However, $\mu_n$ is not a joining of the two systems since it is only invariant under the action of the subgroup

$$\{(h, g_n h g_n^{-1}) \mid h \in \mathrm{SL}_2(\mathbb{R})\} \tag{11.12}$$

obtained from conjugation of the diagonal subgroup by $(I_2, g_n)$.

If $\mu_n$ does not converge to $\mu \times \mu$, then we may choose a subsequence which converges to a different limit $\nu$ say.

We claim first that $\nu$ is still a probability measure with $(\pi_j)_*(\nu) = \mu$ for $j = 1, 2$. If $X$ is compact this is clear. If $X$ is not compact, then for any $\varepsilon > 0$ we may choose a compact set $K \subseteq X$ of measure $\mu(K) > 1 - \varepsilon$, and a function $f_K$ in $C_c(X)$ with $0 \leqslant f_K \leqslant 1$ and $f_K(x) = 1$ for all $x \in K$. Then for any non-negative function $f \in C_c(X)$, the function

$$(x_1, x_2) \mapsto f(x_1) f_K(x_2)$$

is in $C_c(X \times X)$, and satisfies

$$0 \leqslant f(x_1) - f(x_1) f_K(x_2) \leqslant \|f\|_\infty (1 - f_K(x_2))$$

and so for any $n$ we have

$$\int_{X \times X} f(x_1) f_K(x_2) \, \mathrm{d}\mu_n(x_1, x_2) \leqslant \int_{X \times X} f(x_1) \, \mathrm{d}\mu_n(x_1, x_2)$$

$$= \int_X f \, \mathrm{d}\mu \leqslant \int_{X \times X} f(x_1) f_K(x_2) \, \mathrm{d}\mu_n + \varepsilon \|f\|_\infty.$$

In the limit as $n \to \infty$, the same property holds for $\nu$. Decreasing $\varepsilon$ and increasing both $K$ and $f_K$ monotonically gives

$$\int_{X \times X} f(x_1) \, \mathrm{d}\nu(x_1, x_2) = \int_X f \, \mathrm{d}\mu$$

for all $f \in C_c(X)$. Therefore, $(\pi_1)_*\nu = \mu$ and similarly $(\pi_2)_*\nu = \mu$.

Next we claim that $\nu$ has some invariance properties (inherited from (11.12)). More precisely, let $\alpha = (g_n)$ and let $u \in S(\alpha)$ be a non-trivial unipotent element (which exists by Lemma 11.24). Then we claim that $\nu$ is invariant under $(I, u)$. So suppose without loss of generality that a sequence $(h_n)$ in $\mathrm{SL}_2(\mathbb{R})$ satisfies $h_n \to I$ as $n \to \infty$, and has $g_n h_n g_n^{-1} \to u$. Let $f \in C_c(X \times X)$. Then, for any $n$,

$$\int f(x_1, x_2) \, \mathrm{d}\mu_n(x_1, x_2) = \int f(h_n \cdot x_1, (g_n h_n g_n^{-1}) \cdot x_2) \, \mathrm{d}\mu_n(x_1, x_2)$$

by invariance under the subgroup in (11.12). For $n \to \infty$ we have

$$\int f(x_1, x_2)\, \mathrm{d}\mu_n(x_1, x_2) \to \int f(x_1, x_2)\, \mathrm{d}\nu(x_1, x_2)$$

and

$$\int f(h_n{\cdot}x_1, (g_n h_n g_n^{-1}){\cdot}x_2)\, \mathrm{d}\mu_n(x_1, x_2) \to \int f(x_1, u{\cdot}x_2)\, \mathrm{d}\nu(x_1, x_2),$$

where the latter follows by uniform continuity of $f$ as $h_n \to e$, $g_n h_n g_n^{-1} \to u$, and so $h_n{\cdot}x_1 \to x_1$, $g_n h_n g_n^{-1}{\cdot}x_2 \to u{\cdot}x_2$ as $n \to \infty$ uniformly for $(x_1, x_2)$ in a neighborhood of the support of $f$.

To summarize: $(\pi_j)_* \nu = \mu$ for $j = 1, 2$ and $\nu$ is invariant under $(I_2, u)$ for some non-trivial unipotent $u \in \mathrm{SL}_2(\mathbb{R})$. We claim that this implies that

$$\nu = \mu \times \mu.$$

Clearly the $\sigma$-algebra

$$\mathscr{A} = \mathscr{B}_X \times \{\varnothing, X\}$$

consists of $(I_2, u)$-invariant sets, so that for almost every $(x_1, x_2)$ the conditional measures $\mu^{\mathscr{A}}_{(x_1,x_2)}$ are supported on $[(x_1, x_2)]_{\mathscr{A}} = \{x_1\} \times X$, and are invariant under $(I_2, u)$ (cf. Corollary 5.24). Writing $\nu_{(x_1,x_2)} = (\pi_2)_* \mu^{\mathscr{A}}_{(x_1,x_2)}$, we see that

$$\mu = (\pi_2)_* \nu = \int \nu_{(x_1,x_2)}\, \mathrm{d}\mu(x_1, x_2)$$

represents $\mu$ as an integral over $u$-invariant measures on $X$. Since $\mu$ is ergodic with respect to $u$ by Corollary 11.21 we conclude that

$$\mu^{\mathscr{A}}_{(x_1,x_2)} = \delta_{x_1} \times \nu_{(x_1,x_2)} = \delta_{x_1} \times \mu$$

for $\nu$-almost every $(x_1, x_2)$. Together with the fact that $(\pi_1)_* \nu = \mu$, this implies that $\nu = \mu \times \mu$ (since $\mathscr{A} = \mathscr{B}_X \times \{\varnothing, X\}$).  $\square$

The proof of the general case of Theorem 11.26 proceeds along similar lines, using induction on the parameter $r$.

PROOF OF THEOREM 11.26. First notice that it is enough* to consider continuous functions of compact support in $C_c(X)$. To see this, let $F_j \in L^\infty(X)$ and choose functions $f_j \in C_c(X)$ with $\|f_j - F_j\|_2 < \varepsilon$ and $\|f_j\|_\infty \leqslant \|F_j\|_\infty$ for $j = 0, 1, \ldots, r-1$. Then

$$\left| \int F_0(x) F_1(g^{(1)}{\cdot}x) \cdots F_{r-1}(g^{(r-1)}{\cdot}x)\, \mathrm{d}\mu(x) \right.$$

$$\left. - \int f_0(x) F_1(g^{(1)}{\cdot}x) \cdots F_{r-1}(g^{(r-1)}{\cdot}x)\, \mathrm{d}\mu(x) \right| \leqslant \varepsilon \|F_1\|_\infty \cdots \|F_{r-1}\|_\infty,$$

---

* These reductions are used frequently in Chap. 7, where higher-order expressions play an important role.

so replacing $F_0$ by $f_0$ comes at the cost of a fixed multiple of $\varepsilon$ only. Repeating as necessary shows that it is enough to prove Theorem 11.26 for functions in $C_c(X)$.

As a result of formulating the result using continuous functions, we can again phrase the required conclusion in terms of weak*-convergence of a sequence of measures. Define the diagonal measure $\Delta = \Delta_r$ on $X^r$ by

$$\int_{X^r} F(x_0, \ldots, x_{r-1}) \, \mathrm{d}\Delta(x_0, \ldots, x_{r-1}) = \int_X F(x, \ldots, x) \, \mathrm{d}\mu(x)$$

for all $F \in C_c(X^r)$. Then

$$\int_{X^r} f_0(x_0) f_1(x_1) \cdots f_{r-1}(x_{r-1}) \, \mathrm{d}(I, g^{(1)}, \ldots, g^{(r-1)})_* \Delta$$

$$= \int_X f_0(x) f_1(g^{(1)} \cdot x) \cdots f_{r-1}(g^{(r-1)} \cdot x) \, \mathrm{d}\mu(x)$$

for $f_j \in C_c(X)$ and $(I, g^{(1)}, \ldots, g^{(r-1)}) \in (\mathrm{SL}_2(\mathbb{R}))^r$ acting on $X^r$.

Choose a sequence $(g_n^{(1)}, \ldots, g_n^{(r-1)})$ of $(r-1)$-tuples of elements of $\mathrm{SL}_2(\mathbb{R})$ with $g_n^{(i)} \to \infty$ and $g_n^{(i)}(g_n^{(j)})^{-1} \to \infty$ as $n \to \infty$ for $i \neq j$. We wish to show that

$$\mu_n = (I, g_n^{(1)}, \ldots, g_n^{(r-1)})_* \Delta$$

converges to $\mu \times \mu \times \cdots \times \mu$ in the weak*-topology as $n \to \infty$. Choosing a subsequence if necessary we can assume that $\mu_n$ converges in the weak*-topology to some limit $\nu$.

Just as in the proof for $r = 2$ above, we have that $\nu$ is a probability measure with $(\pi_j)_* \nu = \mu$ for any of the coordinate projections

$$\pi_j(x_0, x_1, \ldots, x_{r-1}) = x_j.$$

By induction on $r$ we may assume that the theorem holds for $r-1$ functions. This in fact translates to the refined statement that $(\pi_J)_* \nu = \mu^{|J|}$ on $X^{|J|}$ for any proper subset $J \subseteq \{0, 1, \ldots, r-1\}$. Here $\pi_J = \prod_{j \in J} \pi_j$ is the projection onto the coordinates corresponding to the subset $J$. To see this, fix some proper subset $J \subseteq \{0, 1, \ldots, r-1\}$ and functions $f_j \in C_c(X)$ for $j \in J$. Then (by definition, and in the case that $X$ is not compact by the approximation argument used in the case $r = 2$)

$$\int_X \prod_{j \in J} f_j(x_j) \, \mathrm{d}\nu = \lim_{n \to \infty} \int \prod_{j \in J} f_j(g_n^{(j)} \cdot x_j) \, \mathrm{d}\mu$$

$$= \prod_{j \in J} \int f_j \, \mathrm{d}\mu$$

by the assumption on the sequences $(g_n^{(j)})$ and the inductive assumption.

It is also clear that $\mu_n$ is invariant under the group

$$\left\{ \left( h, g_n^{(1)} h (g_n^{(1)})^{-1}, \ldots, g_n^{(r-1)} h (g_n^{(r-1)})^{-1} \right) \mid h \in \mathrm{SL}_2(\mathbb{R}) \right\},$$

from which we again want to derive a non-trivial invariance property of $\nu$. This requires the following generalization of Lemma 11.24. There exists a subsequence of the $g_n^{(j)}$ (passing to this subsequence is suppressed in the notation for simplicity), a sequence $(h_n)$ in $\mathrm{SL}_2(\mathbb{R})$, and a subset $J \subseteq \{1, \ldots, r-1\}$ such that $h_n \to I$ and $g_n^{(j)} h_n (g_n^{(j)})^{-1} \to I$ for $j \notin J$, while $g_n^{(j)} h_n (g_n^{(j)})^{-1} \to u_j$ for $j \in J$, where $u_j \in \mathrm{SL}_2(\mathbb{R})$ is non-trivial and unipotent for each $j \in J$ (see Exercise 11.4.1).

For notational simplicity we assume that $J = \{s, \ldots, r-1\}$, and we write $u = (u_j \mid j \in J)$ for the $|J|$-tuple of unipotent elements constructed as limits. By the same argument as that used in the case $r = 2$, this implies that $\nu$ is invariant under the action of $u$ on the last $(r-s)$ coordinates of $X^r = Z$, namely

$$u \cdot (x_0, x_1, \ldots, x_{r-1}) = (x_0, \ldots, x_{s-1}, u_s \cdot x_s, \ldots, u_{r-1} \cdot x_{r-1})$$

for $(x_0, \ldots, x_{r-1}) \in X^r$. We define the $\sigma$-algebra

$$\mathscr{A} = \mathscr{B}_X \otimes \cdots \otimes \mathscr{B}_X \times \{\varnothing, X\}^{r-s}$$

which consists of $u$-invariant sets. Then the conditional measures $\nu_z^{\mathscr{A}}$ are, for $\nu$-almost every $z \in X^r$, invariant under $u$ by Corollary 5.24. This implies that

$$\mu^{r-s} = (\pi_J)_* \nu = \int_Z (\pi_J)_* \nu_z^{\mathscr{A}} \, \mathrm{d}\nu(z) \tag{11.13}$$

expresses $\mu^{r-s}$ as an integral of $u$-invariant probability measures $(\pi_J)_* \nu_z^{\mathscr{A}}$. However, as the action of $u_j$ for $j = s, \ldots, r-1$ is mixing (and so, in particular weak-mixing) with respect to $\mu$ by Theorem 11.22, it follows from Theorem 2.36 that the action of $u$ on $X^{r-s}$ is ergodic with respect to $\mu^{r-s}$. Therefore, (11.13) implies that

$$\nu_z^{\mathscr{A}} = \delta_{(z_0, \ldots, z_{s-1})} \times \mu^{r-s}$$

for $\nu$-almost every $z \in X^r$ by Theorem 4.4. Furthermore, $\pi_{\{0, \ldots, s-1\}} \nu = \mu^s$ on $X^s$, so that we get $\nu = \mu^r$ on $X^r$ as desired. This concludes the proof of the inductive step, and hence the theorem. $\square$

# Exercises for Sect. 11.4

**Exercise 11.4.1.** Prove the generalization of Lemma 11.24 used in the proof of Theorem 11.26 by analyzing the sequence of linear maps $\mathrm{Ad}_{g_n^{(j)}}$ on $\mathfrak{sl}_2(\mathbb{R})$.

## 11.5 Rigidity of Invariant Measures
## for the Horocycle Flow

In this section we will discuss the set of probability measures on the homogeneous space $X = \Gamma \backslash \operatorname{SL}_2(\mathbb{R})$ that are invariant under the horocycle flow. Recall that

$$U^- = \left\{ u^-(s) = \begin{pmatrix} 1 & s \\ & 1 \end{pmatrix} \mid s \in \mathbb{R} \right\}$$

and that the horocycle flow is defined by

$$h(s) \cdot x = R_{u^-(s)}(x) = x \begin{pmatrix} 1 & -s \\ & 1 \end{pmatrix}$$

for any $x \in \Gamma \backslash \operatorname{SL}_2(\mathbb{R})$. If $\Gamma$ is a lattice, which we will assume here, then $m_X$ is an invariant and ergodic probability measure for the horocycle flow by Theorem 11.15. If $\Gamma$ is cocompact (that is, if $X$ is compact) then the horocycle flow is uniquely ergodic[99] by a theorem of Furstenberg [101]. For the general case we do not have unique ergodicity. For example, if $\Gamma = \operatorname{SL}_2(\mathbb{Z})$ then our reference vector $x_0 = (i, i)$ (the vector pointing upwards based at $i \in \mathbb{H}$) has a periodic orbit under the horocycle flow—and a periodic orbit supports an invariant measure, showing that the flow is not uniquely ergodic. In algebraic terms $x_0$ corresponds to $I \in \operatorname{SL}_2(\mathbb{R})$, and the identity

$$h(1) \cdot \operatorname{SL}_2(\mathbb{Z}) = \operatorname{SL}_2(\mathbb{Z}) \begin{pmatrix} 1 & -1 \\ 0 & 1 \end{pmatrix} = \operatorname{SL}_2(\mathbb{Z})$$

shows that the orbit consists of a periodic cycle. Figure 11.12 shows the periodic cycle for the reference vector $x_0$, which consists of all vectors pointing upwards based at all points in the fundamental domain of the form $i + t$ with $-\frac{1}{2} \leqslant t \leqslant \frac{1}{2}$ since the horocycle flow moves vectors that point upwards horizontally without changing their direction.

Even though in general the horocycle is not uniquely ergodic, it is possible to describe all probability measures that are invariant and ergodic under the horocycle flow, as shown by Dani [61]. The proof we present is different from the original proofs of Furstenberg and Dani, and is due to Margulis.

**Theorem 11.27.** *Let $\Gamma \subseteq \operatorname{SL}_2(\mathbb{R})$ be a lattice and let $X = \Gamma \backslash \operatorname{SL}_2(\mathbb{R})$. Let $\mu$ be a probability measure that is invariant and ergodic under the horocycle flow $h(s)$ for $s \in \mathbb{R}$. Then either $\mu = m_X$ or $\mu$ is the unique invariant measure supported on a periodic orbit* for $U^-$. If $X$ is compact the only*

---

* As a measure-preserving dynamical system, in this case the system $(X, \mu, R_{u^-})$ for $u^- \in U^-$ is conjugate to the rotation flow $(\mathbb{T} = \mathbb{R}/\mathbb{Z}, m_{\mathbb{T}}, R_t)$ for $t \in \mathbb{R}$.

**Fig. 11.12** The periodic cycle for the reference vector $x_0$

possibility is $\mu = m_X$ *(that is, there is unique ergodicity) since there are no periodic orbits for $U^-$.*

Our method of proof requires the assumption that $\Gamma$ is a lattice, but this is not necessary for the results.

### 11.5.1 Existence of Periodic Orbits; Geometric Characterization

We begin by explaining the relationship between compactness of $X$ and the existence of periodic orbits for $U^-$.

**Lemma 11.28.** *Let $\Gamma \subseteq \mathrm{SL}_2(\mathbb{R})$ be a discrete subgroup. Assume that the point $x_0 \in X = \Gamma\backslash\mathrm{SL}_2(\mathbb{R})$ is periodic for $U^-$. Then $R_{a_t}(x_0)$ diverges to infinity in $X$ (that is, leaves any compact subset of $X$). In particular, if $X$ is compact then there are no periodic orbits for $U^-$.*

PROOF. As discussed in Sect. 9.3.3, the space $X$ is locally isomorphic to $\mathrm{SL}_2(\mathbb{R})$. That is, for any $x \in X$ there is an injectivity radius $r_x > 0$ such that the map

$$B_{r_x}^{\mathrm{SL}_2(\mathbb{R})} \longrightarrow X$$
$$g \longmapsto xg$$

is an isometry. Moreover, if $K \subseteq X$ is compact then there exists some uniform $r > 0$ that serves as an injectivity radius across all points of $K$. This puts an immediate constraint on the length $\ell_0$ of a possible periodic orbit of $x_0 \in K$ for the horocycle flow as follows. There exists some $s > 0$ (depending only on $r$) such that for $x \in K$ and $\ell \in \mathbb{R}\backslash\{0\}$ with $|\ell| < s$ we have $h(\ell)\cdot x \neq x$, so that $|\ell_0| \geqslant s$.

Suppose that $x_0 \in X$ is a periodic cycle of length $\ell_0$, so

$$h(\ell_0) \cdot x_0 = x_0 u^-(-\ell_0) = x_0.$$

Then $R_{a_t}(x_0)$ satisfies

$$h(e^{-t}\ell_0) \cdot R_{a_t}(x_0) = x_0 a_t^{-1} u^-(-e^{-t}\ell_0) = x_0 u^-(-\ell_0) a_t^{-1} = R_{a_t}(x_0),$$

so $R_{a_t}(x_0)$ is periodic with period length $e^{-t}\ell_0$ and so $R_{a_t}(x_0) \notin K$ for large enough $t$. This shows that $R_{a_t}(x_0) \to \infty$ for $t \to \infty$, and hence that there cannot be any periodic points if $X$ is compact. □

The converse for lattices relies on the geometry of the Dirichlet region discussed in Sect. 11.1.

**Lemma 11.29.** *If $X = \Gamma \backslash \mathrm{SL}_2(\mathbb{R})$ is not compact but is of finite volume, then there are periodic orbits for the horocycle flow in $X$. In fact, to every cusp of $X$ there corresponds precisely a one-parameter family of periodic $U^-$-orbits in $X$ parameterized by the action of the diagonal subgroup $A$. More precisely, for one point $x$ with periodic $U^-$-orbit we get precisely one element, namely $R_{a_t}(x)$, of all other periodic $U^-$-orbits that are associated to the same cusp, by letting $t \in \mathbb{R}$ vary. Moreover, $x \in X$ is periodic for $U^-$ if and only if $R_{a_t}(x) \to \infty$ for $t \to \infty$.*

As the proof of this lemma will show, the result is easily verified for the case $\Gamma = \mathrm{SL}_2(\mathbb{Z})$.

PROOF OF LEMMA 11.29. Recall from Sect. 11.1 that the Dirichlet region $D$ defined by $p \in \mathbb{H}$ consists of all points $y \in \mathbb{H}$ with

$$\mathsf{d}(p, y) = \min_{\gamma \in \Gamma} \mathsf{d}(p, \gamma(y)),$$

that it is a hyperbolic convex $n$-gon for some $n$, that it represents a fundamental domain for the action of $\Gamma$, and that the $n$-gon $D$ has at least one point on the boundary $\overline{\mathbb{R}} = \partial \mathbb{H}$ since $X$ is not compact by Lemma 11.9. We will refer to the points of $\overline{D}^{\mathbb{H}} \cap \partial \mathbb{H}$ as boundary vertices, while cusps are as before equivalence classes of boundary vertices.

We claim that for every boundary vertex $r$ there is a non-trivial unipotent $\gamma \in \Gamma$ fixing $r$. To do this, we first show that it is sufficient that there be a non-central element $\gamma \in \Gamma$ fixing $r$. By conjugating $\Gamma$ with some element of $\mathrm{SL}_2(\mathbb{R})$ we may assume that $r = \infty$. Now any matrix $\gamma \in \Gamma \subseteq \mathrm{SL}_2(\mathbb{R})$ with $\gamma(\infty) = \infty$ has the form $\gamma = \left(\begin{smallmatrix} a & b \\ 0 & d \end{smallmatrix}\right)$ with $ad = 1$ and $b \in \mathbb{R}$. For such an element, $\gamma(p) = \frac{a}{d}p + \frac{b}{d}$ has imaginary part $\Im(\gamma(p)) = \frac{a}{d}\Im(p)$ with $\frac{a}{d} > 0$. Replacing $\gamma$ by $\gamma^{-1}$ if necessary, we may assume that $\frac{a}{d} \geqslant 1$. If $\frac{a}{d} > 1$, then $\Im(\gamma(p)) > \Im(p)$. In this case, the geodesic line

$$L_\gamma = \{z \in \mathbb{H} \mid \mathsf{d}(z, p) = \mathsf{d}(z, \gamma(p))\}$$

as in Lemma 11.5 is a half-circle in $\mathbb{C}$ rather than a vertical line. In particular, in this case any $z \in \mathbb{H}$ with big enough imaginary part is closer to $\gamma(p)$ than it is to $p$, which contradicts the assumption that $\infty$ is a boundary vertex of the Dirichlet region defined by $p$. We must therefore have $\frac{a}{d} = 1$, so $a = d = \pm 1$. Thus if a non-central element $\gamma \in \Gamma$ fixes the boundary vertex $r$, then $\gamma^2$ is a non-trivial unipotent element fixing $r$.

To prove the claim that for any boundary vertex $r$ there is a non-trivial element $\gamma \in \Gamma$ fixing $r$, we start by considering an edge $E$ which makes up one of the pieces of the boundary of $D$ near $r$. Then $E \subseteq L_{\gamma_1}$ for some $\gamma_1 \in \Gamma$, with $L_{\gamma_1}$ as in Lemma 11.5. We claim that $E = \overline{D} \cap L_{\gamma_1}$ may also be written

$$E = \overline{\gamma_1(D)} \cap L_{\gamma_1};$$

that is, the two hyperbolic convex $n$-gons $D$ and $\gamma_1(D)$ meet full edge to full edge in $E$. To see this, consider a point $y \in \overline{\gamma_1(D)} \cap L_{\gamma_1}$. Then

$$\mathsf{d}(y,p) = \mathsf{d}(y, \gamma_1(p)) = \min_{\gamma \in \Gamma} \mathsf{d}(y, \gamma(p)).$$

We claim that every point on the geodesic from $y$ to $p$ belongs to $D$. If not, then there is some $z$ in that geodesic segment with $\mathsf{d}(y,p) = \mathsf{d}(y,z) + \mathsf{d}(z,p)$ (since $z$ lies on the geodesic joining $y$ to $p$) and with $\mathsf{d}(z, \gamma(p)) \leqslant \mathsf{d}(z,p)$. This would imply that

$$\mathsf{d}(y, \gamma(p)) \leqslant \mathsf{d}(y,z) + \mathsf{d}(z, \gamma(p)) \leqslant \mathsf{d}(y,z) + \mathsf{d}(z,p) = \mathsf{d}(y,p) \leqslant \mathsf{d}(y, \gamma(p)),$$

and, moreover, that the path from $y$ to $z$ and on to $\gamma(p)$ (via the pieces of geodesics) is in fact a length minimizing path. By Proposition 9.4 this can only happen if $y, z$ and $\gamma(p)$ all belong to the same geodesic, which implies that $\gamma(p) = p$ and therefore $\gamma = \pm I$. This implies that $y$ lies in $\overline{D} \cap L_{\gamma_1}$, and so $\overline{D} \cap L_{\gamma_1} \subseteq \overline{\gamma_1(D)} \cap L_{\gamma_1}$. The same argument implies the reversed inclusion.

If there are no other boundary vertices (as, for example, in the case $r = \infty$ and $\Gamma = \mathrm{SL}_2(\mathbb{Z})$) or if none of the other boundary vertices are equivalent to $r$ (as, for example, in the case $r = \infty$ in the example of Sect. 11.2.2), then we claim that $\gamma_1$ fixes $r$. From the previous paragraph, we know that $D$ meets $\gamma_1(D)$ full edge to full edge in $E$, so there is an edge $E'$ of $D$ that is mapped under $\gamma_1$ to $E$. As $E'$ has the boundary vertex $r_1 = \gamma_1^{-1}(r)$, we must have $r_1 = r$, and so $\gamma_1(r) = r$ as claimed.

In general, the boundary vertex $r_1$ satisfying $\gamma_1(r_1) = r$ might not coincide with $r$ (as, for example, in the case $r = 1$ and $r_1 = -1$ arising in Sect. 11.2.2). In the following argument we will make use of pairs $(E', r')$, where $E'$ is an edge of $D$ and $r'$ is both a boundary vertex of $D$ and an endpoint of $E'$. We may think of these pairs as *directed edges*. Starting with the directed edge $(E, r)$, there is an edge $E_1'$ of $D$ that is mapped under $\gamma_1$ to $E$ (that is, the directed edge $(E_1', r_1)$ is mapped to $(E, r)$).

Let $(E_1, r_1)$ be that directed edge with the property that $E_1$ is the other edge of $D$ meeting $E_1'$ at the vertex $r_1$. If $r_1 \neq r$, then there is some $\gamma_2 \in \Gamma$ with $E_1 \subseteq L_{\gamma_2}$. In particular, $D$ meets $\gamma_2(D)$ in the edge $E_1$. Hence there is a directed edge $(E_2', r_2)$ that is mapped under $\gamma_2$ to $(E_1, r_2)$. Let $(E_2, r_2)$ be the other directed edge meeting $E_2'$ in $r_2$, and so on. Geometrically speaking we move from $D$ to the next copy $\gamma_1(D)$ of $D$ near $r$ and then move through $\gamma_1(D)$ to the next copy $\gamma_1\gamma_2(D)$ and so on. Formally, we obtain a sequence of directed edges $(E, r), (E_1', r_1), (E_1, r_1), (E_2', r_2), (E_2, r_2), \ldots$, and elements $\gamma_1, \gamma_2, \cdots \in \Gamma$ with the properties that

- The directed edge $(E_{j+1}', r_{j+1})$ is mapped under $\gamma_{j+1}$ to $(E_j, r_j)$; that is, $\gamma_{j+1}(r_{j+1}) = r_j$ and $\gamma_{j+1}(E_{j+1}') = E_j$ for $j \geqslant 1$. Similarly, $\gamma_1$ maps $(E_1', r_1)$ to $(E, r)$.
- The directed edges $(E_j', r_j)$ and $(E_j, r_j)$ are precisely the two directed edges meeting in the boundary vertex $r_j$ for $j \geqslant 1$.

Note that $E_{j+1}' \subseteq L_{\gamma_{j+1}^{-1}}$ and that $E_j$ is the edge of $D$ that is mapped under $\gamma_{j+1}^{-1}$ to $E_{j+1}$. In particular, the directed edge $(E_{j+1}, r_{j+1})$ determines $(E_j, r_j)$ in the same way that $(E_j, r_j)$ determines $(E_{j+1}, r_{j+1})$. Therefore, there exists some $n \geqslant 1$ for which $E_n = E$ and $r_n = r$. In symbols, we have

$$r = \gamma_1\gamma_2\cdots\gamma_n(r_n) = \gamma(r_n) = \gamma(r)$$

where $\gamma = \gamma_1\gamma_2\cdots\gamma_n$. It remains to show that $\gamma$ is non-trivial. Again assume that $r = \infty$, so that $E$ is a vertical line, and that $\gamma_1(D)$ is to the right of $E$. Then $E_1$ is the right edge of $D$ rising vertically to $\infty$. By induction, we deduce that $\gamma_1\gamma_2\cdots\gamma_n(D)$ is to the right of $D$, so $\gamma$ is non-central. This completes the proof of the claim that every boundary vertex of $D$ is fixed by a non-trivial unipotent $\gamma$.

Given a non-trivial unipotent $\gamma \in \Gamma$ there is an element $g \in \mathrm{SL}_2(\mathbb{R})$ with $g^{-1}\gamma g = u^-(s) \in U^-$. Therefore $h(s)\cdot\Gamma g = \Gamma gu^-(-s) = \Gamma g$ is periodic for the horocycle flow.

It remains to demonstrate the correspondence between one-parameter families of periodic orbits for the horocycle flow and cusps. For this, notice that if we remove from $D$ its intersection with a big compact ball we are left with as many connected components as there are boundary vertices (indeed, what is left will be a union of disjoint open neighborhoods of the boundary vertices). However, as discussed, boundary vertices are identified under the action of $\Gamma$ and correspondingly some of the edges meeting those vertices are identified under the natural map $D \to \Gamma\backslash\mathbb{H}$. In other words, the cusps correspond to connected components of $\Gamma\backslash\mathbb{H}$ after removing from the latter a big compact subset $\Omega$. By Lemma 11.28, any point $x$ which is periodic for $U^-$ diverges under $R_{a_t}$ as $t \to \infty$, so $R_{a_t}(x)$ must approach one and only one of the cusps as $t \to \infty$.

Now consider the point $z \in D$ and the vector $v \in \mathrm{T}_z \mathbb{H}$ corresponding to $R_{a_{t_0}}(x)$ for some large enough $t_0$ for which $R_{a_t}(x)$ stays in the given neighborhood of the appropriate cusp for all $t \geq t_0$. Without loss of generality $z = x + iy$ belongs to the neighborhood of the boundary vertex $r = \infty$ and we may assume that $y > 1$. We claim that $v$ points straight up, which then shows that the corresponding element of $\mathrm{SL}_2(\mathbb{R})$ has the form $\left(\begin{smallmatrix} a & b \\ 0 & d \end{smallmatrix}\right)$. Since we are interested in parameterizing the periodic orbits, it then follows that any two such periodic orbits are on the same orbit of the subgroup

$$A = \left\{ \begin{pmatrix} a & b \\ & d \end{pmatrix} \in \mathrm{SL}_2(\mathbb{R}) \right\}.$$

To prove the claim, suppose that $v$ does not point straight up. Then the associated geodesic line is a semi-circle, and some future point $R_{a_t}(x)$ will have imaginary part equal to 1. Define $K$ to be the compact segment of the line $y = 1$ in $\mathbb{C}$ starting at i and ending at $\gamma(\mathrm{i}) \in \mathbb{R} + \mathrm{i}$, where $\gamma \in \Gamma$ is a unipotent element fixing $\infty$. Since $K$ is compact, we may assume that $\Omega$ contains the image of $K$ in $\Gamma \backslash \mathbb{H}$. This is a contradiction since $R_{a_t}(x)$ by hypothesis does not return to $\Omega$ for $t > t_0$.

The argument above also shows that if $R_{a_t}(x) \to \infty$ for any $x \in X$ then this orbit must eventually get close to a single cusp, and this happens (assuming that $\infty \in \partial \mathbb{H}$ represents this cusp) again only if $v$ points upwards. In this case, however, the horocycle flow moves horizontally and the unipotent element $\gamma$ fixing $\infty$ shows that $x$ is a periodic point for $U^-$. $\qquad \square$

### 11.5.2 Proof of Measure Rigidity for the Horocycle Flow

Just as in the proof of Lemma 11.28 we will use the geodesic flow to study stretches of $U^-$-orbits in the proof of Theorem 11.27. For this our main tool will be the mixing property of the geodesic flow $R_{a_t}$ on $X$.

PROOF OF THEOREM 11.27. Let $\mu$ be an invariant and ergodic probability measure for the horocycle flow $x \mapsto h(s) \cdot x = R_{u^-(s)}(x)$ for $s \in \mathbb{R}$ and $x \in X$. Recall that $x_0 \in X$ is generic for $\mu$ if for all $f \in C_c(X)$ the time averages converge in the sense that

$$\frac{1}{S} \int_0^S f\left(h(s) \cdot x_0\right) \mathrm{d}s \longrightarrow \int f \, \mathrm{d}\mu \qquad (11.14)$$

as $S \to \infty$, and that $\mu$-almost every $x_0 \in X$ is generic for $\mu$. Let $x_0$ be such a generic point. If $x_0$ is periodic for the horocycle flow, then $\mu$ must be the image of the one-dimensional Lebesgue measure on this periodic orbit.

So assume now that $x_0$ is not periodic. We will study the time averages
as in (11.14) and show for some subsequence of times $S$ that the averages
converge to $\int f \, dm_X$. Since for a point $x_0$ that is generic for $\mu$ the time
averages converge to $\int f \, d\mu$ we deduce that $\int f \, d\mu = \int f \, dm_X$ for all $f$
in $C_c(X)$, and so $\mu = m_X$.

Since we may assume that $x_0 \in X$ is not periodic for $U^-$, we know from
Lemma 11.29 that there exists a sequence of times $t_n \to \infty$ for the geodesic
flow and a fixed compact subset $K \subseteq X$ such that $R_{a_{t_n}}(x_0) \in K$ for $n \geqslant 1$.
Proposition 11.30 therefore finishes the proof.                                        □

**Proposition 11.30.** *Let $K \subseteq X$ be a compact set. Then there exists some
constant $\eta > 0$ with the following property for all $x_0 \in X$. Suppose that $(t_n)$
is a sequence in $\mathbb{R}$ with $t_n \to \infty$, and with $R_{a_{t_n}}(x_0) \in K$ for all $n$. Then*

$$\frac{1}{\eta e^{t_n}} \int_0^{\eta e^{t_n}} f\left(h(s) \cdot x_0\right) ds \longrightarrow \int f \, dm_X \qquad (11.15)$$

*as $n \to \infty$ for all $f \in C_c(X)$.*

The basic idea of the proof—ignoring for the moment the slightly mys-
terious constant $\eta$—is as follows. Since $f$ is uniformly continuous, we may
replace the left-hand side of (11.15) by an integral over a slightly thickened
tubular neighborhood $B_n$ of the piece

$$\left\{x_0 u^-(-s) \mid s \in [0, e^{t_n}]\right\} = x_0 u^-\left(-[0, e^{t_n}]\right) \qquad (11.16)$$

of the $U^-$-orbit of $x_0$. We wish to do this in such a way that the image of $B_n$
under $R_{a_{t_n}}$ is easy to describe. Note that the piece of the $U^-$-orbit in (11.16)
is mapped under $R_{a_{t_n}}$ to the set

$$R_{a_{t_n}}(x_0) u^-\left(-[0, 1]\right),$$

Below we will define a set $Q_\delta \subseteq \mathrm{PSL}_2(\mathbb{R})$ which contains $u^-(-[0, 1])$ and
which may be described as a cube with one of the sides being $u^-(-[0, 1])$.
Define

$$B_n = R_{a_{t_n}}^{-1}\left(R_{a_{t_n}}(x_0) Q_\delta\right) = x_0 \left(a_{t_n}^{-1} Q_\delta a_{t_n}\right).$$

We will show that $B_n$ is a slight thickening of the set defined in (11.16). Using
this we will be able to show

$$\frac{1}{e^{t_n}} \int f\left(h(s) \cdot x_0\right) ds \underset{\varepsilon}{\approx} \frac{1}{m_X(B_n)} \int_{B_n} f(x) \, dx = \frac{1}{m_X(B_n)} \langle f, \chi_{B_n} \rangle.$$

Moreover, $B_n$ was defined as a pre-image under $R_{a_{t_n}}$, and together with the
mixing property we expect

$$\langle f, \chi_{B_n} \rangle \longrightarrow \int f \, dm_X \cdot m_X \left( R_{a_{t_n}}(x_0) Q_\delta \right) = \int f \, dm_X \cdot m_X(B_n). \quad (11.17)$$

However, $B_n$ is not defined as the pre-image of a fixed set in $X$, so the mixing statement does not apply directly. Roughly speaking, the set $B_n$ is defined as the pre-image of a set whose "shape" $Q_\delta$ is fixed but whose position $R_{a_{t_n}}(x_0)$ is allowed to vary. Here the assumption that $R_{a_{t_n}}(x_0) \in K$ is crucial—it will allow us to use mixing to prove (11.17).

To make the above outline formal, we need a basic decomposition lemma for the Haar measure[100].

**Lemma 11.31.** *Let $G$ be a $\sigma$-compact unimodular group and let $S, T \subseteq G$ be closed subgroups with the property that $S \cap T = \{e\}$ and the product set $ST$ contains a neighborhood of $e \in G$. Let $\phi : S \times T \to G$ be the product map $\phi(s,t) = st \in ST \subseteq G$. Then the Haar measure $m_G$ restricted to $ST$ is proportional to the push-forward $\phi_* \left( m_S^\ell \times m_T^r \right)$ where $m_S^\ell$ is the left Haar measure on $S$ and $m_T^r$ is the right Haar measure on $T$.*

Clearly if $G \backslash ST$ has Haar measure zero, then the above lemma gives a complete description of $m_G$ in the coordinate system defined by the subgroups $S$ and $T$.

We will apply the lemma in the case where $G = \mathrm{PSL}_2(\mathbb{R})$, $S = U^-$, and

$$T = U^+ A = \left\{ \begin{pmatrix} a & \\ t & a^{-1} \end{pmatrix} \middle| t \in \mathbb{R}, a \in \mathbb{R}^\times \right\}.$$

Notice that $S = K$ and $T = U^+A$ is another choice, that was already discussed implicitly in Lemma 9.16.

PROOF OF LEMMA 11.31. Since $S \cap T = \{e\}$, an element $g \in G$ has at most one decomposition as $g = st$ with $s \in S$ and $t \in T$. Therefore for compact subsets $K_S \subseteq S$ and $K_T \subseteq T$ the map $\phi$ restricted to $K_S \times K_T$ is a homeomorphism, so $\phi^{-1} : ST \to S \times T$ is measurable. The same applies to the map $\psi : S \times T \to ST$ defined by $\psi(s,t) = \phi(s,t^{-1}) = st^{-1}$.

Let $\nu = \left( \psi^{-1} \right)_* m_G$. We consider $S \times T$ as a $\sigma$-compact group by using coordinatewise multiplication. Then, for $B \subseteq S \times T$ and a point $(s,t) \in S \times T$, we have

$$\nu \left( (s,t)B \right) = m_G \left( \psi \left( (s,t)B \right) \right) = m_G \left( s \psi(B) t^{-1} \right) = m_G \left( \psi(B) \right) = \nu(B),$$

since $G$ is unimodular. It follows that $\nu$ is a left Haar measure on $S \times T$ and so must be proportional to $m_S^\ell \times m_T^\ell$. Now $\phi$ and $\psi$ differ only by the inverse in the second component, and the inverse map sends $m_T^\ell$ to a measure proportional to $m_T^r$, so the lemma follows. $\square$

As mentioned above, we are interested in the case $G = \mathrm{PSL}_2(\mathbb{R})$, $S = U^-$, and $T = U^+A$. Clearly, in this case $S \cap T = \{e\}$ and $S, T$ are closed subgroups of $G$. All that needs to be checked is that $ST$ contains a neighborhood of $e$.

This may be seen[*] from the inverse function theorem, since both $S \times T$ and $G$ are three-dimensional, and the derivative of the multiplication map has full rank at the identity $(e, e)$. We choose the Haar measure $m_{U^-}^{\ell}$ to be the usual Lebesgue measure on $\mathbb{R}$ under the identification of $s \in \mathbb{R}$ with $u^-(s) \in U^-$, and we choose the right Haar measure $m_T^r$ on $T$ so that $m_G$ restricted to $U^-T$ coincides with the product measure of $m_{U^-}^{\ell}$ and $m_T^r$.

PROOF OF PROPOSITION 11.30. Let $\eta = \eta(K) > 0$ be chosen so that

$$ u^- \left( -[0, \eta] \right) B_\eta^T \ni g \longmapsto yg $$

is injective for all $y \in K$.

Let $f$ be a function in $C_c(X)$. It is sufficient to prove (11.15) for non-negative functions, so assume that $f(x) \geqslant 0$ for all $x \in X$. Fix $\varepsilon > 0$ and choose by uniform continuity of $f$ some $\delta \in (0, \delta)$ such that

$$ \mathsf{d}(x, y) < \delta \implies |f(x) - f(y)| < \varepsilon. $$

Recall that $T = U^+A$ and define

$$ Q_\delta = u^- \left( -[0, \eta] \right) B_\delta^T. $$

Then by choice of $\eta$, for any $y \in K$ the map $g \mapsto yg$ is injective on $Q_\delta$. Let

$$ B_n = R_{a_{t_n}}^{-1} \left( R_{a_{t_n}}(x_0) Q_\delta \right) = x_0 \left( a_{t_n}^{-1} Q_\delta a_{t_n} \right) \subseteq x_0 \left( u^- \left( -[0, \eta e^{t_n}] \right) B_\delta^T \right), $$

where the last inclusion may be seen by noting that conjugation by $a_{t_n}^{-1}$ contracts $U^+$ and expands $U^-$. Now for any $s \in [0, \eta e^{t_n}]$ and $h \in B_\delta^T$ we have

$$ \mathsf{d}_X \left( x_0 u^-(-s), x_0 u^-(-s)h \right) \leqslant \mathsf{d}_G(e, h) < \delta $$

and so

$$ |f(x_0 u^-(-s)) - f(x_0 u^-(-s)h)| < \varepsilon. $$

By Lemma 11.31 and the discussion after it, we deduce that

$$ \frac{1}{m(B_n)} \int_{B_n} f(x) \, \mathrm{d}m_X = \frac{1}{m(B_n)} \int_{a_{t_n}^{-1} Q_\delta a_{t_n}} f(x_0 g) \, \mathrm{d}m_G(g) $$

$$ = \frac{1}{\eta e^{t_n}} \int_0^{\eta e^{t_n}} \frac{1}{m_T^r(a_{t_n}^{-1} B_\delta^T a_{t_n})} \int_{a_{t_n}^{-1} B_\delta^T a_{t_n}} f(x_0 u^-(-s)h) \, \mathrm{d}m_T^r(h) \, \mathrm{d}s $$

---

[*] More formally, we can consider the Lie algebras $\mathfrak{u}^-$ of $U^-$ and $\mathfrak{t}$ of $T = U^+A$ and the map $\mathfrak{u}^- \times \mathfrak{t} \to \mathfrak{g}$ defined on a neighborhood of 0 by

$$ (v, w) \mapsto \log(\exp(v)\exp(w)). $$

It may be checked that the derivative of this map at 0 is the embedding $\mathfrak{u}^- \times \mathfrak{t} \to \mathfrak{g}$ defined by $(v, w) \mapsto v + w$.

is within $\varepsilon$ of

$$\frac{1}{\eta e^{t_n}} \int_0^{\eta e^{t_n}} f(x_0 u^-(-s)) \, ds.$$

Next we are going to construct finitely many subsets of $X$ to which the mixing property can be applied. Note that

$$\overline{Q}_\delta = u^-(-[0,\eta]) \overline{B_\delta^T}$$

is compact, and the set $Q_\delta^o = u^-(-(0,\eta)) B_\delta^T$ has

$$m_G(Q_\delta) = m_G(\overline{Q}_\delta) = m_G(Q_\delta^o).$$

It follows by regularity that there is a compact subset $P_\delta \subseteq Q_\delta^o$ and an open set $R_\delta \supseteq \overline{Q}_\delta$ with

$$m_G\left(R_\delta \smallsetminus P_\delta\right) < \frac{\varepsilon}{m_G(Q_\delta)}. \tag{11.18}$$

This implies that $B_\kappa^G P_\delta \subseteq Q_\delta$ and $B_\kappa^G Q_\delta \subseteq R_\delta$ for a sufficiently small $\kappa > 0$. By compactness we may choose finitely many points $y_1, \ldots, y_\ell \in K$ with

$$K \subseteq y_1 B_\kappa^G \cup \cdots \cup y_\ell B_\kappa^G.$$

Since the geodesic flow is mixing, we have

$$\frac{m_G(P_\delta)}{m_G(Q_\delta)} \int f \, dm_G - \varepsilon \leqslant \left\langle f, \frac{1}{m_G(Q_\delta)} \chi_{y_i P_\delta} \circ R_{a_{t_n}} \right\rangle \tag{11.19}$$

and

$$\left\langle f, \frac{1}{m_G(Q_\delta)} \chi_{y_i R_\delta} \circ R_{a_{t_n}} \right\rangle \leqslant \frac{m_G(R_\delta)}{m_G(Q_\delta)} \int f \, dm_G + \varepsilon \tag{11.20}$$

for $i = 1, \ldots, \ell$ and all large enough $n$.

Therefore, if $x = R_{a_{t_n}}(x_0) \in y_i B_\eta^G$ then $y_i P_\delta \subseteq x Q_\delta \subseteq y_i R_\delta$, and since $f$ is non-negative this shows that

$$\left\langle f, \frac{1}{m_G(Q_\delta)} \chi_{y_i P_\delta} \circ R_{a_{t_n}} \right\rangle \leqslant \left\langle f, \frac{1}{m_G(Q_\delta)} \chi_{B_n} \right\rangle \leqslant \left\langle f, \frac{1}{m_G(Q_\delta)} \chi_{y_i R_\delta} \circ R_{a_{t_n}} \right\rangle$$

and so, by the inequalities (11.19)–(11.20) and (11.18),

$$(1-\varepsilon) \int f \, dm_X - \varepsilon \leqslant \left\langle f, \frac{1}{m_G(Q_\delta)} \chi_{B_n} \right\rangle \leqslant (1+\varepsilon) \int f \, dm_X + \varepsilon.$$

By combining this with the earlier statement that

$$\left\langle f, \frac{1}{m_G(Q_\delta)} \chi_{B_n} \right\rangle$$

and

$$\frac{1}{\eta e^{t_n}} \int_0^{\eta e^{t_n}} f(x_0 u^-(-s)) \, ds$$

are $\varepsilon$-close, the proposition and Theorem 11.27 follow. ☐

While the proof of Proposition 11.30 shows that either $x$ is periodic for the horocycle flow or that certain long ergodic averages of a function $f \in C_c(X)$ are close to the integral for the Haar measure, it does not establish that in the latter case $x$ is in fact generic for $m_X$ (unless $X$ is compact). This will be proved in Sect. 11.7.

## Exercises for Sect. 11.5

**Exercise 11.5.1.** Suppose that $X = \Gamma \backslash \mathrm{SL}_2(\mathbb{R})$ is compact. Let $u \in \mathrm{SL}_2(\mathbb{R})$ be a nontrivial unipotent matrix. Prove that the map $R_u : X \to X$ is uniquely ergodic. Generalize the statement and proof to non-compact quotients by lattices. (Note that Theorem 11.27 deals with the case of the $\mathbb{R}$-flow only.)

**Exercise 11.5.2.** Suppose that $X = \Gamma \backslash \mathrm{SL}_2(\mathbb{R})$ is not compact. Show that

$$\frac{1}{t(y_n)} \int_0^{t(y_n)} f(h(t) \cdot y_n) \, dt \longrightarrow \int f \, dm_X$$

as $n \to \infty$ for any $f \in C_c(X)$, if each $y_n \in X$ is periodic with least period $t(y_n)$ for the horocycle flow, and $t(y_n) \to \infty$ as $n \to \infty$.

**Exercise 11.5.3.** Let $X = \Gamma \backslash \mathrm{SL}_2(\mathbb{R})$ be the quotient by a lattice $\Gamma$. Prove (without invoking ergodic decomposition Theorem 6.2) that a probability measure $\mu$ on $X$, invariant under the horocycle flow $h(s)$ for all $s \in \mathbb{R}$, that gives zero measure to the set of all periodic orbits of the horocycle flow must be the Haar measure of $X$.

## 11.6 Non-escape of Mass for Horocycle Orbits

We have seen that the geodesic flow $R_{a_t}$ and the horocycle flow $h(s) = R_{u^-(s)}$ have fundamentally different types of behavior. For example, we briefly discussed in Sect. 9.7.2 the fact that orbits for the geodesic flow can be quite erratic; on the other hand we will show in the next section that an orbit for the horocycle flow will either be periodic (and hence compact) or will be equidistributed in the ambient space. Recall that the latter property means

$$\frac{1}{T} \int_0^T f(h(t) \cdot x) \, dt \longrightarrow \int_X f \, dm_X$$

as $T \to \infty$. In order to prove this, we first wish to show that any limit of a sequence of measures of the form

$$\frac{1}{T_n} \int_0^{T_n} \delta_{xu^-(-t)} \, \mathrm{d}t \tag{11.21}$$

with $T_n \to \infty$ as $n \to \infty$, is indeed a probability measure. This property— that limit points of sequences of uniform measures on long orbits are probability measures—is often called *quantitative non-divergence* or *non-escape of mass*[101]. Clearly this does not hold for the geodesic flow: for example, the orbit of the point $(i, i)$ in $\mathrm{SL}_2(\mathbb{Z}) \backslash \mathrm{SL}_2(\mathbb{R})$ is strictly divergent (see Fig. 9.1 and the discussion in Sect. 9.7.2) and so any limit measure along this orbit of the geodesic flow must be zero.

On a positive note, this kind of strict divergence seen in the geodesic orbit of $(i, i)$ is clearly impossible for the horocycle flow. More precisely, for any $x \in \mathrm{SL}_2(\mathbb{Z}) \backslash \mathrm{SL}_2(\mathbb{R})$ there exists a compact set $L \subseteq \mathrm{SL}_2(\mathbb{Z}) \backslash \mathrm{SL}_2(\mathbb{R})$ and a sequence $t_n \nearrow \infty$ such that $h(t_n) \cdot x \in L$ for all $n \geqslant 1$. This property is often called *non-divergence* (and is due to Margulis in much more general situations). To see why this property holds for the horocycle flow, recall that horocycle orbits can be drawn in the upper half-plane $\mathbb{H}$ as circles touching the real axis, or as horizontal lines. This is easy to see since the latter is the orbit of $(i, i)$, and Möbius transformations map horizontal lines either to horizontal lines or to circles tangent to the real axis (see Fig. 9.3 and Sect. 9.2). We conclude that given any $x \in \mathrm{SL}_2(\mathbb{Z}) \backslash \mathrm{SL}_2(\mathbb{R})$, either $x$ is periodic for the horocycle flow, in which case the orbit is compact and the non-divergence statement is trivial, or $x$ is represented in the fundamental domain $F$ from Fig. 9.5 by a vector not pointing straight upwards. In the latter case we choose $L$ to be the closure of the set of all vectors with base point $z \in F$ satisfying $\Im(z) = 1$. As the circle comes back (in both directions) to $\Im(z) = 1$, we can find some $t_1 > 0$ with $h(t_1) \cdot x \in L$. Applying the same argument with $h(t_1 + 1) \cdot x$ in place of $x$ leads to some $t_2 > t_1$ for which $h(t_2) \cdot x \in L$, and so on.

We next state the quantitative non-divergence theorem, which (in a more general setting) is Dani's refinement of Margulis' non-divergence theorem.

**Theorem 11.32.** *For every lattice $\Gamma \subseteq \mathrm{SL}_2(\mathbb{R})$, every compact subset $K$ in $X = \Gamma \backslash \mathrm{SL}_2(\mathbb{R})$, and every $\varepsilon > 0$, there is a compact subset $L = L(K, \varepsilon)$ of $X$ such that*

$$m_{\mathbb{R}} \left( \{ t \mid 0 \leqslant t \leqslant T, h(t) \cdot x \notin L \} \right) \leqslant \varepsilon T \tag{11.22}$$

*for all $T > 0$ and all $x \in K$. Moreover, there is a compact set $L = L(\varepsilon) \subseteq X$ (independent of $K$ and of $x$) such that for any $x \in X$ either $x$ is periodic or there exists some $T_x > 0$ such that (11.22) holds for all $T \geqslant T_x$.*

Notice that the first claim gives (in a more uniform way than needed here) the earlier claim, that for any $x \in X$ and any convergent sequence of measures of the form (11.21), the limit $\mu$ is a probability measure. Indeed, to show that $\mu(X) > 1 - \varepsilon$ one only needs to choose a continuous function $f \in C_c(X)$ with $\chi_L \leqslant f \leqslant 1$ and apply the definition of weak*-convergence, where $L$ is chosen as in Theorem 11.32 for $K = \{x\}$.

Before starting the proof we record the following fundamental difference between the behavior of polynomials (which model some aspects of the horocycle flow) and the exponential function (which models some aspects of the geodesic flow).

Let $p \in \mathbb{R}[t]$ be a polynomial with small coefficients and with $p(T) = 1$. Then a fixed positive proportion of $t \in [0, T]$ has $p(t) > \frac{1}{2}$. Moreover, the Lebesgue measure of the set

$$\{t \in [0, T] \mid |p(t)| < \varepsilon\}$$

is (for small $\varepsilon$) small compared to $T$. Here it is important to note that the quality of these statements is independent of the size of $T$, which may be seen by rescaling the polynomial $p(t)$ on the interval $[0, T]$ to give the polynomial $q(t) = p(tT)$ on $[0, 1]$. Neither of these properties hold for an exponential function $g(t) = ae^{\pm t}$.

### 11.6.1 The Space of Lattices and the Proof of Theorem 11.32 for $X_2 = \mathrm{SL}_2(\mathbb{Z})\backslash \mathrm{SL}_2(\mathbb{R})$

In proving this theorem it will be helpful to think of $X_2 = \mathrm{SL}_2(\mathbb{Z})\backslash \mathrm{SL}_2(\mathbb{R})$ and, more generally, $X_d = \mathrm{SL}_d(\mathbb{Z})\backslash \mathrm{SL}_d(\mathbb{R})$, in a slightly different way. A lattice $\Lambda \subseteq \mathbb{R}^d$ is called *unimodular* if the quotient $\mathbb{R}^d/\Lambda$ has a fundamental domain of Lebesgue volume 1; equivalently if it has *covolume* 1. Any unimodular lattice has the form $\Lambda_g = g^{-1}\mathbb{Z}^d \subseteq \mathbb{R}^d$ for some $g \in \mathrm{SL}_d(\mathbb{R})$. When we wish to emphasize the meaning of a point $x \in X_d$ in the sense of a lattice in $\mathbb{R}^d$, then we will simply use the symbol $\Lambda \in X_d$.

Moreover, matrices $g_1, g_2 \in \mathrm{SL}_d(\mathbb{R})$ define the same lattice if and only if $g_2 g_2^{-1} \in \mathrm{SL}_d(\mathbb{Z})$. Thus $X_d$ can be identified with the space of unimodular lattices. This gives a geometrical interpretation of the space $X_d$; in particular Mahler's compactness criterion [241], which says that a sequence of elements of $X_d$ diverge to infinity if and only if the distance from the origin in $\mathbb{R}^d$ to the set of non-trivial lattice elements converges to zero.

**Theorem 11.33.** [MAHLER COMPACTNESS CRITERION] *A set $K \subseteq X_d$ has compact closure if and only if there is an $s > 0$ with the property that*

$$\Lambda \cap B_s(0) = \{0\}$$

*for all $\Lambda \in K$.*

One way of formulating this result is as follows. Lattices $\Lambda \subseteq \mathbb{R}^d$ are clearly discrete, and so compact subsets of lattices in $\mathbb{R}^d$ should be uniformly discrete.

PROOF OF THEOREM 11.33. We start by showing that a compact subset $K$ of $X_d$ must have the uniform discreteness property. Suppose therefore that $K$ has compact closure but for every $n \geqslant 1$ there is some $\Lambda_n \in K$ with

$$\Lambda_n \cap B_{1/n}(0) \neq \{0\}.$$

By compactness there exists some $\Lambda \in X_d$ with $\Lambda_n \to \Lambda$. By definition, this means that $\Lambda_n = g_n^{-1}\mathbb{Z}^d$, $\Lambda = g^{-1}\mathbb{Z}^d$, and we can choose $g_n$ (which is only unique up to left multiplication by $\mathrm{SL}_d(\mathbb{Z})$) such that $\mathrm{d}(g_n, g) \to 0$ as $n \to \infty$. Equivalently, $g_n^{-1} \to g^{-1}$ as $n \to \infty$, which in terms of the lattices means that one can choose a basis $\{\mathbf{b}_j^{(n)} = g_n^{-1}\mathbf{e}_j \mid 1 \leqslant j \leqslant d\}$ of $\Lambda_n$ which converges to a basis $\{\mathbf{b}_j = g^{-1}\mathbf{e}_j \mid 1 \leqslant j \leqslant d\}$ of $\Lambda$. However, as $(g_n)$ and $(g_n^{-1})$ both converge, it follows that the lattices $\Lambda_n = g_n^{-1}\mathbb{Z}^d$ cannot contain arbitrarily small elements of $\mathbb{R}^d$, which contradicts our choice and proves the easier half of the theorem.

We now claim that for every lattice $\Lambda$ with $\Lambda \cap B_s(0) = \{0\}$ one can find a basis of vectors $\mathbf{b}_1, \ldots, \mathbf{b}_n \in \Lambda$ that belong to a given ball of radius depending on $s$. Choose $\mathbf{b}_1 \in \Lambda$ with the property that

$$\|\mathbf{b}_1\| = \min\{\|\mathbf{b}\| \mid \mathbf{b} \in \Lambda \smallsetminus \{0\}\}, \tag{11.23}$$

so $\|\mathbf{b}_1\| \geqslant s$, and $\mathbf{b}_1$ generates $\Lambda \cap \mathbb{R}\mathbf{b}_1$. Moreover, there is a constant $C_d$ depending only on the dimension $d$ with $\|\mathbf{b}_1\| \leqslant C_d$, since if all the vectors in $\Lambda \smallsetminus \{0\}$ are very long, then it cannot be unimodular by Minkowski's convex body theorem (which simply relies on the argument that if $B_{2r}(0)$ does not contain a lattice element of $\Lambda$, then $B_r(0)$ is mapped injectively onto $\mathbb{R}^d/\Lambda$, which makes the volume of the fundamental domain of $\Lambda$ at least as large as the volume of $B_r(0)$). Define

$$W = (\mathbb{R}\mathbf{b}_1)^{\perp}$$

and

$$\Lambda_W = \pi_W(\Lambda)$$

where $\pi_W : \mathbb{R}^d \to W$ is the projection along the line $\mathbb{R}\mathbf{b}_1$. The $(d-1)$-dimensional lattice $\Lambda_W$ need not be unimodular, but the covolume is $\frac{1}{\|\mathbf{b}_1\|}$, which is uniformly bounded away from 0 and infinity. So after rescaling by a bounded scalar, we may assume that the lattice $\Lambda_W$ is unimodular. We claim that all the non-zero vectors in $\Lambda_W$ have length bounded away from 0 by a uniform amount. For, if not, a very short vector $\pi_W(\mathbf{x}) \in \Lambda_W$ with $\mathbf{x} \in \Lambda$

would have the property that $\mathbf{x} + n\mathbf{b}_1$ is closer to $0$ than $\mathbf{b}_1$ for some $n \in \mathbb{Z}$, contradicting (11.23) (see Fig. 11.13).

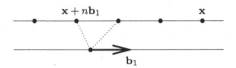

**Fig. 11.13** Choosing the vector $\mathbf{x} + n\mathbf{b}_1$

By induction, this shows that $\Lambda_W$ has a basis consisting of vectors whose length is bounded uniformly away from $0$ and infinity that can be lifted to complete the basis $\mathbf{b}_1, \mathbf{b}_2, \ldots, \mathbf{b}_n$. The length of $\mathbf{b}_j$ for $j \geqslant 2$ can be chosen to be less than $\sqrt{\|\pi_W(\mathbf{b}_j)\|^2 + \|\mathbf{b}_1\|^2}$, which is bounded by a number depending on $d$. It follows that for any sequence $\Lambda_n \in K$ we can write $\Lambda_n = g_n^{-1}\mathbb{Z}^d$ where the columns of $g_n^{-1}$ are uniformly bounded (only depending on $s$). However, as $\det(g_n) = 1$ we know that $g_n$ belongs to a fixed compact set of $\mathrm{SL}_d(\mathbb{R})$, so $K \subseteq X_d$ is contained in the compact image of that set, giving the result. $\square$

We now have the background needed to proceed with the proof of quantitative non-divergence in the case $X_2 = \mathrm{SL}_2(\mathbb{Z})\backslash\mathrm{SL}_2(\mathbb{R})$. As the proof will show, the theorem simply relies on the polynomial (in fact, linear) behavior of orbits of the horocycle flow, together with the simple observation that a unimodular lattice in $\mathbb{R}^2$ cannot contain two linearly independent vectors of norm less than one.

PROOF OF THEOREM 11.32. By Theorem 11.33 we have

$$K \subseteq \Omega_\delta = \{\Lambda \in X_2 \mid \Lambda \cap B_\delta(0) = \{0\}\}$$

for some $\delta \in (0,1)$. We need to choose some constant $\eta \in (0, \frac{\delta}{4})$ such that the set $L = \Omega_\eta$ satisfies (11.22) for all $\Lambda \in K$ and $T > 0$.

Note that an element $h(t)\cdot x = xu^-(-t)$ of the horocycle flow maps the lattice $\Lambda = g^{-1}\mathbb{Z}^2$ corresponding to $x = \Gamma g \in X_2$ to $u^-(t)\Lambda$. In other words, the horocycle flow corresponds to application of the matrix $\begin{pmatrix} 1 & t \\ & 1 \end{pmatrix}$, which fixes the $x$-axis and shears the $y$-axis towards the direction of the $x$-axis. Specifically, the image of a vector $\begin{pmatrix} \alpha_1 \\ \alpha_2 \end{pmatrix}$ under $\begin{pmatrix} 1 & t \\ & 1 \end{pmatrix}$ is $\begin{pmatrix} \alpha_1 + t\alpha_2 \\ \alpha_2 \end{pmatrix}$ for any $t \in \mathbb{R}$, so the horocycle flow moves this point at linear speed determined by $\alpha_2$ through the plane.

When applying the horocycle flow to $\Lambda$ there could be different lattice elements $v \in \Lambda$ that at some time $t$ give rise to a short vector $u^-(t)v$ in $u^-(t)\Lambda$, which prevents $u^-(t)\Lambda$ from belonging to $L = \Omega_\eta$. For this reason we let $\{v_1, v_2, \ldots\} \subseteq \Lambda$ be a maximal set of mutually non-proportional primitive* elements of $\Lambda$. Notice that if a vector $u^-(t)v \in u^-(t)\Lambda \setminus \{0\}$ has norm less than $\eta$, then $v$ is a multiple of a primitive vector $v = nv_j$, and so $u^-(t)v_j$ will have norm less than $\eta$. Thus it is enough to consider only the orbits of the primitive vectors $v_1, v_2, \ldots$.

Recall that $\Lambda \in \Omega_\delta$ by assumption, and so $\|v_i\| \geqslant \delta$ for $i = 1, 2, \ldots$. Fix some $T > 0$. Then for each $i \geqslant 1$, we define

$$B_i = \{t \in [0, T] \mid u^-(t)v_i \in B_\eta^{\mathbb{R}^2}(0)\}$$

(the set of *bad times* in $[0, T]$) and

$$P_i = \{t \in [0, T] \mid u^-(t)v_i \in B_\delta^{\mathbb{R}^2}(0)\}$$

(the set of *protecting times* in $[0, T]$). By assumption $\eta < \delta$, so $B_i \subseteq P_i$. If $i \neq j$ then $P_i \cap P_j = \varnothing$, since if $t \in P_i \cap P_j$ then $u^-(t)v_i$ and $u^-(t)v_j$ both lie in the unimodular lattice $u^-(t)\Lambda$ and have norm less than $\delta < 1$, so they are linearly dependent[†], and hence $i = j$ by construction.

Finally, we claim that

$$m_{\mathbb{R}}(B_i) \leqslant \frac{8\eta}{\delta} m_{\mathbb{R}}(P_i). \tag{11.24}$$

This implies the first claim in the theorem. Indeed, summing over $i$ and using disjointness of the sets $P_i$ the inequality (11.24) gives

$$m_{\mathbb{R}}\left(\bigsqcup_{i \geqslant 1} B_i\right) \leqslant \frac{8\eta}{\delta} m_{\mathbb{R}}\left(\bigsqcup_{i \geqslant 1} P_i\right) = \frac{8\eta}{\delta} \sum_{i \geqslant 1} m_{\mathbb{R}}(P_i) \leqslant \frac{8\eta}{\delta} T.$$

However, as discussed above,

$$m_{\mathbb{R}}\left(\{t \mid u^-(t)\Lambda \notin \Omega_\eta\}\right) = m_{\mathbb{R}}\left(\{t \mid \|u^-(t)v_i\| < \eta \text{ for some } i\}\right)$$

$$= m_{\mathbb{R}}\left(\bigcup_{i \geqslant 1} B_i\right).$$

Choosing $\eta \leqslant \frac{\varepsilon\delta}{8}$ gives the estimate needed.

---

* A vector $v \in \Lambda$ is called primitive if the equation $v = nw$ with $n \in \mathbb{Z}$ and $w \in \Lambda$ implies that $n = \pm 1$.

† The determinant of a $2 \times 2$ matrix $(w_1, w_2)$ formed by the column vectors $w_1, w_2$ is bounded by the product of their Euclidean norms, since it is equal to $\|w_1\|\|w_2\|\cos\phi$, where $\phi$ is the angle between the vectors. Moreover, if $w_1, w_2$ lie in the lattice

$$\Lambda = \mathbf{b}_1\mathbb{Z} + \mathbf{b}_2\mathbb{Z},$$

then covolume$(\Lambda) = |\det(\mathbf{b}_1, \mathbf{b}_2)| \leqslant |\det(w_1, w_2)|$.

To prove the inequality (11.24), let $v_i = \left(\begin{smallmatrix} \alpha_1 \\ \alpha_2 \end{smallmatrix}\right)$ so that

$$u^-(t)v_i = \begin{pmatrix} \alpha_1 + t\alpha_2 \\ \alpha_2 \end{pmatrix}.$$

Then $u^-(t)v_i$ moves at linear speed along a horizontal line, as in Fig. 11.14.

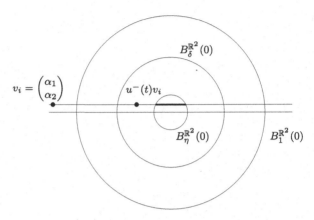

**Fig. 11.14** The linear speed in the horocycle flow makes it easy to describe the lengths of the vectors in the orbit

Clearly $B_i$ and $P_i$ are both intervals. To prove the inequality (11.24), we have to relate the lengths of these intervals to the speed $\alpha_2$ of the vector $u^-(t)v_i$ and to the radii $\eta$ and $\delta$. If $B_i \subseteq [0, T]$ is empty, then there is nothing to prove. This holds, in particular, if $\alpha_2 = 0$ or if $|\alpha_2| \geqslant \frac{\delta}{4} \geqslant \eta$. Otherwise, $P_i = [T_1, T_2]$ must contain some $T'$ with $|\alpha_1 + T'\alpha_2| < \eta \leqslant \frac{\delta}{4}$ while $\|u^-(T_1)v\| = \delta$ gives $|\alpha_1 + T_1\alpha_2| \geqslant \frac{\delta}{2}$ since $|\alpha_2| \leqslant \frac{\delta}{2}$. Thus

$$m_{\mathbb{R}}(P_i) = (T_2 - T_1) \geqslant (T' - T_1) \geqslant |\alpha_2|^{-1}\frac{\delta}{4}.$$

On the other hand, $t \in B_i = [T_3, T_4]$ implies $|\alpha_1 + t\alpha_2| \leqslant \eta$, so

$$m_{\mathbb{R}}(B_i) = T_4 - T_3 < \frac{2\eta}{|\alpha_2|}.$$

Together, this gives the inequality (11.24).

For the final claim let $K = \Omega_{1/2}$ and apply the first statement of the theorem to $\varepsilon/2$ to define the set $L$. Now let $x \in X_2$, and notice that $x$ is periodic for the horocycle flow if and only if the corresponding lattice $\Lambda \subseteq \mathbb{R}^2$ intersects the $x$-axis. Now assume that $x \in X_2$ is not periodic, so that $\Lambda$ may have a very short primitive vector $v \in \Lambda$ that is not fixed. Therefore we may choose $T_1 > 0$ such that $u^-(T_1)v$ has norm between $\frac{1}{2}$ and 1. As in the argument above, this shows that $\pm u^-(T_1)v$ are the only non-trivial vectors of norm less than 1 for $u^-(T_1)\Lambda$, so that $xu^-(-T_1) \in K$. We choose $T_x$ so that $T_1 = \frac{\varepsilon}{2}T_x$, and it is easy to check that (11.22) holds for all $T \geqslant T_x$. $\square$

## 11.6.2 Extension to the General Case

Notice first that Theorem 11.32, in the case of quotients $\Gamma\backslash\operatorname{PSL}_2(\mathbb{R})$ by lattices $\Gamma$ in $\operatorname{PSL}_2(\mathbb{R})$, also implies the same statement for quotients $\Gamma\backslash\operatorname{SL}_2(\mathbb{R})$ by lattices in $\operatorname{SL}_2(\mathbb{R})$. Indeed, if $\Gamma \subseteq \operatorname{SL}_2(\mathbb{R})$ is a lattice, then

$$\overline{\Gamma} = \Gamma \cdot \{\pm I\} \subseteq \operatorname{SL}_2(\mathbb{R})/\{\pm I\} = \operatorname{PSL}_2(\mathbb{R})$$

is also a lattice. Moreover, the natural map

$$\Gamma\backslash\operatorname{SL}_2(\mathbb{R}) \longrightarrow \overline{\Gamma}\backslash\operatorname{PSL}_2(\mathbb{R})$$

is proper (that is, every compact set $\overline{L} \subseteq \overline{\Gamma}\backslash\operatorname{PSL}_2(\mathbb{R})$ has a compact pre-image $L$ in $\Gamma\backslash\operatorname{SL}_2(\mathbb{R})$. From this the earlier claim follows quickly.

With this reduction, we may use the geometry of $\mathbb{H}$ and Dirichlet domains. Indeed, we now state a description of $\overline{\Gamma}\backslash\operatorname{PSL}_2(\mathbb{R})$ which we essentially proved together with Lemma 11.29.

**Proposition 11.34.** *Let $\Gamma \subseteq \operatorname{PSL}_2(\mathbb{R})$ be a lattice. Then either $\Gamma$ is uniform and $\Gamma\backslash\operatorname{PSL}_2(\mathbb{R})$ is compact, or there exists a compact subset*

$$\Omega_{cp} \subseteq X = \Gamma\backslash\operatorname{PSL}_2(\mathbb{R})$$

*such that $X\backslash\Omega_{cp}$ has finitely many connected components, each of which can be identified with a special finite volume subset of the infinite volume quotient*

$$\mathscr{T} = \left\{ \begin{pmatrix} 1 & n \\ & 1 \end{pmatrix} \right\} \backslash \operatorname{PSL}_2(\mathbb{R}).$$

*Indeed, for each component $C$ of $X\backslash\Omega_{cp}$ we find some $y_C > 0$ such that the subset $\mathscr{T}_C \subseteq \mathscr{T}$ is the image of $\{(z,v) \in \mathrm{T}^1\mathbb{H} \mid \Im(z) > y_C\}$, and the identification $\iota_C : C \to \mathscr{T}_C$ has the property that for a sufficiently small element $g \in \operatorname{PSL}_2(\mathbb{R})$ and any $x \in C$ with $x{\cdot}g \in C$ we have $\iota_C(x{\cdot}g) = \iota_C(x)g$.*

Roughly speaking, the components of $X\backslash\Omega_{cp}$ are the cusps of $X$.

PROOF OF PROPOSITION 11.34. We will leave some of the details of the proof as an exercise, but indicate how the proof of Lemma 11.29 applies to Proposition 11.34. We defined a cusp of $X = \Gamma\backslash\operatorname{PSL}_2(\mathbb{R})$ to be an equivalence class of boundary vertices of a Dirichlet domain $D$ for $\Gamma$, and showed in the proof of Lemma 11.29 that for every boundary vertex $r$ there exists a non-trivial unipotent element of $\Gamma$ fixing the boundary vertex $r$. Without loss of generality, we may assume that $r = \infty$. While proving this we used several other elements $\gamma_i \in \Gamma$, and the respective images $\gamma_i(D)$ of the Dirichlet domain. At the end of this argument we had obtained the copies $\gamma_1(D), \gamma_1\gamma_2(D), \ldots, \gamma_1\gamma_2 \cdots \gamma_n(D)$ of $D$ which were all located to the right of $D$ meeting each other full-edge to full-edge on vertical geodesics. Let

$$F = D \cup \gamma_1(D) \cup \gamma_1\gamma_2(D) \cup \cdots \cup \gamma_1\gamma_2 \cdots \gamma_n(D)$$

be the union of these domains. By construction, the quotient map

$$\mathrm{T}^1\mathbb{H} \to \Gamma\backslash\mathrm{PSL}_2(\mathbb{R})$$

restricted to $\{(z,v) \in \mathrm{T}^1\mathbb{H} \mid z \in F, \Im(z) > y_\infty\}$ is injective if $y_\infty > 0$ is chosen sufficiently large. Denote this injective image by $C_\infty \subseteq X$. By applying the Möbius transformation $\left(\begin{smallmatrix} a & \\ & a^{-1} \end{smallmatrix}\right)$ for some $a > 0$ we can ensure that the unipotent element $\gamma = \gamma_1 \cdots \gamma_n$ has the form $\left(\begin{smallmatrix} 1 & 1 \\ & 1 \end{smallmatrix}\right)$.

If necessary we may repeat this construction for the remaining cusps, and can ensure disjointness by choosing the associated parameters $y$ sufficiently big. After completion of this argument for every cusp $D$, the pre-images of the various sets give a set whose image in $X$ has compact closure. $\qquad\square$

SKETCH PROOF OF THEOREM 11.32 FOR GENERAL $\Gamma$. As mentioned before, it is enough to treat the case of a lattice $\Gamma \subseteq \mathrm{PSL}_2(\mathbb{R})$. Applying Proposition 11.34, we can find a compact subset $\Omega_{cp} \subseteq X = \Gamma\backslash\mathrm{PSL}_2(\mathbb{R})$ and finitely many tube-like sets $C_1, \ldots, C_\ell$ which together give a partition

$$X = \Omega_{cp} \sqcup C_1 \sqcup \cdots \sqcup C_\ell.$$

Enlarging $\Omega_{cp}$ if necessary, we may assume that $K \subseteq \Omega_{cp}$ and that each $C_i$ is naturally identified with a tube

$$\left\{ \begin{pmatrix} 1 & n \\ & 1 \end{pmatrix} \mid n \in \mathbb{Z} \right\} \backslash \left\{ (z,v) \in \mathrm{T}^1\mathbb{H} \mid \Im(z) > y_i \right\} = \mathscr{T}(y_i)$$

with $y_i > 1$. We claim that for each $C_i$ there is a subset $L_i \subseteq C_i$ with compact closure such that

$$L = \Omega_{cp} \sqcup L_1 \sqcup \cdots \sqcup L_\ell$$

satisfies the first statement of Theorem 11.32.

Let us now describe the form of non-divergence we will prove (more precisely, that we have already proved) for the tube-like sets $\mathscr{T}(y)$ with $y > 1$. There exists a compact set $L \subseteq \overline{\mathscr{T}(y)}$ such that for any $x \in \partial\mathscr{T}(y)$ and any $T > 0$, either $h(T){\cdot}x \notin \overline{\mathscr{T}(y)}$ or

$$m_{\mathbb{R}}\left( \{ t \in [0,T] \mid h(t){\cdot}x \notin L \} \right) < \varepsilon T. \tag{11.25}$$

Notice that $x \in \overline{\mathscr{T}(y)}$ and $h(T) \cdot x \in \overline{\mathscr{T}(y)}$ implies that $h(t) \cdot x \in \overline{\mathscr{T}(y)}$ for $t$ in $[0, T]$, because of the geometry of horocycle orbits in $\mathrm{T}^1 \mathbb{H}$.

Applying the above claim for each of the sets $C_i$, we obtain the partition

$$L = \Omega_{cp} \sqcup L_1 \sqcup \cdots \sqcup L_\ell.$$

If $x \in K \subseteq \Omega_{cp}$ and $T > 0$, then the interval $[0, T]$ is naturally decomposed into subintervals $I = [T_0, T_1]$ of times where

$$h(t) \cdot x \in \Omega_{cp} \text{ for all } t \in I^o, \text{ or}$$

$$h(t) \cdot x \in C_1 \text{ for all } t \in I^o, \text{ or}$$

$$\vdots$$

$$h(t) \cdot x \in C_\ell \text{ for all } t \in I^o.$$

For a subinterval of the first type there is nothing to show, while for an interval of the other types we can apply the claim in (11.25) to the initial point in $\mathscr{T}_i$ corresponding to $xu^-(-T_0) \in \partial C_i$ and time $T_1 - T_0$. Adding these estimates together gives the result.

To see that we already know (11.25), notice that $\mathscr{T}(y)$ maps injectively to $X_2 = \mathrm{SL}_2(\mathbb{Z}) \backslash \mathrm{SL}_2(\mathbb{R})$ under the natural quotient map $\pi$. Apply Theorem 11.32 to the image $K$ of $\partial \mathscr{T}(y)$ in $X_2$, to obtain a compact subset in $X_2$. Let $L \subseteq \mathscr{T}(y)$ be its pre-image under the quotient map, which has compact closure in $\mathscr{T}$.

Choose $x \in \partial \mathscr{T}(y)$ and $T > 0$ such that $h(T) \cdot x \in \overline{\mathscr{T}(y)}$, then $h(t) \cdot x \in L$ if and only if $h(t) \cdot \pi(x) \in \pi(L)$ for any $t \in [0, T]$, which gives the claim.

The last statement in the theorem also follows from the special case of $X_2 = \mathrm{SL}_2(\mathbb{Z}) \backslash \mathrm{SL}_2(\mathbb{R})$ in the same way. □

# Exercises for Sect. 11.6

**Exercise 11.6.1.** As discussed above, $\mathrm{SL}_2(\mathbb{Z}) \backslash \mathrm{SL}_2(\mathbb{R})$ can be identified both with the unit tangent bundle of the modular surface (and so with vectors attached to points in the fundamental domain $F$ shown in Fig. 9.5 on p. 308), and with the space of unimodular lattices in $\mathbb{R}^2$. Make this correspondence explicit, and in particular explain the meaning of the imaginary coordinate in terms of properties of the associated lattice. Use this to deduce an independent proof of Theorem 11.33 in the case $d = 2$.

**Exercise 11.6.2.** In this exercise we show how special algebraic groups defined over $\mathbb{Q}$ give rise to closed orbits in $\mathrm{SL}_n(\mathbb{Z}) \backslash \mathrm{SL}_n(\mathbb{R})$. Let

$$\rho : \mathrm{SL}_n(\mathbb{R}) \hookrightarrow \mathrm{SL}_N(\mathbb{R})$$

be a linear representation such that for $1 \leqslant i, j \leqslant N$ the $(i,j)$th matrix entry of $\rho(g)$ is a polynomial in the matrix entries of $g$ with rational coefficients (that is, $\rho$ is an algebraic representation defined over $\mathbb{Q}$). Let $v \in \mathbb{Q}^N$ be a vector, and define*

$$\mathbb{G} = \{ g \in \mathrm{SL}_n \mid \rho(g)v = v \}$$

to be the stabilizer of $v$. Prove that the orbit $\mathrm{SL}_n(\mathbb{Z})\mathbb{G}(\mathbb{R})$ in $\mathrm{SL}_n(\mathbb{Z})\backslash \mathrm{SL}_n(\mathbb{R})$ under the group $\mathbb{G}(\mathbb{R})$ is closed.

**Exercise 11.6.3.** In this exercise we show how special quaternion division algebras over $\mathbb{Q}$ give rise to uniform (and, by definition, also) arithmetic lattices in $\mathrm{SL}_2(\mathbb{R})$. A quaternion division algebra over $\mathbb{Q}$ is an algebra $D$ over the field $\mathbb{Q}$ such that $D$ has dimension four over $\mathbb{Q}$, $D$ has a unit $1_D$, $\mathbb{Q}$ (identified with $\mathbb{Q} \cdot 1_D$) is the center of $D$, and every non-zero element of $D$ has a multiplicative inverse. The best-known example is the algebra of rational Hamiltonian quaternions defined by $\mathbb{Q} + \mathbb{Q}i + \mathbb{Q}j + \mathbb{Q}k$ with $i^2 = j^2 = k^2 = -1$, and $ij = -ji = k \ldots$: We will not be able to use this particular algebra to construct a lattice in $\mathrm{SL}_2(\mathbb{R})$, so we begin by constructing a different quaternion algebra, and then show how it gives rise to a lattice.
(a) Show that for $a, b, c, d \in \mathbb{Z}$ the sum $a^2 + b^2 + c^2 + d^2$ is not divisible by 8 unless $a, b, c, d$ are all even.
(b) Use (a) to show that

$$D = \left\{ \begin{pmatrix} a + \sqrt{7}b & c + \sqrt{7}d \\ -(c - \sqrt{7}d) & a - \sqrt{7}b \end{pmatrix} \mid a, b, c, d \in \mathbb{Q} \right\}$$

is a quaternion division algebra over $\mathbb{Q}$.
(c) Show that $D \otimes_{\mathbb{R}} \mathbb{R} = \mathrm{Mat}_{22}(\mathbb{R})$ (that is, show that $D$ is $\mathbb{R}$-split).
(d) Embed $D$ into $\mathrm{Mat}_{44}(\mathbb{Q})$ by using the basis

$$1_D = \begin{pmatrix} 1 & \\ & 1 \end{pmatrix}, i_D = \begin{pmatrix} \sqrt{7} & \\ & -\sqrt{7} \end{pmatrix}, j_D = \begin{pmatrix} & 1 \\ -1 & \end{pmatrix}, k_D = \begin{pmatrix} & -\sqrt{7} \\ \sqrt{7} & \end{pmatrix}$$

of $D$, and identifying $d \in D$ with the linear map $D \ni f \mapsto df \in D$. The image of $D$ is a four-dimensional linear subspace $W$ of $\mathrm{Mat}_{44}(\mathbb{Q})$ defined by 12 linear equations. On $W$ the original coordinates of $a, b, c, d$ with respect to the chosen basis can be represented by rational linear combinations of the matrix entries of $g \in W$.
(e) Show that

$$\mathbb{G} = \{ g \in \mathrm{SL}_4 \mid g \in W, a^2 - 7b^2 + c^2 - 7d^2 = 1 \}$$

---

* Here we intentionally omit the field or ring after $\mathbb{G}$ and $\mathrm{SL}_n$, but for any of the choices $\mathbb{Z}, \mathbb{Q}, \mathbb{R}$ we obtain definitions of $\mathbb{G}(\mathbb{Z}), \mathbb{G}(\mathbb{Q}), \mathbb{G}(\mathbb{R})$ by requiring that the elements belong to $\mathrm{SL}_n(\mathbb{Z}), \mathrm{SL}_n(\mathbb{Q}), \mathrm{SL}_n(\mathbb{R})$ respectively.

has $\mathbb{G}(\mathbb{R}) \cong \mathrm{SL}_2(\mathbb{R})$. Here $\mathbb{G}(\mathbb{R})$ is defined to consist of all $g \in W$ (that is, $g$ satisfying the 12 linear equations) for which the pre-image in $D$ satisfies

$$a^2 - 7b^2 + c^2 - 7d^2 = 1.$$

(f) Show that $\mathrm{SL}_4(\mathbb{Z})\mathbb{G}(\mathbb{R})$ is closed by using Exercise 11.6.2.
(g) Show that $\mathrm{SL}_4(\mathbb{Z})\mathbb{G}(\mathbb{R}) \cong \Gamma \backslash \mathrm{SL}_2(\mathbb{R})$ is compact by using the Mahler compactness criterion (Theorem 11.33) on $\mathrm{SL}_4(\mathbb{Z}) \backslash \mathrm{SL}_4(\mathbb{R})$.

## 11.7 Equidistribution of Horocycle Orbits

We are now ready to prove the theorem promised earlier, due to Dani and Smillie [63]. The argument used below is due to Ratner [305, Lem. 2.1].

**Theorem 11.35.** *Let $\Gamma \subseteq \mathrm{SL}_2(\mathbb{R})$ be a lattice, and let $x \in X = \Gamma \backslash \mathrm{SL}_2(\mathbb{R})$. Then either $x$ is periodic for the horocycle flow (that is, $h(t) \cdot x = x$ for some $t > 0$), or the horocycle orbit of $x$ is equidistributed with respect to the Haar measure $m_X$ of $X$:*

$$\frac{1}{T} \int_0^T f(h(t) \cdot x) \, dt \longrightarrow \int f \, dm_X$$

*as $T \to \infty$.*

If $X$ is compact, then there are no periodic orbits by Lemma 11.28, and the statement of Theorem 11.35 is already known by Theorem 11.27.

PROOF. Suppose that the point $x_0 \in X$ is not periodic for the horocycle flow, let $T_n \nearrow \infty$ be any sequence, and define a sequence $(\mu_n)$ of probability measures by

$$\int f \, d\mu_n = \frac{1}{T_n} \int_0^{T_n} f(h(t) \cdot x) \, dt$$

for $f \in C_c(X)$. By passing to a subsequence, we may assume that $\mu_n \to \mu$ in the weak*-topology, and by Theorem 11.32 the limit $\mu$ is a probability measure. To show the theorem we need to show that $\mu = m_X$.

By the version of Theorem 4.1 for flows (which is contained in the proof of Theorem 8.10), we know that $\mu$ is invariant under the horocycle flow. By the ergodic decomposition theorem* (Theorem 6.2) extended to flows (which is contained in the more general Theorem 8.20) we can write

$$\mu = \int \mu_y \, d\nu(y)$$

---

* By using Exercise 11.5.3, this dependence on material from Chap. 6 can be avoided.

as a generalized convex combination of probability measures that are invariant and ergodic under the horocycle flow. By Theorem 11.27, each such measure is either the Haar measure $m_X$ or is the Lebesgue measure on a periodic orbit of a point $y = h(t_y) \cdot y$ with $t_y > 0$. Therefore, to show that $\mu = m_X$ we only have to show

$$\mu\left(\{y \in X \mid h(t_y) \cdot y = y \text{ for some } t_y > 0\}\right) = 0. \qquad (11.26)$$

Let us point out a few complications at this point. The set of periodic points as in (11.26) is dense (in fact long periodic orbits become equidistributed by Exercise 11.5.2). Moreover, to show that $\mu(B) = 0$ for a given measurable set $B$ it is not sufficient to show that $\mu_i(B) \to 0$ as $i \to \infty^*$. However, if $B$ is compact then, in order to prove that $\mu(B) = 0$, it is enough to find, for every $\varepsilon > 0$, an open set $O \supseteq B$ with

$$\limsup_{i \to \infty} \mu_i(O) \leqslant \varepsilon.$$

This criterion for vanishing of $\mu(B)$ follows easily from the definition of the weak*-topology and the existence of a continuous function $f \in C_c(X)$ with $\chi_B \leqslant f \leqslant \chi_O$. In order to apply this criterion, we need to write the set of periodic points as a countable union of compact sets, for which we have already developed all the necessary tools. By Lemma 11.29 there are finitely many one-parameter families of periodic orbits (one for each cusp). Fix a periodic point $x \in X$, and restrict the parameter $t \in \mathbb{R}$ from Lemma 11.29 to a compact set $I \subseteq \mathbb{R}$. Allowing $s \in [0, t_x]$ to vary, we obtain a compact set $B = \{xu^-(s)a(-t) \mid s \in [0, t_x], t \in I\}$ comprising periodic orbits. Varying $x$ through a finite list and increasing $I$, we can write the set of periodic points as in (11.26) as a countable union of such compact sets $B$. Thus it is sufficient to show that $\mu(B) = 0$ for one such set $B$. Now fix $\varepsilon > 0$ and let $L = L(\varepsilon) \subseteq X$ be a compact set constructed as in the final conclusion of Theorem 11.32. Recall that the period of $xa(-t)$ with respect to the horocycle flow is $e^{-2t}t_x$ if $t_x > 0$ is the period of $x$, and so $xa(-t)$ diverges to one of the cusps as $t \to \infty$ by Lemma 11.29. Since $I$ is compact, there exists some $t_\varepsilon \in \mathbb{R}$ with $Ba(-t_\varepsilon) \subseteq X \smallsetminus L$. We define an open set $O = (X \smallsetminus L)a(t_\varepsilon) \supseteq B$ and claim that $\mu_i(O) \leqslant \varepsilon$ for all large enough $i$. However,

$$\mu_i(O) = \frac{1}{T_i} m_{\mathbb{R}}\left(\{s \in [0, T_i] \mid h(s) \cdot x \in O = (X \smallsetminus L)a(t_\varepsilon)\}\right)$$

$$= \frac{1}{T_i} m_{\mathbb{R}}\left(\{s \in [0, T_i] \mid xa(-t_\varepsilon)u^-(-e^{-2t}s) \notin L\}\right)$$

$$= \frac{1}{e^{-2t_\varepsilon}T_i} m_{\mathbb{R}}\left(\{s \in [0, e^{-2t_\varepsilon}T_i] \mid xa(-t_\varepsilon)u^-(s) \notin L\}\right)$$

---

* For example, if $B = X \smallsetminus x_0 U^-$ then $\mu_i(B) = 0$ for all $i \geqslant 1$, but we cannot have $\mu(B) = 0$ as $x_0 U^- \cong U^-$ cannot have a $U^-$-invariant probability measure.

is the expression discussed in (11.22) for the initial point $xa(-t_\varepsilon)$. Since $e^{-t_\varepsilon}T_i \to \infty$ as $i \to \infty$, and $x$ (equivalently, $xa(-t_\varepsilon)$) is not periodic by assumption, the final statement in Theorem 11.32 shows that $\mu_i(O) \leqslant \varepsilon$ for large enough $i$. This shows that $\mu(B) = 0$ and hence (11.26), which gives the theorem.                                                                                     □

As discussed in Chap. 1, Raghunathan and Dani formulated far-reaching conjectures. Firstly Raghunathan conjectured that orbit closures under unipotent actions are always orbits of closed subgroups, generalizing the classification of orbits for the horocycle flow (that is, the statement that a horocycle orbit is either dense or periodic, which follows trivially from the stronger equidistribution result in Theorem 11.35). Dani conjectured measure rigidity of unipotent flows other than the horocycle flow (which is handled in Theorem 11.27). Ratner proved these conjectures in full, and in addition proved the generalization of equidistribution for horocycle orbits (Theorem 11.35) to other unipotent flows, leading to numerous applications especially in number theory.

# Notes to Chap. 11

[93](Page 347) Some of the striking ergodic and dynamical properties satisfied by the horocycle flow are the following. It is a highly non-trivial example of a flow that is mixing of all orders, as shown by Marcus [244]; Ratner has classified joinings between horocycle flows [300], measurable factors of horocycle flows [299], and many other measurable rigidity properties (see Ratner's survey article [301]); Marcus has also shown a form of rigidity for topological conjugacy of horocycle flows [245].

[94](Page 348) The material in this section follows the book of Bekka and Mayer [21, Chap. 2] closely.

[95](Page 361) This *Klein model* was used by both Klein [198] and Poincaré; the upper half-plane model is usually called the *Poincaré model* because of the influential paper [287] exploring its properties. Both models were used earlier by Beltrami [23] in order to show that hyperbolic geometry is as consistent as Euclidean geometry.

[96](Page 362) The uniformization theorem generalizes the Riemann mapping theorem, which states that if $U$ is a simply connected open proper subset of $\mathbb{C}$, then there exists a bijective holomorphic map from $U$ onto the open unit disk. This was stated in Riemann's thesis [309], and the first complete proof was given by Carathéodory [48]. The uniformization theorem was finally proved by Koebe [207] and by Poincaré [289]; a convenient account may be found in the monograph of Farkas and Kra [89, Chap. IV] or Stillwell [353, Chap. 5].

[97](Page 364) The work of Mautner clarified earlier results of Gel'fand and Fomin [112]. A useful overview of its use in ergodic theory may be found in Starkov's monograph [350].

[98](Page 372) A different approach comes from work of Ryzhikov, using joinings. For the case of horocycle flows, Ryzhikov [327] showed mixing of all orders as a consequence of a new criterion for mixing of all orders, giving a new proof of the theorem of Marcus [244] (this is also described in the article of Thouvenot [359]).

[99](Page 378) For a compact quotient, Hedlund showed that the horocycle flow is minimal [145] (every point has a dense orbit), and this was later generalized by Veech to nilpotent flows on semi-simple Lie groups [368].

[100](Page 385) There are more general results concerning decompositions of Haar measure in locally compact groups—see Knapp [204, Sect. 8.3].

[101](Page 389) The original work on non-divergence is due to Margulis [246]; subsequent refinements include work of Dani [62] and Kleinbock and Margulis [199] on quantitative statements; Kleinbock and Tomanov [200] on extensions to the $S$-arithmetic setting; Ghosh [114] on the positive characteristic case.

# Appendix A
# Measure Theory

Complete treatments of the results stated in this appendix may be found in any measure theory book; see for example Parthasarathy [280], Royden [320] or Kingman and Taylor [195]. A similar summary of measure theory without proofs may be found in Walters [374, Chap. 0]. Some of this appendix will use terminology from Appendix B.

## A.1 Measure Spaces

Let $X$ be a set, which will usually be infinite, and denote by $\mathbb{P}(X)$ the collection of all subsets of $X$.

**Definition A.1.** A set $\mathscr{S} \subseteq \mathbb{P}(X)$ is called a *semi-algebra* if

(1) $\varnothing \in \mathscr{S}$,
(2) $A, B \in \mathscr{S}$ implies that $A \cap B \in \mathscr{S}$, and
(3) if $A \in \mathscr{S}$ then the complement $X \smallsetminus A$ is a finite union of pairwise disjoint elements in $\mathscr{S}$;

if in addition

(4) $A \in \mathscr{S}$ implies that $X \smallsetminus A \in \mathscr{S}$,

then it is called an *algebra*. If $\mathscr{S}$ satisfies the additional property

(5) $A_1, A_2, \cdots \in \mathscr{S}$ implies that $\bigcup_{n=1}^{\infty} A_n \in \mathscr{S}$,

then $\mathscr{S}$ is called a $\sigma$-algebra. For any collection of sets $\mathscr{A}$, write $\sigma(\mathscr{A})$ for the smallest $\sigma$-algebra containing $\mathscr{A}$ (this is possible since the intersection of $\sigma$-algebras is a $\sigma$-algebra).

*Example A.2.* The collection of intervals in $[0, 1]$ forms a semi-algebra.

M. Einsiedler, T. Ward, *Ergodic Theory*, Graduate Texts in Mathematics 259,
DOI 10.1007/978-0-85729-021-2, © Springer-Verlag London Limited 2011

**Definition A.3.** A collection $\mathscr{M} \subseteq \mathbb{P}(X)$ is called a *monotone class* if

$$A_1 \subseteq A_2 \subseteq \cdots \text{ and } A_n \in \mathscr{M} \text{ for all } n \geqslant 1 \implies \bigcup_{n=1}^{\infty} A_n \in \mathscr{M}$$

and

$$B_1 \supseteq B_2 \supseteq \cdots \text{ and } B_n \in \mathscr{M} \text{ for all } n \geqslant 1 \implies \bigcap_{n=1}^{\infty} B_n \in \mathscr{M}.$$

The intersection of two monotone classes is a monotone class, so there is a well-defined smallest monotone class $\mathscr{M}(\mathscr{A})$ containing any given collection of sets $\mathscr{A}$. This gives an alternative characterization of the $\sigma$-algebra generated by an algebra.

**Theorem A.4.** *Let $\mathscr{A}$ be an algebra. Then the smallest monotone class containing $\mathscr{A}$ is $\sigma(\mathscr{A})$.*

A function $\mu : \mathscr{S} \to \mathbb{R}_{\geqslant 0} \cup \{\infty\}$ is *finitely additive* if $\mu(\varnothing) = 0$ and*

$$\mu(A \cup B) = \mu(A) + \mu(B) \tag{A.1}$$

for any disjoint elements $A$ and $B$ of $\mathscr{S}$ with $A \sqcup B \in \mathscr{S}$, and is *countably additive* if

$$\mu \left( \bigcup_{n=1}^{\infty} A_n \right) = \sum_{n=1}^{\infty} \mu(A_n)$$

if $\{A_n\}$ is a collection of disjoint elements of $\mathscr{S}$ with $\bigsqcup_{n=1}^{\infty} A_n \in \mathscr{S}$.

The main structure of interest in ergodic theory is that of a *probability space* or *finite measure space*.

**Definition A.5.** A triple $(X, \mathscr{B}, \mu)$ is called a finite measure space if $\mathscr{B}$ is a $\sigma$-algebra and $\mu$ is a countably additive measure defined on $\mathscr{B}$ with $\mu(X) < \infty$. A triple $(X, \mathscr{B}, \mu)$ is called a $\sigma$-finite measure space if $X$ is a countable union of elements of $\mathscr{B}$ of finite measure. If $\mu(X) = 1$ then a finite measure space is called a probability space.

A probability measure $\mu$ is said to be *concentrated* on a measurable set $A$ if $\mu(A) = 1$.

**Theorem A.6.** *If $\mu : \mathscr{S} \to \mathbb{R}_{\geqslant 0}$ is a countably additive measure defined on a semi-algebra, then there is a unique countably additive measure defined on $\sigma(\mathscr{S})$ which extends $\mu$.*

---

* The conventions concerning the symbol $\infty$ in this setting are that $\infty + c = \infty$ for any $c$ in $\mathbb{R}_{\geqslant 0} \cup \{\infty\}$, $c \cdot \infty = \infty$ for any $c > 0$, and $0 \cdot \infty = 0$.

**Theorem A.7.** *Let $\mathscr{A} \subseteq \mathscr{B}$ be an algebra in a probability space $(X, \mathscr{B}, \mu)$. Then the collection of sets $B$ with the property that for any $\varepsilon > 0$ there is an $A \in \mathscr{A}$ with $\mu(A \triangle B) < \varepsilon$ is a $\sigma$-algebra.*

As discussed in Sect. 2.1, the basic objects of ergodic theory are measure-preserving maps (see Definition 2.1). The next result gives a convenient way to check whether a transformation is measure-preserving.

**Theorem A.8.** *Let $(X, \mathscr{B}_X, \mu)$ and $(Y, \mathscr{B}_Y, \nu)$ be probability spaces, and let $\mathscr{S}$ be a semi-algebra which generates $\mathscr{B}_Y$. A measurable map $\phi : X \to Y$ is measure-preserving if and only if*

$$\mu(\phi^{-1}B) = \nu(B)$$

*for all $B \in \mathscr{S}$.*

PROOF. Let

$$\mathscr{S}' = \{B \in \mathscr{B}_Y \mid \phi^{-1}(B) \in \mathscr{B}_X, \mu(\phi^{-1}B) = \nu(B)\}.$$

Then $\mathscr{S} \subseteq \mathscr{S}'$, and (since each member of the algebra generated by $\mathscr{S}$ is a finite disjoint union of elements of $\mathscr{S}$) the algebra generated by $\mathscr{S}$ lies in $\mathscr{S}'$. It is clear that $\mathscr{S}'$ is a monotone class, so Theorem A.4 shows that $\mathscr{S}' = \mathscr{B}_Y$ as required. □

The next result is an important lemma from probability; what it means is that if the sum of the probabilities of a sequence of events is finite, then the probability that infinitely many of them occur is zero.

**Theorem A.9 (Borel–Cantelli[102]).** *Let $(X, \mathscr{B}, \mu)$ be a probability space, and let $(A_n)_{n \geqslant 1}$ be a sequence of measurable sets with $\sum_{n=1}^{\infty} \mu(A_n) < \infty$. Then*

$$\mu\left(\limsup_{n \to \infty} A_n\right) = \mu\left(\bigcap_{n=1}^{\infty} \left(\bigcup_{m=n}^{\infty} A_m\right)\right) = 0.$$

*If the sequence of sets are pairwise independent, that is if*

$$\mu(A_i \cap A_j) = \mu(A_i)\mu(A_j)$$

*for all $i \neq j$, then $\sum_{n=1}^{\infty} \mu(A_n) = \infty$ implies that*

$$\mu\left(\limsup_{n \to \infty} A_n\right) = \mu\left(\bigcap_{n=1}^{\infty} \left(\bigcup_{m=n}^{\infty} A_m\right)\right) = 1.$$

The elements of a $\sigma$-algebra are typically very complex, and it is often enough to approximate sets by a convenient smaller collection of sets.

**Theorem A.10.** *If $(X, \mathscr{B}, \mu)$ is a probability space and $\mathscr{A}$ is an algebra which generates $\mathscr{B}$ (that is, with $\sigma(\mathscr{A}) = \mathscr{B}$), then for any $B \in \mathscr{B}$ and $\varepsilon > 0$ there is an $A \in \mathscr{A}$ with $\mu(A \triangle B) < \varepsilon$.*

A measure space is called *complete* if any subset of a null set is measurable. If $X$ is a topological space, then there is a distinguished collection of sets to start with, namely the open sets. The $\sigma$-algebra generated by the open sets is called the *Borel $\sigma$-algebra*. If the space is second countable, then the *support* of a measure is the largest closed set with the property that every open neighborhood of every point in the set has positive measure; equivalently the support of a measure is the complement of the largest open set of zero measure.

If $X$ is a metric space, then any Borel probability measure $\mu$ on $X$ (that is, any probability measure defined on the Borel $\sigma$-algebra $\mathscr{B}$ of $X$) is *regular*[103]: for any Borel set $B \subseteq X$ and $\varepsilon > 0$ there is an open set $O$ and a closed set $C$ with $C \subseteq B \subseteq O$ and $\mu(O \smallsetminus C) < \varepsilon$.

## A.2 Product Spaces

Let $I \subseteq \mathbb{Z}$ and assume that for each $i \in I$ a probability space $\mathsf{X}_i = (X_i, \mathscr{B}_i, \mu_i)$ is given. Then the product space $X = \prod_{i \in I} X_i$ may be given the structure of a probability space $(X, \mathscr{B}, \mu)$ as follows. Any set of the form

$$\prod_{i \in I, i < \min(F)} X_i \times \prod_{i \in F} A_i \times \prod_{i \in I, i > \max(F)} X_i,$$

or equivalently of the form

$$\{x = (x_i)_{i \in I} \in X \mid x_i \in A_i \text{ for } i \in F\},$$

for some finite set $F \subseteq I$, is called a *measurable rectangle*. The collection of all measurable rectangles forms a semi-algebra $\mathscr{S}$, and the product $\sigma$-algebra is $\mathscr{B} = \sigma(\mathscr{S})$. The product measure $\mu$ is obtained by defining the measure of the measurable rectangle above to be $\prod_{i \in F} \mu_i(A_i)$ and then extending to $\mathscr{B}$.

The main extension result in this setting is the Kolmogorov consistency theorem, which allows measures on infinite product spaces to be built up from measures on finite product spaces.

**Theorem A.11.** *Let $X = \prod_{i \in I} X_i$ with $I \subseteq \mathbb{Z}$ and each $X_i$ a probability space. Suppose that for every finite subset $F \subseteq I$ there is a probability measure $\mu_F$ defined on $X_F = \prod_{i \in F} X_i$, and that these measures are consistent in the sense that if $E \subseteq F$ then the projection map*

$$\left( \prod_{i \in F} X_i, \mu_F \right) \longrightarrow \left( \prod_{i \in E} X_i, \mu_E \right)$$

*is measure-preserving. Then there is a unique probability measure $\mu$ on the probability space $\prod_{i \in I} X_i$ with the property that for any $F \subseteq I$ the projection* *map*

$$\left(\prod_{i\in I} X_i, \mu\right) \longrightarrow \left(\prod_{i\in F} X_i, \mu_F\right)$$

*is measure-preserving.*

In the construction of an infinite product $\prod_{i\in I} \mu_i$ of probability measures above, the finite products $\mu_F = \prod_{i\in F} \mu_i$ satisfy the compatibility conditions needed in Theorem A.11.

In many situations each $X_i = (X_i, d_i)$ is a fixed compact metric space with $0 < \operatorname{diam}(X_i) < \infty$. In this case the product space $X = \prod_{n\in\mathbb{Z}} X_n$ is also a compact metric space with respect to the metric

$$\mathsf{d}(x,y) = \sum_{n\in\mathbb{Z}} \frac{d_n(x_n, y_n)}{2^n \operatorname{diam}(X_n)},$$

and the Borel $\sigma$-algebra of $X$ coincides with the product $\sigma$-algebra defined above.

## A.3 Measurable Functions

Let $(X, \mathscr{B}, \mu)$ be a probability space. Natural classes of measurable functions on $X$ are built up from simpler functions, just as the $\sigma$-algebra $\mathscr{B}$ may be built up from simpler collections of sets.

A function $f : X \to \mathbb{R}$ is called *simple* if

$$f(x) = \sum_{j=1}^{m} c_j \chi_{A_j}(x)$$

for constants $c_j \in \mathbb{R}$ and disjoint sets $A_j \in \mathscr{B}$. The integral of $f$ is then defined to be

$$\int f \, d\mu = \sum_{j=1}^{m} c_j \mu(A_j).$$

A function $g : X \to \mathbb{R}$ is called *measurable* if $g^{-1}(A) \in \mathscr{B}$ for any (Borel) measurable set $A \subseteq \mathbb{R}$. The basic approximation result states that for any measurable function $g : X \to \mathbb{R}_{\geq 0}$ there is a pointwise increasing sequence of simple functions $(f_n)_{n\geq 1}$ with $f_n(x) \nearrow g(x)$ for each $x \in X$. This allows us to define

$$\int g \, d\mu = \lim_{n\to\infty} \int f_n \, d\mu,$$

which is guaranteed to exist since

$$f_n(x) \leq f_{n+1}(x)$$

for all $n \geqslant 1$ and $x \in X$ (in contrast to the usual terminology from calculus, we include the possibility that the integral and the limit are infinite). It may be shown that this is well-defined (independent of the choice of the sequence of simple functions).

A measurable function $g : X \to \mathbb{R}_{\geqslant 0}$ is *integrable* if $\int g \, d\mu < \infty$. In general, a measurable function $g : X \to \mathbb{R}$ has a unique decomposition into $g = g^+ - g^-$ with $g^+(x) = \max\{g(x), 0\}$; both $g^+$ and $g^-$ are measurable. The function $g$ is said to be integrable if both $g^+$ and $g^-$ are integrable, and the integral is defined by $\int g \, d\mu = \int g^+ \, d\mu - \int g^- \, d\mu$. If $f$ is integrable and $g$ is measurable with $|g| \leqslant f$, then $g$ is integrable. The integral of an integrable function $f$ over a measurable set $A$ is defined by

$$\int_A f \, d\mu = \int f \chi_A \, d\mu.$$

For $1 \leqslant p < \infty$, the space $\mathscr{L}_\mu^p$ (or $\mathscr{L}^p(X)$, $\mathscr{L}^p(X, \mu)$ and so on) comprises the measurable functions $f : X \to \mathbb{R}$ with $\int |f|^p \, d\mu < \infty$. Define an equivalence relation on $\mathscr{L}_\mu^p$ by $f \sim g$ if $\int |f - g|^p \, d\mu = 0$ and write $L_\mu^p = \mathscr{L}_\mu^p / \sim$ for the space of equivalence classes. Elements of $L_\mu^p$ will be described as functions rather than equivalence classes, but it is important to remember that this is an abuse of notation (for example, in the construction of conditional measures on page 138). In particular the value of an element of $L_\mu^p$ at a specific point does not make sense, unless that point itself has positive $\mu$-measure. The function $\| \cdot \|_p$ defined by

$$\|f\|_p = \left( \int |f|^p \, d\mu \right)^{1/p}$$

is a norm (see Appendix B), and under this norm $L^p$ is a Banach space.

The case $p = \infty$ is distinguished: the *essential supremum* is the generalization to measurable functions of the supremum of a continuous function, and is defined by

$$\|f\|_\infty = \inf \left\{ \alpha \mid \mu \left( \{ x \in X \mid f(x) > \alpha \} \right) = 0 \right\}.$$

The space $\mathscr{L}_\mu^\infty$ is then defined to be the space of measurable functions $f$ with $\|f\|_\infty < \infty$, and once again $L_\mu^\infty$ is defined to be $\mathscr{L}_\mu^\infty / \sim$. The norm $\| \cdot \|_\infty$ makes $L_\mu^\infty$ into a Banach space. For $1 \leqslant p < q \leqslant \infty$ we have $L^p \supseteq L^q$ for any finite measure space, with strict inclusion except in some degenerate cases.

In practice we will more often use $\mathscr{L}^\infty$, which denotes the bounded functions.

An important consequence of the Borel–Cantelli lemma is that norm convergence in $L^p$ forces pointwise convergence along a subsequence.

**Corollary A.12.** *If $(f_n)$ is a sequence convergent in $L_\mu^p$ ($1 \leqslant p \leqslant \infty$) to $f$, then there is a subsequence $(f_{n_k})$ converging pointwise almost everywhere to $f$.*

PROOF. Choose the sequence $(n_k)$ so that

$$\|f_{n_k} - f\|_p^p < \frac{1}{k^{2+p}}$$

for all $k \geqslant 1$. Then

$$\mu\left(\left\{x \in X \,\Big|\, |f_{n_k}(x) - f(x)| > \frac{1}{k}\right\}\right) < \frac{1}{k^2}.$$

It follows by Theorem A.9 that for almost every $x$, $|f_{n_k}(x) - f(x)| > \frac{1}{k}$ for only finitely many $k$, so $f_{n_k}(x) \to f(x)$ for almost every $x$. □

Finally we turn to integration of functions of several variables; a measure space $(X, \mathscr{B}, \mu)$ is called $\sigma$-finite if there is a sequence $A_1, A_2, \ldots$ of measurable sets with $\mu(A_n) < \infty$ for all $n \geqslant 1$ and with $X = \bigcup_{n \geqslant 1} A_n$.

**Theorem A.13 (Fubini–Tonelli[104]).** *Let $f$ be a non-negative integrable function on the product of two $\sigma$-finite measure spaces $(X, \mathscr{B}, \mu)$ and $(Y, \mathscr{C}, \nu)$. Then, for almost every $x \in X$ and $y \in Y$, the functions*

$$h(x) = \int_Y f(x, y) \, \mathrm{d}\nu, \quad g(y) = \int_X f(x, y) \, \mathrm{d}\mu$$

*are integrable, and*

$$\int_{X \times Y} f \, \mathrm{d}(\mu \times \nu) = \int_X h \, \mathrm{d}\mu = \int_Y g \, \mathrm{d}\nu. \qquad (A.2)$$

*This may also be written in a more familiar form as*

$$\int_{X \times Y} f(x, y) \, \mathrm{d}(\mu \times \nu)(x, y) = \int_X \left(\int_Y f(x, y) \, \mathrm{d}\nu(y)\right) \mathrm{d}\mu(x)$$

$$= \int_Y \left(\int_X f(x, y) \, \mathrm{d}\mu(x)\right) \mathrm{d}\nu(y).$$

We note that integration makes sense for functions taking values in some other spaces as well, and this will be discussed further in Sect. B.7.

# A.4 Radon–Nikodym Derivatives

One of the fundamental ideas in measure theory concerns the properties of a probability measure viewed from the perspective of a given measure. Fix a $\sigma$-finite measure space $(X, \mathscr{B}, \mu)$ and some measure $\nu$ defined on $\mathscr{B}$.

- The measure $\nu$ is *absolutely continuous* with respect to $\mu$, written $\nu \ll \mu$, if $\mu(A) = 0 \implies \nu(A) = 0$ for any $A \in \mathscr{B}$.

- If $\nu \ll \mu$ and $\mu \ll \nu$ then $\mu$ and $\nu$ are said to be *equivalent*.
- The measures $\mu$ and $\nu$ are mutually *singular*, written $\mu \perp \nu$, if there exist disjoint sets $A$ and $B$ in $\mathscr{B}$ with $A \cup B = X$ and with $\mu(A) = \nu(B) = 0$.

These notions are related by two important theorems.

**Theorem A.14 (Lebesgue decomposition).** *Given $\sigma$-finite measures $\mu$ and $\nu$ on $(X, \mathscr{B})$, there are measures $\nu_0$ and $\nu_1$ with the properties that*

(1) $\nu = \nu_0 + \nu_1$;
(2) $\nu_0 \ll \mu$; and
(3) $\nu_1 \perp \mu$.

*The measures $\nu_0$ and $\nu_1$ are uniquely determined by these properties.*

**Theorem A.15 (Radon–Nikodym derivative**[105]**).** *If $\nu \ll \mu$ then there is a measurable function $f \geqslant 0$ on $X$ with the property that*

$$\nu(A) = \int_A f \, \mathrm{d}\mu$$

*for any set $A \in \mathscr{B}$.*

By analogy with the fundamental theorem of calculus (Theorem A.25), the function $f$ is written $\frac{\mathrm{d}\nu}{\mathrm{d}\mu}$ and is called the *Radon–Nikodym derivative* of $\nu$ with respect to $\mu$. Notice that for any two measures $\mu_1, \mu_2$ we can form a new measure $\mu_1 + \mu_2$ simply by defining $(\mu_1 + \mu_2)(A) = \mu_1(A) + \mu_2(A)$ for any measurable set $A$. Then $\mu_i \ll \mu_1 + \mu_2$, so there is a Radon–Nikodym derivative of $\mu_i$ with respect to $\mu_1 + \mu_2$ for $i = 1, 2$.

## A.5 Convergence Theorems

The most important distinction between integration on $L^p$ spaces as defined above and Riemann integration on bounded Riemann-integrable functions is that the $L^p$ functions are closed under several natural limiting operations, allowing for the following important convergence theorems.

**Theorem A.16 (Monotone Convergence Theorem).** *If $f_1 \leqslant f_2 \leqslant \cdots$ is a pointwise increasing sequence of integrable functions on the probability space $(X, \mathscr{B}, \mu)$, then $f = \lim_{n \to \infty} f_n$ satisfies*

$$\int f \, \mathrm{d}\mu = \lim_{n \to \infty} \int f_n \, \mathrm{d}\mu.$$

*In particular, if $\lim_{n \to \infty} \int f_n \, \mathrm{d}\mu < \infty$, then $f$ is finite almost everywhere.*

**Theorem A.17 (Fatou's Lemma).** *Let* $(f_n)_{n \geqslant 1}$ *be a sequence of measurable real-valued functions on a probability space, all bounded below by some integrable function. If* $\liminf_{n \to \infty} \int f_n \, \mathrm{d}\mu < \infty$ *then* $\liminf_{n \to \infty} f_n$ *is integrable, and*

$$\int \liminf_{n \to \infty} f_n \, \mathrm{d}\mu \leqslant \liminf_{n \to \infty} \int f_n \, \mathrm{d}\mu.$$

**Theorem A.18 (Dominated Convergence Theorem).** *If* $h : X \to \mathbb{R}$ *is an integrable function and* $(f_n)_{n \geqslant 1}$ *is a sequence of measurable real-valued functions which are dominated by* $h$ *in the sense that* $|f_n| \leqslant h$ *for all* $n \geqslant 1$, *and* $\lim_{n \to \infty} f_n = f$ *exists almost everywhere, then* $f$ *is integrable and*

$$\int f \, \mathrm{d}\mu = \lim_{n \to \infty} \int f_n \, \mathrm{d}\mu.$$

## A.6 Well-Behaved Measure Spaces

It is convenient to slightly extend the notion of a Borel probability space as follows (cf. Definition 5.13).

**Definition A.19.** Let $X$ be a dense Borel subset of a compact metric space $\overline{X}$, with a probability measure $\mu$ defined on the restriction of the Borel $\sigma$-algebra $\mathscr{B}$ to $X$. The resulting probability space $(X, \mathscr{B}, \mu)$ is a *Borel probability space*\*.

For our purposes, this is the most convenient notion of a measure space that is on the one hand sufficiently general for the applications needed, while on the other has enough structure to permit explicit and convenient proofs.

A circle of results called Lusin's theorem [237] (or Luzin's theorem) show that measurable functions are continuous off a small set. These results are true in almost any context where continuity makes sense, but we state a form of the result here in the setting needed.

**Theorem A.20 (Lusin).** *Let* $(X, \mathscr{B}, \mu)$ *be a Borel probability space and let* $f : X \to \mathbb{R}$ *be a measurable function. Then, for any* $\varepsilon > 0$, *there is a continuous function* $g : X \to \mathbb{R}$ *with the property that*

$$\mu \left( \{ x \in X \mid f(x) \neq g(x) \} \right) < \varepsilon.$$

As mentioned in the endnote to Definition 5.13, there is a slightly different formulation of the standard setting for ergodic theory, in terms of Lebesgue spaces.

---

\* Commonly the $\sigma$-algebra $\mathscr{B}$ is enlarged to its *completion* $\mathscr{B}_\mu$, which is the smallest $\sigma$-algebra containing both $\mathscr{B}$ and all subsets of null sets with respect to $\mu$. It is also standard to allow any probability space that is isomorphic to $(X, \mathscr{B}_\mu, \mu)$ in Definition A.19 as a measure space to be called a Lebesgue space.

**Definition A.21.** A probability space is a *Lebesgue space* if it is isomorphic as a measure space to

$$\left( [0,s] \sqcup A, \mathscr{B}, m_{[0,s]} + \sum_{a \in A} p_a \delta_a \right)$$

for some countable set $A$ of *atoms* and numbers $s, p_a > 0$ with

$$s + \sum_{a \in A} p_a = 1,$$

where $\mathscr{B}$ comprises unions of Lebesgue measurable sets in $[0,s]$ and arbitrary subsets of $A$, $m_{[0,s]}$ is the Lebesgue measure on $[0,s]$, and $\delta_a$ is the Dirac measure defined by $\delta_a(B) = \chi_B(a)$.

The next result shows, *inter alia*, that this notion agrees with that used in Definition A.19 (a proof of this may be found in the book of Parthasarathy [280, Chap. V]) up to completion of the measure space (a measure space is complete if all subsets of a null set are measurable and null). We will not use this result here.

**Theorem A.22.** *A probability space is a Lebesgue space in the sense of Definition* A.21 *if and only if it is isomorphic to* $(X, \mathscr{B}_\mu, \mu)$ *for some probability measure* $\mu$ *on the completion* $\mathscr{B}_\mu$ *of the Borel* $\sigma$-*algebra* $\mathscr{B}$ *of a complete separable metric space* $X$.

The function spaces from Sect. A.3 are particularly well-behaved for Lebesgue spaces.

**Theorem A.23 (Riesz–Fischer**[106]**).** *Let* $(X, \mathscr{B}, \mu)$ *be a Lebesgue space. For any* $p$, $1 \leqslant p < \infty$, *the space* $L_\mu^p$ *is a separable Banach space with respect to the* $\| \cdot \|_p$-*norm. In particular,* $L_\mu^2$ *is a separable Hilbert space.*

# A.7 Lebesgue Density Theorem

The space $\mathbb{R}$ together with the usual metric and Lebesgue measure $m_{\mathbb{R}}$ is a particularly important and well-behaved special case, and here it is possible to say that a set of positive measure is thick in a precise sense.

**Theorem A.24 (Lebesgue**[107]**).** *If* $A \subseteq \mathbb{R}$ *is a measurable set, then*

$$\lim_{\varepsilon \to 0} \frac{1}{2\varepsilon} m_{\mathbb{R}} \left( A \cap (a - \varepsilon, a + \varepsilon) \right) = 1$$

*for* $m_{\mathbb{R}}$-*almost every* $a \in A$.

A point $a$ with this property is said to be a *Lebesgue density point* or a *point with Lebesgue density* 1. An equivalent and more familiar formulation of the result is a form of the fundamental theorem of calculus.

**Theorem A.25.** *If $f : \mathbb{R} \to \mathbb{R}$ is an integrable function then*

$$\lim_{\varepsilon \to 0} \frac{1}{\varepsilon} \int_s^{s+\varepsilon} f(t)\,\mathrm{d}t = f(s)$$

*for $m_{\mathbb{R}}$-almost every $s \in [0, \infty)$.*

The equivalence of Theorem A.24 and A.25 may be seen by approximating an integrable function with simple functions.

## A.8 Substitution Rule

Let $O \subseteq \mathbb{R}^n$ be an open set, and let $\phi : O \to \mathbb{R}^n$ be a $C^1$-map with Jacobian $J_\phi = |\det \mathrm{D}\,\phi|$. Then for any measurable function $f \geq 0$ (or for any integrable function $f$) defined on $\phi(O) \subseteq \mathbb{R}^n$ we have[108].

$$\int_O f(\phi(\mathbf{x}))J_\phi(\mathbf{x})\,\mathrm{d}m_{\mathbb{R}^n}(\mathbf{x}) = \int_{\phi(O)} f(\mathbf{y})\,\mathrm{d}m_{\mathbb{R}^n}(\mathbf{y}). \tag{A.3}$$

We recall the definition of the push-forward of a measure. Let $(X, \mathscr{B}_X)$ and $(Y, \mathscr{B}_Y)$ be two spaces equipped with $\sigma$-algebras. Let $\mu$ be a measure on $X$ defined on $\mathscr{B}_X$, and let $\phi : X \to Y$ be measurable. Then the push-forward $\phi_*\mu$ is the measure on $(Y, \mathscr{B}_Y)$ defined by $(\phi_*\mu)(B) = \mu(\phi^{-1}(B))$ for all $B \in \mathscr{B}_Y$.

The substitution rule allows us to calculate the push-forward of the Lebesgue measure under smooth maps as follows.

**Lemma A.26.** *Let $O \subseteq \mathbb{R}^n$ be open, let $\phi : O \to \mathbb{R}^n$ be a smooth injective map with non-vanishing Jacobian $J_\phi = |\det \mathrm{D}\,\phi|$. Then the push-forward $\phi_*m_O$ of the Lebesgue measure $m_O = m_{\mathbb{R}^n}|_O$ restricted to $O$ is absolutely continuous with respect to $m_{\mathbb{R}^n}$ and is given by*

$$\mathrm{d}\phi_*m_O = J_\phi^{-1} \circ \phi^{-1}\,\mathrm{d}m_{\phi(O)}.$$

*Moreover, if we consider a measure $\mathrm{d}\mu = F\,\mathrm{d}m_O$ absolutely continuous with respect to $m_O$, then similarly*

$$\mathrm{d}\phi_*\mu = F \circ \phi^{-1} J_\phi^{-1} \circ \phi^{-1}\,\mathrm{d}m_{\phi(O)}.$$

PROOF. Recall that under the assumptions of the lemma, $\phi^{-1}$ is smooth and $J_{\phi^{-1}} = J_\phi^{-1} \circ \phi^{-1}$. Therefore, by (A.3) and the definition of the push-forward,

$$\int_{\phi(O)} f(x) J_\phi^{-1}\left(\phi^{-1}(x)\right) \mathrm{d}m_{\mathbb{R}^n}(x) = \int_{\phi(O)} f\left(\phi(\phi^{-1}(x))\right) J_{\phi^{-1}}(x) \,\mathrm{d}m_{\mathbb{R}^n}(x)$$

$$= \int_O f(\phi(y)) \,\mathrm{d}m_{\mathbb{R}^n}(y)$$

$$= \int_{\phi(O)} f(x) \,\mathrm{d}\phi_* m_O(x)$$

for any characteristic function $f = \chi_B$ of a measurable set $B \subseteq \phi(O)$. This implies the first claim. Moreover, for any measurable functions $f \geqslant 0, F \geqslant 0$ defined on $\phi(O), O$ respectively,

$$\int_{\phi(O)} f(x) F(\phi^{-1}(x)) J_\phi^{-1}(\phi^{-1}(x)) \,\mathrm{d}m_{\mathbb{R}^n}(x) = \int_O f(\phi(y)) F(y) \,\mathrm{d}m_{\mathbb{R}^n},$$

which implies the second claim.                                                       □

# Notes to Appendix A

[102](Page 405) This result was stated by Borel [40, p. 252] for independent events as part of his study of normal numbers, but as pointed out by Barone and Novikoff [18] there are some problems with the proofs. Cantelli [46] noticed that half of the theorem holds without independence; this had also been noted by Hausdorff [142] in a special case. Erdős and Rényi [84] showed that the result holds under the much weaker assumption of pairwise independence.

[103](Page 406) This is shown, for example, in Parthasarathy [280, Th. 1.2]: defining a Borel set $A$ to be regular if, for any $\varepsilon > 0$, there is an open set $O_\varepsilon$ and a closed set $C_\varepsilon$ with $C_\varepsilon \subseteq A \subseteq O_\varepsilon$ and $\mu(O_\varepsilon \setminus C_\varepsilon) < \varepsilon$, it may be shown that the collection of all regular sets forms a $\sigma$-algebra and contains the closed sets.

[104](Page 409) A form of this theorem goes back to Cauchy for continuous functions on the reals, and this was extended by Lebesgue [220] to bounded measurable functions. Fubini [97] extended this to integrable functions, showing that if $f : [a, b] \times [c, d] \rightarrow \mathbb{R}$ is integrable then $y \mapsto f(x, y)$ is integrable for almost every $x$, and proving (A.2). Tonelli [362] gave the formulation here, for non-negative functions on products of $\sigma$-finite spaces. Complete proofs may be found in Royden [320] or Lieb and Loss [229, Th. 1.12]. While the result is robust and of central importance, some hypotheses are needed: if the function is not integrable or the spaces are not $\sigma$-finite, the integrals may have different values. A detailed treatment of the minimal hypotheses needed for a theorem of Fubini type, along with counterexamples and applications, is given by Fremlin [96, Sect. 252].

[105](Page 410) This result is due to Radon [297] when $\mu$ is Lebesgue measure on $\mathbb{R}^n$, and to Nikodym [272] in the general case.

[106](Page 412) This result emerged in several notes of Riesz and two notes of Fischer [91, 92], with a full treatment of the result that $L^2(\mathbb{R})$ is complete appearing in a paper of Riesz [311].

[107](Page 412) This is due to Lebesgue [220], and a convenient source for the proof is the monograph of Oxtoby [276]. Notice that Theorem A.24 expresses how constrained measurable sets are: it is impossible, for example, to find a measurable subset $A$ of $[0, 1]$ with the property that $m_{\mathbb{R}}(A \cap [a, b]) = \frac{1}{2}(b - a)$ for all $b > a$. While a measurable subset

of measure $\frac{1}{2}$ may have an intricate structure, it cannot occupy only half of the space on all possible scales.

[108] (Page 413) The usual hypotheses are that the map $\phi$ is injective and the Jacobian non-vanishing; these may be relaxed considerably, and the theorem holds in very general settings both measurable (see Hewitt and Stromberg [152]) and smooth (see Spivak [349]).

# Appendix B
# Functional Analysis

Functional analysis abstracts the basic ideas of real and complex analysis in order to study spaces of functions and operators between them[109]. A *normed space* is a vector space $E$ over a field $\mathbb{F}$ (either $\mathbb{R}$ or $\mathbb{C}$) equipped with a map $\| \cdot \|$ from $E \to \mathbb{R}$ satisfying the properties

- $\|x\| \geqslant 0$ for all $x \in E$ and $\|x\| = 0$ if and only if $x = 0$;
- $\|\lambda x\| = |\lambda| \|x\|$ for all $x \in E$ and $\lambda \in \mathbb{F}$; and
- $\|x + y\| \leqslant \|x\| + \|y\|$.

If $(E, \| \cdot \|)$ is a normed space, then $\mathsf{d}(x, y) = \|x - y\|$ defines a metric on $E$. A *semi-norm* is a map with the first property weakened to

- $\|x\| \geqslant 0$ for all $x \in E$.

A normed space is a *Banach space* if it is complete as a metric space: that is, the condition that the sequence $(x_n)$ is Cauchy (for all $\varepsilon > 0$ there is some $N$ for which $m > n > N$ implies $\|x_m - x_n\| < \varepsilon$) is equivalent to the condition that the sequence $(x_n)$ converges (there is some $y \in E$ with the property that for all $\varepsilon > 0$ there is some $N$ for which $n > N$ implies $\|x_n - y\| < \varepsilon$).

As discussed in Sect. A.3, for any probability space $(X, \mathscr{B}, \mu)$, the norm $\| \cdot \|_p$ makes the space $L^p_\mu$ into a Banach space.

## B.1 Sequence Spaces

For $1 \leqslant p < \infty$ and a countable set $\Gamma$ (in practice this will be $\mathbb{N}$ or $\mathbb{Z}$) we denote by $\ell^p(\Gamma)$ the space

$$\{x = (x_\gamma) \in \mathbb{R}^\Gamma \mid \sum_{\gamma \in \Gamma} |x_\gamma|^p < \infty\},$$

and for $p = \infty$ write

$$\ell^\infty(\Gamma) = \{x = (x_\gamma) \in \mathbb{R}^\Gamma \mid \sup_{\gamma \in \Gamma} |x_\gamma| < \infty\}.$$

The norms $\|x\|_p = (\sum_{\gamma \in \Gamma} |x_\gamma|^p)^{1/p}$ and $\|x\|_\infty = \sup_{\gamma \in \Gamma} |x_\gamma|$ make $\ell_p(\Gamma)$ into a complete space for $1 \leqslant p \leqslant \infty$.

## B.2 Linear Functionals

A vector space $V$ over a normed field $\mathbb{F}$, equipped with a topology $\tau$, and with the property that each point of $V$ is closed and the vector space operations (addition of vectors and multiplication by scalars) are continuous is called a *topological vector space*. Any topological vector space is Hausdorff. If $0 \in V$ has an open neighborhood with compact closure, then $V$ is said to be locally compact.

Let $\lambda : V \to W$ be a linear map between topological vector spaces. Then the following properties are equivalent:

(1) $\lambda$ is continuous;
(2) $\lambda$ is continuous at $0 \in V$;
(3) $\lambda$ is uniformly continuous in the sense that for any neighborhood $O_W$ of $0 \in W$ there is a neighborhood $O_V$ of $0 \in V$ for which $v - v' \in O_V$ implies $\lambda(v) - \lambda(v') \in O_W$ for all $v, v' \in V$.

Of particular importance are linear maps into the ground field. For a linear map $\lambda : V \to \mathbb{F}$, the following properties are equivalent:

(1) $\lambda$ is continuous;
(2) the kernel $\ker(\lambda) = \{v \in V \mid \lambda(v) = 0\}$ is a closed subset of $V$;
(3) $\ker(\lambda)$ is not dense in $V$;
(4) $\lambda$ is bounded on some neighborhood of $0 \in V$.

Continuous linear maps $\lambda : V \to \mathbb{F}$ are particularly important: they are called *linear functionals* and the collection of all linear functionals is denoted $V^*$. If $V$ has a norm $\|\cdot\|$ defining the topology $\tau$, then $V^*$ is a normed space under the norm

$$\|\lambda\|_{\text{operator}} = \sup_{\|v\|=1} \{|\lambda(v)|_\mathbb{F}\}$$

where $|\cdot|_\mathbb{F}$ is the norm on the ground field $\mathbb{F}$. The normed space $V^*$ is complete if $\mathbb{F}$ is complete. The next result asserts that there are many linear functionals, and allows them to be constructed in a flexible and controlled way.

**Theorem B.1 (Hahn–Banach[110]).** *Let $\lambda : U \to \mathbb{F}$ be a linear functional defined on a subspace $U \subseteq V$ of a normed linear space and let*

$$p : V \to \mathbb{R}_{\geqslant 0}$$

*be a semi-norm. If* $|f(u)| \leqslant p(u)$ *for* $u \in U$, *then there is a linear functional* $\lambda' : V \to \mathbb{F}$ *that extends* $\lambda$ *in the sense that* $\lambda'(u) = \lambda(u)$ *for all* $u \in U$, *and* $|\lambda'(v)| \leqslant p(v)$ *for all* $v \in V$.

# B.3 Linear Operators

It is conventional to call maps between normed spaces *operators*, because in many of the applications the elements of the normed spaces are themselves functions. A map $f : E \to F$ between normed vector spaces $(E, \| \cdot \|_E)$ and $(F, \| \cdot \|_F)$ is *continuous at* $a$ if for any $\varepsilon > 0$ there is some $\delta > 0$ for which

$$\|x - a\|_E < \delta \implies \|f(x) - f(a)\|_F < \varepsilon,$$

is *continuous* if it is continuous at every point, and is *bounded* if there is some $R$ with $\|f(x)\|_F \leqslant R\|x\|_E$ for all $x \in E$. If $f : E \to F$ is linear, then the following are equivalent:

- $f$ is continuous;
- $f$ is bounded;
- $f$ is continuous at $0 \in E$.

A linear map $f : E \to F$ is an *isometry* if $\|f(x)\|_F = \|x\|_E$ for all $x \in E$, and is an *isomorphism of normed spaces* if $f$ is a bijection and both $f$ and $f^{-1}$ are continuous.

Norms $\| \cdot \|_1$ and $\| \cdot \|_2$ on $E$ are *equivalent* if the identity map

$$(E, \| \cdot \|_1) \to (E, \| \cdot \|_2)$$

is an isomorphism of normed spaces; equivalently, if there are positive constants $r, R$ for which

$$r\|x\|_1 \leqslant \|x\|_2 \leqslant R\|x\|_1$$

for all $x \in E$. If $E, F$ are finite-dimensional, then all norms on $E$ are equivalent and all linear maps $E \to F$ are continuous.

**Theorem B.2 (Open Mapping Theorem).** *If* $f : E \to F$ *is a continuous bijection of Banach spaces, then* $f$ *is an isomorphism.*

The space of all bounded linear maps from $E$ to $F$ is denoted $B(E, F)$; this is clearly a vector space. Defining

$$\|f\|_{\text{operator}} = \sup_{\|x\|_E \leqslant 1} \{\|f(x)\|_F\}$$

makes $B(E, F)$ into a normed space, and if $F$ is a Banach space then $B(E, F)$ is a Banach space. An important special case is the space of linear functionals, $E^* = B(E, \mathbb{F})$.

Assume now that $E$ and $F$ are Banach spaces. An operator $f : E \to F$ is *compact* if the image $f(U)$ of the open unit ball $U = \{x \in E \mid \|x\|_E < 1\}$ has compact closure in $F$. Equivalently, an operator is compact if and only if every bounded sequence $(x_n)$ in $E$ contains a subsequence $(x_{n_j})$ with the property that $\big(f(x_{n_j})\big)$ converges in $F$. Many operators that arise naturally in the study of integral equations, for example the Hilbert–Schmidt integral operators $T$ defined on $L^2_\mu(X)$ by

$$(Tf)(s) = \int_X K(s,t)\, \mathrm{d}\mu(t)$$

for some *kernel* $K \in L^2_{\mu \times \mu}(X \times X)$, are compact operators.

Now assume that $E$ is a Banach space. Then $B(E) = B(E,E)$ is not only a Banach space but also an algebra: if $S, T \in B(E)$ then $ST \in B(E)$ where $(ST)(x) = S(T(x))$, and $\|ST\| \leqslant \|S\|\|T\|$. Write $I$ for the identity operator, and define the *spectrum* of an operator $T \in B(E)$ to be

$$\sigma_{\mathrm{operator}}(T) = \{\lambda \in \mathbb{F} \mid (T - \lambda I) \text{ does not have a continuous inverse}\}.$$

**Theorem B.3.** *Let $E$ and $F$ be Banach spaces.*

(1) *If $T \in B(E,E)$ is compact and $\lambda \neq 0$, then the kernel of $T - \lambda I$ is finite-dimensional.*

(2) *If $E$ is not finite-dimensional and $T \in B(E)$ is compact, then $\sigma_{\mathrm{operator}}(T)$ contains $0$.*

(3) *If $S, T \in B(E)$ and $T$ is compact, then $ST$ and $TS$ are compact.*

Functional analysis on Hilbert space is particularly useful in ergodic theory, because each measure-preserving system $(X, \mathscr{B}, \mu, T)$ has an associated Koopman operator $U_T : L^2_\mu \to L^2_\mu$ defined by $U_T(f) = f \circ T$.

An invertible measure-preserving transformation $T$ is said to have *continuous spectrum* if $1$ is the only eigenvalue of $U_T$ and any eigenfunction of $U_T$ is a constant.

**Theorem B.4 (Spectral Theorem).** *Let $U$ be a unitary operator on a complex Hilbert space $\mathscr{H}$.*

(1) *For each element $f \in \mathscr{H}$ there is a unique finite Borel measure $\mu_f$ on $\mathbb{S}^1$ with the property that*

$$\langle U^n f, f \rangle = \int_{\mathbb{S}^1} z^n \, \mathrm{d}\mu_f(z) \tag{B.1}$$

*for all $n \in \mathbb{Z}$.*

(2) *The map*

$$\sum_{n=-N}^{N} c_n z^n \mapsto \sum_{n=-N}^{N} c_n U^n f$$

*extends by continuity to a unitary isomorphism between $L^2(\mathbb{S}^1, \mu_f)$ and the smallest $U$-invariant subspace in $\mathcal{H}$ containing $f$.*

*(3) If $T$ has continuous spectrum and $f \in L^2_\mu$ has $\int_X f \, d\mu = 0$, then the spectral measure $\mu_f$ associated to the unitary operator $U_T$ is non-atomic.*

We will also need two fundamental compactness results due to Alaoglu, Banach and Tychonoff[111].

**Theorem B.5 (Tychonoff).** *If $\{X_\gamma\}_{\gamma \in \Gamma}$ is a collection of compact topological spaces, then the product space $\prod_{\gamma \in \Gamma} X_\gamma$ endowed with the product topology is itself a compact space.*

**Theorem B.6 (Alaoglu).** *Let $X$ be a topological vector space with $U$ a neighborhood of $0$ in $X$. Then the set of linear operators $x^* : X \to \mathbb{R}$ with $\sup_{x \in U} |x^*(x)| \leqslant 1$ is weak\*-compact.*

# B.4 Continuous Functions

Let $(X, \mathsf{d})$ be a compact metric space. The space $C_\mathbb{C}(X)$ of continuous functions $f : X \to \mathbb{C}$ is a metric space with respect to the uniform metric

$$\mathsf{d}(f, g) = \sup_{x \in X} |f(x) - g(x)|;$$

defining $\|f\|_\infty = \sup_{x \in X} |f(x)|$ makes $C_\mathbb{C}(X)$ into a normed space.

It is often important to know when a subspace of a normed space of functions is dense.

**Theorem B.7 (Stone–Weierstrass Theorem[112]).** *Let $(X, \mathsf{d})$ be a compact metric space, and let $\mathscr{A} \subseteq C_\mathbb{C}(X)$ be a linear subspace with the following properties:*

- *$\mathscr{A}$ is closed under multiplication (that is, $\mathscr{A}$ is a subalgebra);*
- *$\mathscr{A}$ contains the constant functions;*
- *$\mathscr{A}$ separates points (for $x \neq y$ there is a function $f \in \mathscr{A}$ with $f(x) \neq f(y)$); and*
- *for any $f \in \mathscr{A}$, the complex conjugate $\overline{f} \in \mathscr{A}$.*

*Then $\mathscr{A}$ is dense in $C_\mathbb{C}(X)$.*

**Lemma B.8.** *The spaces $C_\mathbb{C}(X)$ and $C(X)$ are separable metric spaces with respect to the metric induced by the uniform norm.*

PROOF. Let $\{x_1, x_2, \dots\}$ be a dense set in $X$, and define a set

$$F = \{f_1, f_2, \dots\}$$

of continuous functions by $f_n(x) = \mathsf{d}(x, x_n)$ where $\mathsf{d}$ is the given metric on $X$. The set $F$ separates points since the set $\{x_1, x_2, \dots\}$ is dense. It follows that the algebra generated by $F$ is dense in $C(X)$ by the Stone–Weierstrass Theorem (Theorem B.7). The same holds for the $\mathbb{Q}$-algebra generated by $F$ (that is, for the set of finite linear combinations $\sum_{i=1}^{m} c_i h_i$ with $c_i \in \mathbb{Q}$ and $h_i = \prod_{k=1}^{K_i} g_{k,i}$ with $g_{k,i} \in F$ and $K_i \in \mathbb{N}$). However, this set is countable, which shows the lemma for real-valued functions. The same argument using the $\mathbb{Q}(i)$-algebra gives the complex case.                                    □

The next lemma is a simple instance of a more general result of Urysohn that characterizes normal spaces[113].

**Theorem B.9 (Tietze–Urysohn extension).** *Any continuous real-valued function on a closed subspace of a normal topological space may be extended to a continuous real-valued function on the entire space.*

We will only need this in the metric setting, and any metric space is normal as a topological space.

**Corollary B.10.** *If $(X, \mathsf{d})$ is a metric space, then for any non-empty closed sets $A, B \subseteq X$ with $A \cap B = \varnothing$, there is a continuous function $f : X \to [0,1]$ with $f(A) = \{0\}$ and $f(B) = \{1\}$.*

# B.5 Measures on Compact Metric Spaces

The material in this section deals with measures and linear operators. It is standard; a convenient source is Parthasarathy [280].

Let $(X, \mathsf{d})$ be a compact metric space, with Borel $\sigma$-algebra $\mathscr{B}$. Denote by $\mathscr{M}(X)$ the space of Borel probability measures on $X$. The dual space $C(X)^*$ of continuous real functionals on the space $C(X)$ of continuous functions $X \to \mathbb{R}$ can be naturally identified with the space of finite signed measures on $X$. A functional $F : C(X) \to \mathbb{C}$ is called *positive* if $f \geqslant 0$ implies that $F(f) \geqslant 0$, and the *Riesz representation theorem* states that any continuous positive functional $F$ is defined by a unique measure $\mu \in \mathscr{M}(X)$ via

$$F(f) = \int_X f \, \mathsf{d}\mu.$$

The main properties of $\mathscr{M}(X)$ needed are the following. Recall that a set $\mathscr{M}$ of measures is said to be *convex* if the convex combination

$$s\mu_1 + (1 - s)\mu_2$$

lies in $\mathscr{M}$ for any $\mu_1, \mu_2 \in \mathscr{M}(X)$ and $s \in [0,1]$.

**Theorem B.11.** (1) $\mathscr{M}(X)$ *is convex.*

(2) *For $\mu_1, \mu_2 \in \mathcal{M}(X)$,*

$$\int f \, d\mu_1 = \int f \, d\mu_2 \tag{B.2}$$

*for all $f \in C(X)$ if and only if $\mu_1 = \mu_2$.*

(3) *The weak\*-topology on $\mathcal{M}(X)$ is the weakest topology making each of the evaluation maps*

$$\mu \mapsto \int f \, d\mu$$

*continuous for any $f \in C(X)$; this topology is metrizable and in this topology $\mathcal{M}(X)$ is compact.*

(4) *In the weak\*-topology, $\mu_n \to \mu$ if and only if any of the following conditions hold:*

- $\int f \, d\mu_n \to \int f \, d\mu$ *for every $f \in C(X)$;*
- *for every closed set $C \subseteq X$, $\limsup_{n \to \infty} \mu_n(C) \leqslant \mu(C)$;*
- *for every open set $O \subseteq X$, $\liminf_{n \to \infty} \mu_n(O) \geqslant \mu(O)$;*
- *for every Borel set $A$ with $\mu(\partial(A)) = 0$, $\mu_n(A) \to \mu(A)$.*

PROOF OF PART (3). Recall that by the Riesz representation theorem the dual space $C(X)^*$ of continuous linear real functionals $C(X) \to \mathbb{R}$ with the operator norm coincides with the space of finite signed measures, with the functional being given by integration with respect to the measure. Moreover, by the Banach–Alaoglu theorem the unit ball $B_1$ in $C(X)^*$ is compact in the weak\*-topology. It follows that

$$\mathcal{M}(X) = \left\{ \mu \in C(X) \mid \int 1 \, d\mu = 1, \int f \, d\mu \geqslant 0 \text{ for } f \in C(X) \text{ with } f \geqslant 0 \right\}$$

is a weak\*-closed subset of $B_1$ and is therefore compact in the weak\*-topology.

To show that the weak\*-topology is metrizable on $\mathcal{M}(X)$ we use the fact that $C(X)$ is separable by Lemma B.8. Suppose that $\{f_1, f_2, \dots\}$ is a dense set in $C(X)$. Then the weak\*-topology on $\mathcal{M}(X)$ is generated by the intersections of the open neighborhoods of $\mu \in \mathcal{M}(X)$ defined by

$$V_{\varepsilon,n}(\mu) = \left\{ \nu \in \mathcal{M}(X) \mid \left| \int f_n \, d\nu - \int f_n \, d\mu \right| < \varepsilon \right\}.$$

This holds since for any $f \in C(X)$ and neighborhood

$$V_{\varepsilon,f}(\mu) = \left\{ \nu \in \mathcal{M}(X) \mid \left| \int f \, d\nu - \int f \, d\mu \right| < \varepsilon \right\}$$

we can find some $n$ with $\|f_n - f\| < \frac{\varepsilon}{3}$ and it is easily checked that

$$V_{\varepsilon/3,n}(\mu) \subseteq V_{\varepsilon,f}(\mu).$$

Define

$$\mathsf{d}_{\mathscr{M}}(\mu, \nu) = \sum_{n=1}^{\infty} \frac{1}{2^n} \frac{|\int f_n \, \mathrm{d}\mu - \int f_n \, \mathrm{d}\nu|}{1 + |\int f_n \, \mathrm{d}\mu - \int f_n \, \mathrm{d}\nu|} \qquad (\text{B.3})$$

for $\mu, \nu \in \mathscr{M}(X)$. A calculation shows that $\mathsf{d}_{\mathscr{M}}$ is a metric on $\mathscr{M}(X)$.

We finish the proof by comparing the metric neighborhoods $B_\delta(\mu)$ defined by $\mathsf{d}_{\mathscr{M}}$ with the neighborhoods $V_{\varepsilon,n}(\mu)$. Fix $\delta > 0$ and choose $K$ such that $\sum_{n=K+1}^{\infty} \frac{1}{2^n} < \frac{\delta}{2}$. Then, for sufficiently small $\varepsilon > 0$, any measure

$$\nu \in V_{\varepsilon, f_1}(\mu) \cap \cdots \cap V_{\varepsilon, f_K}(\mu)$$

will satisfy

$$\sum_{n=1}^{K} \frac{1}{2^n} \frac{|\int f_n \, \mathrm{d}\mu - \int f_n \, \mathrm{d}\nu|}{1 + |\int f_n \, \mathrm{d}\mu - \int f_n \, \mathrm{d}\nu|} < \frac{\delta}{2},$$

showing that $\nu \in B_\delta(\mu)$. Similarly, if $n \geqslant 1$ and $\varepsilon > 0$ are given, we may choose $\delta$ small enough to ensure that $\frac{1}{2^n} \frac{s}{1+s} < \delta$ implies that $s < \varepsilon$. Then for any $\nu \in B_\delta(\mu)$ we will have $\nu \in V_{\varepsilon,n}$. It follows that the metric neighborhoods give the weak*-topology.                                                                                        $\square$

A continuous map $T : X \to X$ induces a map $T_* : \mathscr{M}(X) \to \mathscr{M}(X)$ defined by $T_*(\mu)(A) = \mu(T^{-1}A)$ for any Borel set $A \subseteq X$. Each $x \in X$ defines a measure $\delta_x$ by

$$\delta_x(A) = \begin{cases} 1 \text{ if } x \in A; \\ 0 \text{ if } x \notin A. \end{cases},$$

and $T_*(\delta_x) = \delta_{T(x)}$ for any $x \in X$.

For $f \geqslant 0$ a measurable map and $\mu \in \mathscr{M}(X)$,

$$\int f \, \mathrm{d}T_*\mu = \int f \circ T \, \mathrm{d}\mu. \qquad (\text{B.4})$$

This may be seen by the argument used in the first part of the proof of Lemma 2.6. In particular, (B.4) holds for all $f \in C(X)$, and from this it is easy to check that the map $T_* : \mathscr{M}(X) \to \mathscr{M}(X)$ is continuous with respect to the weak*-topology on $\mathscr{M}(X)$.

**Lemma B.12.** *Let $\mu$ be a measure in $\mathscr{M}(X)$. Then $\mu \in \mathscr{M}^T(X)$ if and only if $\int f \circ T \, \mathrm{d}\mu = \int f \, \mathrm{d}\mu$ for all $f \in C(X)$.*

The map $T_*$ is continuous and affine, so the set $\mathscr{M}^T(X)$ of $T$-invariant measures is a closed convex subset of $\mathscr{M}(X)$.

## B.6 Measures on Other Spaces

Our emphasis is on compact metric spaces and finite measure spaces, but we are sometimes forced to consider larger spaces. As mentioned in Definition A.5, a measure space is called $\sigma$-finite if it is a countable union of measurable sets with finite measure. Similarly, a metric space is called $\sigma$-compact if it is a countable union of compact subsets. A measure defined on the Borel sets of a metric space is called *locally finite* if every point of the space has an open neighborhood of finite measure.

**Theorem B.13.** *Let $\mu$ be a locally finite measure on the Borel sets of a $\sigma$-compact metric space. Then $\mu$ is* regular, *meaning that*

$$\mu(B) = \sup\{\mu(K) \mid K \subseteq B, K \ compact\} = \inf\{U \mid B \subseteq U, U \ open\}$$

*for any Borel set $B$.*

## B.7 Vector-valued Integration

It is often useful to integrate functions taking values in the space of measures (for example, in Theorem 6.2, in Sect. 6.5, and in Theorem 8.10). It is also useful to integrate functions $f : X \to V$ defined on a measure space $(X, \mathscr{B}, \mu)$ and taking values in a topological vector space $V$. The goal is to define $\int_X f \, d\mu$ as an element of $V$ that behaves like an integral: for example, if $\lambda : V \to \mathbb{R}$ is a continuous linear functional on $V$, then we would like

$$\lambda\left(\int_X f \, d\mu\right) = \int_X (\lambda f) \, d\mu \tag{B.5}$$

to hold whenever $\int_X f \, d\mu$ is defined. One (of many[(114)]) approaches to defining integration in this setting is to use the property in (B.5) to *characterize* the integral; in order for this to work we need to restrict attention to topological vector spaces in which there are enough functionals. We say that $V^*$ *separates points* in $V$ if for any $v \neq v'$ in $V$ there is a $\lambda \in V^*$ with $\lambda(v) \neq \lambda(v')$.

**Definition B.14.** Let $V$ be a topological vector space on which $V^*$ separates points, and let $f : X \to V$ be a function defined on a measure space $(X, \mathscr{B}, \mu)$ with the property that the scalar functions $\lambda(f) : X \to \mathbb{F}$ lie in $L^1_\mu(X)$ for every $\lambda \in V^*$. If there is an element $v \in V$ for which

$$\lambda(v) = \int_X (\lambda f) \, d\mu$$

for every $\lambda \in V^*$, then we define

$$\int_X f \, d\mu = v.$$

We start with the simplest example of integration for functions taking values in a Hilbert space.

*Example B.15.* If $V$ is a Hilbert space $\mathscr{H}$ then the characterization in Definition B.14 takes the form

$$\left\langle \int_X f \, d\mu, h \right\rangle = \int_X \langle f(x), h \rangle \, d\mu(x) \tag{B.6}$$

for all $h \in V$. Note that in this setting the right-hand side of (B.6) defines a continuous linear functional on $\mathscr{H}$. It follows that the integral $\int_X f \, d\mu$ exists by the Riesz representation theorem (see p. 422).

We now describe two more situations in which the existence of the integral can be established quite easily.

*Example B.16.* Let $V = L^p_\nu(Y)$ for a probability space $(Y, \nu)$ with $1 \leqslant p < \infty$, and let $F : X \times Y \to \mathbb{C}$ be an element of $L^p_{\mu \times \nu}(X \times Y)$. In this case we define

$$f : (X, \mu) \to V$$

by defining $f(x)$ to be the equivalence class of the function

$$F(x, \cdot) : y \longmapsto F(x, y).$$

We claim that $v = \int_X f \, d\mu$ exists and is given by the equivalence class of

$$\nu(y) = \int F(x, y) \, d\mu(x),$$

which is well-defined by the Fubini–Tonelli Theorem (Theorem A.13), since

$$L^p_{\mu \times \nu}(X \times Y) \subseteq L^1_{\mu \times \nu}(X \times Y).$$

To see this claim, recall that $V^* = L^q_\nu(Y)$ where $\frac{1}{p} + \frac{1}{q} = 1$, and let $w \in L^q_\nu(Y)$. Then $Fw \in L^1_{\mu \times \nu}(X \times Y)$ and so

$$\int_X \langle f(x), w \rangle \, d\mu = \int_X \int_Y F(x, y) w(y) \, d\nu \, d\mu$$

$$= \int_Y \int_X F(x, y) \, d\mu \cdot w(y) \, d\nu = \langle v, w \rangle$$

by Fubini, as required (notice that the last equation also implies that $v$ lies in $L^p_\nu(Y)$).

*Example B.17.* Suppose now that $V$ is a Banach space, and that $f : X \to V^*$ takes values in the dual space $V^*$ of $V$. Assume moreover that $\|f(x)\|$ is integrable and for any $v \in V$ the map $x \mapsto \langle v, f(x) \rangle$ is measurable (and hence automatically integrable, since $|\langle v, f(x) \rangle| \leqslant \|v\| \cdot \|f(x)\|$). Then

$$\int_X f(x) \, \mathrm{d}\mu(x) \in V^*$$

exists if we equip $V^*$ with the weak*-topology: In fact, we may let $\int_X f \, \mathrm{d}\mu$ be the map

$$V \ni v \longmapsto \int_X \langle v, f(x) \rangle \, \mathrm{d}\mu,$$

which depends linearly and continuously on $v$. Moreover, with respect to the weak*-topology on $V^*$ all continuous functionals on $V^*$ are evaluation maps on $V$.

The last example includes (and generalizes) the first two examples above, but also includes another important case. A similar construction is used in Sect. 5.3, in the construction of conditional measures.

*Example B.18.* Let $V = C(Y)$ for a compact metric space $Y$, so that $V^*$ is the space of signed finite measures on $Y$. Hence, for any probability-valued function

$$\Theta : X \to \mathscr{M}(Y)$$

with the property that $\int f(y) \, \mathrm{d}\Theta_x(y)$ depends measurably on $x \in X$, there exists a measure $\int_X \Theta_x \, \mathrm{d}\mu(x)$ on $Y$.

The next result gives a general criterion that guarantees existence of integrals in this sense (see Folland [94, App. A]).

**Theorem B.19.** *If $(X, \mathscr{B}, \mu)$ is a Borel probability space, $V^*$ separates points of $V$, $f : X \to V$ is measurable, and the smallest closed convex subset $I$ of $V$ containing $f(X)$ is compact, then the integral $\int_X f \, \mathrm{d}\mu$ in the sense of Definition B.14 exists, and lies in $I$.*

A second approach is to generalize Riemann integration to allow continuous functions defined on a compact metric space equipped with a Borel probability measure and taking values in a Banach space. If $V$ is a Banach space with norm $\| \cdot \|$, $(X, \mathrm{d})$ is a compact metric space with a finite Borel measure $\mu$, and $f : X \to V$ is continuous, then $f$ is uniformly continuous since $X$ is compact. Given a finite partition $\xi$ of $X$ into Borel sets and a choice $x_P \in X$ of a point $x_P \in P \in \xi$ for each atom $P$ of $\xi$, define the associated Riemann sum

$$R_\xi(f) = \sum_{P \in \xi} f(g_P) \mu(P).$$

It is readily checked that $R_\xi(f)$ converges as

$$\mathrm{diam}(\xi) = \max_{P \in \xi} \mathrm{diam}(P) \to 0,$$

and we define

$$\int_X f \, \mathrm{d}\mu = \lim_{\mathrm{diam}(\xi) \to 0} R_\xi(f)$$

to be the (Riemann) integral of $f$ with respect to $\mu$. It is clear from the definition that

$$\left\| \int_X f \, \mathrm{d}\mu \right\| \leqslant \int_X \|f\| \, \mathrm{d}\mu,$$

where the integral on the right-hand side has the same definition for the continuous function $x \mapsto \|f(x)\|$ taking values in $\mathbb{R}$ (and therefore coincides with the Lebesgue integral).

# Notes to Appendix B

[109](Page 417) Convenient sources for most of the material described here include Rudin [321] and Folland [94]; many of the ideas go back to Banach's monograph [17], originally published in 1932.

[110](Page 418) The Hahn–Banach theorem is usually proved using the Axiom of Choice (though it is not equivalent to it), and is often the most convenient form of the Axiom of Choice for functional analysis arguments. Significant special cases were found by Riesz [312, 313] in connection with extending linear functionals on $L^q$, and by Helly [147] who gave a more abstract formulation in terms of operators on normed sequence spaces. Hahn [132] and Banach [16] formulated the theorem as it is used today, using transfinite induction in a way that became a central tool in analysis.

[111](Page 421) Tychonoff's original proof appeared in 1929 [364]; the result requires and implies the Axiom of Choice. Alaoglu's theorem appeared in 1940 [4], clarifying the treatment of weak topologies by Banach [17].

[112](Page 421) Weierstrass proved that the polynomials are dense in $C[a, b]$ (corresponding to the algebra of real functions generated by the constants and the function $f(t) = t$). Stone [355] proved the result in great generality.

[113](Page 422) Urysohn [366] shows that a topological space is *normal* (that is, Hausdorff and with the property that disjoint closed sets have disjoint open neighborhoods) if and only if it has the extension property in Theorem B.9. A simple example of a non-normal topological space is the space of all functions $\mathbb{R} \to \mathbb{R}$ with the topology of pointwise convergence. Earlier, Tietze [361] had shown the same extension theorem for metric spaces, and in particular Corollary B.10, which for normal spaces is usually called Urysohn's lemma.

[114](Page 425) Integration can also be defined by emulating the real-valued case using partitions of the domain to produce a theory of vector-valued Riemann integration, or by using the Borel $\sigma$-algebra in $V$ to produce a theory of vector-valued Lebesgue integration: the article of Hildebrandt [153] gives an overview.

# Appendix C
# Topological Groups

Many groups arising naturally in mathematics have a topology with respect to which the group operations are continuous. Abstracting this observation has given rise to the important theory described here. We give a brief overview, but note that most of the discussions and examples in this volume concern concrete groups, so knowledge of the general theory summarized in this appendix is useful but often not strictly necessary.

## C.1 General Definitions

**Definition C.1.** A *topological group* is a group $G$ that carries a topology with respect to which the maps $(g, h) \mapsto gh$ and $g \mapsto g^{-1}$ are continuous as maps $G \times G \to G$ and $G \to G$ respectively.

Any topological group can be viewed as a uniform space in two ways: the *left uniformity* renders each left multiplication $g \mapsto hg$ into a uniformly continuous map while the *right uniformity* renders each right multiplication $g \mapsto gh$ into a uniformly continuous map. As a uniform space, any topological group is completely regular, and hence[115] is Hausdorff if it is $T_0$. Since the topological groups we need usually have a natural metric giving the topology, we will not need to develop this further.

The topological and algebraic structure on a topological group interact in many ways. For example, in any topological group $G$:

- the connected component of the identity is a closed normal subgroup;
- the inverse map $g \mapsto g^{-1}$ is a homeomorphism;
- for any $h \in G$ the left multiplication map $g \mapsto hg$ and the right multiplication map $g \mapsto gh$ are homeomorphisms;
- if $H$ is a subgroup of $G$ then the closure of $H$ is also a subgroup;
- if $H$ is a normal subgroup of $G$, then the closure of $H$ is also a normal subgroup.

M. Einsiedler, T. Ward, *Ergodic Theory*, Graduate Texts in Mathematics 259, DOI 10.1007/978-0-85729-021-2, © Springer-Verlag London Limited 2011

A topological group is called *monothetic* if it is Hausdorff and has a dense cyclic subgroup; a monothetic group is automatically abelian. Any generator of a dense subgroup is called a *topological generator*. Monothetic groups arise in many parts of dynamics.

A subgroup of a topological group is itself a topological group in the subspace topology. If $H$ is a subgroup of a topological group $G$ then the set of left (or right) cosets $G/H$ (or $H\backslash G$) is a topological space in the quotient topology (the smallest topology which makes the natural projection $g \mapsto gH$ or $Hg$ continuous). The quotient map is always open. If $H$ is a normal subgroup of $G$, then the quotient group becomes a topological group. However, if $H$ is not closed in $G$, then the quotient group will not be $T_0$ even if $G$ is. It is therefore natural to restrict attention to the category of Hausdorff topological groups, continuous homomorphisms and closed subgroups, which is closed under many natural group-theoretic operations.

If the topology on a topological group is metrizable[116], then there is a compatible metric defining the topology that is invariant under each of the maps $g \mapsto hg$ (a left-invariant metric) and there is similarly a right-invariant metric.

**Lemma C.2.** *If $G$ is compact and metrizable, then $G$ has a compatible metric invariant under all translations (that is, a bi-invariant metric).*

PROOF. Choose a basis $\{U_n\}_{n \geqslant 1}$ of open neighborhoods of the identity $e \in G$, with $\cap_{n \geqslant 1} U_n = \{e\}$, and for each $n \geqslant 1$ choose (by Theorem B.9) a continuous function $f_n : G \to [0,1]$ with $\|f_n\| = 1$, $f_n(e) = 1$ and $f_n(G \backslash U_n) = \{0\}$. Let

$$f(g) = \sum_{n=1}^{\infty} f_n(g)/2^n,$$

so that $f$ is continuous, $f^{-1}(\{1\}) = e$, and define

$$\mathrm{d}(x,y) = \sup_{a,b \in G} \{|f(axb) - f(ayb)|\}.$$

Then d is bi-invariant and compatible with the topology on $G$. $\qquad \square$

*Example C.3.* [117] The group $\mathrm{GL}_n(\mathbb{C})$ carries a natural norm

$$\|x\| = \max \left\{ \left( \sum_{i=1}^{n} \left| \sum_{j=1}^{n} x_{ij} v_j \right|^2 \right)^{1/2} \mid \sum_{i=1}^{n} |v_i|^2 = 1 \right\}$$

from viewing a matrix $x = (x_{ij})_{1 \leqslant i,j \leqslant n} \in \mathrm{GL}_n(\mathbb{C})$ as a linear operator on $\mathbb{C}^n$. Then the function

$$\mathrm{d}(x,y) = \log \left( 1 + \|x^{-1}y - I_n\| + \|y^{-1}x - I_n\| \right)$$

is a left-invariant metric compatible with the topology. For $n \geqslant 2$, there is no bi-invariant metric on $\mathrm{GL}_n(\mathbb{C})$. To see this, notice that for such a metric conjugation would be an isometry, while

$$\begin{pmatrix} m & 1 \\ 0 & 1 \end{pmatrix} \begin{pmatrix} \frac{1}{m} & \frac{1}{m^2} \\ 0 & 1 \end{pmatrix} = \begin{pmatrix} 1 & 1 + \frac{1}{m} \\ 0 & 1 \end{pmatrix} \longrightarrow \begin{pmatrix} 1 & 1 \\ 0 & 1 \end{pmatrix}$$

as $m \to \infty$, and

$$\begin{pmatrix} \frac{1}{m} & \frac{1}{m^2} \\ 0 & 1 \end{pmatrix} \begin{pmatrix} m & 1 \\ 0 & 1 \end{pmatrix} = \begin{pmatrix} 1 & \frac{1}{m} + \frac{1}{m^2} \\ 0 & 1 \end{pmatrix} \longrightarrow \begin{pmatrix} 1 & 0 \\ 0 & 1 \end{pmatrix}$$

as $m \to \infty$.

## C.2 Haar Measure on Locally Compact Groups

Further specializing to locally compact topological groups (that is, topological groups in which every point has a neighborhood containing a compact neighborhood) produces a class of particular importance in ergodic theory for the following reason.

**Theorem C.4 (Haar[118]).** *Let $G$ be a locally compact group.*

(1) *There is a measure $m_G$ defined on the Borel subsets of $G$ that is invariant under left translation, is positive on non-empty open sets, and is finite on compact sets.*

(2) *The measure $m_G$ is unique in the following sense: if $\mu$ is any measure with the properties of (1) then there is a constant $C$ with $\mu(A) = Cm_G(A)$ for all Borel sets $A$.*

(3) *$m_G(G) < \infty$ if and only if $G$ is compact.*

The measure $m_G$ is called (a) *left Haar measure* on $G$; if $G$ is compact it is usually normalized to have $m_G(G) = 1$. There is a similar right Haar measure. If $m_G$ is a left Haar measure on $G$, then for any $g \in G$ the measure defined by $A \mapsto m_G(Ag)$ is also a left Haar measure. By Theorem C.4, there must therefore be a unique function mod, called the *modular function* or *modular character* with the property that

$$m_G(Ag) = \mathrm{mod}(g)m_G(A)$$

for all Borel sets $A$. The modular function is the continuous homomorphism $\mathrm{mod} : G \to \mathbb{R}_{>0}$. A group in which the left and right Haar measures coincide (equivalently, whose modular function is identically 1) is called *unimodular*: examples include all abelian groups, all compact groups (since there are no non-trivial compact subgroups of $\mathbb{R}_{>0}$), and semi-simple Lie groups.

There are several different proofs of Theorem C.4. For compact groups, it may be shown using fixed-point theorems from functional analysis. A particularly intuitive construction, due to von Neumann, starts by assigning measure one to some fixed compact set $K$ with non-empty interior, then uses translates of some small open set to efficiently cover $K$ and any other compact set $L$. The Haar measure of $L$ is then approximately the number of translates needed to cover $L$ divided by the number needed to cover $K$ (see Sect. 8.3 for more details). Remarkably, Theorem C.4 has a converse: under some technical hypotheses, a group with a Haar measure must be locally compact[119].

Haar measure produces an important class of examples for ergodic theory: if $\phi : G \rightarrow G$ is a surjective homomorphism and $G$ is compact, then $\phi$ preserves[120] the Haar measure on $G$. Haar measure also connects[121] the topology and the algebraic structure of locally compact groups.

*Example C.5.* In many situations, the Haar measure is readily described.

(1) The Lebesgue measure $\lambda$ on $\mathbb{R}^n$, characterized by the property that

$$\lambda\left([a_1, b_1] \times \cdots \times [a_n, b_n]\right) = \prod_{i=1}^{n} (b_i - a_i)$$

for $a_i < b_i$, is translation invariant and so is a Haar measure for $\mathbb{R}^n$ (unique up to multiplication by a scalar).

(2) The Lebesgue measure $\lambda$ on $\mathbb{T}^n$, characterized in the same way by the measure it gives to rectangles, is a Haar measure (unique if we choose to normalize so that the measure of the whole group $\mathbb{T}^n$ is 1).

As we have seen, a measure can be described in terms of how it integrates integrable functions. For the remaining examples, we will describe a Haar measure $m_G$ by giving a 'formula' for $\int f \, dm_G$. Thus the statement about the Haar measure $m_{\mathbb{R}^n}$ in (1) above could be written somewhat cryptically as

$$\int_{\mathbb{R}^n} f(\mathbf{x}) \, dm_{\mathbb{R}^n}(\mathbf{x}) = \int_{\mathbb{R}^n} f(\mathbf{x}) \, dx_1 \dots dx_n$$

for all functions $f$ for which the right-hand side is finite. Evaluating a Haar measure on a group with explicit coordinates often amounts to computing a Jacobian.

(3) Let $G = \mathbb{R} \setminus \{0\} = \mathrm{GL}_1(\mathbb{R})$, the real multiplicative group. The transformation $x \mapsto ax$ has Jacobian $a$: it can be readily checked that

$$\int f(ax) \frac{dx}{|x|} = \int f(x) \frac{dx}{|x|}$$

for any integrable $f$ and $a \neq 0$. Hence a Haar measure $m_G$ is defined by

$$\int_G f(x)\,dm_G(x) = \int_G \frac{f(x)}{|x|}\,dx$$

for any integrable $f$. Similarly, if $G = \mathbb{C}\setminus\{0\} = \mathrm{GL}_1(\mathbb{C})$, then

$$\int_{\mathbb{C}\setminus\{0\}} f(z)\,dm_G(z) = \iint_{\mathbb{R}^2\setminus\{(0,0)\}} \frac{f(x+iy)}{x^2+y^2}\,dx\,dy.$$

(4) Let $G = \{\left(\begin{smallmatrix} a & b \\ 0 & 1 \end{smallmatrix}\right) \mid a \in \mathbb{R}\setminus\{0\}, b \in \mathbb{R}\}$, and identify elements of $G$ with pairs $(a,b)$. Then

$$\int_G f(a,b)\,dm_G^{(\ell)} = \int_{\mathbb{R}}\int_{\mathbb{R}\setminus\{0\}} \frac{f(a,b)}{a^2}\,da\,db$$

defines a left Haar measure, while

$$\int_G f(a,b)\,dm_G^{(r)} = \int_{\mathbb{R}}\int_{\mathbb{R}\setminus\{0\}} \frac{f(a,b)}{|a|}\,da\,db$$

defines a right Haar measure. As $G$ is isomorphic to the group of affine transformations $x \mapsto ax + b$ under composition, it is called the '$ax + b$' group. It is an example of a non-unimodular group, with $\mathrm{mod}(a,b) = \frac{1}{|a|}$.

(5) Let $G = \mathrm{GL}_2(\mathbb{R})$, and identify the element $(x_{ij})_{1\leqslant i,j\leqslant 2}$ with

$$(x_{11}, x_{12}, x_{21}, x_{22}) \in A = \{\mathbf{x} \in \mathbb{R}^4 \mid x_{11}x_{22} - x_{12}x_{21} \neq 0\}.$$

Then

$$\int_G f\,dm_G = \iiiint_A \frac{f(x_{11}, x_{12}, x_{21}, x_{22})}{(x_{11}x_{22} - x_{12}x_{21})^2}\,dx_{11}\,dx_{12}\,dx_{21}\,dx_{22}$$

defines a left and a right Haar measure on $G$, which is therefore unimodular.

## C.3 Pontryagin Duality

Specializing yet further brings us to the class of *locally compact abelian groups* (LCA groups) which have a very powerful theory[122] generalizing Fourier analysis on the circle. Throughout this section, $L^p(G)$ denotes $L^p_{m_G}(G)$ for some Haar measure $m_G$ on $G$.

A *character* on a LCA group $G$ is a continuous homomorphism

$$\chi : G \to \mathbb{S}^1 = \{z \in \mathbb{C} \mid |z| = 1\}.$$

The set of all continuous characters on $G$ forms a group under pointwise multiplication, denoted $\widehat{G}$ (this means the operation on $\widehat{G}$ is defined by

$$(\chi_1 + \chi_2)(g) = \chi_1(g)\chi_2(g)$$

for all $g \in G$, and the trivial character $\chi(g) = 1$ is the identity). The image of $g \in G$ under $\chi \in \widehat{G}$ will also be written $\langle g, \chi \rangle$ to emphasize that this is a pairing between $G$ and $\widehat{G}$. For compact $K \subseteq G$ and $\varepsilon > 0$ the sets

$$N(K, \varepsilon) = \{\chi \mid |\chi(g) - 1| < \varepsilon \text{ for } g \in K\}$$

and their translates form a basis for a topology on $\widehat{G}$, the topology of *uniform convergence on compact sets*.

**Theorem C.6.** *In the topology described above, the character group of a LCA group is itself a LCA group. A subgroup of the character group that separates points is dense.*

A subset $E \subseteq \widehat{G}$ is said to *separate points* if for $g \neq h$ in $G$ there is some $\chi \in E$ with $\chi(g) \neq \chi(h)$.

Using the Haar measure on $G$ the usual $L^p$ function spaces may be defined. For $f \in L^1(G)$ the *Fourier transform of* $f$, denoted $\widehat{f}$ is the function on $\widehat{G}$ given by

$$\widehat{f}(\chi) = \int_G f(g)\overline{\langle g, \chi \rangle}\, dm_G.$$

Some of the basic properties of the Fourier transform are as follows.

- The image of the map $f \mapsto \widehat{f}$ is a separating self-adjoint algebra in $C_0(\widehat{G})$ (the continuous complex functions vanishing at infinity) and hence is dense in $C_0(\widehat{G})$ in the uniform metric.
- The Fourier transform of the convolution $f * g$ is the product $\widehat{f} \cdot \widehat{g}$.
- The Fourier transform satisfies $\|\widehat{f}\|_\infty \leqslant \|f\|_1$ and so is a continuous operator from $L^1(G)$ to $L^\infty(\widehat{G})$.

**Lemma C.7.** *If $G$ is discrete, then $\widehat{G}$ is compact, and if $G$ is compact then $\widehat{G}$ is discrete.*

We prove the second part of this lemma to illustrate how Fourier analysis may be used to study these groups. Assume that $G$ is compact, so that the constant function $\chi_0 \equiv 1$ is in $L^1(G)$.

Also under the assumption of compactness of $G$, we have the following orthogonality relations. Let $\chi \neq \eta$ be characters on $G$. Then we may find an element $h \in G$ with $(\chi\eta^{-1})(h) \neq 1$. On the other hand,

$$\int_G (\chi\eta^{-1})(g)\, dm_G = \int_G (\chi\eta^{-1})(g + h)\, dm_G = (\chi\eta^{-1})(h) \int (\chi\eta^{-1})(g)\, dm_G,$$

so $\int_G (\chi \eta^{-1})(g)\,\mathrm{d}m_G = 0$ and the characters $\chi$ and $\eta$ are orthogonal with respect to the inner-product

$$\langle f_1, f_2 \rangle = \int_G f_1 \overline{f_2}\,\mathrm{d}m_G$$

on $\widehat{G}$. Thus distinct characters are orthogonal as elements of $L^2(G)$.

Finally, note that the Fourier transform of any $L^1$ function is continuous on the dual group, and the orthogonality relations mean that $\widehat{\chi_0}(\chi) = 1$ if $\chi$ is the trivial character $\chi_0$, and $\widehat{\chi_0}(\chi) = 0$ if not. It follows that $\{\chi_0\}$ is an open subset of $\widehat{G}$, so $\widehat{G}$ is discrete.

The Fourier transform is defined on $L^1(G) \cap L^2(G)$, and maps into a dense linear subspace of $L^2(\widehat{G})$ as an $L^2$ isometry. It therefore extends uniquely to an isometry $L^2(G) \to L^2(\widehat{G})$, known as the *Fourier* or *Plancherel transform* and also denoted by $f \mapsto \widehat{f}$. We note that this map is surjective.

Recall that there is a natural inner-product structure on $L^2(G)$.

**Theorem C.8 (Parseval Formula).** *Let $f$ and $g$ be functions in $L^2(G)$. Then*

$$\langle f, g \rangle_G = \int_G f(x)\overline{g(x)}\,\mathrm{d}m_G = \int_{\widehat{G}} \widehat{f}(\chi)\overline{\widehat{g}(\chi)}\,\mathrm{d}m_{\widehat{G}} = \langle \widehat{f}, \widehat{g} \rangle_{\widehat{G}}.$$

Given a finite Borel measure $\mu$ on the dual group $\widehat{G}$ of a locally compact abelian group $G$, the inverse Fourier transform of $\mu$ is the function $\check{\mu} : G \to \mathbb{C}$ defined by

$$\check{\mu}(x) = \int_{\widehat{G}} \chi(x)\,\mathrm{d}\mu(\chi).$$

A function $f : G \to \mathbb{C}$ is called *positive-definite* if for any $a_1, \ldots, a_r \in \mathbb{C}$ and $x_1, \ldots, x_r \in G$,

$$\sum_{i=1}^{r} \sum_{j=1}^{r} a_i \overline{a_j} f(x_i x_j^{-1}) \geqslant 0. \tag{C.1}$$

**Theorem C.9 (Herglotz–Bochner[123]).** *Let $G$ be an abelian locally compact group. A function $f : G \to \mathbb{C}$ is positive-definite if and only if it is the Fourier transform of a finite positive Borel measure.*

Denote by $B(G)$ the set of all functions $f$ on $G$ which have a representation in the form

$$f(x) = \int_{\widehat{G}} \langle x, \chi \rangle\,\mathrm{d}\mu(\chi)$$

for $x \in G$ and a finite positive Borel measure $\mu$ on $\widehat{G}$. A consequence of the Herglotz–Bochner theorem (Theorem C.9) is that $B(G)$ coincides with the set of finite linear combinations of continuous positive-definite functions on $G$ (see (C.1)).

**Theorem C.10 (Inversion Theorem).** *Let $G$ be a locally compact group. If $f \in L^1(G) \cap B(G)$, then $\widehat{f} \in L^1(\widehat{G})$. Having chosen a Haar measure on $G$, the Haar measure $m_{\widehat{G}}$ on $\widehat{G}$ may be normalized to make*

$$f(g) = \int_{\widehat{G}} \widehat{f}(\chi)\langle g, \chi \rangle \, dm_{\widehat{G}} \qquad (C.2)$$

*for $g \in G$ and any $f \in L^1(G) \cap B(G)$.*

We will usually use Theorem C.10 for a compact metric abelian group $G$. In this case the Haar measure is normalized to make $m(G) = 1$, and the measure on the discrete countable group $\widehat{G}$ is simply counting measure, so that the right-hand side of (C.2) is a series.

In particular, for the case $G = \mathbb{T} = \mathbb{R}/\mathbb{Z}$ we find $\widehat{G} = \{\chi_k \mid k \in \mathbb{Z}\}$ where $\chi_k(t) = e^{2\pi i k t}$. Theorem C.10 then says that for any $f \in L^2(\mathbb{T})$ we have the Fourier expansion

$$f(t) = \sum_{k \in \mathbb{Z}} \widehat{f}(\chi_k) e^{2\pi i k t}$$

for almost every $t$.

Similarly, for any compact $G$, the set of characters of $G$ forms an orthonormal basis of $L^2(G)$. We already showed the orthonormality property in the discussion after Lemma C.7; here we indicate briefly how the completeness of the set of characters can be established, both for concrete groups and in general.

Let $\mathscr{A}$ denote the set of finite linear combinations of the form

$$p(g) = \sum_{i=1}^{n} c_i \chi_i$$

with $c_i \in \mathbb{C}$. Then $\mathscr{A}$ is a subalgebra of $C_{\mathbb{C}}(X)$ which is closed under conjugation. If we know in addition that $\mathscr{A}$ separates points in $G$, then by the Stone–Weierstrass Theorem (Theorem B.7) we have that $\mathscr{A}$ is dense in $C_{\mathbb{C}}(X)$. Moreover, in that case $\mathscr{A}$ is also dense in $L^2(G)$, and so the set of characters forms an orthonormal basis for $L^2(G)$. That $\mathscr{A}$ separates points can be checked explicitly for many compact abelian groups; in particular for $G = \mathbb{R}^d/\mathbb{Z}^d$ the characters are of the form

$$\chi_{\mathbf{n}}(\mathbf{x}) = e^{2\pi i (n_1 x_1 + \cdots + n_d x_d)} \qquad (C.3)$$

with $\mathbf{n} \in \mathbb{Z}^d$, and this explicit presentation may be used to show that the set of characters separates points on the $d$-torus. In general, one can prove that $\widehat{G}$ separates points by showing that the functions in $B(G)$ separate points, and then applying the Herglotz–Bochner theorem (Theorem C.9).

**Theorem C.11.** *For any compact abelian group $G$, the set of characters separates points and therefore forms a complete orthonormal basis for $L^2(G)$.*

The highlight of this theory is *Pontryagin duality*, which directly links the algebraic structure of LCA groups to their (Fourier-)analytic structure. If $G$ is an LCA group, then $\Gamma = \widehat{G}$ is also an LCA group, which therefore has a character group $\widehat{\Gamma}$, which is again LCA. Any element $g \in G$ defines a character $\chi \mapsto \chi(g)$ on $\Gamma$.

**Theorem C.12 (Pontryagin Duality).** *The map $\alpha : G \to \widehat{\Gamma}$ defined by*

$$\langle g, \chi \rangle = \langle \chi, \alpha(g) \rangle$$

*is a continuous isomorphism of LCA groups.*

The Pontryagin duality theorem relates to the subgroup structure of an LCA group as follows.

**Theorem C.13.** *If $H \subseteq G$ is a closed subgroup, then $G/H$ is also an LCA group. The set*

$$H^\perp = \{\chi \in \widehat{G} \mid \chi(h) = 1 \text{ for all } h \in H\},$$

*the annihilator of $H$, is a closed subgroup of $\widehat{G}$. Moreover,*

- $\widehat{G/H} \cong H^\perp$;
- $\widehat{G}/H^\perp \cong \widehat{H}$;
- *if $H_1, H_2$ are closed subgroups of $G$ then*

$$H_1^\perp + H_2^\perp \cong \widehat{X}$$

*where $X = G/(H_1 \cap H_2)$;*
- $H^{\perp\perp} \cong H$.

The dual of a continuous homomorphism $\theta : G \to H$ is a homomorphism

$$\widehat{\theta} : \widehat{H} \to \widehat{G}$$

defined by $\widehat{\theta}(\chi)(g) = \chi(\theta(g))$. There are simple dualities for homomorphisms, for example $\theta$ has dense image if and only if $\widehat{\theta}$ is injective. Pontryagin duality expresses topological properties in algebraic terms. For example, if $G$ is compact then $\widehat{G}$ is torsion if and only if $G$ is zero-dimensional (that is, has a basis for the topology comprising sets that are both closed and open), and $\widehat{G}$ is torsion-free if and only if $G$ is connected. Duality also gives a description of monothetic groups: if $G$ is a compact abelian group with a countable basis for its topology then $G$ is monothetic if and only if the dual group $\widehat{G}$ is isomorphic as an abstract group to a countable subgroup of $\mathbb{S}^1$. If $G$ is monothetic, then any such isomorphism is given by choosing a topological generator $g \in G$ and then sending $\chi \in \widehat{G}$ to $\chi(g) \in \mathbb{S}^1$.

*Example C.14.* As in the case of Haar measure in Example C.5, the character group of many groups can be written down in a simple way.

(1) If $G = \mathbb{Z}$ with the discrete topology, then any character $\chi \in \widehat{\mathbb{Z}}$ is determined by the value $\chi(1) \in \mathbb{S}^1$, and any choice of $\chi(1)$ defines a character. It follows that the map $z \mapsto \chi_z$, where $\chi$ is the unique character on $\mathbb{Z}$ with $\chi_z(1) = z$, is an isomorphism $\mathbb{S}^1 \to \widehat{\mathbb{Z}}$.

(2) Consider the group $\mathbb{R}$ with the usual topology. Then for any $s \in \mathbb{R}$ the map $\chi_s : t \mapsto e^{\mathrm{i}st}$ is a character on $\mathbb{R}$, and any character has this form. In other words, the map $s \mapsto \chi_s$ is an isomorphism $\mathbb{R} \to \widehat{\mathbb{R}}$.

(3) More generally, let $\mathbb{K}$ be any locally compact non-discrete field, and assume that $\chi_0 : \mathbb{K} \to \mathbb{S}^1$ is a non-trivial character on the additive group structure of $\mathbb{K}$. Then the map $a \mapsto \chi_a$, where $\chi_a(x) = \chi_0(ax)$, defines an isomorphism $\mathbb{K} \to \widehat{\mathbb{K}}$.

(4) An important example of (3) concerns the field of $p$-adic numbers $\mathbb{Q}_p$. For each prime number $p$, the field $\mathbb{Q}_p$ is the set of formal power series $\sum_{n \geqslant k} a_n p^n$ where $a_n \in \{0, 1, \ldots, p-1\}$ and $k \in \mathbb{Z}$ and we always choose $a_k \neq 0$, with the usual addition and multiplication. The metric $\mathsf{d}(x, y) = |x - y|_p$, where $|\sum_{n \geqslant k} a_n p^n|_p = p^{-k}$ and $|0|_p = 0$, makes $\mathbb{Q}_p$ into a non-discrete locally compact field. By (3) an isomorphism $\widehat{\mathbb{Q}_p} \to \mathbb{Q}_p$ is determined by any non-trivial character on $\mathbb{Q}_p$, for example the map

$$\sum_{n \geqslant k} a_n p^n \mapsto \exp\left( 2\pi\mathrm{i} \sum_{n=k}^{-1} a_n p^{-n} \right).$$

(5) Consider the additive group $\mathbb{Q}$ with the discrete topology. Then the group of characters is compact. Any element of $\widehat{\mathbb{R}}$ restricts to a character of $\mathbb{Q}$, so there is an embedding $\mathbb{R} \hookrightarrow \widehat{\mathbb{Q}}$ (injective because a continuous character on $\mathbb{R}$ is defined by its values on the dense set $\mathbb{Q}$). The group $\widehat{\mathbb{Q}}$ is an example of a *solenoid*, and there is a detailed account of its structure in terms of adeles in the monograph of Weil [378].

**Lemma C.15 (Riemmann–Lebesgue[124]).** *Let $G$ be a locally compact abelian group, and let $\mu$ be a measure on $G$ absolutely continuous with respect to Haar measure $m_G$. Then*

$$\widehat{\mu}(\chi) = \int_G \chi(g) \, \mathrm{d}\mu(t) \to 0$$

*as $\chi \to \infty^*$.*

---

* A sequence $\chi_n \to \infty$ if for any compact set $K \subseteq \widehat{G}$ there exists $N = N(K)$ for which

$$n \geqslant N \implies \chi_n \notin K.$$

The Riemann–Lebesgue lemma generalizes to absolutely continuous measures with respect to any sufficiently smooth measure.

**Lemma C.16.** *Let $\nu$ be a finite measure on $\mathbb{S}^1$, and assume that*

$$\int e^{2\pi i n t}\, d\nu(t) \to 0$$

*as $|n| \to \infty$. Then for any finite measure $\mu$ that is absolutely continuous with respect to $\nu$,*

$$\int e^{2\pi i n t}\, \frac{d\mu}{d\nu}\, d\nu(t) \to 0$$

*as $|n| \to \infty$.*

# Notes to Appendix C

[115](Page 429) Given a topological space $(X, \mathcal{T})$, points $x$ and $y$ are said to be topologically indistinguishable if for any open set $U \in \mathcal{T}$ we have $x \in U$ if and only if $y \in U$ (they have the same neighborhoods). The space is said to be $T_0$ or Kolmogorov if distinct points are always topologically distinguishable. This is the weakest of a hierarchy of topological separation axioms; for topological groups many of these collapse to the following natural property: the space is $T_2$ or Hausdorff if distinct points always have some distinct neighborhoods.

[116](Page 430) A topological group is metrizable if and only if every point has a countable basis of neighborhoods (this was shown by Kakutani [170] and Birkhoff [34]) and has a metric invariant under all translations if there is a countable basis $\{V_n\}$ at the identity with $x V_n x^{-1} = V_n$ for all $n$ (see Hewitt and Ross [151, p. 79]).

[117](Page 430) This explicit construction of a left-invariant metric on $\mathrm{GL}_n(\mathbb{C})$ is due to Kakutani [170] and von Dantzig [64].

[118](Page 431) Haar's original proof appears in his paper [130]; more accessible treatments may be found in the books of Folland [94], Weil [377] or Hewitt and Ross [151]. The important lecture notes of von Neumann from 1940–41, when he developed much of the theory from a new perspective, have now been edited and made available by the American Mathematical Society [269].

[119](Page 432) This result was announced in part in a note by Weil [376] and then complete proofs were given by Kodaira [206]; these results were later sharpened by Mackey [239].

[120](Page 432) This observation is due to Halmos [134], who determined when Haar measure is ergodic, and accounts for the special role of compact group automorphisms as distinguished examples of measure-preserving transformations in ergodic theory. The proof is straightforward: the measure defined by $\mu(A) = m_G(\phi^{-1} A)$ is also a translation-invariant probability measure defined on the Borel sets, so $\mu = m_G$.

[121](Page 432) For example, if $G$ and $H$ are locally compact groups and $G$ has a countable basis for its topology then any measurable homomorphism $\phi : H \to G$ is continuous (Mackey [240]); in any locally compact group, for any compact set $A$ with positive Haar measure, the set $AA^{-1}$ contains a neighborhood of the identity; if $H \subseteq G$ is closed under multiplication and conull then $H = G$.

[122](Page 433) The theory described in this section is normally called Pontryagin duality or Pontryagin–von Kampen duality; the original sources are the book of Pontryagin [293] and the papers of van Kampen [181]. More accessible treatments may be found in Folland [94], Weil [377], Rudin [322] or Hewitt and Ross [151].

(123)(Page 435) This result is due to Herglotz [148] for functions on $\mathbb{Z}$, to Bochner [37] for $\mathbb{R}$, and to Weil [377] for locally compact abelian groups; accessible sources include the later translation [38] and Folland [94].

(124)(Page 438) Riemann [310] proved that the Fourier coefficients of a Riemann integrable periodic function converge to zero, and this was extended by Lebesgue [219]. The finite Borel measures on $\mathbb{T}$ with $\widehat{\mu}(n) \to 0$ as $|n| \to \infty$ are the *Rajchman* measures; all absolutely continuous measures are Rajchman measures but not conversely. Menshov, in his construction of a Lebesgue null set of multiplicity, constructed a singular Rajchman measure in 1916 by modifying the natural measure on the Cantor middle-third set (though notice that the Cantor–Lebesgue measure $\nu$ on the middle-third Cantor set has $\widehat{\nu}(n) = \widehat{\nu}(3n)$, so is a continuous measure that is not Rajchman). Riesz raised the question of whether a Rajchman measure must be continuous, and this was proved by Neder in 1920. Wiener gave a complete characterization of continuous measures by showing that $\nu$ is continuous if and only if $\frac{1}{2n+1} \sum_{k=-n}^{n} |\widehat{\mu}(k)| \to 0$ as $n \to \infty$. A convenient account is the survey by Lyons [238].

# Hints for Selected Exercises

**Exercise 2.1.5** (p. 20): For (a) use a one-sided full shift.

**Exercise 2.4.2** (p. 32): Recall from Exercise 2.1.1 that the spaces themselves are isomorphic. Try to do this directly, but if all else fails look at it again using the material from Sect. 2.7.

**Exercise 2.4.4** (p. 32): Use the fact that the kernel of $A^n - I$ on the torus only contains points with rational coordinates.

**Exercise 2.5.5** (p. 36): Apply the uniform mean ergodic theorem to the inner product $\langle U_T^n \chi_B, \chi_B \rangle$ and notice that $\int P_T(\chi_B) \, d\mu > 0$ since the projection onto the constants already has this property.

**Exercise 2.6.3** (p. 48): This is an easy consequence of a later formulation of the ergodic theorem, described in Theorem 6.1. Try to prove it directly.

**Exercise 2.7.1** (p. 52): Fix some $B$ with $0 < \mu(B) < 1$, and use the Baire category theorem to find $A$.

**Exercise 2.7.8** (p. 53): Prove, and then use, the polarization identity

$$4 \langle U_T^n f, g \rangle = \langle U_T^n(f + g), f + g \rangle + i \langle U_T^n(f + ig), f + ig \rangle$$
$$- \langle U_T^n(f - g), f - g \rangle - i \langle U_T^n(f - ig), f - ig \rangle .$$

**Exercise 2.7.10** (p. 53): Fix $A \in \mathscr{B}$ and consider the closed linear subspace $M$ of $L_\mu^2$ containing the constant functions and $\{U_T^n \chi_A \mid n \in \mathbb{Z}\}$. Prove that

$$\langle U_T^n \chi_A, U_T^k \chi_A \rangle \to \mu(A)^2$$

as $n \to \infty$, and then decompose each function $f \in L_\mu^2$ into $f_1 + f_2$ with $f_1 \in M$ and $f_2 \in M^\perp$.

**Exercise 2.7.13** (p. 54): For (a) and (b) recall that smoothness corresponds to polynomially rapid decay of Fourier coefficients. For (b) diagonalize the

M. Einsiedler, T. Ward, *Ergodic Theory*, Graduate Texts in Mathematics 259, 441
DOI 10.1007/978-0-85729-021-2, © Springer-Verlag London Limited 2011

matrix $A = \begin{pmatrix} 0 & 1 \\ 1 & 1 \end{pmatrix}$ defining the automorphism, and show that for any integer point in $\mathbb{Z}^2$ the product of the coordinates when expressed in the diagonalizing coordinates is bounded from below. Then argue as in part (a).

**Exercise 3.2.1** (p. 86): You will need to formulate the ergodic theorem for the system $(B, \mathscr{B}|_B, \frac{1}{\mu(B)}\mu, T|_B)$ for a $T$-invariant measurable set $B$ of positive measure.

**Exercise 3.3.3** (p. 91): If $|u - \frac{p}{q}| > 1$ the statement is clear. Assume therefore that $|u - \frac{p}{q}| \leqslant 1$ and try to find upper and lower bounds for the size of

$$|f(u) - f(\tfrac{p}{q})|,$$

where $f \in \mathbb{Z}[t]$ is the minimal polynomial of $u$.

**Exercise 4.2.1** (p. 104): Use Zorn's lemma for (c).

**Exercise 4.4.1** (p. 117): For part (c), notice that it is enough to find a point $x \in \mathbb{T}$ with the property that $\left(\frac{1}{N} \sum_{n=0}^{N-1} f(T_2^n x)\right)_{N \geqslant 1}$ does not converge for some $f \in C(\mathbb{T})$.

**Exercise 5.3.2** (p. 144): Assume there is no such set, consider what that implies about the collection of $\varepsilon$-balls around a dense set of points in $X$, and deduce a contradiction of aperiodicity.

**Exercise 6.5.2** (p. 168): Let $(X, \mathscr{B}_X, \mu, T)$ be ergodic, let $(Y, \mathscr{B}_Y, \nu, S)$ have $S = I_Y$, and let $\lambda \in J(\mathsf{X}, \mathsf{Y})$. Write $P_\lambda : L_\mu^2 \to L_\nu^2$ for the operator defined by

$$f \mapsto P_{L_\nu^2}(f \otimes 1)$$

where $(f \otimes g)(x, y) = f(x)g(y)$, and show that for $f \in L_\mu^2$, $P_\lambda(f)$ is constant almost everywhere. Then show that for $g \in L_\nu^2$,

$$\int f \otimes g \, d\lambda = \int f \, d\mu \int g \, d\nu,$$

and deduce that $\lambda$ is product measure.

**Exercise 6.5.4** (p. 168): Let $\rho$ be a joining, and notice that

$$L_\nu^2(Y) \subseteq L_\rho^2(X \times Y).$$

If $f$ is an eigenfunction for $S$, then $E_\rho(f \,|\, \mathscr{B}_X)$ is an eigenfunction for $T$.

**Exercise 7.1.1** (p. 174): This is a simple instance of a wide-ranging compactness principle in Ramsey theory. One direction is immediate; for the reverse assume that there are $r$-colorings of $[0, N]$ with no monochrome arithmetic progression of length $\ell$. Extend each of these colorings to all of $\mathbb{N}$ arbitrarily, and then show that a limit point of those $r$-colorings of $\mathbb{N}$ (viewed as maps $\mathbb{N} \to \{1, \ldots, r\}$) defines an $r$-coloring of $\mathbb{N}$ with no monochrome arithmetic progression of length $\ell$.

**Exercise 7.1.2** (p. 174): This is discussed in detail in [125, Chap. 2], where it is shown that we may take $N(3,2) = 325$.

**Exercise 7.2.2** (p. 178): This is a difficult result, and you should expect to need to consult the references (see, for example the book of Petersen [282, Sec. 4.3]). Prove it first for $k = 1$, and start by using Zorn's Lemma to show that $(X, T)$ must contain a non-empty minimal set (see Exercise 4.2.1 on p. 104). Then use induction on the length $k$, and express the property in terms of the action of the map $T \times T^2 \times \cdots \times T^k$ on the diagonal in $X \times \cdots \times X$.

**Exercise 8.6.2** (p. 266): If the group is not unimodular, show that there is an element $a$ of $G$ for which the Haar measure of $B_1(e)a^n$ grows exponentially in $n$. On the other hand, use property (D) and the inclusion

$$B_1(e)a^n \subseteq B_{1+n d(a,e)}(e)$$

to derive a contradiction.

**Exercise 8.8.2** (p. 274): Define $\overline{X} = X \times \overline{G}^{\mathbb{N}}$, where $\overline{G} = G \cup \{\infty\}$ is the one-point compactification of $G$, use $\mu_0 \times \nu^{\mathbb{N}}$ and the transformation

$$T\left(x, (g_n)_{n \geqslant 1}\right) = \left(g_1 \cdot x, (g_{n+1})_{n \geqslant 1}\right).$$

**Exercise 9.1.1** (p. 282): Given $y_1 > y_0 > 0$ show that there are constants $c_0, c_1$ with

$$c_0 \| \cdot \|_2 \leqslant \| \cdot \|_z \leqslant c_1 \| \cdot \|_2$$

for $z = x + iy$ with $y_0 < y < y_1$.

**Exercise 9.3.4** (p. 305): For part (c) use Furstenberg's theorem (Theorem 4.21).

**Exercise 9.4.2** (p. 313): Use Proposition 9.20.

**Exercise 9.4.3** (p. 313): Use the one-parameter subgroups that were mentioned on p. 311.

**Exercise 9.4.4** (p. 314): This is equivalent to a careful interpretation of Fig. H.1 (and the argument is similar to the proof of Lemma 11.11).

**Exercise 11.2.1** (p. 362): Let $\tau_1, \ldots, \tau_4$ be the four rotations by an angle of $\pi$ around the centers of the four edges of the regular four-gon $D$, and let $\gamma_1, \ldots, \gamma_4$ be the four counter-clockwise rotations by an angle of $\frac{\pi}{3}$ at the four vertices of $D$. The existence of the tiling (and the corresponding lattice) reduces to the claim that the interior of $D$ does not intersect the image $\eta(D)$, where $\eta$ is a word in $\tau_1, \ldots, \tau_4, \gamma_1, \ldots, \gamma_4$, unless $\eta(D) = D$.

To prove this claim, associate to the word $\eta$ the closed path that starts at the center of the $D$ and moves from there to the center of an image

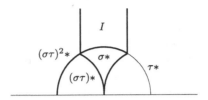

**Fig. H.1** Using the action to show that $\mathrm{PSL}_2(\mathbb{Z})$ is a free product

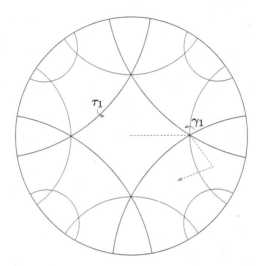

**Fig. H.2** The tiling constructed in Lemma 11.12

of $D$ under $\tau_i$ or $\gamma_j$ and so on, until it reaches the center of $\eta(D)$. If, for example, the word $\eta$ starts with $\gamma_1^2$ we let the path go from the center of $D$ to the center of $\gamma_1^2(D)$ in such a way that the path stays in the union of the interior of $D$, the interior of $\gamma_1^2(D)$, and the vertex where these two four-gons touch (see Fig. H.2). Depending on how the word $\eta$ continues, we then let the path continue in a similar fashion. We can also reverse this procedure, for a piecewise analytic path connecting the center of $D$ to itself we can attach a word $\eta'$ and a corresponding image $\eta'(D)$ of $D$. Now show that the image $\eta'(D)$ remains unchanged under homotopies, and conclude that $\eta(D) = D$ as $\mathbb{H}$ is simply connected.

**Exercise 11.2.3** (p. 363): Identify the geometric consequences of the fact that $V_x$ acts via the map $z \mapsto \frac{z}{t_x z + 1}$ conjugated by the translation $z \mapsto z + x$.

**Exercise 11.4.1** (p. 377): Choose $j_0 \in \{1, \ldots, r-1\}$ with

$$\| \mathrm{Ad}_{g_{j_0,n}} \| \geqslant \| \mathrm{Ad}_{g_{j,n}} \|$$

for $j = 1, \ldots, r - 1$, and choose $v_n$ of size about

$$\| \mathrm{Ad}_{g_{j_0,n}} \|^{-1}$$

such that $\mathrm{Ad}_{g_{j_0,n}} v_n$ converges to a non-zero element.

**Exercise 11.5.1** (p. 388): Suppose that $u = u^-(1) \in U^-$. Observe that for an $R_u$-invariant measure $\mu$ the measure $\int_0^1 \left( R_{u^-(s)} \right)_* \mu \, ds$ is invariant under $U^-$.

**Exercise 11.5.2** (p. 388): Use Lemma 11.29 and the argument in the proof of Proposition 11.30.

**Exercise 11.5.3** (p. 388): Analyze the proof of Theorem 11.27.

**Exercise 11.6.2** (p. 397): Suppose that $g_n \in \mathbb{G}(\mathbb{R})$ and $g \in \mathrm{SL}_n(\mathbb{R})$ have the property that $\mathrm{SL}_n(\mathbb{Z})g_n \to \mathrm{SL}_n(\mathbb{Z})g$ as $n \to \infty$. Then there exists $\gamma_n \in \mathrm{SL}_n(\mathbb{Z})$ such that $\gamma_n g_n \to g$. Now study the rational vectors $\rho(\gamma_n)v = \rho(\gamma_n g_n)v \subseteq \mathbb{Q}^N$, their denominators, and their convergence properties.

**Exercise 11.6.3** (p. 398): For (g), use the fact that $\mathbb{G}(\mathbb{R})$ does not change the quadratic form $a^2 - 7b^2 + c^2 - 7d^2$ for $(a, b, c, d) \in \mathbb{R}^4$ and that for an integer vector $(a, b, c, d)$ the form vanishes only if $a = b = c = d = 0$.

# References

1. J. Aaronson, *An introduction to infinite ergodic theory*, in *Mathematical Surveys and Monographs* **50** (American Mathematical Society, Providence, RI, 1997).
2. R. L. Adler, M. Keane, and M. Smorodinsky, 'A construction of a normal number for the continued fraction transformation', *J. Number Theory* **13** (1981), no. 1, 95–105.
3. E. Akin, *The general topology of dynamical systems*, in *Graduate Studies in Mathematics* **1** (American Mathematical Society, Providence, RI, 1993).
4. L. Alaoglu, 'Weak topologies of normed linear spaces', *Ann. of Math. (2)* **41** (1940), 252–267.
5. H. Anzai, 'Ergodic skew product transformations on the torus', *Osaka Math. J.* **3** (1951), 83–99.
6. L. Arenas-Carmona, D. Berend, and V. Bergelson, 'Ledrappier's system is almost mixing of all orders', *Ergodic Theory Dynam. Systems* **28** (2008), no. 2, 339–365.
7. V. I. Arnol'd and A. Avez, *Ergodic problems of classical mechanics*, Translated from the French by A. Avez (W. A. Benjamin, Inc., New York-Amsterdam, 1968).
8. E. Artin, 'Ein mechanisches System mit quasiergodischen Bahnen', *Hamb. Math. Abh.* **3** (1924), 170–177.
9. J. Auslander, *Minimal flows and their extensions*, in *North-Holland Mathematics Studies* **153** (North-Holland Publishing Co., Amsterdam, 1988). Notas de Matemática [Mathematical Notes], 122.
10. L. Auslander, L. Green, and F. Hahn, *Flows on homogeneous spaces, With the assistance of L. Markus and W. Massey, and an appendix by L. Greenberg*, in *Annals of Mathematics Studies*, No. **53** (Princeton University Press, Princeton, N.J., 1963).
11. J. Avigad, 'The metamathematics of ergodic theory', *Ann. Pure Appl. Logic* **157** (2009), no. 2–3, 64–76.
12. J. Avigad, P. Gerhardy, and H. Towsner, 'Local stability of ergodic averages', *Trans. Amer. Math. Soc.* **362** (2010), 261–288.
13. J. Avigad and K. Simic, 'Fundamental notions of analysis in subsystems of second-order arithmetic', *Ann. Pure Appl. Logic* **139** (2006), no. 1–3, 138–184.
14. D. H. Bailey, J. M. Borwein, and R. E. Crandall, 'On the Khintchine constant', *Math. Comp.* **66** (1997), no. 217, 417–431.
15. V. Baladi, 'Decay of correlations', in *Smooth ergodic theory and its applications (Seattle, WA, 1999)*, in *Proc. Sympos. Pure Math.* **69**, pp. 297–325 (Amer. Math. Soc., Providence, RI, 2001).
16. S. Banach, 'Sur les fonctionelles linéaires. I, II', *Studia Mathematica* **1** (1929), 211–216, 223–239.
17. S. Banach, *Théorie des óperations linéaires* (Chelsea Publishing Co., New York, 1955).

18. J. Barone and A. Novikoff, 'A history of the axiomatic formulation of probability from Borel to Kolmogorov. I', *Arch. History Exact Sci.* **18** (1977/78), no. 2, 123–190.

19. L. Barreira and Y. B. Pesin, *Lyapunov exponents and smooth ergodic theory*, in *University Lecture Series* **23** (American Mathematical Society, Providence, RI, 2002).

20. V. Becher and S. Figueira, 'An example of a computable absolutely normal number', *Theoret. Comput. Sci.* **270** (2002), no. 1–2, 947–958.

21. M. B. Bekka and M. Mayer, *Ergodic theory and topological dynamics of group actions on homogeneous spaces*, in *London Mathematical Society Lecture Note Series* **269** (Cambridge University Press, Cambridge, 2000).

22. A. Bellow and H. Furstenberg, 'An application of number theory to ergodic theory and the construction of uniquely ergodic models', *Israel J. Math.* **33** (1979), no. 3–4, 231–240 (1980). A collection of invited papers on ergodic theory.

23. E. Beltrami, 'Teoria fondamentale degli spazii di curvatura costante', *Brioschi Ann. (2)* **2** (1868), 232–255.

24. Y. Benoist and J.-F. Quint, 'Mesures stationnaires et fermés invariants des espaces homogènes', *C. R. Math. Acad. Sci. Paris* **347** (2009), no. 1, 9–13.

25. V. Bergelson, 'Some historical remarks and modern questions around the ergodic theorem', in *Dynamics of Complex Systems*, pp. 1–11 (Research Institute for Math. Sciences, Kyoto, 2004).

26. V. Bergelson, 'Combinatorial and Diophantine applications of ergodic theory', in *Handbook of dynamical systems. Vol. 1B*, pp. 745–869 (Elsevier B. V., Amsterdam, 2006). Appendix A by A. Leibman and Appendix B by Anthony Quas and Máté Wierdl.

27. V. Bergelson and A. Gorodnik, 'Weakly mixing group actions: a brief survey and an example', in *Modern dynamical systems and applications*, pp. 3–25 (Cambridge Univ. Press, Cambridge, 2004).

28. V. Bergelson and A. Gorodnik, 'Ergodicity and mixing of noncommuting epimorphisms', *Proc. London Math. Soc. (3)* **95** (2007), 329–359.

29. V. Bergelson and A. Leibman, 'Polynomial extensions of van der Waerden's and Szemerédi's theorems', *J. Amer. Math. Soc.* **9** (1996), no. 3, 725–753.

30. S. Bhattacharya, 'Higher order mixing and rigidity of algebraic actions on compact abelian groups', *Israel J. Math.* **137** (2003), 211–221.

31. P. Billingsley, *Ergodic theory and information* (John Wiley & Sons Inc., New York, 1965).

32. G. D. Birkhoff, *Dynamical systems* (American Mathematical Society, Providence, R.I., 1927).

33. G. D. Birkhoff, 'Proof of the ergodic theorem', *Proc. Nat. Acad. Sci. U.S.A.* **17** (1931), 656–660.

34. G. D. Birkhoff, 'A note on topological groups', *Compos. Math.* **3** (1936), 427–430.

35. G. D. Birkhoff and P. A. Smith, 'Structure analysis of surface transformations', *Journ. de Math.* **7** (1928), 345–379.

36. E. Bishop, *Foundations of constructive analysis* (McGraw-Hill Book Co., New York, 1967).

37. S. Bochner, *Vorlesungen über Fouriersche Integrale* (Leipzig, 1932).

38. S. Bochner, *Lectures on Fourier integrals. With an author's supplement on monotonic functions, Stieltjes integrals, and harmonic analysis*, in *Annals of Mathematics Studies*, No. **42** (Princeton University Press, Princeton, N.J., 1959).

39. P. Bohl, 'Über ein in der Theorie der säkularen Störungen vorkommendes Problem', *J. für Math.* **135** (1909), 189–283.

40. É. Borel, 'Les probabilités dénombrables et leurs applications arithmétiques', *Rend. Circ. Mat. Palermo* **27** (1909), 247–271.

41. J. Bourgain, 'An approach to pointwise ergodic theorems', in *Geometric aspects of functional analysis (1986/87)*, in *Lecture Notes in Math.* **1317**, pp. 204–223 (Springer, Berlin, 1988).

42. J. Bourgain, A. Furman, E. Lindenstrauss, and S. Mozes, 'Invariant measures and stiffness for non-abelian groups of toral automorphisms', *C. R. Math. Acad. Sci. Paris* **344** (2007), no. 12, 737–742.

43. A. Brauer, 'Review of *"Three pearls of number theory"* by A. Y. Khinchin', *Bull. Amer. Math. Soc.* **61** (1955), no. 4, 351–353.

44. M. Brin and G. Stuck, *Introduction to dynamical systems* (Cambridge University Press, Cambridge, 2002).

45. A. P. Calderon, 'A general ergodic theorem', *Ann. of Math. (2)* **58** (1953), 182–191.

46. F. P. Cantelli, 'Sulla probabilità come limite della frequenza', *Rom. Acc. L. Rend. (5)* **26** (1917), 39–45.

47. G. Cantor, 'Über eine Eigenschaft des Ingebriffes aller reelen algebraischen Zahlen', *J. Reine Angew. Math.* **77** (1874), 258–262.

48. C. Carathéodory, 'Untersuchungen über die konformen Abbildungen von festen und veränderlichen Gebieten', *Math. Ann.* **72** (1912), 107–144.

49. C. Carathéodory, 'Über den Wiederkehrsatz von Poincaré', *S. B. Preuss. Akad. Wiss.* (1919), 580–584.

50. J. W. S. Cassels and H. P. F. Swinnerton-Dyer, 'On the product of three homogeneous linear forms and the indefinite ternary quadratic forms', *Philos. Trans. Roy. Soc. London. Ser. A.* **248** (1955), 73–96.

51. R. V. Chacon, 'Transformations having continuous spectrum', *J. Math. Mech.* **16** (1966), 399–415.

52. R. V. Chacon, 'A geometric construction of measure preserving transformations', in *Proc. Fifth Berkeley Sympos. Math. Statist. and Probability (Berkeley, Calif., 1965/66), Vol. II: Contributions to Probability Theory, Part 2*, pp. 335–360 (Univ. California Press, Berkeley, Calif., 1967).

53. G. J. Chaitin, 'A theory of program size formally identical to information theory', *J. Assoc. Comput. Mach.* **22** (1975), 329–340.

54. D. G. Champernowne, 'The construction of decimals normal in the scale of ten', *J. London Math. Soc.* **8** (1933), 254–260.

55. G. Choquet, 'Existence et unicité des représentations intégrales au moyen des points extrémaux dans les cônes convexes', in *Séminaire Bourbaki, Vol. 4*, pp. Exp. No. 139, 33–47 (Soc. Math. France, Paris, 1959).

56. G. Choquet, 'Les cônes convexes faiblement complets dans l'analyse', in *Proc. Internat. Congr. Mathematicians (Stockholm, 1962)*, pp. 317–330 (Inst. Mittag-Leffler, Djursholm, 1963).

57. H. Chu, 'Some results on affine transformations of compact groups', *Invent. Math.* **28** (1975), 161–183.

58. J.-P. Conze and E. Lesigne, 'Sur un théorème ergodique pour des mesures diagonales', *C. R. Acad. Sci. Paris Sér. I Math.* **306** (1988), no. 12, 491–493.

59. J.-P. Conze and E. Lesigne, 'Sur un théorème ergodique pour des mesures diagonales', in *Probabilités*, in *Publ. Inst. Rech. Math. Rennes* **1987**, pp. 1–31 (Univ. Rennes I, Rennes, 1988).

60. I. P. Cornfeld, S. V. Fomin, and Y. G. Sinaï, *Ergodic theory* (Springer-Verlag, New York, 1982).

61. S. G. Dani, 'Invariant measures and minimal sets of horospherical flows', *Invent. Math.* **64** (1981), no. 2, 357–385.

62. S. G. Dani, 'On orbits of unipotent flows on homogeneous spaces. II', *Ergodic Theory Dynam. Systems* **6** (1986), no. 2, 167–182.

63. S. G. Dani and J. Smillie, 'Uniform distribution of horocycle orbits for Fuchsian groups', *Duke Math. J.* **51** (1984), no. 1, 185–194.

64. D. v. Dantzig, 'Zur topologischen Algebra. I. Komplettierungstheorie', *Math. Ann.* **107** (1932), 587–626.

65. M. M. Day, 'Amenable semigroups', *Illinois J. Math.* **1** (1957), 509–544.

66. M. M. Day, 'Amenability and equicontinuity', *Studia Math.* **31** (1968), 481–494.

67. A. del Junco and J. Rosenblatt, 'Counterexamples in ergodic theory and number theory', *Math. Ann.* **245** (1979), no. 3, 185–197.

68. A. del Junco and R. Yassawi, 'Multiple mixing and rank one group actions', *Canad. J. Math.* **52** (2000), no. 2, 332–347.

69. M. Denker, 'On strict ergodicity', *Math. Z.* **134** (1973), 231–253.

70. M. Denker and E. Eberlein, 'Ergodic flows are strictly ergodic', *Advances in Math.* **13** (1974), 437–473.

71. W. Doeblin, 'Remarques sur la théorie métrique des fractions continues', *Compositio Math.* **7** (1940), 353–371.

72. J. L. Doob, 'Regularity properties of certain families of chance variables', *Trans. Amer. Math. Soc.* **47** (1940), 455–486.

73. J. L. Doob, *Stochastic processes* (John Wiley & Sons Inc., New York, 1953).

74. T. Downarowicz, 'The Choquet simplex of invariant measures for minimal flows', *Israel J. Math.* **74** (1991), no. 2–3, 241–256.

75. M. Drmota and R. F. Tichy, *Sequences, discrepancies and applications*, in *Lecture Notes in Mathematics* **1651** (Springer-Verlag, Berlin, 1997).

76. W. Duke, Z. Rudnick, and P. Sarnak, 'Density of integer points on affine homogeneous varieties', *Duke Math. J.* **71** (1993), no. 1, 143–179.

77. F. J. Dyson, 'The approximation to algebraic numbers by rationals', *Acta Math., Uppsala* **79** (1947), 225–240.

78. P. Ehrenfest and T. Ehrenfest, *The conceptual foundations of the statistical approach in mechanics* (Cornell University Press, Ithaca, N.Y., 1959). Translated by M. J. Moravcsik.

79. M. Einsiedler, A. Katok, and E. Lindenstrauss, 'Invariant measures and the set of exceptions to Littlewood's conjecture', *Ann. of Math. (2)* **164** (2006), no. 2, 513–560.

80. M. Einsiedler and E. Lindenstrauss, Diagonal actions on locally homogeneous spaces, in *Homogeneous flows, moduli spaces and arithmetic*, in *Clay Math. Proc.* (Amer. Math. Soc., Providence, RI, 2010).

81. M. Einsiedler and T. Ward, *Ergodic theory: supplementary material.* www.mth.uea.ac.uk/ergodic.

82. N. D. Elkies and C. T. McMullen, 'Gaps in $\sqrt{n}$ mod 1 and ergodic theory', *Duke Math. J.* **123** (2004), no. 1, 95–139.

83. J. S. Ellenberg and A. Venkatesh, 'Local-global principles for representations of quadratic forms', *Invent. Math.* **171** (2008), no. 2, 257–279.

84. P. Erdős and A. Rényi, 'On Cantor's series with convergent $\sum 1/q_n$', *Ann. Univ. Sci. Budapest. Eötvös. Sect. Math.* **2** (1959), 93–109.

85. P. Erdős and P. Turán, 'On some sequences of integers', *J. London Math. Soc.* **11** (1936), 261–264.

86. A. Eskin and C. McMullen, 'Mixing, counting, and equidistribution in Lie groups', *Duke Math. J.* **71** (1993), no. 1, 181–209.

87. G. Everest and T. Ward, *An introduction to number theory*, in *Graduate Texts in Mathematics* **232** (Springer-Verlag London Ltd., London, 2005).

88. J.-H. Evertse, 'On sums of $S$-units and linear recurrences', *Compositio Math.* **53** (1984), no. 2, 225–244.

89. H. M. Farkas and I. Kra, *Riemann surfaces*, in *Graduate Texts in Mathematics* **71** (Springer-Verlag, New York, second ed., 1992).

90. R. Feres, *Dynamical systems and semisimple groups: an introduction*, in *Cambridge Tracts in Mathematics* **126** (Cambridge University Press, Cambridge, 1998).

91. E. Fischer, 'Applications d'un théorème sur la convergence en moyenne', *Comptes Rendus* **144** (1907), 1148–1151.

92. E. Fischer, 'Sur la convergence en moyenne', *Comptes Rendus* **144** (1907), 1022–1024.

93. N. P. Fogg, *Substitutions in dynamics, arithmetics and combinatorics*, in *Lecture Notes in Mathematics* **1794** (Springer-Verlag, Berlin, 2002). Edited by V. Berthé, S. Ferenczi, C. Mauduit and A. Siegel.

94. G. B. Folland, *A course in abstract harmonic analysis*, in *Studies in Advanced Mathematics* (CRC Press, Boca Raton, FL, 1995).

95. N. Frantzikinakis, 'The structure of strongly stationary systems', *J. Anal. Math.* **93** (2004), 359–388.

96. D. H. Fremlin, *Measure theory. Vol. 2* (Torres Fremlin, Colchester, 2003). Broad foundations, Corrected second printing of the 2001 original.

97. G. Fubini, 'Sugli integrali multipli', *Rom. Acc. L. Rend. (5)* **16** (1907), 608–614.

98. H. Furstenberg, *Stationary processes and prediction theory*, in *Annals of Mathematics Studies*, No. **44** (Princeton University Press, Princeton, N.J., 1960).

99. H. Furstenberg, 'Strict ergodicity and transformation of the torus', *Amer. J. Math.* **83** (1961), 573–601.

100. H. Furstenberg, 'Disjointness in ergodic theory, minimal sets, and a problem in Diophantine approximation', *Math. Systems Theory* **1** (1967), 1–49.

101. H. Furstenberg, 'The unique ergodicity of the horocycle flow', in *Recent advances in topological dynamics (Proc. Conf., Yale Univ., New Haven, Conn., 1972; in honor of Gustav Arnold Hedlund)*, pp. 95–115. Lecture Notes in Math., Vol. **318** (Springer, Berlin, 1973).

102. H. Furstenberg, 'Ergodic behavior of diagonal measures and a theorem of Szemerédi on arithmetic progressions', *J. Analyse Math.* **31** (1977), 204–256.

103. H. Furstenberg, *Recurrence in ergodic theory and combinatorial number theory* (Princeton University Press, Princeton, N.J., 1981). M. B. Porter Lectures.

104. H. Furstenberg, 'Stiffness of group actions', in *Lie groups and ergodic theory (Mumbai, 1996)*, in *Tata Inst. Fund. Res. Stud. Math.* **14**, pp. 105–117 (Tata Inst. Fund. Res., Bombay, 1998).

105. H. Furstenberg and Y. Katznelson, 'An ergodic Szemerédi theorem for commuting transformations', *J. Analyse Math.* **34** (1978), 275–291 (1979).

106. H. Furstenberg and Y. Katznelson, 'A density version of the Hales-Jewett theorem', *J. Anal. Math.* **57** (1991), 64–119.

107. H. Furstenberg, Y. Katznelson, and D. Ornstein, 'The ergodic theoretical proof of Szemerédi's theorem', *Bull. Amer. Math. Soc. (N.S.)* **7** (1982), no. 3, 527–552.

108. H. Furstenberg and B. Weiss, 'The finite multipliers of infinite ergodic transformations', in *The structure of attractors in dynamical systems (Proc. Conf., North Dakota State Univ., Fargo, N.D., 1977)*, in *Lecture Notes in Math.* **668**, pp. 127–132 (Springer, Berlin, 1978).

109. H. Furstenberg and B. Weiss, 'Topological dynamics and combinatorial number theory', *J. Analyse Math.* **34** (1978), 61–85 (1979).

110. H. Furstenberg and B. Weiss, 'A mean ergodic theorem for $(1/N) \sum_{n=1}^{N} f(T^n x) g(T^{n^2} x)$', in *Convergence in ergodic theory and probability (Columbus, OH, 1993)*, in *Ohio State Univ. Math. Res. Inst. Publ.* **5**, pp. 193–227 (de Gruyter, Berlin, 1996).

111. C. F. Gauss, *Disquisitiones arithmeticae* (Springer-Verlag, New York, 1986). Translated and with a preface by Arthur A. Clarke, Revised by William C. Waterhouse, Cornelius Greither and A. W. Grootendorst, and with a preface by Waterhouse.

112. I. M. Gel'fand and S. V. Fomin, 'Geodesic flows on manifolds of constant negative curvature', *Uspehi Matem. Nauk (N.S.)* **7** (1952), no. 1(47), 118–137.

113. A. Gelfond, 'Sur le septieme problème de Hilbert', *Bull. Acad. Sci. URSS, VII. Ser.* **1934** (1934), no. 4, 623–630.

114. A. Ghosh, 'Metric Diophantine approximation over a local field of positive characteristic', *J. Number Theory* **124** (2007), no. 2, 454–469.

115. J. Gillis, 'Centrally biased discrete random walk', *Quart. J. Math. Oxford Ser. (2)* **7** (1956), 144–152.

116. E. Glasner, *Ergodic theory via joinings*, in *Mathematical Surveys and Monographs* **101** (American Mathematical Society, Providence, RI, 2003).

117. D. A. Goldston, Y. Motohashi, J. Pintz, and C. Y. Yıldırım, 'Small gaps between primes exist', *Proc. Japan Acad. Ser. A Math. Sci.* **82** (2006), no. 4, 61–65.

118. D. A. Goldston, J. Pintz, and C. Y. Yıldırım, 'Primes in tuples. III. On the difference $p_{n+\nu} - p_n$', *Funct. Approx. Comment. Math.* **35** (2006), 79–89.

119. D. A. Goldston, J. Pintz, and C. Y. Yıldırım, 'The path to recent progress on small gaps between primes', in *Analytic number theory*, in *Clay Math. Proc.* **7**, pp. 129–139 (Amer. Math. Soc., Providence, RI, 2007).

120. A. Gorodnik and A. Nevo, *The ergodic theory of lattice subgroups*, in *Annals of Math. Studies* **172** (Princeton University Press, Princeton, 2009).

121. W. H. Gottschalk and G. A. Hedlund, *Topological dynamics*, in *American Mathematical Society Colloquium Publications*, Vol. **36** (American Mathematical Society, Providence, R. I., 1955).

122. W. T. Gowers, 'A new proof of Szemerédi's theorem', *Geom. Funct. Anal.* **11** (2001), no. 3, 465–588.

123. W. T. Gowers, 'Hypergraph regularity and the multidimensional Szemerédi theorem', *Ann. of Math. (2)* **166** (2007), no. 3, 897–946.

124. R. L. Graham and B. L. Rothschild, 'A short proof of van der Waerden's theorem on arithmetic progressions', *Proc. Amer. Math. Soc.* **42** (1974), 385–386.

125. R. L. Graham, B. L. Rothschild, and J. H. Spencer, *Ramsey theory*, in *Wiley-Interscience Series in Discrete Mathematics and Optimization* (John Wiley & Sons Inc., New York, second ed., 1990). A Wiley-Interscience Publication.

126. I. Grattan-Guinness (ed.), *Landmark writings in western mathematics 1640–1940* (Elsevier B. V., Amsterdam, 2005).

127. B. Green and T. Tao, 'The primes contain arbitrarily long arithmetic progressions', *Ann. of Math. (2)* **167** (2008), no. 2, 481–547.

128. F. P. Greenleaf, *Invariant means on topological groups and their applications*, in *Van Nostrand Mathematical Studies*, No. **16** (Van Nostrand Reinhold Co., New York, 1969).

129. G. Greschonig and K. Schmidt, 'Ergodic decomposition of quasi-invariant probability measures', *Colloq. Math.* **84/85** (2000), no. 2, 495–514.

130. A. Haar, 'Der Massbegriff in der Theorie der kontinuierlichen Gruppen', *Ann. of Math. (2)* **34** (1933), no. 1, 147–169.

131. F. Hahn and W. Parry, 'Minimal dynamical systems with quasi-discrete spectrum', *J. London Math. Soc.* **40** (1965), 309–323.

132. H. Hahn, 'Über lineare Gleichungssysteme in linearen Räumen', *J. Reine Angew. Math.* **157** (1927), 214–229.

133. A. W. Hales and R. I. Jewett, 'Regularity and positional games', *Trans. Amer. Math. Soc.* **106** (1963), 222–229.

134. P. R. Halmos, 'On automorphisms of compact groups', *Bull. Amer. Math. Soc.* **49** (1943), 619–624.

135. P. R. Halmos, 'In general a measure preserving transformation is mixing', *Ann. of Math. (2)* **45** (1944), 786–792.

136. P. R. Halmos, 'A nonhomogeneous ergodic theorem', *Trans. Amer. Math. Soc.* **66** (1949), 284–288.

137. P. R. Halmos, *Measure Theory* (D. Van Nostrand Company, Inc., New York, N.Y., 1950).

138. P. R. Halmos, *Lectures on ergodic theory* (Chelsea Publishing Co., New York, 1960).

139. P. R. Halmos and J. v. Neumann, 'Operator methods in classical mechanics. II', *Ann. of Math. (2)* **43** (1942), 332–350.

140. G. Hansel and J. P. Raoult, 'Ergodicity, uniformity and unique ergodicity', *Indiana Univ. Math. J.* **23** (1973/74), 221–237.

141. P. Hartman and A. Wintner, 'On the law of the iterated logarithm', *Amer. J. Math.* **63** (1941), 169–176.

142. F. Hausdorff, *Grundzüge der Mengenlehre* (Leipzig: Veit & Comp. VIII u. 476 S. gr. 8°, 1914).

143. G. A. Hedlund, 'On the metrical transitivity of the geodesics on closed surfaces of constant negative curvature', *Ann. of Math. (2)* **35** (1934), no. 4, 787–808.

144. G. A. Hedlund, 'A metrically transitive group defined by the modular groups', *Amer. J. Math.* **57** (1935), no. 3, 668–678.

145. G. A. Hedlund, 'Fuchsian groups and transitive horocycles', *Duke Math. J.* **2** (1936), no. 3, 530–542.

146. S.-M. Heinemann and O. Schmitt, 'Rokhlin's lemma for non-invertible maps', *Dynam. Systems Appl.* **10** (2001), no. 2, 201–213.

147. E. Helly, 'Über lineare Funktionaloperationen', *Wien. Ber.* **121** (1912), 265–297.

148. G. Herglotz, 'Über Potenzreihen mit positivem, reellem Teil im Einheitskreis', *Leipz. Ber.* **63** (1911), 501–511.

149. C. Hermite, 'Sur la fonction exponentielle', *C. R. Acad. Sci. Paris* **77** (1873), 18–24, 74–79, 226–233.

150. I. N. Herstein and I. Kaplansky, *Matters mathematical* (Chelsea Publishing Co., New York, 1978).

151. E. Hewitt and K. A. Ross, *Abstract harmonic analysis. Vol. I*, in *Grundlehren der Mathematischen Wissenschaften* **115** (Springer-Verlag, Berlin, second ed., 1979).

152. E. Hewitt and K. Stromberg, *Real and abstract analysis. A modern treatment of the theory of functions of a real variable* (Springer-Verlag, New York, 1965).

153. T. H. Hildebrandt, 'Integration in abstract spaces', *Bull. Amer. Math. Soc.* **59** (1953), 111–139.

154. E. Hlawka, *The theory of uniform distribution* (A B Academic Publishers, Berkhamsted, 1984). With a foreword by S. K. Zaremba, Translated from the German by Henry Orde.

155. C. Hoffman and D. Rudolph, 'Uniform endomorphisms which are isomorphic to a Bernoulli shift', *Ann. of Math. (2)* **156** (2002), no. 1, 79–101.

156. E. Hopf, 'Statistik der geodätischen Linien in Mannigfaltigkeiten negativer Krümmung', *Ber. Verh. Sächs. Akad. Wiss. Leipzig* **91** (1939), 261–304.

157. E. Hopf, 'Ergodic theory and the geodesic flow on surfaces of constant negative curvature', *Bull. Amer. Math. Soc.* **77** (1971), 863–877.

158. B. Host, 'Mixing of all orders and pairwise independent joinings of systems with singular spectrum', *Israel J. Math.* **76** (1991), no. 3, 289–298.

159. B. Host and B. Kra, 'Nonconventional ergodic averages and nilmanifolds', *Ann. of Math. (2)* **161** (2005), no. 1, 397–488.

160. R. E. Howe and C. C. Moore, 'Asymptotic properties of unitary representations', *J. Funct. Anal.* **32** (1979), no. 1, 72–96.

161. M. Iosifescu and C. Kraaikamp, *Metrical theory of continued fractions*, in *Mathematics and its Applications* **547** (Kluwer Academic Publishers, Dordrecht, 2002).

162. H. Iwaniec and E. Kowalski, *Analytic number theory*, in *American Mathematical Society Colloquium Publications* **53** (American Mathematical Society, Providence, RI, 2004).

163. K. Jacobs, *Neuere Methoden und Ergebnisse der Ergodentheorie (Ergebnisse der Mathematik und ihrer Grenzgebiete. N. F., Heft* **29**. Springer-Verlag, Berlin-Göttingen-Heidelberg, 1960).

164. K. Jacobs, 'Lipschitz functions and the prevalence of strict ergodicity for continuous-time flows', in *Contributions to Ergodic Theory and Probability (Proc. Conf., Ohio State Univ., Columbus, Ohio, 1970)*, pp. 87–124 (Springer, Berlin, 1970).

165. M. Jerison, 'Martingale formulation of ergodic theorems', *Proc. Amer. Math. Soc.* **10** (1959), 531–539.

166. R. I. Jewett, 'The prevalence of uniquely ergodic systems', *J. Math. Mech.* **19** (1969/1970), 717–729.

167. R. L. Jones, 'New proofs for the maximal ergodic theorem and the Hardy-Littlewood maximal theorem', *Proc. Amer. Math. Soc.* **87** (1983), no. 4, 681–684.

168. M. Kac, 'On the notion of recurrence in discrete stochastic processes', *Bull. Amer. Math. Soc.* **53** (1947), 1002–1010.

169. A. G. Kachurovskiĭ, 'General theories unifying ergodic averages and martingales', *Tr. Mat. Inst. Steklova* **256** (2007), no. Din. Sist. i Optim., 172–200.

170. S. Kakutani, 'Über die Metrisation der topologischen Gruppen', *Proc. Imp. Acad. Jap.* **12** (1936), 82–84.

171. S. Kakutani, 'Two fixed-point theorems concerning bicompact convex sets', *Proc. Imp. Acad. Jap.* **14** (1938), 242–245.

172. S. Kakutani, 'Induced measure preserving transformations', *Proc. Imp. Acad. Tokyo* **19** (1943), 635–641.

173. S. Kakutani, 'Ergodic theory', in *Proceedings of the International Congress of Mathematicians, Cambridge, Mass., 1950, vol. 2* (Amer. Math. Soc., Providence, R. I., 1952), 128–142.

174. S. Kakutani, 'Examples of ergodic measure preserving transformations which are weakly mixing but not strongly mixing', in *Recent advances in topological dynamics (Proc. Conf., Yale Univ., New Haven, Conn., 1972; in honor of Gustav Arnold Hedlund)*, pp. 143–149. Lecture Notes in Math., Vol. **318** (Springer, Berlin, 1973).

175. S. Kakutani and W. Parry, 'Infinite measure preserving transformations with "mixing"', *Bull. Amer. Math. Soc.* **69** (1963), 752–756.

176. S. A. Kalikow, 'Twofold mixing implies threefold mixing for rank one transformations', *Ergodic Theory Dynam. Systems* **4** (1984), no. 2, 237–259.

177. T. Kamae, 'A simple proof of the ergodic theorem using nonstandard analysis', *Israel J. Math.* **42** (1982), no. 4, 284–290.

178. T. Kamae and M. Mendès France, 'van der Corput's difference theorem', *Israel J. Math.* **31** (1978), no. 3–4, 335–342.

179. B. Kamiński, 'Mixing properties of two-dimensional dynamical systems with completely positive entropy', *Bull. Acad. Polon. Sci. Sér. Sci. Math.* **28** (1980), no. 9–10, 453–463 (1981).

180. J. W. Kammeyer and D. J. Rudolph, *Restricted orbit equivalence for actions of discrete amenable groups*, in *Cambridge Tracts in Mathematics* **146** (Cambridge University Press, Cambridge, 2002).

181. E. R. v. Kampen, 'Locally bicompact abelian groups and their character groups', *Ann. of Math. (2)* **36** (1935), no. 2, 448–463.

182. A. Katok and B. Hasselblatt, *Introduction to the modern theory of dynamical systems*, in *Encyclopedia of Mathematics and its Applications* **54** (Cambridge University Press, Cambridge, 1995). With a supplementary chapter by Anatole Katok and Leonardo Mendoza.

183. A. Katok and R. J. Spatzier, 'Invariant measures for higher-rank hyperbolic abelian actions', *Ergodic Theory Dynam. Systems* **16** (1996), no. 4, 751–778.

184. A. Katok and R. J. Spatzier, 'Corrections to: "Invariant measures for higher-rank hyperbolic abelian actions" [Ergodic Theory Dynam. Systems **16** (1996), no. 4, 751–778; MR1406432 (97d:58116)]', *Ergodic Theory Dynam. Systems* **18** (1998), no. 2, 503–507.

185. A. Katok and A. M. Stepin, 'Approximations in ergodic theory', *Uspehi Mat. Nauk* **22** (1967), no. 5 (137), 81–106.

186. Y. Katznelson and B. Weiss, 'A simple proof of some ergodic theorems', *Israel J. Math.* **42** (1982), no. 4, 291–296.

187. M. Keane, 'A continued fraction titbit', *Fractals* **3** (1995), no. 4, 641–650. Symposium in Honor of Benoit Mandelbrot (Curaçao, 1995).

188. M. Keane and P. Michel, 'Généralisation d'un lemme de "découpage" de Rokhlin', *C. R. Acad. Sci. Paris Sér. A-B* **274** (1972), 926–928.

189. A. Y. Khinchin, 'Zu Birkhoffs Lösung des Ergodenproblems', *Math. Ann.* **107** (1933), 485–488.

190. A. Y. Khinchin, 'Metrische Kettenbruchprobleme', *Compos. Math.* **1** (1935), 361–382.

191. A. Y. Khinchin, *Continued fractions* (The University of Chicago Press, Chicago, Ill.-London, 1964).

192. A. Y. Khinchin, *Three pearls of number theory* (Dover Publications Inc., Mineola, NY, 1998). Translated from the Russian by F. Bagemihl, H. Komm, and W. Seidel, Reprint of the 1952 translation.

193. J. L. King, 'Ergodic properties where order 4 implies infinite order', *Israel J. Math.* **80** (1992), no. 1–2, 65–86.

194. J. L. King, 'Three problems in search of a measure', *Amer. Math. Monthly* **101** (1994), no. 7, 609–628.

195. J. F. C. Kingman and S. J. Taylor, *Introduction to measure and probability* (Cambridge University Press, London, 1966).

196. B. Kitchens and K. Schmidt, 'Mixing sets and relative entropies for higher-dimensional Markov shifts', *Ergodic Theory Dynam. Systems* **13** (1993), no. 4, 705–735.

197. B. P. Kitchens, *Symbolic dynamics*, in *Universitext* (Springer-Verlag, Berlin, 1998).

198. F. J. Klein, 'Über die sogenannte Nicht-Euklidische Geometrie', *Math. Ann.* **4** (1871), 573–625.

199. D. Kleinbock and G. A. Margulis, 'Flows on homogeneous spaces and Diophantine approximation on manifolds', *Ann. of Math. (2)* **148** (1998), no. 1, 339–360.

200. D. Kleinbock and G. Tomanov, 'Flows on $S$-arithmetic homogeneous spaces and applications to metric Diophantine approximation', *Comment. Math. Helv.* **82** (2007), no. 3, 519–581.

201. H. D. Kloosterman, 'On the representation of numbers in the form $ax^2 + by^2 + cz^2 + dt^2$', *Proc. London Math. Soc.* **25** (1926), no. 2, 143–173.

202. H. D. Kloosterman, 'On the representation of numbers in the form $ax^2 + by^2 + cz^2 + dt^2$', *Acta Math.* **49** (1927), no. 3–4, 407–464.

203. M. Kmošek, *Rozwinięcie niektórych liczb niewymiernych na ułamki łańcuchowe* (Master's thesis, University of Warsaw, 1979).

204. A. W. Knapp, *Lie groups beyond an introduction*, in *Progress in Mathematics* **140** (Birkhäuser Boston Inc., Boston, MA, 2002).

205. K. Knopp, 'Mengentheoretische Behandlung einiger Probleme der diophantischen Approximationen und der transfiniten Wahrscheinlichkeiten', *Math. Ann.* **95** (1926), no. 1, 409–426.

206. K. Kodaira, 'Über die Beziehung zwischen den Massen und den Topologien in einer Gruppe', *Proc. Phys.-Math. Soc. Japan (3)* **23** (1941), 67–119.

207. P. Koebe, 'Über die Uniformisierung der algebraischen Kurven. I, II, III, IV', *Math. Ann.* **67, 69, 72, 75** (1909, 1910, 1912, 1914), 145–224, 1–81, 437–516, 42–129.

208. B. O. Koopman, 'Hamiltonian systems and transformations in Hilbert space', *Proc. Nat. Acad. Sci. USA* **17** (1931), 315–318.

209. B. O. Koopman and J. v. Neumann, 'Dynamical systems of continuous spectra', *Proc. Nat. Acad. Sci. U.S.A.* **18** (1932), 255–263.

210. U. Krengel, 'On the speed of convergence in the ergodic theorem', *Monatsh. Math.* **86** (1978/79), no. 1, 3–6.

211. U. Krengel, *Ergodic theorems*, in *de Gruyter Studies in Mathematics* **6** (Walter de Gruyter & Co., Berlin, 1985). With a supplement by Antoine Brunel.

212. W. Krieger, 'On entropy and generators of measure-preserving transformations', *Trans. Amer. Math. Soc.* **149** (1970), 453–464.

213. W. Krieger, 'On unique ergodicity', in *Proceedings of the Sixth Berkeley Symposium on Mathematical Statistics and Probability (Univ. California, Berkeley, Calif., 1970/1971), Vol. II: Probability theory* (Univ. California Press, Berkeley, Calif., 1972), 327–346.

214. N. Kryloff and N. Bogoliouboff, 'La théorie générale de la mesure dans son application à l'étude des systèmes dynamiques de la mécanique non linéaire', *Ann. of Math. (2)* **38** (1937), no. 1, 65–113.

215. L. Kuipers and H. Niederreiter, *Uniform distribution of sequences* (Wiley-Interscience [John Wiley & Sons], New York, 1974). *Pure and Applied Mathematics.*

216. A. G. Kurosh, *The theory of groups* (Chelsea Publishing Co., New York, 1960). Translated from the Russian and edited by K. A. Hirsch. 2nd English ed. 2 volumes.

217. R. Kuz'min, 'Sur un problème de Gauss', *C. R. Acad. Sc. URSS* **1928** (1928), 375–380 (Russian).

218. J. L. Lagrange, 'Additions au mémoire sur la résolution des équations numériques', *Mém. Acad. Royale soc. et belles-lettres Berlin* **24** (1770).

219. H. Lebesgue, 'Sur les séries trigonométriques', *Ann. Sci. École Norm. Sup.* (**3**) 20 (1903), 453–485.

220. H. Lebesgue, *Leçons sur l'intégration et la recherche de fonctions primitives* (Paris: Gauthier-Villars. VII u. 136 S. 8°, 1904).

221. F. Ledrappier, 'Un champ markovien peut être d'entropie nulle et mélangeant', *C. R. Acad. Sci. Paris Sér. A-B* **287** (1978), no. 7, 561–563.

222. E. Lehrer, 'Topological mixing and uniquely ergodic systems', *Israel J. Math.* **57** (1987), no. 2, 239–255.

223. A. Leibman, 'Pointwise convergence of ergodic averages for polynomial sequences of translations on a nilmanifold', *Ergodic Theory Dynam. Systems* **25** (2005), no. 1, 201–213.

224. A. Leibman, 'Orbits on a nilmanifold under the action of a polynomial sequence of translations', *Ergodic Theory Dynam. Systems* **27** (2007), no. 4, 1239–1252.

225. M. Lemańczyk, 'Canonical factors on a Lebesgue space', *Bull. Polish Acad. Sci. Math.* **36** (1988), no. 9–10, 541–544.

226. E. Lesigne, 'Sur une nil-variété, les parties minimales associées à une translation sont uniquement ergodiques', *Ergodic Theory Dynam. Systems* **11** (1991), no. 2, 379–391.

227. P. Lévy, 'Sur quelques points de la théorie des probabilités dénombrables', *Ann. Inst. H. Poincaré* **6** (1936), no. 2, 153–184.

228. P. Lévy, *Processus stochastiques et mouvement brownien*, in *Les Grands Classiques Gauthier-Villars. [Gauthier-Villars Great Classics]* (Éditions Jacques Gabay, Sceaux, 1992). Followed by a note by M. Loève, Reprint of the second (1965) edition.

229. E. H. Lieb and M. Loss, *Analysis*, in *Graduate Studies in Mathematics* **14** (American Mathematical Society, Providence, RI, second ed., 2001).

230. D. A. Lind and B. Marcus, *An introduction to symbolic dynamics and coding* (Cambridge University Press, Cambridge, 1995).

231. D. A. Lind and J.-P. Thouvenot, 'Measure-preserving homeomorphisms of the torus represent all finite entropy ergodic transformations', *Math. Systems Theory* **11** (1977/78), no. 3, 275–282.

232. F. Lindemann, 'Über die Zahl π', *Math. Ann.* **20** (1882), 213–225.

233. E. Lindenstrauss, 'Pointwise theorems for amenable groups', *Invent. Math.* **146** (2001), no. 2, 259–295.

234. J. Liouville, 'Nouvelle demonstration d'un théorème sur les irrationnelles algébriques inséré dans le compte rendu de la dernière séance', *C.R. Acad. Sci. Paris* **18** (1844), 910–911.

235. J. Liouville, 'Sur des classes très-étendues de quantités dont la valeur n'est ni algébrique, ni même réductible à des irrationelles algébriques', *C. R. Acad. Sci., Paris* **18** (1844), 883–885.

236. J. Liouville, 'Sur des classes très-étendues de quantités dont la valeur n'est ni algébrique, ni même réductible à des irrationelles algébriques', *J. Math. Pures Appl.* **16** (1851), 133–142.

237. N. Lusin, *Leçons sur les ensembles analytiques et leurs applications* (Chelsea Publishing Co., New York, 1972). With a note by W. Sierpiński, preface by Henri Lebesgue, reprint of the 1930 edition.

238. R. Lyons, 'Seventy years of Rajchman measures', *J. Fourier Anal. Appl.* (1995), 363–377. Special Issue.

239. G. W. Mackey, 'Borel structure in groups and their duals', *Trans. Amer. Math. Soc.* **85** (1957), 134–165.

240. G. W. Mackey, 'Point realizations of transformation groups', *Illinois J. Math.* **6** (1962), 327–335.

241. K. Mahler, 'On lattice points in *n*-dimensional star bodies. I. Existence theorems', *Proc. Roy. Soc. London. Ser. A.* **187** (1946), 151–187.

242. A. Manning, 'Dynamics of geodesic and horocycle flows on surfaces of constant nega-tive curvature', in *Ergodic theory, symbolic dynamics, and hyperbolic spaces (Trieste, 1989)*, in *Oxford Sci. Publ.*, pp. 71–91 (Oxford Univ. Press, New York, 1991).
243. R. Mansuy, 'Histoire de martingales', *Math. Sci. Hum. Math. Soc. Sci.* (2005), no. 169, 105–113.
244. B. Marcus, 'The horocycle flow is mixing of all degrees', *Invent. Math.* **46** (1978), no. 3, 201–209.
245. B. Marcus, 'Topological conjugacy of horocycle flows', *Amer. J. Math.* **105** (1983), no. 3, 623–632.
246. G. A. Margulis, 'On the action of unipotent groups in the space of lattices', in *Lie groups and their representations (Proc. Summer School, Bolyai, János Math. Soc., Budapest, 1971)*, pp. 365–370 (Halsted, New York, 1975).
247. G. A. Margulis, 'Discrete subgroups and ergodic theory', in *Number theory, trace formulas and discrete groups (Oslo, 1987)*, pp. 377–398 (Academic Press, Boston, MA, 1989).
248. G. A. Margulis, *Discrete subgroups of semisimple Lie groups*, in *Ergebnisse der Math-ematik und ihrer Grenzgebiete (3)* **17** (Springer-Verlag, Berlin, 1991).
249. G. A. Margulis, 'Dynamical and ergodic properties of subgroup actions on homoge-neous spaces with applications to number theory', in *Proceedings of the International Congress of Mathematicians, Vol. I, II (Kyoto, 1990)* (Math. Soc. Japan, Tokyo, 1991), 193–215.
250. G. A. Margulis, 'Oppenheim conjecture', in *Fields Medallists' lectures*, in *World Sci. Ser. 20th Century Math.* **5**, pp. 272–327 (World Sci. Publishing, River Edge, NJ, 1997).
251. G. A. Margulis, 'Problems and conjectures in rigidity theory', in *Mathematics: fron-tiers and perspectives*, pp. 161–174 (Amer. Math. Soc., Providence, RI, 2000).
252. G. A. Margulis, *On some aspects of the theory of Anosov systems*, in *Springer Mono-graphs in Mathematics* (Springer-Verlag, Berlin, 2004). With a survey by Richard Sharp: Periodic orbits of hyperbolic flows, Translated from the Russian by Valentina Vladimirovna Szulikowska.
253. G. A. Margulis and G. M. Tomanov, 'Invariant measures for actions of unipotent groups over local fields on homogeneous spaces', *Invent. Math.* **116** (1994), no. 1–3, 347–392.
254. M. H. Martin, 'Metrically transitive point transformations', *Bull. Amer. Math. Soc.* **40** (1934), no. 8, 606–612.
255. G. Maruyama, 'The harmonic analysis of stationary stochastic processes', *Mem. Fac. Sci. Kyūsyū Univ. A.* **4** (1949), 45–106.
256. D. W. Masser, 'Mixing and linear equations over groups in positive characteristic', *Israel J. Math.* **142** (2004), 189–204.
257. F. I. Mautner, 'Geodesic flows on symmetric Riemann spaces', *Ann. of Math. (2)* **65** (1957), 416–431.
258. A. L. Mayhew, 'On some etymologies of English words', *The Modern Language Re-view* **7** (1912), no. 4, 499–507.
259. C. T. McMullen, 'Uniformly Diophantine numbers in a fixed real quadratic field', *Compositio Math.* **145** (2009), 827–844.
260. C. C. Moore, 'The Mautner phenomenon for general unitary representations', *Pacific J. Math.* **86** (1980), no. 1, 155–169.
261. G. Morris and T. Ward, 'A note on mixing properties of invertible extensions', *Acta Math. Univ. Comenian. (N.S.)* **66** (1997), no. 2, 307–311.
262. S. Mozes, 'Mixing of all orders of Lie groups actions', *Invent. Math.* **107** (1992), no. 2, 235–241. Erratum: **119** (1995), no. 2, 399.
263. M. G. Nadkarni, *Basic ergodic theory*, in *Birkhäuser Advanced Texts: Basler Lehrbücher* (Birkhäuser Verlag, Basel, 1998).
264. M. G. Nadkarni, *Spectral theory of dynamical systems*, in *Birkhäuser Advanced Texts: Basler Lehrbücher* (Birkhäuser Verlag, Basel, 1998).

265. B. Nagle, V. Rödl, and M. Schacht, 'The counting lemma for regular $k$-uniform hypergraphs', *Random Structures Algorithms* **28** (2006), no. 2, 113–179.

266. J. v. Neumann, 'Allgemeine Eigenwerttheorie Hermitescher Funktionaloperatoren', *Math. Ann.* **102** (1929), 49–131.

267. J. v. Neumann, 'Proof of the quasi-ergodic hypothesis', *Proc. Nat. Acad. Sci. U.S.A.* **18** (1932), 70–82.

268. J. v. Neumann, 'Zur Operatorenmethode in der klassischen Mechanik', *Ann. of Math. (2)* **33** (1932), 587–642.

269. J. v. Neumann, *Invariant measures* (American Mathematical Society, Providence, RI, 1999).

270. J. Neveu, 'Relations entre la théorie des martingales et la théorie ergodique', *Ann. Inst. Fourier (Grenoble)* **15** (1965), no. fasc. 1, 31–42.

271. A. Nevo, 'Pointwise ergodic theorems for actions of groups', in *Handbook of dynamical systems. Vol. 1B*, pp. 871–982 (Elsevier B. V., Amsterdam, 2006).

272. O. Nikodym, 'Sur une généralisation des intégrales de M. J. Radon', *Fundamenta Mathematicae* **15** (1930), 131–179.

273. A. Novikoff and J. Barone, 'The Borel law of normal numbers, the Borel zero-one law, and the work of Van Vleck', *Historia Math.* **4** (1977), 43–65.

274. D. S. Ornstein and B. Weiss, 'The Shannon-McMillan-Breiman theorem for a class of amenable groups', *Israel J. Math.* **44** (1983), no. 1, 53–60.

275. D. S. Ornstein and B. Weiss, 'Entropy and isomorphism theorems for actions of amenable groups', *J. Analyse Math.* **48** (1987), 1–141.

276. J. C. Oxtoby, *Measure and category. A survey of the analogies between topological and measure spaces*, in *Graduate Texts in Mathematics* **2** (Springer-Verlag, New York, 1971).

277. W. Parry, *Entropy and generators in ergodic theory* (W. A. Benjamin, Inc., New York-Amsterdam, 1969).

278. W. Parry, 'Ergodic properties of affine transformations and flows on nilmanifolds', *Amer. J. Math.* **91** (1969), 757–771.

279. W. Parry, *Topics in ergodic theory*, in *Cambridge Tracts in Mathematics* **75** (Cambridge University Press, Cambridge, 1981).

280. K. R. Parthasarathy, *Probability measures on metric spaces* (AMS Chelsea Publishing, Providence, RI, 2005). Reprint of the 1967 original.

281. A. L. T. Paterson, *Amenability*, in *Mathematical Surveys and Monographs* **29** (American Mathematical Society, Providence, RI, 1988).

282. K. Petersen, *Ergodic theory*, in *Cambridge Studies in Advanced Mathematics* **2** (Cambridge University Press, Cambridge, 1989). Corrected reprint of the 1983 original.

283. R. R. Phelps, *Lectures on Choquet's theorem*, in *Lecture Notes in Mathematics* **1757** (Springer-Verlag, Berlin, second ed., 2001).

284. J.-P. Pier, *Amenable locally compact groups*, in *Pure and Applied Mathematics* (New York) (John Wiley & Sons Inc., New York, 1984).

285. I. I. Pjateckiĭ-Šapiro, 'On the distribution of the fractional parts of the exponential function', *Moskov. Gos. Ped. Inst. Uč. Zap.* **108** (1957), 317–322.

286. V. Platonov and A. Rapinchuk, *Algebraic groups and number theory*, in *Pure and Applied Mathematics* **139** (Academic Press Inc., Boston, MA, 1994).

287. H. Poincaré, 'Théorie des groupes Fuchsiens', *Acta Math.* **1** (1882), 1–62.

288. H. Poincaré, 'Sur le problème des trois corps et les équations de la dynamique', *Acta Math.* **13** (1890), 1–270.

289. H. Poincaré, 'Sur l'uniformisation des fonctions analytiques', *Acta Math.* **31** (1908), no. 1, 1–63.

290. A. D. Pollington and S. L. Velani, 'On a problem in simultaneous Diophantine approximation: Littlewood's conjecture', *Acta Math.* **185** (2000), no. 2, 287–306.

291. D. H. J. Polymath, *A new proof of the density Hales–Jewett theorem*. arXiv:math/0910.3926.

292. J.-V. Poncelet, *Traité des propriétés projectives des figures; ouvrange utile a ceux qui s'occupent des applications de la geometrie descriptive et d'operations geometriques sûr le terrain* (Gauthier-Villars, Paris, 1866). Second ed.; First ed. published 1822.

293. L. S. Pontrjagin, *Topological Groups*, in *Princeton Mathematical Series*, v. 2 (Princeton University Press, Princeton, 1939). Translated from the Russian by Emma Lehmer.

294. A. J. v. d. Poorten, 'Notes on continued fractions and recurrence sequences', in *Number theory and cryptography (Sydney, 1989)*, in *London Math. Soc. Lecture Note Ser.* **154**, pp. 86–97 (Cambridge Univ. Press, Cambridge, 1990).

295. A. J. v. d. Poorten and H. P. Schlickewei, 'Additive relations in fields', *J. Austral. Math. Soc. Ser. A* **51** (1991), no. 1, 154–170.

296. R. Rado, 'Note on combinatorial analysis', *Proc. London Math. Soc. (2)* **48** (1943), 122–160.

297. J. Radon, 'Theorie und Anwendungen der absolut additiven Mengenfunktionen', *Wien. Ber.* **122** (1913), 1295–1438.

298. M. M. Rao, 'Abstract martingales and ergodic theory', in *Multivariate analysis, III (Proc. Third Internat. Sympos., Wright State Univ., Dayton, Ohio, 1972)*, pp. 45–60 (Academic Press, New York, 1973).

299. M. Ratner, 'Factors of horocycle flows', *Ergodic Theory Dynam. Systems* **2** (1982), no. 3–4, 465–489 (1983).

300. M. Ratner, 'Horocycle flows, joinings and rigidity of products', *Ann. of Math. (2)* **118** (1983), no. 2, 277–313.

301. M. Ratner, 'Ergodic theory in hyperbolic space', in *Conference in modern analysis and probability (New Haven, Conn., 1982)*, in *Contemp. Math.* **26**, pp. 309–334 (Amer. Math. Soc., Providence, RI, 1984).

302. M. Ratner, 'On measure rigidity of unipotent subgroups of semisimple groups', *Acta Math.* **165** (1990), no. 3–4, 229–309.

303. M. Ratner, 'Strict measure rigidity for unipotent subgroups of solvable groups', *Invent. Math.* **101** (1990), no. 2, 449–482.

304. M. Ratner, 'On Raghunathan's measure conjecture', *Ann. of Math. (2)* **134** (1991), no. 3, 545–607.

305. M. Ratner, 'Raghunathan's topological conjecture and distributions of unipotent flows', *Duke Math. J.* **63** (1991), no. 1, 235–280.

306. M. Ratner, 'Raghunathan's conjectures for Cartesian products of real and $p$-adic Lie groups', *Duke Math. J.* **77** (1995), no. 2, 275–382.

307. H. Reiter, *Classical harmonic analysis and locally compact groups* (Clarendon Press, Oxford, 1968).

308. A. Rényi, 'On mixing sequences of sets', *Acta Math. Acad. Sci. Hungar.* **9** (1958), 215–228.

309. B. Riemann, *Grundlagen für eine allgemeine Theorie der Functionen einer veränderlichen complexen Grösse* (Ph.D. thesis, Göttingen, 1851).

310. B. Riemann, 'Über die Darstellbarkeit einer Function durch eine trigonometrische Reihe (Habilitationsschrift, 1854)', *Abhandlungen der Königlichen Gesellschaft der Wissenschaften zu Göttingen* **13** (1868).

311. F. Riesz, 'Integrálható függvények sorozatai', *Matematikai és Physikai Lapok* **19** (1910), 165–182, 228–243.

312. F. Riesz, 'Sur certains systèmes d'équations fonctionnelles et l'approximation des fonctions continues', *Comptes Rendus* **150** (1910), 674–677.

313. F. Riesz, 'Sur certains systèmes singuliers d'équations intégrales', *Ann. Sci. École Norm. Sup.* **28** (1911), 33–62.

314. F. Riesz, 'Some mean ergodic theorems', *J. London Math. Soc.* **13** (1938), 274–278.

315. V. A. Rokhlin, 'A "general" measure-preserving transformation is not mixing', *Doklady Akad. Nauk SSSR (N.S.)* **60** (1948), 349–351.

316. V. A. Rokhlin, 'On endomorphisms of compact commutative groups', *Izvestiya Akad. Nauk SSSR. Ser. Mat.* **13** (1949), 329–340.

317. A. Rosenthal, 'Strictly ergodic models for noninvertible transformations', *Israel J. Math.* **64** (1988), no. 1, 57–72.

318. K. F. Roth, 'Sur quelques ensembles d'entiers', *C. R. Acad. Sci. Paris* **234** (1952), 388–390.

319. K. F. Roth, 'Rational approximations to algebraic numbers', *Mathematika* **2** (1955), 1–20.

320. H. L. Royden, *Real analysis* (Macmillan Publishing Company, New York, third ed., 1988).

321. W. Rudin, *Functional analysis* (McGraw-Hill Book Co., New York, 1973). *McGraw-Hill Series in Higher Mathematics*.

322. W. Rudin, *Fourier analysis on groups*, in *Wiley Classics Library* (John Wiley & Sons Inc., New York, 1990). Reprint of the 1962 original.

323. D. J. Rudolph, 'An example of a measure preserving map with minimal self-joinings, and applications', *J. Analyse Math.* **35** (1979), 97–122.

324. D. J. Rudolph, *Fundamentals of measurable dynamics*, in *Oxford Science Publications* (The Clarendon Press Oxford University Press, New York, 1990).

325. D. J. Rudolph and B. Weiss, 'Entropy and mixing for amenable group actions', *Ann. of Math. (2)* **151** (2000), no. 3, 1119–1150.

326. C. Ryll-Nardzewski, 'On the ergodic theorems. II. Ergodic theory of continued fractions', *Studia Math.* **12** (1951), 74–79.

327. V. V. Ryzhikov, 'On a connection between the mixing properties of a flow with an isomorphism entering into its transformations', *Mat. Zametki* **49** (1991), no. 6, 98–106, 159.

328. V. V. Ryzhikov, 'Mixing, rank and minimal self-joining of actions with invariant measure', *Mat. Sb.* **183** (1992), no. 3, 133–160.

329. A. Sárközy, 'On difference sets of sequences of integers. III', *Acta Math. Acad. Sci. Hungar.* **31** (1978), no. 3–4, 355–386.

330. H. P. Schlickewei, '*S*-unit equations over number fields', *Invent. Math.* **102** (1990), no. 1, 95–107.

331. E. Schmidt, 'Zur Theorie der linearen und nichtlinearen Integralgleichungen. III', *Math. Ann.* **65** (1908), 370–399.

332. K. Schmidt, *Dynamical systems of algebraic origin* (Birkhäuser Verlag, Basel, 1995).

333. K. Schmidt and T. Ward, 'Mixing automorphisms of compact groups and a theorem of Schlickewei', *Invent. Math.* **111** (1993), no. 1, 69–76.

334. T. Schneider, 'Transzendenzuntersuchungen periodischer Funktionen. I. Transzendenz von Potenzen', *J. Reine Angew. Math.* **172** (1934), 65–69.

335. R. Schulze-Pillot, 'Representation by integral quadratic forms—a survey', in *Algebraic and arithmetic theory of quadratic forms*, in *Contemp. Math.* **344**, pp. 303–321 (Amer. Math. Soc., Providence, RI, 2004).

336. C. Series, 'The modular surface and continued fractions', *J. London Math. Soc. (2)* **31** (1985), no. 1, 69–80.

337. C. Series, 'Geometrical methods of symbolic coding', in *Ergodic theory, symbolic dynamics, and hyperbolic spaces (Trieste, 1989)*, in *Oxford Sci. Publ.*, pp. 125–151 (Oxford Univ. Press, New York, 1991).

338. J.-P. Serre, *A course in arithmetic*, in *Graduate Texts in Mathematics* **7** (Springer-Verlag, New York, 1973).

339. J. Shallit, 'Simple continued fractions for some irrational numbers', *J. Number Theory* **11** (1979), no. 2, 209–217.

340. J. Shallit, 'Real numbers with bounded partial quotients: a survey', *Enseign. Math. (2)* **38** (1992), no. 1–2, 151–187.

341. S. Shelah, 'Primitive recursive bounds for van der Waerden numbers', *J. Amer. Math. Soc.* **1** (1988), no. 3, 683–697.

342. P. C. Shields, *The ergodic theory of discrete sample paths*, in *Graduate Studies in Mathematics* **13** (American Mathematical Society, Providence, RI, 1996).

343. C. L. Siegel, 'Approximation algebraischer Zahlen', *Jahrbuch d. philos. Fakultät Göttingen* (1921), 291–296.
344. W. Sierpiński, 'Sur la valeur asymptotique d'une certaine somme', *Bull Intl. Acad. Polonmaise des Sci. et des Lettres (Cracovie)* (1910), 9–11.
345. W. Sierpiński, 'Démonstration élémentaire du théorème de M. Borel sur, les nombres absolument normaux et détermination effective d'une tel nombre', *Bull. Soc. Math. France* **45** (1917), 125–132.
346. K. Simic, 'The pointwise ergodic theorem in subsystems of second-order arithmetic', *J. Symbolic Logic* **72** (2007), no. 1, 45–66.
347. A. Soifer, *The mathematical coloring book* (Springer, New York, 2009). Mathematics of coloring and the colorful life of its creators, With forewords by Branko Grünbaum, Peter D. Johnson, Jr. and Cecil Rousseau.
348. B. Spitters, 'A constructive view on ergodic theorems', *J. Symbolic Logic* **71** (2006), no. 2, 611–623. Corrigendum: **71** (2006), no. 4, 1431–1432.
349. M. Spivak, *Calculus on manifolds. A modern approach to classical theorems of advanced calculus* (W. A. Benjamin, Inc., New York-Amsterdam, 1965).
350. A. N. Starkov, *Dynamical systems on homogeneous spaces*, in *Translations of Mathematical Monographs* **190** (American Mathematical Society, Providence, RI, 2000). Translated from the 1999 Russian original by the author.
351. J. M. Steele, 'Covering finite sets by ergodic images', *Canad. Math. Bull.* **21** (1978), no. 1, 85–91.
352. J. Steinig, 'A proof of Lagrange's theorem on periodic continued fractions', *Arch. Math. (Basel)* **59** (1992), no. 1, 21–23.
353. J. Stillwell, *Geometry of surfaces*, in *Universitext* (Springer-Verlag, New York, 1992). Corrected reprint of the 1992 original.
354. M. H. Stone, 'Linear transformation in Hilbert space. I: Geometrical aspects; II: Analytical aspects; III: Operational methods and group theory', *Proc. Nat. Acad. Sci. U.S.A.* **15** (1929), 198–200; 423–425; **16** (1930) 172–175.
355. M. H. Stone, 'Applications of the theory of Boolean rings to general topology', *Trans. Amer. Math. Soc.* **41** (1937), 375–481.
356. E. Szemerédi, 'On sets of integers containing no four elements in arithmetic progression', *Acta Math. Acad. Sci. Hungar.* **20** (1969), 89–104.
357. E. Szemerédi, 'On sets of integers containing no $k$ elements in arithmetic progression', *Acta Arith.* **27** (1975), 199–245.
358. A. A. Tempel'man, *Ergodic theorems for group actions*, in *Mathematics and its Applications* **78** (Kluwer Academic Publishers Group, Dordrecht, 1992). Informational and thermodynamical aspects, Translated and revised from the 1986 Russian original.
359. J.-P. Thouvenot, 'Some properties and applications of joinings in ergodic theory', in *Ergodic theory and its connections with harmonic analysis (Alexandria, 1993)*, in *London Math. Soc. Lecture Note Ser.* **205**, pp. 207–235 (Cambridge Univ. Press, Cambridge, 1995).
360. A. Thue, 'Über Annäherungswerte algebraischer Zahlen', *J. Reine Angew. Math.* **135** (1909), 284–305.
361. H. Tietze, 'Über funktionen, die auf einer abgeschlossenen Menge stetig sind', *J. Reine Angew. Math.* **145** (1914), 9–14.
362. L. Tonelli, 'Sull' integrazione per parti', *Rom. Acc. L. Rend. (5)* **18** (1909), 246–253.
363. H. Towsner, *A model theoretic proof of Szemerédi's theorem.* arXiv:math/1002.4456.
364. A. Tychonoff, 'Über die topologische Erweiterung von Räumen', *Math. Ann.* **102** (1929), 544–561.
365. S. M. Ulam, *Adventures of a mathematician* (Charles Scribner's Sons, New York, 1976).
366. P. Urysohn, 'Über die Mächtigkeit der zusammenhängenden Mengen', *Math. Ann.* **94** (1925), 262–295.

367. V. Vatsal, 'Special values of anticyclotomic $L$-functions', *Duke Math. J.* **116** (2003), no. 2, 219–261.

368. W. A. Veech, 'Minimality of horospherical flows', *Israel J. Math.* **21** (1975), no. 2–3, 233–239.

369. G. Vitali, 'Sui gruppi di punti e sulle funzioni di variabili reali', *Torino Atti* **43** (1908), 229–246.

370. E. B. v. Vleck, 'On non-measurable sets of points, with an example', *Trans. Amer. Math. Soc.* **9** (1908), no. 2, 237–244.

371. B. L. v. d. Waerden, 'Beweis einer Baudet'schen Vermutung', *Nieuw. Arch. Wisk.* **15** (1927), 212–216.

372. B. L. v. d. Waerden, 'How the proof of Baudet's conjecture was found', in *Studies in Pure Mathematics (Presented to Richard Rado)*, pp. 251–260 (Academic Press, London, 1971).

373. S. Wagon, *The Banach-Tarski paradox* (Cambridge University Press, Cambridge, 1993). With a foreword by Jan Mycielski, Corrected reprint of the 1985 original.

374. P. Walters, *An introduction to ergodic theory*, in *Graduate Texts in Mathematics* **79** (Springer-Verlag, New York, 1982).

375. T. Ward, 'Three results on mixing shapes', *New York J. Math.* **3A** (1997/98), 1–10.

376. A. Weil, 'Sur les groupes topologiques et les groupes mesures', *C. R. Acad. Sci., Paris* **202** (1936), 1147–1149.

377. A. Weil, *L'intègration dans les groupes topologiques et ses applications*, in *Actual. Sci. Ind.*, no. **869** (Hermann et Cie., Paris, 1940).

378. A. Weil, *Basic number theory*, in *Die Grundlehren der mathematischen Wissenschaften*, Band **144** (Springer-Verlag New York, Inc., New York, 1967).

379. B. Weiss, 'Strictly ergodic models for dynamical systems', *Bull. Amer. Math. Soc. (N.S.)* **13** (1985), no. 2, 143–146.

380. H. Weyl, 'Über die *Gibbs*sche Erscheinung und verwandte Konvergenzphänomene', *Rendiconti del Circolo Matematico di Palermo* **30** (1910), 377–407.

381. H. Weyl, 'Uber die Gleichverteilung von Zahlen mod Eins', *Math. Ann.* **77** (1916), 313–352.

382. N. Wiener, 'The ergodic theorem', *Duke Math. J.* **5** (1939), no. 1, 1–18.

383. N. Wiener and A. Wintner, 'Harmonic analysis and ergodic theory', *Amer. J. Math.* **63** (1941), 415–426.

384. S. M. J. Wilson, 'Limit points in the Lagrange spectrum of a quadratic field', *Bull. Soc. Math. France* **108** (1980), no. 2, 137–141.

385. D. Witte Morris, *Introduction to arithmetic groups*. In preparation.

386. D. Witte Morris, 'Rigidity of some translations on homogeneous spaces', *Invent. Math.* **81** (1985), no. 1, 1–27.

387. D. Witte Morris, *Ratner's theorems on unipotent flows*, in *Chicago Lectures in Mathematics* (University of Chicago Press, Chicago, IL, 2005).

388. A. C. Woods, 'The Markoff spectrum of an algebraic number field', *J. Austral. Math. Soc. Ser. A* **25** (1978), no. 4, 486–488.

389. F. B. Wright, 'Mean least recurrence time', *J. London Math. Soc.* **36** (1961), 382–384.

390. R. Yassawi, 'Multiple mixing and local rank group actions', *Ergodic Theory Dynam. Systems* **23** (2003), no. 4, 1275–1304.

391. K. Yosida, 'Mean ergodic theorem in Banach spaces', *Proc. Imp. Acad., Tokyo* **14** (1938), 292–294.

392. K. Yosida and S. Kakutani, 'Birkhoff's ergodic theorem and the maximal ergodic theorem', *Proc. Imp. Acad., Tokyo* **15** (1939), 165–168.

393. T. Ziegler, 'Universal characteristic factors and Furstenberg averages', *J. Amer. Math. Soc.* **20** (2007), no. 1, 53–97.

394. R. J. Zimmer, *Ergodic theory and semisimple groups*, in *Monographs in Mathematics* **81** (Birkhäuser Verlag, Basel, 1984).

395. J. D. Zund, 'George David Birkhoff and John von Neumann: a question of priority and the ergodic theorems, 1931–1932', *Historia Math.* **29** (2002), no. 2, 138–156.

# Author Index

Aaronson, 22
Adler, 94
Akin, 274
Alaoglu, 372, 428
Ambrose, 67
Anzai, 230
Arenas-Carmona, 275
Arnol'd, 2
Artin, 171, 277
atom, 133
Auslander, 274, 345
Avez, 2
Avigad, 67

Bailey, 94
Baladi, 67
Banach, 171, 275, 417, 418, 428
Barone, 66, 414
Barreira, viii
Baudet, 171
Becher, 67
Bekka, 12
Bellow, 119
Beltrami, 401
Berend, 275
Bergelson, 66, 179, 227, 230, 274, 275
Bernoulli, 17
Berthé, 65
Bhattacharya, 274
Billingsley, 12, 94
Birkhoff, 11, 32, 44, 66, 104, 229, 439
Bishop, 67
Bochner, 61, 440
Bogoliouboff, 98, 107, 251, 252
Bohl, 119
Boltzmann, 11
Bonnet, 352

Borel, 2, 132, 134, 405, 411, 414
Borwein, 94
Bourgain, 40, 44, 67, 275
Brauer, 229
Brin, vii

Calderon, 275
Cantelli, 132, 405, 414
Cantor, 95
Carathéodory, 65, 401
Cassels, 7
Chacon, 67
Champernowne, 67
Choquet, 103, 154, 272
Chu, 109
Conze, 228
Cornfeld, vii, 12, 94
van der Corput, 184, 187, 195, 229
Crandall, 94

Dani, 10, 378, 389, 399, 401, 402
von Dantzig, 439
Day, 275
Denker, 119
Dirichlet, 75, 94, 380
Doeblin, 94
Doob, 126, 127, 151
Downarowicz, 119
Drmota, 119
Duke, 10
Dyson, 95

Eberlein, 119
Ehrenfest, P., 11
Ehrenfest, T., 11
Einsiedler, vii, 7
Elkies, 10

M. Einsiedler, T. Ward, *Ergodic Theory*, Graduate Texts in Mathematics 259, 463
DOI 10.1007/978-0-85729-021-2, © Springer-Verlag London Limited 2011

Ellenberg, 8, 9
Erdős, 4, 171, 229, 414
Eskin, 9
Evertse, 275

Farkas, 401
Fatou, 177, 411
Ferenczi, 65
Feres, 12, 330
Figuera, 67
Fischer, 412
Folland, 243, 428, 439
Følner, 251
Fomin, vii, 12, 94, 401
Fourier, 161, 434, 435
Frantzikinakis, 230
Fremlin, 414
Frobenius, 240
Fubini, 60, 262, 409, 414
Furman, 275
Furstenberg, 7, 53, 114, 119, 169, 171, 175,
    178, 180, 228–230, 274, 275, 378

Gauss, 9, 76, 77, 352
Gel'fand, 11, 401
Gelfond, 95
Gerhardy, 67
Ghosh, 402
Gillis, 67
Glasner, 12, 169
Goldston, 5, 12
Gorodnik, 274, 275
Gottschalk, 274
Gowers, 4, 230
Graham, 172, 229
Gray, 12
Green, 5, 229, 345
Greenleaf, 275
Greschonig, 266

Haar, 14, 108, 161, 228, 243, 311, 439
Hahn, Frank, 109, 345
Hahn, Hans, 418, 428
Hales, 230
Halmos, 67, 68, 162, 275
Hansel, 119
Hartman, 66
Hasselblatt, vii
Hausdorff, 414
Hecke, 362
Hedlund, 274, 330, 401
Heinemann, 68
Heisenberg, 300, 331
Helly, 428

Herglotz, 61, 440
Hermite, 95
Herstein, 95
Hewitt, 275, 415, 439
Hilbert, 67, 420
Hildebrandt, 428
Hlawka, 119
Hoffman, 68
Hopf, 314
Host, 5, 67, 228
Howe, 370, 372

Iosifescu, 94
Iwaniec, 119

Jacobi, 293
Jacobs, 151
Jakobs, 119
Jerison, 151
Jewett, 119, 230
Jones, 67
del Junco, 67, 257

Kac, 63
Kachurovskiĭ, 151
Kakutani, 61–63, 65–67, 151, 439
Kalikow, 67
Kamae, 67, 229
Kamiński, 274
Kammeyer, 275
van Kampen, 439
Kaplansky, 95
Katok, vii, 7, 67
Katznelson, 67, 171, 230
Keane, 68, 94
Khinchin, 66, 94
King, 12
Kingman, 403
Kitchens, 65, 275
Klein, 401
Kleinbock, 402
Kloosterman, 8
Kmošek, 94
Knapp, 330, 402
Knopp, 94
Kodaira, 439
Koebe, 401
Koopman, 29, 50, 66
Kowalski, 119
Kra, 5, 228, 401
Kraaikamp, 94
Krengel, 66, 67
Krieger, 11, 119, 169
Kronecker, 160, 228

Kryloff, 98, 107, 251, 252
Kuipers, 119
Kurosh, 310
Kuz'min, 94

Lagrange, 88
Lebesgue, 410, 412, 414, 438, 439
Ledrappier, 236
Lehrer, 119
Leibman, 227, 230, 345
Lemańczyk, 162
Lesigne, 228, 345
Lévy, 94, 151
Lie, 228
Lieb, 414
Lind, 65, 119
Lindemann, 95
Lindenstrauss, vii, 7, 257, 275
Liouville, 91, 95, 118
Littlewood, 7
Loss, 414
Lusin, 314, 411
Lyons, 440

Mackey, 439
Mahler, 390
Manning, 330
Mansuy, 151
Marcus, 65, 401
Margulis, 6, 7, 9, 12, 367, 389, 402
Martin, 330
Maruyama, 67
Masser, 274
Mauduit, 65
Mautner, 364, 369
Mayer, 12
Mayhew, 151
McMullen, 9, 10, 95, 330
Mendès France, 229
Menshov, 440
Michel, 68
Minkowski, 391
Möbius, 279, 284
Moore, 369, 370, 372
Morris, 274
Motohashi, 12
Mozes, 9, 274, 275, 372

Nadkarni, vii, 47
Nagle, 230
Neder, 440
von Neumann, 32, 50, 66, 162, 230, 275, 432,
        439
Neveu, 151

Nevo, 275
Niederreiter, 119
Nikodym, 100, 123, 325, 410
Novikoff, 66, 414

Oppenheim, 5
Ornstein, vii, 11, 171, 275
Oxtoby, 414

Parry, 12, 67, 109, 151
Parseval, 435
Parthasarathy, 403, 412, 422
Paterson, 275
Pesin, viii
Petersen, vii, 67, 443
Phelps, 118
Pier, 275
Pintz, 5, 12
Pjateckiǐ-Šapiro, 119
Plancherel, 435
Platonov, 12
Poincaré, 4, 340, 401
Pollington, 7
Polymath, 230
Poncelet, 12
Pontryagin, 161, 437, 439
van der Poorten, 94, 239

Rado, 229
Radon, 100, 123, 325, 410
Raghunathan, 9, 10, 401
Rajchman, 440
Rao, 151
Raoult, 119
Rapinchuk, 12
Ratner, 9, 10, 12, 345, 401
Reiter, 275
Rényi, 53, 414
Riemann, 401, 438, 439
Riesz, 138, 245, 275, 412, 422, 423, 428
Rödl, 230
Rokhlin, 49, 63, 65, 67, 236
Rosenblatt, 257
Rosenthal, 68, 119
Ross, 275, 439
Roth, 95, 171, 230
Rothschild, 172, 229
Royden, 414
Rudin, 67, 417, 428, 439
Rudnick, 10
Rudolph, vii, 12, 68, 169, 275
Ryll-Nardzewski, 94
Ryzhikov, 67, 401

Sárközy, 180
Sarnak, 10
Schacht, 230
Schlickewei, 239
Schmidt, Erhard, 67, 420
Schmidt, Klaus, vii, 235, 266, 274, 275
Schmitt, 68
Schneider, 95
Schreier, 171
Schulze-Pillot, 9
Schur, 171
Series, 330
Serre, 307, 330
Shah, 9
Shallit, 94
Shelah, 229
Shields, 12
Siegel, 65, 95
Sierpiński, 67, 119
Sinaĭ, vii
Smillie, 399
Smith, 66
Smorodinsky, 94
Soifer, 229
Spatzier, 7
Spencer, 229
Spitters, 67
Spivak, 415
Starkov, 12, 401
Steele, 65
Steinig, 95
Stepin, 67
Stillwell, 401
Stone, 66, 421, 428
Stromberg, 415
Stuck, vii
Swinnerton-Dyer, 7
Szemerédi, 4, 171, 178

Tao, 5, 229
Tarski, 275
Taylor, 403

Tempel'man, 275
Thouvenot, 119, 162, 401
Thue, 95
Tichy, 119
Tietze, 428
Tomanov, 9, 12, 402
Tonelli, 414
Towsner, 67, 230
Turán, 4, 171
Tychonoff, 372, 428

Urysohn, 428

Vatsal, 10
.Veech, 401
Velani, 7
Venkatesh, 8, 9
Vitali, 40, 261, 275
van Vleck, 66

van der Waerden, 171, 172
Wagon, 275
Walters, vii
Ward, 274
Weierstrass, 421
Weil, 438–440
Weiss, 67, 119, 178, 228, 229, 275
Weyl, 3, 114, 119, 183, 201
Wiener, 66, 259, 275
Wilson, 95
Wintner, 66
Witte Morris, 12, 345
Woods, 95

Yassawi, 67
Yıldırım, 5
Yosida, 66

Ziegler, 5, 228
Zimmer, 12, 275
Zorn, 442, 443
Zund, 66

# Index of Notation

$\mathbb{N}$, natural numbers, viii

$\mathbb{N}_0$, non-negative integers, viii

$\mathbb{Z}$, integers, viii

$\mathbb{Q}$, rational numbers, viii

$\mathbb{R}$, real numbers, viii

$\mathbb{C}$, complex numbers, viii

$\mathbb{S}^1$, multiplicative circle, viii

$\mathbb{T}$, additive circle, viii

$\Re(\cdot), \Im(\cdot)$, real and imaginary parts, viii

$O(\cdot)$, order of growth, viii

$o(\cdot)$, order of growth, viii

$\sim$, similar growth, viii

$\ll$, relation between growth in functions, viii

$C(X)$, real-valued continuous functions on $X$, viii

$C_{\mathbb{C}}(X)$, complex-valued continuous functions on $X$, viii

$C_c(X)$, compactly supported continuous functions on $X$, viii

$A \smallsetminus B$, difference of two sets, viii

$\chi_A(\cdot)$, indicator or characteristic function of $A$, 2

$m$, Lebesgue measure, 2

a.e., almost everywhere, 2

$\{\cdot\}$, fractional part, 3

$\langle\cdot\rangle$, distance to nearest integer, 7

$R_\alpha$, rotation by $\alpha$, 14

$m_{\mathbb{T}}$, Lebesgue (Haar) measure on the circle, 14

$m_X$, Haar measure on $X$, 14

$f \circ T$, composition $x \mapsto f(Tx)$, 15

$L^1$, integrable functions, 15

$x|_I$, block in a sequence, 18

$\widetilde{X} = (\widetilde{X}, \widetilde{B}, \widetilde{\mu}, \widetilde{T})$, invertible extension, 20

$d(A)$, density of the set $A$ in $\mathbb{N}$ or $\mathbb{Z}$, 22

$\overline{\mathbf{d}}(A)$, upper density of the set $A$ in $\mathbb{N}$ or $\mathbb{Z}$, 23

$\triangle$, $A \triangle B = (A \smallsetminus B) \cup (B \smallsetminus A)$, 23

$U_T$, unitary operator associated to $T$, 28

$U^*$, adjoint of operator $U$, 29

$\xrightarrow[L_\mu^p]{}$, convergence in $L_\mu^p$ norm, 32

$P_T$, projection onto subspace invariant under $U_T$, 32

$A_N^f$, $\frac{1}{N}\sum_{n=0}^{N-1} f \circ T^n$ (with various interpretations), 34

$r_A(\cdot)$, first return time to the set $A$, 61

$\mu|_{.}$, restriction of measure $\mu$, 62

$\bigsqcup$, disjoint union, 62

$(X^{(r)}, \mathscr{B}^{(r)}, \mu^{(r)}, T^{(r)})$, suspension defined by $r$, 64

$\lfloor \cdot \rfloor$, floor function, 76

$C(X)^*$, dual of $C(X)$, 97

$\mathscr{M}(X)$, probability measures on $X$, 97

$T_*$, map on measures induced by $T$, 97

$\delta_.$, point measure, 97

$\mathscr{M}^T(X)$, $T$-invariant probability measures on $X$ (Borel if $X$ is a compact metric space), 98

$\mathscr{E}^T(X)$, ergodic Borel probability measures on $X$, 101

$E(\cdot|\mathscr{A})$, conditional expectation with respect to $\mathscr{A}$, 121

$\sigma(\{A_\gamma \mid \gamma \in \Gamma\})$, smallest $\sigma$-algebra containing the sets $A_\gamma$, 122

$\mu_x^{\mathscr{A}}$, conditional measure, 133

$\subseteq_{\mu}$, inclusion up to $\mu$-null sets, 135

$=_{\mu}$, equality up to $\mu$-null sets, 135

$[x]_{\mathscr{A}}$, atom of $\mathscr{A}$ containing $x$, 135

M. Einsiedler, T. Ward, *Ergodic Theory*, Graduate Texts in Mathematics 259, 467
DOI 10.1007/978-0-85729-021-2, © Springer-Verlag London Limited 2011

$C(X)$, real-valued continuous functions on $X$, 138

$G_\delta$-set, a countable intersection of open sets, 139

$F_\sigma$-set, a countable union of closed sets, 139

$\phi_*$, induced map on measure spaces, 145

$\mathscr{E}$, $\sigma$-algebra of invariant sets, 153

$\mathscr{B}_X \otimes \mathscr{B}_Y$, product $\sigma$-algebra, 158

X, Y, measure-preserving systems, 158

J(X, Y), set of joinings of X and Y, 158

$\pi_X$, projection onto factor $X$, 159

$\mathbb{S}^{\mathbb{N}}$, infinite multiplicative torus, 160

$\widehat{K}$, dual (character) group of $K$, 161

X $\perp$ Y, disjointness of measure-preserving systems, 163

$\mu \times_\lambda \nu$ or $\mu \times_Z \nu$, relatively independent joining, 164

$\overline{\mathbf{d}}_B(A)$, upper Banach density of $A$, 171

SZ, Szemerédi system, 176

$\widetilde{X} = (\widetilde{X}, \widetilde{\mathscr{B}}, \widetilde{\mu}, \widetilde{T})$, invertible extension, 177

$\mathrm{diam}(\cdot)$, diameter of a set in a metric space, 178

$\mathscr{H}_a$, functions invariant under $U_T^a$, 181

$\mathscr{E}_a$, functions ergodic under $T^a$, 181

$\mathscr{H}_{\mathrm{rat}}$, rational spectrum part of $L^2$, 181

$\mathscr{E}_*$, totally ergodic part of $L^2$, 181

$\mathscr{H}$, Hilbert space, 184

$\langle \cdot, \cdot \rangle$, inner product, 184

AP, almost-periodic, 200

$\mathscr{N}_X$, trivial $\sigma$-algebra on $X$, 202

$f \otimes f'$, $(x, x') \to f(x) f(x')$, 221

MPT$(X, \mathscr{B}, \mu)$, group of invertible measure-preserving transformations of $(X, \mathscr{B}, \mu)$, 231

$g_*$, group action on measures, 232

$\mathscr{M}^G(X)$, set of $G$-invariant measures, 232

$g_n \to \infty$, going to infinity in a group, 232

$\mathbf{e}_1, \mathbf{e}_2$, standard basis of $\mathbb{R}^2$, 236

$X_{\bullet\bullet}$, Ledrappier's example, 236

$[K:L]$, minimal number of left translates $gL$ of $L$ needed to cover $K$, 243

$\mathrm{mod}(\cdot)$, modular function on a group, 248

Homeo$(X)$, set of homeomorphisms of a compact topological space $X$, 252

$U_g$, induced unitary representation for a group action, 255

(P), property that a metric on a group is proper, 260

(D), doubling property of metric balls in a group, 260

(F), Følner property of metric balls in a group, 260

$\mathrm{Im}(\cdot)$, image of map or operator, 267

$\mathbb{H}$, hyperbolic plane, 277

T$\mathbb{H}$, tangent bundle to $\mathbb{H}$, 277

T$_z\mathbb{H}$, tangent plane to $\mathbb{H}$ at $z$, 277

D, derivative, 277

$\langle \cdot, \cdot \rangle_z$, inner product at $z$, 278

$L(\phi)$, length of piecewise smooth curve $\phi$, 278

SL$_2(\mathbb{R})$, special linear group, 279

PSL$_2(\mathbb{R})$, projective special linear group, 279

PSO(2), projective special orthogonal group, 280

SO(2), special orthogonal group, 280

T$^1\mathbb{H}$, unit tangent bundle of $\mathbb{H}$, 281

$U^\pm, u^\pm$, horospherical subgroups and elements, 287

Mat$_{dd}(\mathbb{R})$, $d \times d$ matrices over $\mathbb{R}$, 288

SL$_d(\mathbb{R})$, group of real matrices with determinant 1, 289

GL$(V)$, group of automorphisms of vector space $V$, 289

T$G$, tangent bundle to closed linear group $G$, 289

tr$(v)$, trace of matrix $v$, 290

$\mathfrak{g}$, Lie algebra of closed linear group $G$, 291

$G^0$, connected component of the identity in $G$, 294

$\mathfrak{sl}_d(\mathbb{R})$, Lie algebra of SL$_d(\mathbb{R})$, 295

$E_{ij}$, element of standard basis of matrices in Mat$_{dd}(\mathbb{R})$, 295

T$_g G$, tangent space to closed linear group $G$ at $g$, 295

$L_g$, left translation, 296

$R_g$, left translation, 297

$m_X$, measure on quotient space induced by Haar measure, 311

$[g, h]$, commutator of $g$ and $h$, 313

$[G, G]$, commutator subgroup of $G$, 313

$X_2$, the space PSL$_2(\mathbb{Z})\backslash$PSL$_2(\mathbb{R})$, 317

GL$_2(\mathbb{Z})$, integer matrices of determinant $\pm 1$, 327

$H_n$, $(2n+1)$-dimensional real Heisenberg group, 342

$G^{(\mathrm{ab})}$, abelianization of $G$, 344

$\Gamma(N)$, principal congruence lattice, 358

$\Gamma_0(2)$, example of a congruence lattice in SL$_2(\mathbb{Z})$, 358

$L_0^2$, square-integrable functions with
    integral zero, 366

$B_\delta^G, B_\delta^G(I)$, metric open ball around the
    identity in $G$, 366

$U_g^-, U_g^+$, stable and unstable horospherical
    subgroup, 367

$m_G^\ell, m_G^r$, left- and right-invariant Haar
    measure, 385

$X_n$, the space $\mathrm{SL}_n(\mathbb{Z}) \backslash \mathrm{SL}_n(\mathbb{R})$, 390

$\mathbb{P}(X)$, collection of all subsets of $X$, 403

$\mathscr{L}_\mu^p$, $p$-integrable measurable functions, 408

$L_\mu^p$, equivalence classes of elements of $\mathscr{L}_\mu^p$,
    408

$\mathscr{B}_\mu$, completion of $\mathscr{B}$ with respect to $\mu$,
    411

$\ell^p$, space of sequences with finite $p$ norm,
    417

$B(E, F)$, normed space of bounded linear
    maps $E \to F$, 419

$\sigma_{\mathrm{operator}}(T)$, spectrum of an operator, 420

$\mu_f$, spectral measure associated to $f$, 420

$C(X)^*$, dual of $C(X)$, 422

$T_*$, map on measures induced by $T$, 424

$\langle g, \chi \rangle$, pairing between $g \in G$ and
    character $\chi \in \widehat{G}$, 434

$\widehat{f}$, Fourier transform of $f$, 434

$B(G)$, functions on a locally compact
    group with a representation as
    an integral over characters, 435

$H^\perp$, annihilator of closed subgroup, 437

# General Index

abelianization, 344
absolutely continuous measure, 123, 409
action
    continuous, 231
    of a group, 231
    properly discontinuous, 348
adjoint
    operator, 29
    representation, 297
Alaoglu's theorem (Theorem B.6), 268, 421
algebra, 403
almost
    -periodic, 67, 200
    disjoint, 307
    invariant, 24
amenable, 251
    'ax + b'-group, 254
    group, 251
        abelian, 252
        compact, 253
        invariant measure, 252
        mean ergodic theorem, 257
        pointwise ergodic theorem, 257
        SL₂(ℝ) is not, 232
        Heisenberg group, 254
annihilator, 437
AP, 200
aperiodic, 65, 144
approximable
    badly, 87
        golden mean, 88
        quadratic irrational, 90
        very well, 91
arithmetic lattice, 358
associated unitary operator, 29
atom, 136
    conditional measure, 144

ergodic decomposition, 137
    measurability, 142
    null sets, 136
'ax + b'-group
    amenable, 254, 433
    not unimodular, 254, 433

badly approximable, 6, 87
    quadratic irrational (Corollary 3.14), 90
ball model of hyperbolic plane, 361
Banach
    algebra, 420
    space, 408, 417
        mean ergodic theorem, 66
        open mapping theorem, 419
        separable, 412
    upper density, 171
        Erdős–Turán conjecture, 171
Bernoulli
    automorphism, 18
    measure, 103, 329
    shift, 17, 18
    ergodic, 25
    mixing of all orders
        (Exercise 2.7.9), 53
    multiple recurrence, 176
    non-invertible theory, 68
bi-invariant metric, 430
Birkhoff
    ergodic theorem (Theorem 2.30), 44
    recurrence theorem (Exercise 4.2.2), 104
Borel
    σ-algebra, 15, 126, 128, 134, 406, 411
    –Cantelli lemma, 132, 405

M. Einsiedler, T. Ward, *Ergodic Theory*, Graduate Texts in Mathematics 259, 471
DOI 10.1007/978-0-85729-021-2, © Springer-Verlag London Limited 2011

and pointwise convergence along a
    subsequence, 408
normal number theorem, 2, 82
probability measure, 15, 97
probability space, ix, 134, 411
    constraint on complexity
    of $\sigma$-algebras, 141
    space, 134
bundle, 333

ceiling function, 322
character, 312, 434
    modular function, 312, 431
characteristic factor, 228
    degree, 229
    nilmanifold, 228
Choquet
    ergodic decomposition, 154
    representation theorem, 103, 154, 272
    simplex, 103
        all arise, 119
circle, viii
    doubling map, 14
        ergodic, 27
        invertible extension, 20
    rotation, 14
        disjoint (Exercise 6.5.1), 168
        equidistributed, 112
        not weak-mixing, 51
        uniquely ergodic, 2, 107
closed linear group, 288
    adjoint representation, 297
    discrete subgroup, 301, 365
    exponential map, 290
    left-invariant metric, 297
    left-invariant metric on a subgroup,
        300
    logarithm map, 290
    subgroup, 300
    topology, 298
cocompact lattice, 378
    uniform, 307
coloring, 172, 210, 442
    compactness, 442
commutator, 313, 331
    subgroup, 313, 332, 343
compact
    extension, 200
        dichotomy, 202
    group
        action, 254
        amenable, 253
        dual, 161
        endomorphism, 15

rotation, 161
    operator, 197, 204, 420
    kernel, 420
complete
    Banach space, 417
    function space (Theorem A.23), 412
    function spaces
        ergodic theorem, 66
        measure space, 406
        normed space, 418
        orthonormal basis, 437
        sequence space, 418
completely independent, 132
conditional
    expectation, 121
        continuity, 125
        existence, 123
        functoriality, 148
    measure, 133, 164
        existence for Borel probability
            spaces, 135
        geometric characterization
            (Proposition 5.19), 142
congruence lattice, 358
conjugacy, 65
    topological, 102
conjugate, 365
continued fraction, 7, 69
    badly approximable, 87
    convergents, 72, 79
    convergents are optimal, 73
    in terms of $2 \times 2$ matrices, 70
    map, 76
        as a homeomorphism, 86
        as an extension, 94
        ergodic, 79, 323
        invertible extension, 91
        normal number, 94
        partial quotient, 70
        recursion for convergents, 71
        typical behavior of digits, 82
        uniqueness, 75
        very well approximable, 91
continuous
    group action, 231
    map
        ergodic decomposition, 118
        invariant measure, 97, 99
        invertible extension, 102
        minimal, 104, 119
        unique ergodicity, 105
        spectrum, 421
        weak mixing, 51
conull set, 135

convex
   Choquet theorem, 103
   combination
      of measures, 98
      function, 185
      set of measures, 422
      set of joinings, 159
      subset of $\mathscr{M}(X)$, 98
van der Corput lemma, 184, 187, 229
countably-generated $\sigma$-algebra, 135
covolume, 390
cusps, 357
   as equivalence classes, 357
cylinder set, 17, 18, 25, 179, 236, 242

decreasing martingale theorem, 129
density, 22
   Banach upper, 171
   integer sequence, 50
   Lebesgue, 126, 412
   point, 413
   subsets of $\mathbb{Z}^d$, 227
   uniform, 113
   upper, 4, 23, 180
diagonal measure, 373
Dirichlet
   principle, 75
   region, 349, 357, 380
      and uniform lattices, 356
   theorem, 94
discrete spectrum, 161
   group rotation (Exercise 6.4.1), 163
   theorem (Theorem 6.13), 162
disintegration of a measure, 136
disjoint, 163
   circle rotations, 168
displacement, 337
division algebra (quaternion), 398
dominated convergence theorem
      (Theorem A.18), 411
Doob's inequality (Lemma 5.6), 127
doubling property, 260
dual group, 161
   Pontryagin theorem, 161

eigenvalue, 50
elliptic element of $SL_2(\mathbb{R})$, 365
$\varepsilon$-dense, 26, 200, 204
equidistribution, 2, 110
   circle rotation, 112
      rate, 118
   generic point, 113
   horocycle orbit, 347, 388
   unique ergodicity, 114

Weyl's criterion, 111
Weyl's theorem, 4, 114, 183
equivalent measures, 410
ergodic, 23
   average, 34
   Bernoulli shift, 25
   circle rotation, 26, 30
   continued fraction map, 79, 323
   decomposition, 103, 154
      atom, 137
      example, 107
      group actions, 266
   disjoint from identity, 168
   group
      action, 232
   endomorphism, 31
   rotation, 108, 161
   in terms of invariant $L^p$ functions, 28
   in terms of invariant measurable
      functions, 23
   maximal theorem, 37
   measures, 99
      are mutually singular, 101
      are not closed, 102
      as extreme points, 99, 234
      dense in the space of all measures,
      103
      for group actions, 266
   possible etymology, 11
   preserved by isomorphism, 27
   simple eigenvalues, 160
   theorem
      along squares, 118
      Banach space, 66
      Birkhoff (Theorem 2.30), 44
      conditional expectation
         (Theorem 6.1), 153
      flow, 257
      for infinite integral, 86
      for permutations (Exercise 2.6.1),
      47
      local, 259
      mean (Theorem 2.21), 11, 32
      von Neumann (Theorem 2.21), 32
      pointwise (Theorem 2.30), 11, 44
      pointwise for group action, 264
      uniform mean, 36
   topologically, 104
   toral endomorphism, 31
   totally, 181
      spectral characterization, 187
   unique, 105
   unitary property, 29
expectation

conditional, 121
  continuity, 125
  existence, 123
exponential map, 289
  locally invertible, 291
  one-parameter subgroup, 290
extension, 17, 199
  compact, 200
    dichotomy, 202
    SZ property, 208
  relatively weak-mixing
    dichotomy, 202
    non-ergodic, 201
    products, 223
  trivial, 17

factor, 156
  Kronecker, 160
  map, 16, 157
  non-trivial, 17
  from a system to itself, 20
  topological, 102
  trivial, 17
Fatou's lemma (Theorem A.17), 177, 411
fiber, 333
finitely additive, 404
flow, 257, 320
  built under a function
    ergodicity, 322
  built under a function, 321
    ceiling function, 322
    fiber, 322
  ergodic theorem, 257
Følner sequence, 251
forward measurable, 68
Fourier
  coefficients, 161
  transform, 434, 435
    inverse, 435
free product, 310
Fubini's theorem, 409
Fuchsian group, 307
  Dirichlet region, 348, 349
  properly discontinuous, 348
fundamental domain, 14, 307
  Dirichlet region, 349
  open, 348
  strict, 307
  well-defined volume, 312
Furstenberg
  correspondence principle, 178, 180
  joinings, 169

Gauss

–Bonnet formula, 352, 360
map, 76
  ergodic, 79, 323
  invertible extension, 91
  measure, 77
general linear group, 288
generator (existence), 11, 119
generic
  measure-preserving transformation,
    67
  point, 113, 336
geodesic
  flow, 277, 286
  escape of mass, 389
  shadowing lemma, 329
  stable manifold, 287
  unstable manifold, 288
  return time function, 320
group
  action, 231
    continuous, 231, 232
    ergodic, 232
    ergodicity does not descend to
      subgroups, 233
    induced unitary representation, 255
    invariant measure, 232
    invariant set, 233
    maximal ergodic theorem, 262
    maximal inequality, 260
    mixing, 233
    mixing of all orders, 233
    pointwise ergodic theorem, 264
    rigid, 233
    weak-mixing, 232
  endomorphism
    ergodic, 31
  locally compact, 431
    Haar measure, 243
    modular function, 431
  metrizable, 430
  monothetic, 430
  rotation
    discrete spectrum (Exercise 6.4.1),
      163
    uniquely ergodic, 108
  topological, 429
  triangle inequality, 262

Haar measure, 161, 311
  decomposition, 385
  existence, 243
  left, right, unimodular, 431
  modular function, 431
Hahn–Banach theorem (Theorem B.1), 418

Halmos–von-Neumann theorem
    (Theorem 6.13), 162
harmonic analysis, 30
    less amenable to generalization, 31
Hausdorff dimension, 7
Hecke triangle group, 362
Heisenberg group, 300, 331
    $(2n + 1)$-dimensional, 342
    amenable, 254
    center, 332
    commutator subgroup, 332
    discrete subgroup, 303
    nilrotation, 331
    quotient, 331
    unimodular, 254
Herglotz–Bochner theorem (Theorem C.9),
    435
Hilbert–Schmidt operator, 67, 420
homogeneous space, 228
horocycle
    flow, 287
        equidistributed orbit, 388
        invariant measure, 378
        long periodic orbits
            (Exercise 11.5.2), 388
        minimal, 401
        mixing of all orders, 401
        non-divergence, 389
        non-escape of mass, 388, 389
        periodic orbits and compactness,
            378
        quantitative non-divergence, 389
        uniquely ergodic, 378
    stable and unstable flows, 311
horospherical subgroup
    stable, unstable, 367
hyperbolic
    area form, 306
    element of $SL_2(\mathbb{R})$, 365
    metric, 278
    plane, 277
        action of $PSL_2(\mathbb{R})$ is transitive, 280
        ball model, 361
        boundary, 278
        geodesic curve, path, 283
        geodesic flow, 286
        minimizing path, 282
        polygon, 352
        regular, 361
    Riemannian metric, 278
    space
        Hecke triangle group, 362
        tiling, 362
    volume form, 306

increasing martingale theorem
    (Theorem 5.5), 126, 182
independent, 49
    completely, 132
induced transformation, 61
    ergodic, 61
infinite measure, 22
injectivity radius, 302
integrable function, 408
invariant
    $\sigma$-algebra
        Kronecker factor, 160
    eigenspace, 29
    function, 25
    measurable set, 25
    measure, 13, 32
        characterized, 15
        continuous maps, 97
        convex combination, 100
        ergodic decomposition, 103
        ergodicity, 99
        maps without any, 102
        North–South map, 99
        unique, 105
    set for group action, 233
    sets
        $\sigma$-algebra, 153
        sub-$\sigma$-algebra, 20, 156
        subspace, 32
        vectors, 372
inverse Fourier transform, 435
invertible extension, 20, 177, 178
    continuous map, 102
    mixing properties, 178, 235
    universal property, 20
isometry, 29
    between normed spaces, 419
isomorphism
    measurable, 16
    normed spaces, 419
    theorem, 11

Jacobi identity, 293
Jacobian, 325
Jewett-Krieger theorem, 119
joining, 153
    disjoint, 163
    and factors (Exercise 6.3.3), 159
    induced by an isomorphism, 164
    relatively independent, 164, 201, 221
        basic properties, 165
    trivial, 158

Kac's theorem (Theorem 2.44), 63

Kakutani
  induced transformation, 61
  –Rokhlin lemma (Lemma 2.45), 63
    fails for other sequences, 65
  skyscraper, 62
Khinchin
  –Lévy constant, 94
  Three pearls of number theory, 229
  constant, 94
Klein model, 401
Koopman operator, 29, 66, 420
Krieger's theorem on existence of
    generators, 11, 119
Kronecker
  factor, 160, 228
  system, 161, 199, 226
    relatively compact orbits, 199
    SZ property is syndetic, 190
    SZ property, 189, 226
Kryloff–Bogoliouboff theorem
    (Corollary 4.2), 98, 251, 252
  analog for amenable groups, 252

Lagrange's theorem (Theorem 3.13), 88
lattice, 12, 307
  arithmetic, 358
  congruence, 358
  covolume, 390
  forces unimodularity, 312
  modular group, 307
  principal congruence, 358
  uniform, 307
  unimodular, 390
law of the iterated logarithm, 66
Lebesgue
  decomposition (Theorem A.14), 410
  density point, 413
  density theorem, 126, 412
  space, 411, 412
Ledrappier's example
  mixing, 236, 242
  not mixing on 3 sets, 238
Lie
  algebra, 293
    corresponding to subgroup, 300
    determines $G^0$, 294
  bracket, 293
  group, 228, 289
linear
  functional, 418
  group, see closed linear group
  operator
    bounded, continuous, 419
    compact, 420

isometry, 419
spectrum, 420
Liouville
  number
    equidistribution rate, 118
  number (Exercise 4.4.5), 118
  theorem (Exercise 3.3.3), 91
Littlewood's conjecture, 7
local ergodic theorem, 259
locally
  compact abelian group, 433
    annihilator, 437
    character, 434
    Fourier transform, 434
    inversion theorem, 436
    Parseval formula (Theorem C.8),
      435
    Plancherel transform, 435
    Pontryagin duality, 437
    solenoid, 438
  compact group, 431
    Haar measure, 431
    modular function, 431
    unimodular, 431
  finite measure, 425
  isomorphic, 379
logarithm map, 289, 290
lower central series, 343
Lusin's theorem (Theorem A.20), 314, 411

Möbius transformation, 279, 284, 285, 308
Mahler's compactness criterion, 390
martingale, 126
  decreasing, 129
  increasing, 126
  relation to ergodic theorems, 151
matrix
  coefficient, 367
  coefficients, vanishing, 372
  nilpotent, 370
  unipotent, 370
Mautner phenomenon, 364, 369
  unitary, 369
maximal
  ergodic theorem, 37, 38
    for group action, 262
  inequality
    analog for martingales, 127
    for group action, 260
  operator, 39
  transformations, 38
mean ergodic theorem (Theorem 2.21), 32
  uniform, 36
measurable

forward, 68
function, 407
isomorphism, 16
rectangle, 406
measure
    absolutely continuous, 409
    concentrated, 404
    conditional, 135
    diagonal, 373
    disintegration, 136
    equivalent, 410
    ergodic not closed, 102
    Haar, 431
    joining, 153
    Lebesgue decomposition, 410
    locally finite, 425
    regular, 425
    $\sigma$-finite, 404
    singular, 410
    space
        complete, 406
    stationary, 272
        and random walks, 273
    support, 28, 406
measure-preserving
    flow, 257
    system
        disjoint, 163
        ergodic disjoint from identity
            (Exercise 6.5.2), 168
        group action, 231
        transformation, 13
    transformation, 13
        aperiodic, 65, 144
        associated unitary, 28
        Bernoulli shift, 17
        circle rotation, 14
        continued fraction, 76
        continuous map, 98, 252
        continuous spectrum, 51, 421
        discrete spectrum, 161
        ergodic, 23
        extension, 17
        factor, 156
        forward, 68
        group endomorphism, 15
        invertible extension, 20, 177, 178
        isomorphism, 16
        mixing, 49
        mixing on $(k+1)$ sets, $k$-fold, or
            order $k$, 49
        suspension, 64
        universal property of invertible
            extension, 20

weak-mixing, 50, 53
mild-mixing, 49, 274
minimal, 104
    homeomorphism, 104
    set, 104
mixing, 49
    exponential rate for toral
        endomorphisms
        (Exercise 2.7.13), 54
    group action, 233
        descends to subgroups, 233
        of all orders, 233
    Ledrappier's example, 236
    mild, 49, 274
    of all orders, 50, 274
        in positive cones, 242
    $SL_2(\mathbb{R})$-actions, 373
    on $k+1$ sets, $k$-fold, or order $k$, 49
    Rokhlin problem, 67
    semigroup actions, 235
    strong, 49
    $\times 2, \times 3$ system, 242
    weak, 49, 50
        of all orders, 218
    weak but not strong, 50
Möbius transformation, 284
modular
    function, 312, 431
    group, 307
    lattice, 307
monothetic group, 160, 430
    classification, 437
    topological generator, 430
monotone
    class, 404
    class theorem (Theorem A.4), 404
    convergence theorem
        (Theorem A.16), 410
multiple recurrence, 175
    SZ system, 176
    Bernoulli shift, 176
    circle rotation, 175
    topological, 178
mutually singular, 101

von Neumann ergodic theorem
    (Theorem 2.21), 32
nilmanifold, 228
nilpotent
    group
        2-step, 332
        $k$-step, 343
        lower central series, 343
        matrix, 370

nilrotation, 331, 333
    linear drift, 337
    non-ergodic, 341
    uniquely ergodic, 334
nilsystem, 344
non-amenable group, 232
non-divergence, 389
non-escape of mass, 389
normed space, 417
    Banach, 417
    equivalent, isomorphic, 419
North–South map, 99, 232
    in a non-amenable group, 232

one-parameter subgroup, 10
open mapping theorem, 419
Oppenheim's conjecture, 5
orbit, 1
    closure, 179
    dense, 401
    geodesic, 277, 389
    horocycle, 388
    periodic, 378, 379
orthogonality relations
    compact group, 237, 434, 435

pairwise independent, 405
parabolic element of $SL_2(\mathbb{R})$, 365
Parseval formula, 435
partial quotient, 70
partition, 122
paths, 278
periodic, 3
permutation
    cyclic, 47
    ergodic theorem, 47
pigeon-hole principle, 21
ping-pong lemma, 359
Plancherel transform, 435
Poincaré
    model, 401
    recurrence (Theorem 2.11), 4, 21
        finite, finitely additive, 22
pointwise ergodic theorem, 44
    group action (Theorem 8.19), 264
polarization identity, 441
polynomial
    equidistributed, 116
    homogeneous, 5
    horocycle orbit, 390
    irrational, 114
    recurrence, 180, 181
    trigonometric, 109
Pontryagin duality, 437

positive
    operator, 39
    upper density, 4
positive-definite
    function, 435
pre-periodic, 3
primitive vector, 392
principal congruence lattice, 358
pro-nilmanifold, 228
probability space, 134, 404
projective
    space, 279
    special linear group, 279
    special orthogonal group, 280
proper
    action, 288, 357
    map, 289
    metric, 260
properly discontinuous, 348
push-forward, 93, 357, 373

quadratic
    form, 5
        congruence obstruction, 8
        indefinite, 6
        integral, 8
        non-degenerate, 6
        rational, 6
        irrational, 88
quantitative non-divergence, 389
quaternion division algebra, 398
    $\mathbb{R}$-split, 398

Radon–Nikodym
    derivative, 100, 410
    theorem (Theorem A.15), 410
Raghunathan conjecture, 9, 10, 401
rational spectrum, 181
recurrence
    multiple, 175
        SZ system, 176
        Bernoulli shift, 176
        circle rotation, 175
        topological, 178
        Poincaré, 21
recurrent point, 104
regular
    measure, 406, 425
    polygon, 361
relatively
    compact orbits, 199
    independent joining, 164, 201, 221,
        223
        basic properties, 165

weak-mixing extension, 201
 dichotomy, 202
 non-ergodic, 201
 product, 223
return time, 61
 expected (Theorem 2.44), 63
 function, 320
Riemann surface
 genus two, 362
 uniformization theorem, 362
Riemannian metric, 295, 297
Riesz representation theorem, 138, 422
rigid group action, 233
Rokhlin
 problem, 67
 tower, 64
 base, height, residual set, 64
rotation
 circle, 14
  not weak-mixing, 51
  uniquely ergodic, 2, 107
 compact group
  rigid, 235
 quotient of nilpotent group, 331
 torus
  uniquely ergodic, 109
Roth theorem (Theorem 7.14)
 Kronecker system, 194
 orderly and chaotic parts, 193

Sárközy's theorem (Theorem 7.9)
 orderly and chaotic parts, 181
 rational spectrum component, 181
 totally ergodic component, 181
semi-algebra, 52, 403
shadowing lemma for geodesic flow, 329
shift
 action, 236
 Bernoulli, 17
 map, 17
σ-algebra, 403
 Borel, 406
 completion, 412
 countably-generated, 135
 not countably-generated, 136, 156
 product, 406
σ-finite measure, 404
simple
 eigenvalue, 29, 160
 function, 16, 407
simplex
 all Choquet simplexes arise, 119
 Choquet representation theorem, 103
simply transitive, 279

singular, 410
skew-product, 230
 uniquely ergodic, 114
solenoid, 21, 438
special
 flow, 321
 linear group, 279
 projective, 279
spectral
 theorem (Theorem B.4), 59, 183, 420
spectrum
 linear operator, 420
speed of a path, 278
stable
 horospherical subgroup, 367
 manifold, 287
stationary
 measure, 272
 and random walks, 273
stiff action, 275
substitution rule, 413
S-unit theorem, 239
 and mixing, 240
 fails in positive characteristic, 240
 vanishing subsums, 239
support of a measure, 28, 406
suspension, 64
syndetic, 36, 175, 190, 192, 212
SZ property, 176
 Kronecker systems, 189, 226
 maximal factor, 216
 property
  compact extension, 207
  relatively weak-mixing extension,
   218, 224
 system, 176
  invertible extension, 177
  limit of factors, 216
  reduction to Borel probability
   space, 177
  reduction to ergodic case, 178
  transfinite induction, 226
  weak-mixing systems, 191
Szemerédi's theorem, 4
 consequence of multiple recurrence,
  178
 effective (Gowers' theorem), 4
 finite version, 178
 finitistic, 5
 polynomial, 227

tangent bundle, 295
 concrete realization, 297
 modular surface, 397

unit, 281
tangent bundle
    closed linear group, 468
Tietze–Urysohn extension lemma
    (Theorem B.9), 422
tiling, 362
×2, ×3 system is mixing of all orders, 239
topological
    conjugacy, 102
    dynamical system, 102
    conjugate, 102
    minimal, 104
    ergodicity, 104
    factor, 102
    group, 429
        bi-invariant metric, 430
        Hausdorff, 429, 430
        locally compact, 431
        metrizable, 430
        monothetic, 430
        multiple recurrence, 178
    space
        Hausdorff, 428, 439
        Kolmogorov, 439
        normal, 428
        $T_0$, 439
        $T_2$, 439
    vector space, 418
        dual separates points, 425
        linear functional, 418
        linear map, 418
toral
    endomorphism
        ergodicity, 31
        rate of mixing (Exercise 2.7.13), 54
    rotation
        unique ergodicity, 109
totally
    bounded, 199
    ergodic, 36, 181
        component, 181
        spectral characterization, 187
    transitive, 279, 280
trivial
    extension, factor, 17
    joining (product), 158
    $\sigma$-algebra, 202
    ergodicity, 153
Tychonoff's theorem (Theorem B.5), 421

uniformization theorem, 362
    co-compact lattice, 401
uniformly
    continuous, 418

discrete, 301
distributed, see equidistributed
unimodular, 248, 312, 313
    '$ax + b$'-group is not, 254, 433
    forced by presence of a lattice, 312
    $GL_2(\mathbb{R})$, 433
    Heisenberg group, 254
    lattice, 390
        Mahler compactness criterion, 390
    locally compact group, 431
    $SL_d(\mathbb{R})$, 313
unipotent
    matrix, 370
    one-parameter subgroup, 10
unique ergodicity, 4, 105
    circle rotation, 2, 107
    equidistribution, 114
    Furstenberg's theorem, 114
    group rotation, 108
    horocycle flow, 378, 379
    nilrotation, 334
    toral rotation, 109
unitary
    operator, 28, 29
        eigenvalue, 50
        spectral theorem (Theorem B.4),
            420
    property, 29
        ergodicity, 29
    representation, 366, 369
        for group action, 255
        Mautner phenomenon, 369
unstable
    horospherical subgroup, 367
    manifold, 288
Urysohn lemma, 428

vanishing of matrix coefficients
    (Theorem 11.25), 372
vertex, 350
very well approximable, 91
Vitali covering lemma, 40, 261
    integers (Corollary 2.28), 41

van der Waerden theorem (Theorem 7.1),
    171, 172
weak
    convergence, 187
    mixing, 49, 50, 53
        continuous spectrum, 51
        equivalent formulations, 51
        group action, 232
        of all orders, 218, 219
        SZ property, 191

mixing without strong mixing, 50
weak*
-compact, 98
-limit
  of ergodic measures not ergodic,
  102
  of orbit measures, 328
-limit point, 98
-topology, 146
  characterized in terms of functions,
  sets, 423

definition, 423
  metric, 134, 147, 423
Weyl equidistribution
  criterion, 111
  theorem (Theorem 1.4), 4, 114, 183,
  201

zero-dimensional
  group, 437
  groups, and mixing, 274, 275
Zorn's Lemma, 442, 443